Advances in
Carbohydrate Chemistry and Biochemistry

Volume 61

Advances in Carbohydrate Chemistry and Biochemistry

Editor

DEREK HORTON

The American University

Washington, DC

Volume 61

Amsterdam • Boston • Heidelberg • London • New York • Oxford
Paris • San Diego • San Francisco • Singapore • Sydney • Tokyo
Academic Press is an imprint of Elsevier

Academic Press is an imprint of Elsevier
84 Theobald's Road, London WC1X 8RR, UK
Radarweg 29, PO Box 211, 1000 AE Amsterdam, The Netherlands
Linacre House, Jordan Hill, Oxford OX2 8DP, UK
30 Corporate Drive, Suite 400, Burlington, MA 01803, USA
525 B Street, Suite 1900, San Diego, CA 92101-4495, USA

First edition 2008

ISBN: 978-0-12-373920-9
ISSN: 0065-2318

For information on all Academic Press publications
visit our website at books.elsevier.com

Printed and bound in USA

08 09 10 11 12 10 9 8 7 6 5 4 3 2 1

CONTENTS

Anti-Influenza Drug Discovery: Are We Ready for the Next Pandemic?

Tasneem Islam and Mark von Itzstein

Chemoselective Neoglycosylation

FRANCESCO NICOTRA, LAURA CIPOLLA, FRANCESCO PERI, BARBARA LA FERLA AND CRISTINA REDAELLI

LIST OF CONTRIBUTORS

Stephen J. Angyal, School of Chemistry, University of New South Wales, Sydney, NSW 2052, Australia

Laura Cipolla, Department of Biotechnology and Bioscience, University of Milano Bicocca, Piazza della Scienza 2, I-20126 Milano, Italy

Barbara La Ferla, Department of Biotechnology and Bioscience, University of Milano Bicocca, Piazza della Scienza 2, I-20126 Milano, Italy

Juliette Fitremann, Laboratoire des IMRCP (UMR 5623 CNRS-UPS), Bât 2R1, Université Paul Sabatier Toulouse 3, 118 Route de Narbonne, 31062 Toulouse cedex 4, France

Tasneem Islam, Institute for Glycomics, Griffith University (Gold Coast Campus), PMB50, Gold Coast Mail Centre, Queensland 9726, Australia

Slawomir Jarosz, Institute of Organic Chemistry, Polish Academy of Sciences, ul. Kasprzaka 44/52, 01-224 Warsaw, Poland

Yuriy Knirel, Russian Academy of Sciences, N.D. Zelinsky Institute of Organic Chemistry, Leninsky Prospekt 47, 119991 Moscow, Russia

Maria Kochetkova, formerly Russian Academy of Sciences, N.D. Zelinsky Institute of Organic Chemistry, Leninsky Prospekt 47, 119991 Moscow, Russia

Rosa M. de Lederkremer, CIHIDECAR, Departamento de Química Orgánica, Facultad de Ciencias Exactas y Naturales, Universidad de Buenos Aires, Pabellón II, Ciudad Universitaria, 1428 Buenos Aires, Argentina

Bartosz Lewandowski, Institute of Organic Chemistry, Polish Academy of Sciences, ul. Kasprzaka 44/52, 01-224 Warsaw, Poland

Carla Marino, CIHIDECAR, Departamento de Química Orgánica, Facultad de Ciencias Exactas y Naturales, Universidad de Buenos Aires, Pabellón II, Ciudad Universitaria, 1428 Buenos Aires, Argentina

Francesco Nicotra, Department of Biotechnology and Bioscience, University of Milano Bicocca, Piazza della Scienza 2, I-20126 Milano, Italy

Francesco Peri, Department of Biotechnology and Bioscience, University of Milano Bicocca, Piazza della Scienza 2, I-20126 Milano, Italy

Russell Pickford, Department of Chemistry, University of York, Heslington, York YO10 5DD, UK

Yves Queneau, Laboratoire de Chimie Organique; ICBMS; UMR 5246 CNRS; Université Lyon 1; INSA-Lyon; CPE Lyon; Institut National des Sciences Appliquées, Bâtiment Jules Verne, 20 Avenue Albert Einstein, 69621 Villeurbanne Cedex, France

Cristina Redaelli, Department of Biotechnology and Bioscience, University of Milano Bicocca, Piazza della Scienza 2, I-20126 Milano, Italy

James C. Reynolds, Department of Chemistry, University of York, Heslington, York YO10 5DD, UK

João A. Rodrigues, Department of Chemistry, University of York, Heslington, York YO10 5DD, UK

David P. Sumpton, Department of Chemistry, University of York, Heslington, York YO10 5DD, UK

Adrian M. Taylor, Department of Chemistry, University of York, Heslington, York YO10 5DD, UK

Jane Thomas-Oates, Department of Chemistry, University of York, Heslington, York YO10 5DD, UK

Mark von Itzstein, Institute for Glycomics, Griffith University (Gold Coast Campus), PMB50, Gold Coast Mail Centre, Queensland 9726, Australia

PREFACE

The sixty preceding volumes in this Series have recorded significant advances in our knowledge of every aspect of the carbohydrate field. Some of the articles compile information of permanent reference value, while others have documented the interim state of knowledge in a developing area, drawing attention to gaps in our understanding, and providing pointers for future research. This current Volume 61 offers authoritative articles of both types, covering a wide variety of carbohydrate topics.

The field had its early beginnings in empirical studies on those macromolecular biomaterials we call carbohydrates, together with the study of simple sugars as organic compounds that present challenges of stereochemical complexity and multifunctionality. Over time the discipline has burgeoned, from its early bases in food and fiber technology and a specialized area of organic chemistry, onto center stage in its present wide role in chemistry, biochemistry, and glycobiology.

Mass spectrometry has had enormous impact on our ability to analyze carbohydrate structures. Its fundamentals for the analysis of sugars and their derivatives under electron impact were documented by Kochetkov and Chizhov in Volume 21 of this Series, and the advent of Fast Atom Bombardment significantly extended the technique to applications with molecules of considerably greater complexity, as recorded by Dell in Volume 45. Progress in instrumental capabilities in this rapidly evolving field has greatly augmented the reach of mass spectrometry in characterization of glycans and glycoconjugates, warranting an early re-visit to the topic in the present article by Thomas-Oates and her team (York, UK). Interestingly, the traditional techniques for structural analysis of carbohydrates, namely permethylation, peracylation, and labeling of the reducing terminal, remain core methodologies within these mass-spectrometric applications.

Deoxy sugars are of wide occurrence as components of nucleic acids, natural glycosides, and antibiotics, and they were the subject of an earlier article by Hanessian in Volume 21. Here, de Lederkremer and Marino (Buenos Aires) provide a detailed update on the distribution of deoxy sugars in natural products, along with a survey of methods for their synthesis.

The most abundant simple carbohydrate, and indeed the most abundant pure organic chemical, is sucrose. Its central role as a nutritive sweetener is well documented, but its potential as a precursor for organic synthetic manipulation (sucrochemistry) has long been a story of largely unrealized potential. The article here by Queneau (Lyon), Jarosz (Warsaw) and their coauthors provides a comprehensive account of actual and potential uses of sucrose as a cheap precursor for a range of applications as food additives, pharmaceuticals, surfactants, and complexing agents. The authors assess fundamental questions of selective reactivity among the eight functional groups in the sucrose molecule, as well as practical aspects concerning processing conditions and reaction solvents.

The article by Islam and von Itzstein (Brisbane) focuses on the worldwide threat of a possible avian influenza pandemic. The key involvement in the infective process of sialidase, a surface glycoprotein of the influenza virus, provides a pointer to the design of effective drugs that can combat all pathogenic strains, rather than vaccines, which target a single strain. Inhibitors of the sialidase, based on the 1,2-unsaturated analogue (Neu5Ac2en) of *N*-acetylneuraminic acid have led via computational methodology to effective mimetics, most notably Relenza (zanamivir) and Tamiflu (oseltamivir), along with other inhibitors. The chapter details the very extensive synthetic and pharmacological background involved in reaching these optimized targets.

Nicotra's group in Milan contributed the article on chemoselective neoglycosylation, namely the chemical attachment of saccharide chains to various acceptors: proteins, peptides, lipids, and different types of support media, to encode specific carbohydrate determinants for molecular recognition. Synthetic realization of the specific N- or O-glycosyl links present in the glycoprotein structure is difficult, and other modes of ligation must be employed. The utility of such structures is manifestly evident in the clinical success of such conjugates as the *Haemophilus influenzae* type b (Hib) vaccine, the use of neoglycopeptides in cancer diagnosis, and the enhancement of protein-binding affinity by use of glycoarrays and dendrimers.

The development of carbohydrate science recorded in these volumes by noted experts during the past six decades has been closely paralleled by the content of conferences devoted to the field, where many of these specialists have signaled important progress. Most notably, the International Carbohydrate Symposia, beginning in the early 1960s and proceeding on a biennial basis, have reflected the remarkable growth and breadth of the field. The historical account of these symposia presented here by Angyal (Sydney), a frequent contributor to this Series, documents this evolution in detail. In addition to listing the principal themes and contributors, he offers valuable pointers for organizers of future symposia in the most effective use of these meetings for information exchange and the stimulation of new ideas.

The long and distinguished career of N. K. Kochetkov, a towering figure in Russian science, who made major contributions to the study of nucleotide sugars, the methodology of glycosidic coupling, and the mass spectrometry of carbohydrate derivatives, is recognized in the biographical memoir by Knirel and Kochetkova (Moscow).

DEREK HORTON

Washington, DC
July 2007

NIKOLAY KONSTANTINOVICH KOCHETKOV

1915–2005

Professor Nikolay Konstantinovich Kochetkov was a distinguished Russian scientist of the second half of the twentieth century. He is well known for his outstanding contributions to organic and bioorganic chemistry, especially to carbohydrate research.

He was born in Moscow on May 18, 1915 at the time of World War I. Russia was still a monarchy, ruled by Emperor Nicholas II. Nikolay lived through two revolutions, Democratic and Bolshevik, the Civil War, establishment of the USSR, Stalin's dictatorship, and World War II, during which time he spent six years in the army and at the front. He then relaunched his scientific career from level zero, worked through the period of 'developed socialism', Gorbachev's perestroika, the collapse of the Soviet Union and the establishment of a new Russia. His was quite a long and eventful life ...

The family to which Nikolay was born had a small textile business not far from Moscow. His father Konstantin was educated in Russia and Germany as a chemical engineer, and was expected to develop the family business. This did not happen, because all property was confiscated after the Bolshevik revolution. The only thing that his parents could do for Nikolay and his younger brother Alexander (Shura) was to help them to receive the best possible education.

Nikolay finished primary school in 1929. He said his choice for the future was made early, and this was chemistry. However, he had a long and difficult way to go to realize his wish to become a scientist. His middle-class origin was an obstacle for his education, as Soviet authorities gave priority to teaching the children of workers and peasants. A secondary school with a specialized class in chemistry, where Nikolay was accepted after some struggle, was suddenly re-organized and transformed into a technical school preparing technicians for brick factories. Consequently, Nikolay's first job was leading a team of girls at a

ISBN: 978-0-12-373920-9
DOI: 10.1016/S0065-2318(07)61001-4

brick factory. From there he moved to the control laboratory at a chemical plant in Moscow where his father worked as an engineer.

When the time came for higher education, Nikolay's first choice was the chemical faculty of the Moscow State University. Again his 'unacceptable' social origin played its part and he did not get a place; so he joined the Moscow Institute of Fine Chemical Technology (MITKhT). He always had good memories of his student years, despite the fact that they coincided with a difficult time for the country and its people. All his life he remembered his teachers, who were educated before the revolution and who were all prominent scientists of their time. He also had a very high opinion of their lecturing skills. Their names—Professors Alexander Reformatskii (inorganic chemistry), Sergei Nametkin, Abram Berkengeim (both organic chemistry), and Academician Yakov Syrkin (physical chemistry) among others—may not perhaps ring many bells in the West today.

Students spent summer holidays traveling around the country, and the Caucasus, with its highest mountain Elbrus being among their favorite regions. They had no money and very little equipment for mountaineering, and used costumes sewn by their mothers from old clothes. Vera Volodina, Nikolay's fellow student and future wife, recalls other events, such as a trip to Leningrad (now St. Petersburg) and nearby palaces once belonging to the Russian royal family, and a Pushkin ball masquerade in 1937 devoted to the 100th anniversary of the poet's death. Nikolay was dressed as a Hussar, and it was then that Vera lost her heart to him for the next seventy years. But they also saw their fellow students disappearing, with Nikolay's closest friend being one of them. Nikolay's father had a special prison bag ready at home, as every night he feared arrest. That shortened his life, and he died in his early fifties.

The end of Nikolay's student years in 1939 was dramatic for all Europe and for him personally. He was a bright student and wanted to specialize in organic chemistry. By the time of his graduation, he was offered two postgraduate positions. The first was from Professor Alexander Orekhov, a specialist in alkaloid chemistry and then head of the Pharmaceutical Institute in Moscow, and the other from Professor (later Academician) Alexander Nesmeyanov, who chaired the department of organic chemistry in the MITKhT since 1938. Nikolay happily chose the latter and passed his examinations. A topic for his research was selected and a place in the laboratory allocated. But he was a postgraduate for only two weeks, after which, in his own words, the government made the choice for him and conscripted him to the army.

His first posting was a place 5,000 kilometers east of Moscow, not far from Lake Baikal. To the end of his life Nikolay could not forget the dreadful living conditions there. With the winter temperature falling down to $-50°C$, two to three hundred recruits lived in a huge dugout with a small stove. Add to this lack of warm clothes and boots, and it becomes clear why Nikolay used to say that if he was to choose between freezing and hunger (these two words in Russian, *holod* and *golod*, differ by only one letter), he would choose the latter.

However, Nikolay lived through this nightmare. His chemical education helped and he managed to get a job as a technician in the regiment's pharmacy, which made his life a little easier. As his two-year service was coming to an end he was going to go home, but with the German invasion of the USSR in 1941 the war began and he had to stay in the army for another four years, two of them being at the front. During the war he continued using his chemical expertise, and his rank by the end of the war was head of the regiment's chemical reconnaissance.

With his regiment, Nikolay advanced through Latvia, Poland, Silesia, and Bohemia, and he was awarded the medal *Za Otvagu* (*For Bravery*), which forever remained his most favorite award. He came back in 1945 from Prague, at the same time as Shura, his younger brother. They returned too late to see their father, who had died in 1943; their mother was very ill and died shortly afterwards. In 1946 Nikolay married Vera, who was waiting for him, and he began to restore his career in science after six years in the army—from October 21, 1939 to October 20, 1945.

Science and his work were the two most important things for Nikolay, and it was fortunate that Academician Nesmeyanov, then the head of the organic chemistry department at Moscow State University, offered him a position of lecture assistant. His job was to conduct experiments illustrating the topics of lectures in organic chemistry, which Nesmeyanov gave to about 250 students twice a week. Nikolay struggled to bring back his knowledge of organic chemistry, which he had lost while in the army. Demonstrating experiments to students, together with Nesmeyanov's excellent lectures, helped him to restore his chemical competence.

Kochetkov continued as lecture assistant until 1951, combining this work with research on metalloorganic chemistry in Nesmeyanov's laboratory. He studied organo-mercury compounds, in particular their synthesis by the addition of mercuric chloride to acetylenes. In 1948 Nikolay wrote a thesis and received the Ph.D. degree in chemistry.

Even so his life remained difficult, with little money and severe shortages of food and the necessary things for his family, now including a baby girl Maria born in 1947. Vera accepted once and for all that Nikolay's work was the main thing for him, and she was always very understanding and supportive. She took on all responsibilities for running the home, and created a comfortable environment for his work. She also managed to take a postgraduate course in inorganic chemistry, write a thesis, and receive the Ph.D. degree in the late 1940s. In 1952, they had a son, Sergey.

Nesmeyanov became rector of Moscow State University while its new building was being constructed in the Sparrow (then Lenin) Hills, and Nikolay watched him dealing with all kinds of emerging problems. He loved Nesmeyanov's style of communicating, which was direct and informal, with occasional jokes, and he later compared that with the style he saw in Cambridge. Nikolay considered Nesmeyanov his teacher who contributed, perhaps more than anyone else, to his becoming a scientist; and he was always grateful to him for that. Nesmeyanov also helped Nikolay out on many occasions in his later life.

Meanwhile, Nikolay had started his new research on the preparation, reactions, and synthetic applications of β-chlorovinyl ketones and related compounds (β-aminovinyl ketones, β-ketoacetals). Efficient synthetic methods were developed and improved, including C-ketovinylation (introduction of the RCOCH=CH group), and a number of useful heterocyclic compounds (pyrazoles, triazoles, pyridines, and so on) were synthesized. The discovery of enamine–imine tautomerism in β-aminovinyl ketones was another remarkable achievement by Nikolay at that time. For these studies he received in 1953 the degree of Doctor of Science in chemistry. He continued as a lecturer and supervisor of postgraduates at the University, becoming docent (associate professor) in 1951 and full professor in 1955.

Nikolay's scientific interests now moved to compounds with potential physiological activities. Still keeping a position at Moscow University, he accepted an offer to head the chemical department at the Institute of Pharmacology and Chemotherapy of the Academy of Medical Sciences of the USSR. There his research focused on the search for and synthesis of physiologically active compounds—potential drugs. Some of them, such as diazoline, chloracon, and thiamine, were launched and used for many years as antihistamine, anticonvulsant, and antituberculosis medicines. Together with his new coworkers Marat Karpeisky and Radii Khomutov, he developed an industrial-scale technology for synthesis of cycloserine and its analogues. Nikolay's achievements in this field were recognized, and in 1957 he was elected Correspondent Member of the Academy of Medical Sciences. He retained contacts with the Institute when he moved later on to new research areas, and in the 1960s in collaboration with Arkadii Likhosherstov, Viktor Kulakov, and Alexander Vasil'ev he contributed significantly to the synthetic chemistry of pyrrolizidine alkaloids and necinic acids of high biological activity.

After Stalin's death in 1953, contacts between the Soviet Union and the West began developing, and more government officials traveled to Europe. One of these visits had an important implication for Nikolay's future career. The head of the department of organic chemistry at Cambridge University, Professor (later Lord) Alexander Todd, met a Soviet government delegation led by Alexey Kosygin, then the vice chairman of the USSR Council of Ministers. Todd said he could accept two chemists from the Soviet Union to work in his laboratory for six months. However, he put forward three conditions: they had to be good chemists (he said he would send them back if they were not); they would leave politics alone; and the USSR Embassy in London would not supervise them.

On the return of the delegation to Moscow, Todd's proposal was passed to the president of the Academy of Sciences, Academician Nesmeyanov. He suggested that Nikolay should go to England, to which Nikolay agreed, though with some hesitation. At that time traveling to England was like going to the Moon—nobody could be sure that they would return home safely. Many people traveling abroad or communicating with foreigners had been arrested during Stalin's time, which had ended only three years previously. The second candidate appeared to be a brave recent graduate from the chemical faculty at Moscow University, Electron Mistryukov. Seeing Nikolay off to Cambridge, Nesmeyanov said that he would work in a great laboratory and asked him to learn why 'they' (namely in the West) worked so efficiently.

Nikolay wrote later that Professor Todd was very supportive and encouraging, and told him not to worry about anything. He interviewed Nikolay on his arrival about his area of research and sent him to the library to get acquainted with the latest achievements in nucleoside and nucleotide chemistry. Within a week he allocated Nikolay, to his delight, a place in the laboratory and offered a topic involving synthesis of nucleoside glucuronic acid diphosphate. Nikolay worked hard and regretted that he could not finish the work by the end of his term. He said that he met technicians during later trips to Cambridge, and they remembered him as a crazy Russian who worked in the laboratory at weekends.

After years of the Soviet people's isolation from the West, Nikolay suddenly found himself in a leading global scientific center, which many prominent scientists were visiting. He met Sir Robert Robinson, Sir Ian Heilbron, and Professor Christopher Longuet-Higgins among others at meetings and colloquia, where the most complicated scientific problems were discussed in a casual and merry manner. Nikolay learned a lot from these meetings; he came to love their style and adopted it for himself.

He appreciated Professor Todd's guidance not only in research, but also in his new life in the West, which was completely different from his previous experience. For instance, Professor Todd told Nikolay that he had ordered that no journalists should approach Nikolay and he advised him not to speak to any of them either. This proved to be a good idea, as Nikolay's stay in Cambridge coincided with the Suez crisis and the uprising against the Soviet Union in Hungary, both of which events could make journalists very curious about comments of a Russian.

Six months passed, and Nikolay returned to Moscow. His visit paved the way for other Soviet researchers to come to the UK, and afterwards many of them worked in Cambridge and other places. Communications between Nikolay and Lord Todd continued into the late 1980s. They met at international conferences and every time when they traveled to each others' countries.

The work in England boosted Nikolay's career at home. His return to Moscow from Cambridge came at the start of a boom in biochemistry and molecular biology in the Soviet Union, which followed great discoveries made in the West, including the Watson–Crick double helix. To boost research in the life sciences, the USSR Academy of Sciences asked the government to establish two new institutes. A positive decision did not come easily. First the importance of developing of a new 'exotic' scientific field had to be proved and reasons why it would be expensive explained. Two prominent scientists, Academicians Vladimir Engelhardt and Ivan Nazarov, were among those who pursued the process. Finally in 1958, a decree was issued authorizing the organization of the Institute for Chemistry of Natural Products (IKhPS; now M. M. Shemyakin-Yu. A. Ovchinnikov Institute of Bioorganic Chemistry) and the Institute of Radiological and Physico-Chemical Biology (now V. A. Enghelhardt Institute of Molecular Biology).

Academician Nazarov was nominated as director-to-be of the IKhPS, but he died, and Academician Mikhail Shemyakin, a specialist in the chemistry of antibiotics and vitamins, was appointed instead. Nikolay became his deputy and the head of the laboratory of chemistry of carbohydrates and nucleotides. He was an appropriate candidate; he had experience in the chemistry of monosaccharides and nucleosides, which he received in Cambridge and which was rare in the Soviet Union. At that time there was no nucleic acid chemistry going on in the country, and carbohydrate research was limited to industry-oriented technologies for processing cellulose and a few other polysaccharides.

Nikolay was involved in solving problems emerging during the organization of the IKhPS, and was to be one of the major contributors to the project. The first task was to define directions of research and the organizational structure that should be followed in staff recruitment, a difficult process because of the lack of people appropriately educated and qualified for the new research. Slowly, heads of laboratories and team leaders were found, and some of them brought their teams with them. Nikolay invited Varvara Derevitskaya, a specialist in cellulose chemistry, to be head of the glycoproteins group and his deputy in the laboratory. Eduard Budovsky led the nucleic acids group, and Anatolii Khorlin and Izida Zhukova the groups of structures of plant glycosides and glycolipids, respectively. Lev Kudryashov was appointed head of the monosaccharide synthesis group and later the carbohydrate radiochemistry group.

Attention was then transferred to undergraduates and postgraduates of Moscow University and other institutions of higher education. In 1958, Academician Nesmeyanov formed a special group from graduates of the chemical faculty at Moscow University under Nikolay's supervision. The senior staff of the IKhPS, including Nikolay himself, taught this group the chemistry of proteins, nucleotides and nucleic acids, carbohydrates, and steroids; and then many students were given jobs in the IKhPS. Among young people who joined Nikolay's new laboratory were 'Nesmeyanov's recruits' Oleg Chizhov, Viktor Vas'kovsky, and Boris Dmitriev, who were followed by Vladimir Shibaev, Anatolii Usov, Leonid Likhosherstov, Alexei Bochkov, Yury Ovodov, Mikhail Grachev, Sergei Kara-Murza, Leon Backinowsky, Galina Smirnova, and Evgeny Sverdlov. Later, some of them led research groups in the laboratory, and others became heads of their own laboratories or even directors of institutes and members of the Academy of Sciences.

The challenge that Nikolay and his new collaborators faced was even stronger than in some other laboratories, because none of them had a solid background in carbohydrate and nucleotide chemistry. They had to educate themselves, understand the contemporary level of research in these fields, learn the methodology, find their own niche, and only then move forward. All of the staff was enthusiastic and ready to do everything that was needed. Nikolay launched 'anti-illiteracy' seminars in his laboratory where scientific news would be informally discussed once or twice a week, and that proved successful within the year. A professional team, 'the Kochetkov school', was born.

Apart from scientific challenges, Nikolay had many other problems on his plate. The building that was chosen to accommodate the two institutes was unsuitable for either of them and had to be significantly reconstructed. A separate problem was to find equipment, such as chromatographs and spectrometers, and to teach people how to use them; there were not many people available who could do that. Nevertheless, with a help of friends from other institutes, these difficulties were overcome and the IKhPS started working.

Soon, it began developing contacts with foreign academic institutions, and the first of them was the Institute of Organic and Bioorganic Chemistry in Prague, in what was then Czechoslovakia. Nikolay recalled that its director, Professor František Šorm, who was also the president of the National Academy of Sciences, helped the IKhPS a great deal. Later contacts deepened, and it became possible to send researchers for training in leading laboratories in Western Europe and the USA.

The first scientific articles from the IKhPS were submitted for publication in the early 1960s, among them being Nikolay's reports on his work in the new field. His major project in nucleotide chemistry was specific chemical modifications of heterocyclic bases. Reactions of hydroxylamine with cytidine and uridine were studied in detail and a new reagent, *O*-methylhydroxylamine, was proposed for modification of cytidine. These investigations aimed at the development of efficient methods for sequencing and analysis of the secondary structure of polynucleotides. Later, a reaction of chloroacetaldehyde with adenosine and cytidine was discovered and used for preparation of fluorescent polynucleotide derivatives.

Kochetkov's involvement in carbohydrate chemistry began with the development of new methods for synthesis of unusual monosaccharides, including deoxy, amino, thio sugars, and higher monosaccharides. Sugar derivatives of α-amino acids and hydroxy amino acids were prepared as model compounds in degradative studies of glycoprotein structure. Natural nucleoside sugar diphosphates and their analogues were synthesized and assayed in enzymatic reactions to reveal their structure–function relationships. A new method of glycosylation, the orthoster method, was elaborated and it gained general recognition, being applied to the synthesis of a number of natural glycosides and oligosaccharides.

Simultaneously, Kochetkov paid attention to the structural elucidation of natural carbohydrates. Analysis of saponins from Far-Eastern plants of the Araliaceae family, including ginseng, revealed that many of them were glycosides of a new type, having polycyclic triterpene aglycones and long (up to 10 monosaccharide residues) oligosaccharide chains. Nikolay was one of the first to realize the high potential of mass spectrometry in carbohydrate chemistry. His pioneering works with Oleg Chizhov revealed the major electron-impact fragmentation pathways of various monosaccharide and oligosaccharide derivatives, and opened the way to the essential simplification of structural analysis of carbohydrates. This enabled Kochetkov and co-workers to begin studies of more complex carbohydrate-containing products, including animal sphingoglycolipids and mucin-type glycoproteins.

The progress in studies of natural products in the USSR led to the establishment in 1964 of the Institute of Biologically Active Substances (now the Pacific Institute of Bioorganic Chemistry) of the Far-East Branch of the Academy of Sciences in Vladivostok. Its first director and co-founder was Professor (later Academician) Georgy Elyakov, one of Kochetkov's collaborators. Several Ph.D.s from Nikolay's laboratory, including Yury Ovodov and Viktor Vas'kovsky, went to work there, still counting themselves as members of Kochetkov's school.

In 1960 Kochetkov was elected Correspondent Member of the Academy of Sciences. In 1965 he became the coordinator of an international program on the development of carbohydrate chemistry in countries of Eastern Europe, which continued for the next twenty two years. While in the IKhPS, he became a member of the editorial boards of *Carbohydrate Research*, *Advances in Heterocyclic Chemistry,* and a number of national journals. In 1961, in collaboration with Professors Igor' Torgov and Maria Botvinik, specialists in steroid and protein chemistry, Kochetkov wrote *Chemistry of Natural Compounds*, one of the first textbooks in this field in the country.

However, by the mid-1960s, the relationship between Academician Shemyakin and Kochetkov became difficult. Nikolay was dismissed from the post of deputy director and encountered serious problems in his work. Once again, Academician Nesmeyanov helped him. With the support of the then vice president of the Academy of Sciences, Academician Nikolay Semenov, he proposed Kochetkov as a candidate for the directorship of the N. D. Zelinsky Institute of Organic Chemistry (IOKh). The Presidium of the Academy of Sciences ultimately approved him to the post.

Nikolay was more than happy with the appointment, but wanted to take his laboratory with him. After long negotiations the laboratory was split up, with the majority of its carbohydrate unit moving to the IOKh and the nucleotide unit staying in the IKhPS. This was in the snowy winter, and veterans recalled how they used children's sleds as vehicles for transporting the laboratory equipment between the two Institutes, which were about 10 minutes walk apart from each other. During the move some crazy researchers could not interrupt chromatographic separations, and loaded sleds with columns, asking colleagues to pull them and running behind with bottles of solvents!

Nikolay recalled his first day at the IOKh, and how he was in front of the director's office, which he was to occupy for twenty two years. He would often say that he fell in love with the Institute from the first sight. Most of all he loved the people who, as he would note, were intelligent and top specialists in their fields. The directorship in the IOKh became the happiest and most creative period of his life.

Kochetkov's duties included the administration and scientific supervision of the Institute. In line with the Academy of Sciences' practice for directors, he chaired the IOKh's Scientific Council. In one of his last interviews, Nikolay said he thought it would be tragedy for a scientist to become just the manager of an enterprise, and he was able to avoid that. He remained a devoted scientist, while becoming an excellent manager and administrator. He maintained traditions that had developed since the Institute was founded in 1934.

Nikolay was successful in maintaining and extending those fields of organic chemistry in which the IOKh held strong positions. The fundamental research led to results that could be used by various industries. Among them, there were small molecules with potential as human and veterinary pharmaceuticals, insecticides and other plant protectors, and products for the food industry and agriculture. Kochetkov was a pure academic scientist, but he always took pride when he could contribute to the Institute's being able to solve applied problems, posed by the state.

As an enterprise, the IOKh was quite large and employed around 1,200 people—researchers and technicians, engineers of various specialities, pilot production workers, patent experts, librarians, and others. Apart from the academic research facilities, the Institute had a variety of analytical laboratories and services, as well as units for design and construction of new laboratory equipment.

In 1969, Nikolay established a patent department for assessing the value of discoveries and their potential for patenting and licensing. In 1978, the construction of a new laboratory building with modern equipment was completed, almost doubling the Institute's floor space. Nikolay viewed this project as one of the most challenging during the time of his directorship.

Another area of his work at the IOKh was education and publishing. Every year the Institute received a number of postgraduates and graduates, along with trainees from other parts of the Soviet Union and abroad, who had to be taught. There was also a program for the promotion of scientific achievements to the population through special lectures given by scientists and exhibitions. Numerous books, handbooks, monographs, textbooks, and guidelines for applied science institutions and industry were also published.

In 1980, the IOKh won the award of the Academy of Sciences for good work organization and high scientific achievements. In 1981 and 1982 it received similar awards from the Moscow municipal authorities, and in 1984 the Order of the Red Banner of Labour from the Soviet government.

The IOKh's life-science division was developing the chemistry of steroids, terpenoids, proteins, and carbohydrates. Flourishing in the Institute, the Kochetkov's laboratory contributed significantly to the chemistry and biochemistry of carbohydrates and became one of the leading laboratories in this field in the world.

Kochetkov's co-workers who came with him to the IOKh from the IKhPS, along with new people, Tat'yana Druzhinina, Alexander Sviridov, Nargiz Bairamova, Evgenii Klimov, Anatolii Chernyak, Nikolay Arbatskii, Nelly Malysheva, Vitalii Betaneli, Yuriy Knirel, Vladimir Torgov, Leonid Danilov, Andrei Nikolaev, Grigorii Lipkind and others, formed a new team. Alexander Shashkov was invited to develop NMR spectroscopy of carbohydrates. Oleg Chizhov and Anatolii Usov were appointed heads of the IOKh's laboratories of physico-chemical methods of analysis and plant polysaccharides, respectively, and they worked in close cooperation with the Kochetkov's laboratory.

New research projects were launched, and the synthesis of regular homo- and hetero-polysaccharides, including bacterial antigens, was the most challenging. In connection with this task, a number of complex oligosaccharides were synthesized, corresponding to the repeating units of specific polysaccharides of pathogenic bacteria (*Shigella flexneri*, *Salmonella enterica*, *Pseudomonas aeruginosa*, *Streptococcus pneumoniae*). In the course of this work the arsenal of methods for carbohydrate synthesis was developed and improved; however, the synthesis of regular polysaccharides was hampered by the lack of effective stereospecific glycosylation procedures. To solve this problem Kochetkov undertook a systematic study of the glycosylation process, resulting in the development of new, highly stereospecific methods for building 1,2-*trans*- and 1,2-*cis*-glycosidic linkages. Among these, the most useful was found to be the condensation of 1,2-*O*-(1-cyano)ethylidene derivatives with trityl ethers of mono- and oligosaccharides. When both functions occurred in the same molecule, polycondensation took place. Using this approach, a number of linear and branched polysaccharides—structural counterparts of bacterial antigens—were synthesized for the first time.

The trityl-cyanoethylidene polycondensation also permitted the synthesis of polysaccharides having functionalized aglycones for the production of neoglycoconjugates. An alternative approach for preparation of neoglycoconjugates, which could be used for serodiagnosis of infectious diseases (such as salmonellosis), was developed based on copolymerization of allyl or acryloylaminoalkyl glycosides of synthetic oligosaccharides with acrylamide. Yet another significant achievement in the area of neoglycoconjugate synthesis was the elaboration of a useful procedure for preparation of glycosylamines, known worldwide as the Kochetkov method.

Nikolay's team continued structural studies of glycolipids and glycoproteins and began the analysis of bacterial and algal polysaccharides. The structures of the oligosaccharide chains of gangliosides of starfish and sea urchins, blood-group substances, influenza virus hemagglutinin, and riboflavin-binding glycoprotein of egg white, gel-forming polysaccharides of red seaweeds and antigenic polysaccharides of various bacteria, including acute pathogens, were established. The data obtained were important for taxonomy and classification of the organisms and the development of new synthetic and semi-synthetic diagnostic agents and vaccines. In bacterial polymers, a number of unusual monosaccharides were discovered and identified, including new sugar ethers with lactic acid and two novel classes of acidic amino sugars: 2,3-diamino-2,3-dideoxyhexuronic and 5,7-diamino-3,5,7,9-tetradeoxynon-2-ulosonic acids; their structures were confirmed by chemical syntheses.

Kochetkov made major contributions to the methodology for the structure elucidation of complex carbohydrates. With his co-workers he elaborated new methods for chemical fragmentations and modifications of glycopolymers, including specific or selective cleavages of glycuronans and hexosaminoglycans, solvolytic depolymerization of heteropolysaccharides with hydrogen fluoride or trifluoromethanesulfonic acid, mild solvolytic desulfation of sulfated polysaccharides, and selective splitting of amide bonds and *O*-glycosidic linkages for isolation of glycopeptides and oligosaccharides from glycoproteins. An efficient ^{13}C NMR-based computer-assisted approach was devised for structural analysis of regular polysaccharides, including the determination of absolute configurations of constituent monosaccharides.

Nikolay was also interested in other fields of carbohydrate chemistry and biochemistry, such as conformations of oligo- and polysaccharides, synthesis of highly branched manno-oligosaccharides, radiolysis of carbohydrates, chemoenzymatic synthesis of complex polysaccharides, and studies of enzymes and lipids that participate in their biosynthesis. A joint project with Viktor Zhulin, the head of one of IOKh's laboratories, revealed that high pressure could significantly improve the stereospecificity of glycosylation. Kochetkov initiated research into the synthesis of sugar phosphites (with Eduard Nifant'ev of the Moscow State Pedagogical Institute) and phosphates; a so-called glycosyl-*H*-phosphonate method was specially designed for synthesis of oligosaccharides with phosphodiester linkages between subunits. His exploration of monosaccharides as chiral precursors of aglycones of macrolide antibiotics with multiple (up to 10) chiral centers culminated in the synthesis of erythronolides A and B and oleandonolide.

Soon after joining the IOKh, Nikolay published two new books, *Chemistry of Carbohydrates* (1967; with Alexey Bochkov, Boris Dmitriev, Anatolii Usov, Oleg Chizhov, and Vladimir Shibaev) and *Organic Chemistry of Nucleic Acids* (1970; with Eduard Budovskii, Evgeny Sverdlov, Natal'ya Simukova, Mikhail Turchinskii, and Vladimir Shibaev); the latter was then translated into English and Japanese. Another of his books, *Radiation Chemistry of Carbohydrates* (1978; with Lev Kudryashov and Mikhail Chlenov), was also translated into English. Later Kochetkov wrote three more books and a number of comprehensive reviews; his publication list exceeded 1200 scientific articles. He became an editor of the journals *Organic Mass Spectrometry*, *Tetrahedron*, and *Tetrahedron Letters*.

In 1979 Kochetkov was elected full member (Academician) of the Academy of Sciences of the USSR. The Soviet government recognized Nikolay as the Hero of Socialist Labour and awarded him with the Golden Star Order, which was then the highest civil state award. He also received a number of other Soviet Union orders and the highest national scientific awards—the Lenin prize for achievements in synthesis and structural analysis of carbohydrates (1988) and the Demidov prize for outstanding achievements in chemistry and biology (1993). In 1995 he received the Academy of Sciences' highest award for scientific achievements—the Lomonosov Great Gold Medal.

Kochetkov was awarded the Gold Medal of the Slovak Academy of Sciences in 1986 and the Haworth Memorial Medal of the UK Royal Society of Chemistry for achievements in carbohydrate chemistry in 1989. He was a member of the Chemical Society of France (since 1973) and a full member of the Polish Academy of Sciences (1988).

In 1988, the heyday of Gorbachev's perestroika, Nikolay was 73. The government had introduced an age limit for administrators at 70, and Academician Vladimir Tartakovskii was elected as new director of the IOKh. Nikolay continued as director emeritus and also remained the scientific chief of the laboratory.

The collapse of the Soviet Union and the ensuing political events hit science badly, with funding dropping and people leaving to work elsewhere. The situation in the IOKh was no different: scientists, having no particular experience in raising funds for research and commercialization of their scientific achievements, suddenly had to learn 'on the run', struggling also to keep their research going. Although Nikolay had never been a Communist Party member, he could not come to terms with the changes, and they made him upset and angry. He believed that the state's attitude toward science was awful and could destroy it, and that a great country like Russia could not exist without well-developed science.

However, he still remained involved with his laboratory, which was headed by Vladimir Shibaev during 1989–2004, and then by Yuriy Knirel. Nikolay did his best to help the laboratory, to which he had devoted most of his life, to survive, and it has overcome all the challenges during the most difficult years and continues to function. Since 2006, the laboratory has been called the N. K. Kochetkov Laboratory of Carbohydrate Chemistry.

Academician Nikolay Kochetkov died on December 21, 2005 at the age of ninety. He will be remembered not only as a great scientist but also as a tall, handsome, and witty man with charm and a colorful personality, who had a great influence on those who met him.

YURIY KNIREL
MARIA KOCHETKOVA

LIST OF SELECTED PUBLICATIONS BY N. K. KOCHETKOV

Books

1. N. K. Kochetkov, I. V. Torgov, and M. M. Botvinik, *Chemistry of Natural Compounds*, Izdatel. Akad. Nauk SSSR, Moscow, 1961 (in Russian).
2. N. K. Kochetkov, A. F. Bochkov, B. A. Dmitriev, A. I. Usov, O. S. Chizhov, and V. N. Shibaev, *Chemistry of Carbohydrates*, Khimiya, Moscow, 1967 [in Russian].
3. N. K. Kochetkov, E. I. Budowsky, E. D. Sverdlov, N. A. Simukova, M. F. Turchinsky, and V. N. Shibaev, *Organic Chemistry of Nucleic Acids,* Part A, Plenum, London, 1971; Part B, Plenum, London, 1972.
4. N. K. Kochetkov, L. I. Kudrjashov, and M. A. Chlenov, *Radiation Chemistry of Carbohydrates*, Pergamon Press, Oxford, 1979.
5. N. K. Kochetkov, A. F. Sviridov, M. S. Ermolenko, D. V. Yashunsky, and O. S. Chizhov, *Carbohydrates in the Synthesis of Natural Compounds*, Nauka, Moscow, 1984 [in Russian].
6. N. K. Kochetkov, *The Synthesis of Polysaccharides*, Nauka, Moscow, 1994 [in Russian].
7. N. K. Kochetkov, *Essays. Scientific Lectures and Reviews: Selected Works*, Nauka, Moscow, 2006 [in Russian].

Book chapters and review articles

8. N. K. Kochetkov, Chemistry of β-chlorovinyl ketones, *Usp. Khim.*, 24 (1955) 32–51.
9. N. K. Kochetkov and E. E. Nifant'ev, Chemistry of β-ketoacetals, *Usp. Khim.*, 30 (1961) 31–47.

10. N. K. Kochetkov and S. D. Sokolov, Recent developments in isoxazole chemistry, *Adv. Heterocyclic Chem.*, 2 (1963) 365–422.
11. N. K. Kochetkov, E. I. Budowsky, and V. N. Shibaev, Nucleoside diphosphates of sugars: Separation, structure, and biochemical properties, *Usp. Biol. Khim.*, 6 (1964) 108–141.
12. N. K. Kochetkov and A. M. Likhosherstov, Advances in pyrrolizidine chemistry, *Adv. Heterocyclic Chem.*, 5 (1965) 315–367.
13. N. K. Kochetkov and O. S. Chizhov, Mass spectrometry of carbohydrate derivatives, *Adv. Carbohydr. Chem.*, 21 (1966) 39–93.
14. M. I. Rybinskaya, A. N. Nesmeyanov, and N. K. Kochetkov, β-Ketovinylation, *Usp. Khim.*, 38 (1969) 961–1008.
15. N. K. Kochetkov, Chemical modification of nucleotides and its application to the investigation of nucleic acid structure, *Pure Appl. Chem.*, 18 (1969) 257–273.
16. N. K. Kochetkov and E. I. Budowsky, The chemical modification of nucleic acids, *Progr. Nucleic Acid Res. Mol. Biol.*, 9 (1969) 403–438.
17. N. K. Kochetkov and A. F. Bochkov, Sugar orthoesters and their synthetic applications, in R. Bognar, V. Druckner, and S. Szantay (Eds.), *Recent Developments in the Chemistry of Natural Carbon Compounds*, Vol. 4, Akad. Kiadó, Budapest, 1971, pp. 77–192.
18. N. K. Kochetkov and V. N. Shibaev, Glycosyl esters of nucleoside pyrophosphates, *Adv. Carbohydr. Chem. Biochem.*, 28 (1973) 307–399.
19. N. K. Kochetkov, O. S. Chizhov, and A. F. Bochkov, Oligosaccharides: Synthesis and determination of structure, *MTP Intern. Rev. Sci. Org. Chem. Ser. 1*, 7 (1973) 147–190.
20. N. K. Kochetkov, Some new trends in the chemistry of polysaccharides, *Pure Appl. Chem.*, 33 (1973) 53–72.
21. N. K. Kochetkov and V. N. Shibaev, Analogs of glucose uridine diphosphate: Synthesis and reactions with enzymes, *Russian Chem. Bull.*, 23 (1974) 1106–1121.
22. N. K. Kochetkov, Pathways for the synthesis of specific polysaccharides, *Pure Appl. Chem.*, 42 (1975) 327–350.
23. N. K. Kochetkov, O. S. Chizhov, and A. F. Sviridov, in R. L. Whistler and J. N. BeMiller (Eds.), Specific cleavage of polysaccharides containing uronuc acid residues by the Hoffman-Veerman reaction, *Methods in Carbohydrate Chemistry*, Vol. VIII, *General Methods*, Academic Press, New York, 1980, pp. 123–125.
24. N. K. Kochetkov, The synthesis of polysaccharides with a regular structure, *Sov. Sci. Rev. Chem. Rev. Sect. B.*, 4 (1982) 1–69.
25. A. F. Sviridov, A. Y. Shmyrina, O. S. Chizhov, and N. K. Kochetkov, Methods of synthesis of branched monosaccharides, *Bioorg. Khim.*, 8 (1982) 293–325.
26. N. K. Kochetkov, in L. Anderson and F. M. Unger (Eds.), Synthesis of O-antigenic polysaccharides. Pathways for the polymerization of oligosaccharide repeating units, *Bacterial Lipopolysaccharides: Structure, Synthesis and Biological Activities*, ACS Symp. Ser., Vol. 231, Amer. Chem. Soc., New York, 1983, pp. 65–81.
27. N. K. Kochetkov, Synthesis of fragments of bacterial polysaccharides and their application for preparation of synthetic antigens, *Pure Appl. Chem.*, 56 (1984) 923–938.
28. N. K. Kochetkov and G. P. Smirnova, Glycolipids of marine invertebrates, *Adv. Carbohydr. Chem. Biochem.*, 44 (1986) 387–438.
29. Y. A. Knirel and N. K. Kochetkov, 2,3-Diamino-2,3-dideoxyuronic and 5,7-diamino-3,5,7,9-tetradeoxynonulosonic acids: New components of bacterial polysaccharides, *FEMS Microbiol. Rev.*, 46 (1987) 381–385.
30. N. K. Kochetkov, Synthesis of polysaccharides with a regular structure, *Tetrahedron*, 43 (1987) 2389–2436.

31. N. K. Kochetkov and Y. A. Knirel, The structures of O-specific polysaccharides of bacterium *Pseudomonas aeruginosa*, *Sov. Sci. Rev. Chem. Rev. Sect. B*, Part 1, 13 (1989) 1–101.
32. N. K. Kochetkov, A. S. Shashkov, G. M. Lipkind, and Y. A. Knirel, New ^{13}C-NMR approaches to the structural analysis of carbohydrates, *Sov. Sci. Rev. Chem. Rev. Sect. B*, Part 2, 13 (1989) 1–73.
33. N. K. Kochetkov, Microbial polysaccharides: New approaches (Haworth memorial lecture), *Chem. Soc. Rev.*, 19 (1990) 29–54.
34. N. K. Kochetkov and A. F. Sviridov, Stereocontrolled synthesis of the main aglycons of 14-member macrolide antibiotics from carbohydrates, *Mendeleev Khim. Zh.*, 6 (1991) 389–400.
35. N. K. Kochetkov, The synthesis of polysaccharides to 1986, in J. W. Apsimon (Ed.), *The Total Synthesis of Natural Products*, Vol. 8, Wiley, New York, 1992, pp. 245–309.
36. Y. A. Knirel and N. K. Kochetkov, The structure of the lipopolysaccharides of Gram-negative bacteria. I. General characterization of the lipopolysaccharides and the structure of lipid A, *Biochemistry (Moscow)*, 58 (1993) 73–84.
37. Y. A. Knirel and N. K. Kochetkov, The structure of the lipopolysaccharides of Gram-negative bacteria. II. The structure of the core region, *Biochemistry (Moscow)*, 58 (1993) 84–99.
38. Y. A. Knirel and N. K. Kochetkov, The structure of the lipopolysaccharides of Gram-negative bacteria. III. The structure of the O-antigens, *Biochemistry (Moscow)*, 59 (1994) 1325–1383.
39. N. K. Kochetkov, Recent developments in the synthesis of polysaccharides and stereospecificity of glycosylation reactions, in Atta-ur-Rahman (Ed.), *Studies in Natural Products Chemistry*, Vol. 14, Elsevier, Amsterdam, 1994, pp. 201–266.
40. N. K. Kochetkov, Structure and synthesis of polysaccharides (Lomonosov Great Gold Medal winner lecture), *Vestnik Ross. Akad. Nauk*, 65 (1995) 730–738.
41. N. K. Kochetkov, Unusual monosaccharides: Components of O-antigenic polysaccharides of microorganisms, *Russian Chem. Rev.*, 65 (1996) 735–768.
42. N. K. Kochetkov and A. Y. Chernyak, Neoglycoconjugates as artificial antigens: Chemical aspects, *Adv. Macromol. Carbohydr. Res.*, 1 (1997) 93–175.
43. N. K. Kochetkov, Catalytic antibodies: Possible applications in organic synthesis, *Russian Chem. Rev.*, 67 (1998) 999–1029.
44. N. K. Kochetkov, Solid-phase synthesis of oligosaccharides and glycoconjugates, *Russian Chem. Rev.*, 69 (2000) 795–820.

Research articles

45. N. K. Kochetkov, R. M. Khomutov, and M. Ya. Karpeisky, New synthesis of cycloserine, *Dokl. Akad. Nauk SSSR*, 111 (1956) 831–834.
46. U. Dabrowski, Yu. A. Pentin, J. Dabrowski, V. M. Tatevskii, and N. K. Kochetkov, Tautomerism investigation of alkyl β-aminovinyl ketones from infrared absorption spectra, *Zh. Fiz. Khim.*, 32 (1958) 135–140.
47. N. K. Kochetkov, L. I. Kudryashov, and A. I. Usov, Interaction of diisopropylidene-glucose with halogen-triphenylphosphite complexes, *Dokl. Akad. Nauk SSSR*, 133 (1960) 1094–1097.
48. N. K. Kochetkov, V. A. Derevitskaya, and L. M. Likhosherstov, Condensation of monosaccharides and α-amino-acids with carbodiimide, *Chem. Ind.* (1960) 1532–1533.
49. N. K. Kochetkov, A. Ya. Khorlin, and A. F. Bochkov, Synthesis of K-strophanthin-β, *Dokl. Akad. Nauk SSSR*, 136 (1961) 613–616.
50. N. K. Kochetkov, I. G. Zhukova, and I. S. Glukhoded, Thin-layer chromatography of cerebrosides, *Dokl. Akad. Nauk SSSR*, 139 (1961) 608–611.

51. N. K. Kochetkov, A. Ya. Khorlin, V. E. Vaskovsky, and V. E. Zhvirblis, Triterpene saponins. 1. Saponins from roots of *Aralia manschurica*, *Zh. Obsch. Khim.*, 31 (1961) 658–665.
52. N. K. Kochetkov, L. I. Kudryashov, A. I. Usov, and B. A. Dmitriev, Monosaccharides. 1. New synthesis of D-quinovose and D-fucose, *Zh. Obsch. Khim.*, 31 (1961) 3303–3308.
53. N. K. Kochetkov, E. I. Budowsky, and V. N. Shibaev, Synthesis of *N*-methyluridine diphosphate glucose, *Biochim. Biophys. Acta*, 53 (1961) 415–417.
54. N. K. Kochetkov, A. Ya. Khorlin, O. S. Chizhov, and V. I. Sheichenko, Schizandrin—lignan of unusual structure, *Tetrahedron Lett.*, 2 (1961) 730–734.
55. N. K. Kochetkov, A. E. Vasil'ev, and S. N. Levchenko, Synthesis of dihydrosenecic acid, *Russian Chem. Bull.*, 11 (1962) 2141–2142.
56. N. K. Kochetkov, E. I. Budowsky, and N. A. Simukova, The reaction of hydroxylamine with ribonucleic acid, *Biochim. Biophys. Acta*, 55 (1962) 255–257.
57. N. K. Kochetkov, A. Ya. Khorlin, and V. E. Vaskovsky, The structures of aralosides A and B, *Tetrahedron Lett.*, 3 (1962) 713–716.
58. N. K. Kochetkov and A. Ya. Khorlin, Oligosides, a new type of plant glycosides, *Dokl. Akad. Nauk SSSR*, 150 (1963) 1289–1292.
59. N. K. Kochetkov and B. A. Dmitriev, The Wittig reaction in carbohydrate series, *Dokl. Akad. Nauk SSSR*, 151 (1963) 106–109.
60. E. E. Nifant'ev, M. A. Grachev, L. V. Backinowsky, S. G. Kara-Murza, and N. K. Kochetkov, Synthesis of methyl β-chlorovinyl ketone, *Zh. Prikl. Khim.*, 36 (1963) 676–678.
61. N. K. Kochetkov, E. I. Budowsky, and R. P. Shibaeva, The selective reaction of *O*-methylhydroxylamine with the cytidine nucleus, *Biochim. Biophys. Acta*, 68 (1963) 493–496.
62. N. K. Kochetkov, I. G. Zhukova, and I. S. Glukhoded, Sphingoplasmalogens. A new type of sphingolipids, *Biochim. Biophys. Acta*, 70 (1963) 716–719.
63. N. K. Kochetkov and A. I. Usov, The reaction of carbohydrates with triphenyl phosphite methiodide and related compounds. A new synthesis of deoxy sugars, *Tetrahedron*, 19 (1963) 973–983.
64. N. K. Kochetkov, N. S. Wulfson, O. S. Chizhov, and B. M. Zolotarev, Mass spectrometry of carbohydrate derivatives, *Tetrahedron*, 19 (1963) 2209–2224.
65. N. K. Kochetkov, A. Ya. Khorlin, and Yu. S. Ovodov, The structure of gypsoside—triterpenic saponin from *Gypsophila pacifica* Kom., *Tetrahedron Lett.*, 4 (1963) 477–482.
66. N. K. Kochetkov and A. I. Usov, The synthesis of D-chalcose, *Tetrahedron Lett.*, 4 (1963) 519–521.
67. A. Ya. Khorlin, A. F. Bochkov, and N. K. Kochetkov, New synthesis of ortho esters of sugars, *Russian Chem. Bull.*, 13 (1964) 2112–2114.
68. N. K. Kochetkov and O. S. Chizhov, Application of mass spectrometry to methylated monosaccharide identification, *Biochim. Biophys. Acta*, 83 (1964) 134–136.
69. N. K. Kochetkov, E. I. Budowsky, and R. P. Shibaeva, Selective modification of uridine and guanosine, *Biochim. Biophys. Acta*, 87 (1964) 515–518.
70. N. K. Kochetkov, A. Ya. Khorlin, and A. F. Bochkov, New synthesis of glycosides, *Tetrahedron Lett.*, 5 (1964) 289–293.
71. G. B. Elyakov, L. I. Strigina, N. I. Uvarova, V. E. Vaskovsky, A. K. Dzizenko, and N. K. Kochetkov, Glycosides from ginseng roots, *Tetrahedron Lett.*, 5 (1964) 3591–3597.
72. N. K. Kochetkov, E. I. Budowsky, M. F. Turchinsky, and N. A. Simukova, Method for selective chemical degradation of RNA. Basic principle, *Biochem. Biophys. Res. Commun.*, 19 (1965) 49–53.
73. N. K. Kochetkov and B. A. Dmitriev, New route to higher sugars, *Tetrahedron*, 21 (1965) 803–815.
74. N. K. Kochetkov, A. J. Khorlin, and V. J. Chirva, Clematoside C—triterpenic oligoside from *Clematis manshurica*, *Tetrahedron Lett.*, 6 (1965) 2201–2205.

75. E. I. Budowsky, T. N. Druzhinina, G. I. Eliseeva, N. D. Gabrielyan, N. K. Kochetkov, M. A. Novikova, V. N. Shibaev, and G. L. Zhdanov, Synthetic analogues of uridine diphosphate glucose: Biochemical and chemical studies. The secondary structure of nucleoside diphosphate sugars, *Biochim. Biophys. Acta*, 122 (1966) 213–224.
76. N. K. Kochetkov, A. Ya. Khorlin, A. F. Bochkov, and I. G. Yazlovetsky, A new route to the synthesis of polysaccharides, *Carbohydr. Res.*, 2 (1966) 84–85.
77. N. K. Kochetkov, E. E. Nifant'ev, and I. P. Gudkova, Reactions of levoglucosan with hypophosphoric acid. A new type of phosphorus-containing carbohydrate, *Zh. Obsch. Khim.*, 37 (1967) 277.
78. N. K. Kochetkov, A. I. Usov, and L. I. Miroshnikova, Polysaccharides of algae. 1. Water-soluble polysaccharides of the red alga *Laingia pacifica*, *Zh. Obsch. Khim.*, 37 (1967) 792–796.
79. N. E. Broude, E. I. Budovskii, and N. K. Kochetkov, Modification of tRNA by glyoxal, *Mol. Biol.*, 1 (1967) 214–223.
80. N. K. Kochetkov, A. J. Khorlin, and A. F. Bochkov, A new method of glycosylation, *Tetrahedron*, 23 (1967) 693–707.
81. N. K. Kochetkov, E. I. Budovskii, E. D. Sverdlov, R. P. Shibaeva, V. N. Shibaev, and G. S. Monastirskaya, The mechanism of the reaction of hydroxylamine and *O*-methylhydroxylamine with cytidine, *Tetrahedron Lett.*, 8 (1967) 3253–3257.
82. V. A. Derevitskaya, M. G. Vafina, and N. K. Kochetkov, Synthesis and properties of some serine glycosides, *Carbohydr. Res.*, 3 (1967) 377–388.
83. N. K. Kochetkov, V. A. Derevitskaya, and S. G. Kara-Murza, The structure of blood group substance A: General molecular architecture and nature of the carbohydrate-peptide linkage, *Carbohydr. Res.*, 3 (1967) 403–415.
84. N. K. Kochetkov, O. S. Chizhov, and A. F. Sviridov, An approach to the selective degradation of polysaccharide chains, *Carbohydr. Res.*, 4 (1967) 362–363.
85. N. K. Kochetkov, B. A. Dmitriev, and L. V. Bakinovsky, A new approach to the synthesis of higher 3-deoxy-glyculosonic acids, *Carbohydr. Res.*, 5 (1967) 399–405.
86. N. K. Kochetkov and A. I. Usov, Polysaccharides of algae. 2. Polysaccharides of the red alga *Odonthalia corymbifera* (Gmel.) J. Ag. Isolation of 6-*O*-methyl-D-galactose, *Zh. Obsch. Khim.*, 38 (1968) 234–238.
87. N. K. Kochetkov, O. S. Chizhov, and N. V. Molodtsov, Mass spectrometry of oligosaccharides, *Tetrahedron*, 24 (1968) 5587–5593.
88. A. I. Usov, M. A. Rechter, and N. K. Kochetkov, Polysaccharides of algae. 3. Isolation and preliminary investigation of a λ-polysaccharide from *Tichocarpus crinitus* (Gmel.) Rupr., *Zh. Obsch. Khim.*, 39 (1969) 905–911.
89. N. K. Kochetkov, A. M. Likhosherstov, and V. N. Kulakov, The total synthesis of some pyrrolyzidine alkaloids and their absolute configuration, *Tetrahedron*, 25 (1969) 2313–2323.
90. N. K. Kochetkov, V. A. Derevitskaya, and E. M. Klimov, Synthesis of glycosides *via* tert.-butyl ethers of alcohols, *Tetrahedron Lett.*, 10 (1969) 4762–4769.
91. N. K. Kochetkov, B. A. Dmitriev, and L. V. Backinowsky, Application of the Wittig reaction to the synthesis of 3-deoxy-D-*manno*-octulosonic acid, *Carbohydr. Res.*, 11 (1969) 193–197.
92. B. A. Dmitriev, A. V. Kessenich, A. Y. Chernyak, A. D. Naumov, and N. K. Kochetkov, Application of ^{19}F n.m.r. spectroscopy to carbohydrates: *O*-trifluoroacetyl derivatives of methyl glycopyranosides, *Carbohydr. Res.*, 11 (1969) 289–291.
93. A. I. Usov, M. D. Martynova, and N. K. Kochetkov, Discovery of agarase in molluscs of the genus *Littorina*, *Dokl. Acad. Nauk SSSR*, 194 (1970) 455–457.
94. N. K. Kochetkov, O. S. Chizhov, and A. F. Sviridov, Selective cleavage of glycuronosidic linkages, *Carbohydr. Res.*, 14 (1970) 277–285.

95. V. A. Derevitskaya, E. M. Klimov, and N. K. Kochetkov, Synthesis of esters *via* alkyl tert-butyl ethers, *Tetrahedron Lett.*, 11 (1970) 4269–4270.
96. N. K. Kochetkov, A. I. Usov, and K. S. Adamyants, Synthesis and nucleophilic substitution reactions of some iodo-deoxy sugars, *Tetrahedron*, 27 (1971) 549–559.
97. N. K. Kochetkov, A. F. Bochkov, T. A. Sokolovskaya, and V. J. Snyatkova, Modifications of the orthoester method of glycosylation, *Carbohydr. Res.*, 16 (1971) 17–27.
98. N. K. Kochetkov, B. M. Zolotarev, A. I. Usov, and M. A. Rechter, Mass-spectrometric characterization of 3,6-anhydrogalactose derivatives, *Carbohydr. Res.*, 16 (1971) 29–38.
99. A. F. Bochkov, V. M. Dashunin, A. V. Kessenikh, N. K. Kochetkov, A. D. Naumov, and I. V. Obruchnikov, Rigid conformation of tricyclic orthoesters of sugars, *Carbohydr. Res.*, 16 (1971) 497–499.
100. A. I. Usov, K. S. Adamyants, L. I. Miroshnikova, A. A. Shaposhnikova, and N. K. Kochetkov, Solvolytic desulphation of sulphated carbohydrates, *Carbohydr. Res.*, 18 (1971) 336–338.
101. B. A. Dmitriev, L. V. Backinowsky, O. S. Chizhov, B. M. Zolotarev, and N. K. Kochetkov, Gas-liquid chromatography and mass spectrometry of aldononitrile acetates and partially methylated aldononitrile acetates, *Carbohydr. Res.*, 19 (1971) 432–435.
102. N. K. Kochetkov, O. S. Chizhov, N. N. Malysheva, and A. I. Shiyonok, Application of mass spectrometry to oligosaccharide sequencing, *Org. Mass Spectrom.*, 5 (1971) 481–482.
103. N. K. Kochetkov, V. N. Shibaev, and A. A. Kost, Modification of nucleic acid components with α-halo aldehydes, *Dokl. Akad. Nauk SSSR*, 205 (1972) 100–103.
104. B. A. Dmitriev, L. V. Backinowsky, V. L. Lvov, N. K. Kochetkov, and I. L. Hofman, Immunochemical studies on *Shigella dysenteriae* lipopolysaccharides, *Eur. J. Biochem.*, 40 (1973) 355–359.
105. N. K. Kochetkov, I. G. Zhukova, G. P. Smirnova, and I. S. Glukhoded, Isolation and characterization of a sialoglycolipid from the sea urchin *Strongylocentrotus intermedius*, *Biochim. Biophys. Acta*, 326 (1973) 74–83.
106. N. K. Kochetkov, V. A. Derevitskaya, L. M. Likhosherstov, and S. A. Medvedev, A new destruction of blood group substances. Isolation of glycopeptide containing O-glycosidic carbohydrate-peptide linkage, *Biochem. Biophys. Res. Commun.*, 52 (1973) 748–751.
107. N. K. Kochetkov, A. I. Usov, L. I. Miroshnikova, and O. S. Chizhov, Polysaccharides of algae. 12. Partial hydrolysis of a polysaccharide from *Laingia pacifica* Yamada, *Zh. Obsch. Khim.*, 43 (1973) 1832–1839.
108. A. I. Usov, K. S. Adamyants, S. V. Yarotsky, A. A. Anoshina, and N. K. Kochetkov, The isolation of a sulphated mannan and a neutral xylan from the red seaweed *Nemalion vermiculare* Sur., *Carbohydr. Res.*, 26 (1973) 282–283.
109. N. K. Kochetkov, L. I. Kudrjashov, M. A. Chlenov, and T. Y. Livertovskaya, The epimerisation of monosaccharides by γ-irradiation in frozen, aqueous solutions, *Carbohydr. Res.*, 28 (1973) 86–88.
110. N. K. Kochetkov, A. I. Usov, and V. V. Deryabin, General synthetic route to monosaccharide amidosulfates, *Dokl. Acad. Nauk SSSR*, 216 (1974) 97–100.
111. N. K. Kochetkov, B. A. Dmitriev, O. S. Chizhov, E. M. Klimov, N. N. Malysheva, A. Y. Chernyak, N. E. Bayramova, and V. I. Torgov, Synthesis of derivatives of the trisaccharide repeating unit of the O-specific polysaccharide from *Salmonella anatum*, *Carbohydr. Res.*, 33 (1974) c5–c7.
112. V. N. Shibaev, Yu. Yu. Kusov, V. N. Chekunchikov, and N. K. Kochetkov, Chemical conversion of glycosyl phosphates into glycosides, *Dokl. Akad. Nauk SSSR*, 222 (1975) 370–372.
113. T. N. Druzhinina, Y. Y. Kusov, V. N. Shibaev, and N. K. Kochetkov, Interaction of uridine diphosphate glucose with calf liver uridine diphosphate glucose dehydrogenase. Significance of hydroxyl groups at C-3, C-4 and C-6 of hexosyl residue, *Biochim. Biophys. Acta*, 403 (1975) 1–8.

114. A. F. Bochkov, I. V. Obruchnikov, V. M. Kalinevich, and N. K. Kochetkov, Synthesis of β-1 → 6-D-glucan. A new reaction of polycondensation, *Tetrahedron Lett.*, 16 (1975) 3403–3406.
115. A. F. Bochkov and N. K. Kochetkov, A new approach to the synthesis of oligosaccharides, *Carbohydr. Res.*, 39 (1975) 355–357.
116. N. K. Kochetkov, V. N. Shibaev, A. A. Kost, A. P. Razjivin, and A. Y. Borisov, New fluorescent cytidine 5′-phosphate derivatives, *Nucleic Acids Res.*, 3 (1976) 1341–1349.
117. B. A. Dmitriev, Y. A. Knirel, N. K. Kochetkov, and I. L. Hofman, Somatic antigens of *Shigella*. Structural investigation on the O-specific polysaccharide chain of *Shigella dysenteriae* type 1 lipopolysaccharide, *Eur. J. Biochem.*, 66 (1976) 559–566.
118. N. K. Kochetkov, V. A. Derevitskaya, and N. P. Arbatsky, The structure of pentasaccharides and hexasaccharides from blood group substance H, *Eur. J. Biochem.*, 67 (1976) 129–136.
119. N. K. Kochetkov, G. P. Smirnova, and N. V. Chekareva, Isolation and structural studies of a sulfated sialosphingolipid from the sea urchin *Echinocardium cordatum*, *Biochim. Biophys. Acta*, 424 (1976) 274–283.
120. N. K. Kochetkov, B. A. Dmitriev, and L. V. Backinowsky, New sugars from antigenic lipopolysaccharides of bacteria: Identification and synthesis of 3-*O*-[(*R*)-1-carboxyethyl]-L- rhamnose, an acidic component of *Shigella dysenteriae* type 5 lipopolysaccharide, *Carbohydr. Res.*, 51 (1976) 229–237.
121. N. K. Kochetkov, V. N. Shibaev, A. A. Kost, A. P. Razjivin, and S. V. Ermolin, Fluorescent derivatives based on cytosine. Effect of substituents on optical properties of 2-phenylethenocytosines, *Dokl. Akad. Nauk SSSR*, 234 (1977) 227–230.
122. B. A. Dmitriev, Y. A. Knirel, N. K. Kochetkov, B. Jann, and K. Jann, Cell-wall lipopolysaccharide of the *Shigella*-like *Escherichia coli* O58. Structure of the polysaccharide chain, *Eur. J. Biochem.*, 79 (1977) 111–115.
123. N. K. Kochetkov, L. V. Backinowsky, and Y. E. Tsvetkov, Sugar thio-orthoesters as glycosylating agents, *Tetrahedron Lett.*, 18 (1977) 3681–3684.
124. N. K. Kochetkov, L. I. Kudrjashov, M. A. Chlenov, and L. P. Grineva, Radiation-induced dephosphorylation of sugar phosphates, *Carbohydr. Res.*, 53 (1977) 109–116.
125. N. K. Kochetkov, B. A. Dmitriev, and V. L. Lvov, 4-*O*-[(*R*)-1-Carboxyethyl]-D-glucose: A new acidic sugar from *Shigella dysenteriae* type 3 lipopolysaccharide, *Carbohydr. Res.*, 54 (1977) 253–259.
126. Y. V. Vozney and N. K. Kochetkov, Tritylation of secondary hydroxyl groups of sugars by triphenylmethylium salts, *Carbohydr. Res.*, 54 (1977) 300–303.
127. B. A. Dmitriev, V. L. Lvov, and N. K. Kochetkov, Complete structure of the repeating unit of the O-specific polysaccharide chain of *Shigella dysenteriae* type 3 lipopolysaccharide, *Carbohydr. Res.*, 56 (1977) 207–209.
128. G. P. Smirnova, N. V. Chekareva, O. S. Chizhov, B. M. Zolotarev, and N. K. Kochetkov, Mass spectrometry of sialic acids: Peracetylated methyl esters of 5-acylamino-3,5-dideoxy-nononic and -heptonic acids, *Carbohydr. Res.*, 59 (1977) 235–239.
129. V. A. Derevitskaya, N. P. Arbatsky, and N. K. Kochetkov, The structure of carbohydrate chains of blood-group substance. Isolation and elucidation of the structure of higher oligosaccharides from blood-group substance H, *Eur. J. Biochem.*, 86 (1978) 423–437.
130. V. N. Shibaev, A. A. Kost, N. K. Kochetkov, A. P. Razjivin, and S. V. Ermolin, A search for new fluorescent nucleotide derivatives: Properties of 4′-substituted 2-phenylethenocytosines, *Studia biophys.*, 69 (1978) 91–102.
131. B. A. Dmitriev and N. K. Kochetkov, Relation of the structure of the specific polysaccharides of the somatic antigens of the bacteria *Shigella dysenteriae* to immunochemical properties, *Dokl. Akad. Nauk SSSR*, 245 (1979) 765–768.

132. V. N. Shibaev, L. L. Danilov, V. N. Chekunchikov, Yu. Yu. Kusov, and N. K. Kochetkov, New synthesis of polyprenyl glycosyl pyrophosphates, *Bioorg. Khim.*, 5 (1979) 308–310.
133. V. N. Shibaev, T. N. Druzhinina, A. N. Popova, N. K. Kochetkov, S. S. Rozhnova, and V. A. Kilesso, Biosynthesis of the O-specific polysaccharide of *Salmonella senftenberg*, *Bioorg. Khim.*, 5 (1979) 1071–1082.
134. V. I. Betaneli, M. V. Ovchinnikov, L. V. Backinowsky, and N. K. Kochetkov, Synthesis of 1,2-*O*-(1-cyanoalkylidene) sugar derivatives, *Carbohydr. Res.*, 68 (1979) c11–c13.
135. N. K. Kochetkov, A. F. Sviridov, K. A. Arifkhodzhaev, O. S. Chizhov, and A. S. Shashkov, The structure of the extracellular polysaccharide from *Mycobacterium lacticolum* strain 121, *Carbohydr. Res.*, 71 (1979) 193–203.
136. V. I. Betaneli, M. V. Ovchinnikov, L. V. Backinowsky, and N. K. Kochetkov, Glycosylation by 1,2-*O*-cyanoethylidene derivatives of carbohydrates, *Carbohydr. Res.*, 76 (1979) 252–258.
137. N. K. Kochetkov, V. I. Torgov, N. N. Malysheva, and A. S. Shashkov, Synthesis of the pentasaccharide repeating unit of the O-specific polysaccharide from *Salmonella strasbourg*, *Tetrahedron*, 36 (1980) 1099–1105.
138. V. I. Betaneli, M. V. Ovchinnikov, L. V. Backinowsky, and N. K. Kochetkov, Practical synthesis of *O*-β-D-mannopyranosyl-, *O*-α-D-mannopyranosyl-, and *O*-β-D-galactopyranosyl-(1→4)-*O*-α-L-rhamnopyranosyl-(1→3)-D-galactoses, *Carbohydr. Res.*, 84 (1980) 211–224.
139. L. V. Backinowsky, N. F. Balan, A. S. Shashkov, and N. K. Kochetkov, Synthesis and ^{13}C-NMR spectra of β-L-rhamnopyranosides, *Carbohydr. Res.*, 84 (1980) 225–235.
140. L. V. Backinowsky, Y. E. Tsvetkov, N. F. Balan, N. E. Byramova, and N. K. Kochetkov, Synthesis of 1,2-*trans*-disaccharides *via* sugar thio-orthoesters, *Carbohydr. Res.*, 85 (1980) 209–221.
141. Y. A. Knirel, E. V. Vinogradov, B. A. Dmitriev, and N. K. Kochetkov, Determination of location of *O*-acetyl groups in bacterial polysaccharides, *Bioorg. Khim.*, 7 (1981) 463–465.
142. A. S. Shashkov, A. I. Usov, Y. A. Knirel, B. A. Dmitriev, and N. K. Kochetkov, Determination of absolute and anomeric configurations of monosaccharides in oligo- and poly-saccharides by glycosylation effects in ^{13}C-NMR spectra, *Bioorg. Khim.*, 7 (1981) 1364–1371.
143. N. K. Kochetkov, V. I. Betaneli, M. V. Ovchinnikov, and L. V. Backinowsky, Synthesis of the O-antigenic polysaccharide of *Salmonella newington* and of its analogue differing in configuration at the only glycosidic centre, *Tetrahedron*, 37 (1981) 149–156.
144. L. L. Danilov, S. D. Maltsev, V. N. Shibaev, and N. K. Kochetkov, Synthesis of polyprenyl pyrophosphate sugars from unprotected mono- and oligo-saccharide phosphates, *Carbohydr. Res.*, 88 (1981) 203–211.
145. G. M. Lipkind, V. E. Verovskii, and N. K. Kochetkov, Conformational analysis of the carbohydrate chains of blood-group substances, *Bioorg. Khim.*, 8 (1982) 963–970.
146. N. K. Kochetkov, G. P. Smirnova, and I. S. Glukhoded, Gangliosides with sialic acid bound to *N*-acetylgalactosamine from hepatopancreas of the starfish, *Evasterias retifera* and *Asterias amurensis*, *Biochim. Biophys. Acta*, 712 (1982) 650–658.
147. B. A. Dmitriev, N. A. Kocharova, Y. A. Knirel, A. S. Shashkov, N. K. Kochetkov, E. S. Stanislavsky, and G. M. Mashilova, Somatic antigens of *Pseudomonas aeruginosa*. The structure of the polysaccharide chain of *Ps. aeruginosa* O:6 (Lanyi) lipopolysaccharide, *Eur. J. Biochem.*, 125 (1982) 229–237.
148. V. N. Shibaev, L. L. Danilov, T. N. Druzhinina, L. M. Gogilashvili, S. D. Maltsev, and N. K. Kochetkov, Enzymatic synthesis of *Salmonella* O-specific polysaccharide analogs from modified polyprenyl pyrophosphate sugar acceptors, *FEBS Lett.*, 139 (1982) 177–180.
149. B. A. Dmitriev, A. V. Nikolaev, A. S. Shashkov, and N. K. Kochetkov, Block-synthesis of higher oligosaccharides: Synthesis of hexa- and nona-saccharide fragments of the O-antigenic polysaccharide of *Salmonella newington*, *Carbohydr. Res.*, 100 (1982) 195–206.

150. V. I. Betaneli, M. V. Ovchinnikov, L. V. Backinowsky, and N. K. Kochetkov, A convenient synthesis of 1,2-*O*-benzylidene and 1,2-*O*-ethylidene derivatives of carbohydrates, *Carbohydr. Res.*, 107 (1982) 285–291.
151. N. K. Kochetkov, B. A. Dmitriev, A. Y. Chernyak, and A. B. Levinsky, A new type of carbohydrate-containing synthetic antigen: Synthesis of carbohydrate-containing polysaccharide copolymers with the specificity of O:3 and O:4 factors of *Salmonella*, *Carbohydr. Res.*, 110 (1982) c16–c20.
152. V. N. Shibaev, T. N. Druzhinina, L. M. Gogilashvili, N. K. Kochetkov, S. S. Rozhnova, and V. A. Kilesso, Chemico-enzymatic synthesis of glucosylated O-specific polysaccharides of *Salmonella*, *Dokl. Akad. Nauk SSSR*, 270 (1983) 897–899.
153. L. V. Backinowsky, N. E. Nifantiev, and N. K. Kochetkov, Loss of stereospecificity upon glycosylation of methyl 2,3-di-*O*-acetyl-4-*O*-trityl-β-D-xylopyranoside by 1,2-*O*-cyanoalkylidene derivatives of monosaccharides, *Bioorg. Khim.*, 9 (1983) 1089–1096.
154. N. K. Kochetkov and G. P. Smirnova, A disialoglycolipid with two sialic acid residues located in the inner part of the oligosaccharide chain from hepatopancreas of the starfish *Patiria pectinifera*, *Biochim. Biophys. Acta*, 759 (1983) 192–198.
155. B. A. Dmitriev, V. L. Lvov, N. V. Tochtamysheva, A. S. Shashkov, N. K. Kochetkov, B. Jann, and K. Jann, Cell-wall lipopolysaccharide of *Escherichia coli* O114:H2. Structure of the polysaccharide chain, *Eur. J. Biochem.*, 134 (1983) 517–521.
156. V. I. Betaneli, L. V. Backinowsky, N. E. Bayramova, M. V. Ovchinnikov, M. M. Litvak, and N. K. Kochetkov, Glycosylation of 1,2-*O*-cyanoethylidene derivatives of carbohydrates, *Carbohydr. Res.*, 113 (1983) c1–c5.
157. V. A. Derevitskaya, L. M. Likhosherstov, M. D. Martynova, and N. K. Kochetkov, Specific method for the fragmentation of the polypeptide chain of glycoproteins. Distribution of carbohydrate chains on the peptide core of blood-group-specific glycoprotein, *Carbohydr. Res.*, 120 (1983) 85–94.
158. Y. A. Knirel, E. V. Vinogradov, N. A. Kocharova, A. S. Shashkov, B. A. Dmitriev, and N. K. Kochetkov, Synthesis and ^{13}C-n.m.r. spectra of methyl 2,3-diacetamido-2,3-dideoxy-α-D-hexopyranosides and 2,3-diacetamido-2,3-dideoxy-D-hexoses, *Carbohydr. Res.*, 122 (1983) 181–188.
159. N. E. Byramova, M. V. Ovchinnikov, L. V. Backinowsky, and N. K. Kochetkov, Selective removal of *O*-acetyl groups in the presence of *O*-benzoyl groups by acid-catalysed methanolysis, *Carbohydr. Res.*, 124 (1983) c8–c11.
160. L. M. Likhosherstov, O. S. Novikova, V. A. Derevitskaya, and N. K. Kochetkov, New method for cleavage of *N*-linked carbohydrate chains of glycoproteins using alkaline lithium borohydride, *Dokl. Akad. Nauk SSSR*, 274 (1984) 222–225.
161. L. L. Danilov, V. N. Shibaev, and N. K. Kochetkov, Phosphorylation of polyprenols *via* their trichloroacetimidates, *Synthesis* (1984) 404–406.
162. G. M. Lipkind, V. E. Verovsky, and N. K. Kochetkov, Conformational states of cellobiose and maltose in solution: A comparison of calculated and experimental data, *Carbohydr. Res.*, 133 (1984) 1–13.
163. N. K. Kochetkov, O. S. Chizhov, and A. S. Shashkov, Dependence of ^{13}C chemical shifts on the spatial interaction of protons, and its application in structural and conformational studies of oligo- and poly-saccharides, *Carbohydr. Res.*, 133 (1984) 173–185.
164. N. K. Kochetkov, N. E. Byramova, Y. E. Tsvetkov, and L. V. Backinowsky, Synthesis of the O-specific polysaccharide of *Shigella flexneri*, *Tetrahedron*, 41 (1985) 3363–3375.
165. L. V. Backinowsky, S. A. Nepogod'ev, A. S. Shashkov, and N. K. Kochetkov, Stereospecific synthesis of a (1→5)-α-L-arabinan, *Carbohydr. Res.*, 137 (1985) c1–c3.

166. N. E. Byramova, Y. E. Tsvetkov, L. V. Backinowsky, and N. K. Kochetkov, Synthesis of the basic chain of the O-specific polysaccharides of *Shigella flexneri*, *Carbohydr. Res.*, 137 (1985) c8–c13.
167. L. V. Backinowsky, S. A. Nepogod'ev, A. S. Shashkov, and N. K. Kochetkov, 1,2-*O*-Cyanoalkylidene derivatives of furanoses as 1,2-*trans*-glycosylating agents, *Carbohydr. Res.*, 138 (1985) 41–54.
168. L. M. Likhosherstov, O. S. Novikova, V. A. Derevitskaya, and N. K. Kochetkov, A new simple synthesis of amino sugar β-D-glycosylamines, *Carbohydr. Res.*, 146 (1986) c1–c5.
169. G. P. Smirnova, N. K. Kochetkov, and V. L. Sadovskaya, Gangliosides of the starfish *Aphelasterias japonica*, evidence for a new linkage between two N-glycolylneuraminic acid residues through the hydroxy group of the glycolic acid residue, *Biochim. Biophys. Acta*, 920 (1987) 47–55.
170. N. P. Arbatskii, M. D. Martynova, V. A. Derevitskaya, and N. K. Kochetkov, New approach to the structural study of carbohydrate chains of glycoproteins. Application of HPLC for identification of oligosaccharides, *Dokl. Akad. Nauk SSSR*, 297 (1987) 995–999.
171. N. K. Kochetkov, N. E. Nifant'ev, and L. V. Backinowsky, Synthesis of the capsular polysaccharide of *Streptococcus pneumoniae* type 14, *Tetrahedron*, 43 (1987) 3109–3121.
172. N. K. Kochetkov, V. M. Zhulin, E. M. Klimov, N. N. Malysheva, Z. G. Makarova, and A. Y. Ott, The effect of high pressure on the stereospecificity of the glycosylation reaction, *Carbohydr. Res.*, 164 (1987) 241–254.
173. E. V. Vinogradov, Y. A. Knirel, A. S. Shashkov, and N. K. Kochetkov, Determination of the degree of amidation of 2-deoxy-2-formamido-D-galacturonic acid in O-specific polysaccharides of *Pseudomonas aeruginosa* O4 and related strains, *Carbohydr. Res.*, 170 (1987) c1–c4.
174. N. A. Kocharova, Y. A. Knirel, A. S. Shashkov, N. K. Kochetkov, and G. B. Pier, Structure of an extracellular cross-reactive polysaccharide from *Pseudomonas aeruginosa* immunotype 4, *J. Biol. Chem.*, 263 (1988) 11291–11295.
175. G. M. Lipkind, A. S. Shashkov, Y. A. Knirel, E. V. Vinogradov, and N. K. Kochetkov, A computer-assisted structural analysis of regular polysaccharides on the basis of ^{13}C-n.m.r. data, *Carbohydr. Res.*, 175 (1988) 59–75.
176. Y. A. Tsvetkov, A. V. Bukharov, L. V. Backinowsky, and N. K. Kochetkov, Synthesis of homopolysaccharides and block-heteropolysaccharides carrying a spacer arm, *Carbohydr. Res.*, 175 (1988) c1–c4.
177. L. M. Likhosherstov, O. S. Novikova, V. E. Piskarev, E. E. Trusikhina, V. A. Derevitskaya, and N. K. Kochetkov, A method for reductive cleavage of *N*-glycosylamide carbohydrate-peptide bond, *Carbohydr. Res.*, 178 (1988) 155–163.
178. N. P. Arbatsky, V. A. Derevitskaya, A. O. Zheltova, N. K. Kochetkov, L. M. Likhosherstov, S. N. Senchenkova, and D. V. Yurtov, The carbohydrate chains of influenzae virus hemagglutinin, *Carbohydr. Res.*, 178 (1988) 165–181.
179. G. M. Lipkind, A. S. Shashkov, S. S. Mamyan, and N. K. Kochetkov, The nuclear Overhauser effect and structural factors determining the conformations of disaccharide glycosides, *Carbohydr. Res.*, 181 (1988) 1–12.
180. A. S. Shashkov, G. M. Lipkind, Y. A. Knirel, and N. K. Kochetkov, Stereochemical factors determining the effects of glycosylation on the ^{13}C chemical shifts in carbohydrates, *Magn. Reson. Chem.*, 26 (1988) 735–747.
181. N. P. Arbatskii, A. O. Zheltova, D. V. Yurtov, V. A. Derevitskaya, and N. K. Kochetkov, Successive isolation of hemagglutinin and neuraminidase from influenza virus A/Krasnodar/101/59 (H2N2) using bromelain, *Dokl. Akad. Nauk SSSR*, 306 (1989) 1490–1493.

182. V. M. Zhulin, Z. G. Makarova, E. M. Klimov, N. N. Malysheva, and N. K. Kochetkov, Effect of phase transitions of solvents on stereospecificity of glycosylation at high pressures, *Dokl. Akad. Nauk SSSR*, 309 (1989) 641–645.
183. E. V. Vinogradov, D. Krajewska-Pietrasik, W. Kaca, A. S. Shashkov, Y. A. Knirel, and N. K. Kochetkov, Structure of *Proteus mirabilis* O27 O-specific polysaccharide containing amino acids and phosphoethanolamine, *Eur. J. Biochem.*, 185 (1989) 645–650.
184. N. K. Kochetkov, A. F. Sviridov, M. S. Ermolenko, D. V. Yashunsky, and V. S. Borodkin, Stereocontrolled synthesis of erythronolides A and B from 1,6-anhydro-β-D-glucopyranose (levoglucosan). Skeleton assembly in (C_9–C_{13}) + (C_7–C_8) + (C_1–C_6) sequence, *Tetrahedron*, 45 (1989) 5109–5136.
185. N. K. Kochetkov, E. M. Klimov, and N. N. Malysheva, Novel highly stereospecific method of 1,2-*cis*-glycosylation. Synthesis of α-D-glucosyl-D-glucoses, *Tetrahedron Lett.*, 30 (1989) 5459–5462.
186. L. V. Backinowsky, S. A. Nepogod'ev, and N. K. Kochetkov, Formation of cyclo-oligosaccharides by polycondensation of the 3- and 6-*O*-tritylated derivatives of 1,2-*O*-(1-cyanoethylidene)-α-D-galactofuranose, *Carbohydr. Res.*, 185 (1989) c1–c3.
187. N. P. Arbatsky, M. D. Martynova, A. O. Zheltova, V. A. Derevitskaya, and N. K. Kochetkov, Studies on structure and heterogeneity of carbohydrate chains of *N*-glycoproteins by use of liquid chromatography. Oligosaccharide maps of glycoproteins, *Carbohydr. Res.*, 187 (1989) 165–171.
188. Y. A. Knirel, N. A. Kocharova, A. S. Shashkov, N. K. Kochetkov, V. A. Mamontova, and T. F. Solov'eva, Structure of the capsular polysaccharide of *Klebsiella ozaenae* serotype K4 containing 3-deoxy-D-*glycero*-D-*galacto*-nonulosonic acid, *Carbohydr. Res.*, 188 (1989) 145–155.
189. Y. E. Tsvetkov, L. V. Backinowsky, and N. K. Kochetkov, Synthesis of a common polysaccharide antigen of *Pseudomonas aeruginosa* as the 6-aminohexyl glycoside, *Carbohydr. Res.*, 193 (1989) 75–90.
190. N. P. Arbatskii, A. O. Zheltova, D. V. Yurtov, I. G. Kharitonenkov, Yu. P. Abashev, V. A. Derevitskaya, and N. K. Kochetkov, The structure of carbohydrate chains of hemagglutinin and neuraminidase of influenza virus B/Leningrad/179/86, *Bioorg. Khim.*, 16 (1990) 801–807.
191. N. K. Kochetkov, S. A. Nepogod'ev, and L. V. Backinowsky, Synthesis of cyclo-[(l-6)-β-D-galactofurano]-oligosaccharides, *Tetrahedron*, 46 (1990) 139–150.
192. L. M. Likhosherstov, O. S. Novikova, V. A. Derevitskaya, and N. K. Kochetkov, A selective method for sequential splitting of *O*- and *N*-linked glycans from *N*,*O*-glycoproteins, *Carbohydr. Res.*, 199 (1990) 67–76.
193. E. Katzenellenbogen, E. Romanowska, N. A. Kocharova, A. S. Shashkov, G. M. Lipkind, Y. A. Knirel, and N. K. Kochetkov, The structure of the biological repeating unit of the O-antigen of *Hafnia alvei* O39, *Carbohydr. Res.*, 203 (1990) 219–227.
194. A. V. Nikolaev, I. A. Ivanova, V. N. Shibaev, and N. K. Kochetkov, Application of the hydrogenphosphonate approach in the synthesis of glycosyl phosphosugars linked through secondary hydroxyl groups, *Carbohydr. Res.*, 204 (1990) 65–78.
195. L. M. Likhosherstov, V. E. Piskarev, N. F. Sepetov, V. A. Derevitskaya, and N. K. Kochetkov, Structure of the oligosaccharide chains of riboflavin-binding glycoprotein from hen egg white. 4. Neutral oligosaccharides of a hybrid type, *Bioorg. Khim.*, 17 (1991) 246–251.
196. Y. Y. Tendetnik, V. I. Pokrovsky, A. Y. Chernyak, A. B. Levinsky, and N. K. Kochetkov, The use of synthetic O-antigens of *Salmonella* for improvement of serological diagnosis of enteric infections, *FEMS Microbiol. Immunol.*, 76 (1991) 93–98.
197. A. Y. Chernyak, G. V. Sharma, L. O. Kononov, P. Radha Krishna, A. V. Rama Rao, and N. K. Kochetkov, Synthesis of glycuronamides of amino acids, constituents of microbial polysaccharides, and their conversion into neoglycoconjugates of copolymer type, *Glycoconj. J.*, 8 (1991) 82–89.

198. A. F. Sviridov, V. S. Borodkin, M. S. Ermolenko, D. V. Yashunsky, and N. K. Kochetkov, Stereocontrolled synthesis of erythronolides A and B in a (C5–C9) + (C3–C4) + (C1–C2) + (C11–C13) sequence from 1,6-anhydro-β-D-glycopyranose (levoglucosan). Part 1. Synthesis of C1–C10 and C11–C13 segments, *Tetrahedron*, 47 (1991) 2291–2316.
199. A. F. Sviridov, V. S. Borodkin, M. S. Ermolenko, D. V. Yashunsky, and N. K. Kochetkov, Stereocontrolled synthesis of erythronolides A and B in a (C5–C9) + (C3–C4) + (C1–C2) + (C11–C13) sequence from 1,6-anhydro-β-D-glycopyranose (levoglucosan). Part 2, *Tetrahedron*, 47 (1991) 2317–2336.
200. N. K. Kochetkov, E. M. Klimov, N. N. Malysheva, and A. V. Demchenko, A new stereoselective method for 1,2-*cis*-glycosylation, *Carbohydr. Res.*, 212 (1991) 77–91.
201. E. V. Vinogradov, A. S. Shashkov, Y. A. Knirel, N. K. Kochetkov, N. V. Tochtamysheva, S. P. Averin, O. V. Goncharova, and V. S. Khlebnikov, Structure of the O-antigen of *Francisella tularensis* strain 15, *Carbohydr. Res.*, 214 (1991) 289–297.
202. Y. A. Knirel, E. V. Vinogradov, A. S. Shashkov, Z. Sidorczyk, W. Kaca, A. Rozalski, K. Kotelko, and N. K. Kochetkov, The structure of the O-specific polysaccharides of *Proteus*, *Dokl. Akad. Nauk*, 324 (1992) 333–338.
203. N. K. Kochetkov, N. N. Malysheva, E. M. Klimov, and A. V. Demchenko, Synthesis of polysaccharides with 1,2-*cis*-glycosidic linkages by tritylthiocyanate polycondensation. Stereoregular α-(1-6)-D-glucan, *Tetrahedron Lett.*, 33 (1992) 381–384.
204. A. Y. Chernyak, G. V. M. Sharma, L. O. Kononov, P. Radha Krishna, A. B. Levinskii, N. K. Kochetkov, and A. V. Rama Rao, 2-Azidoethyl glycosides: Glycosides potentially useful for the preparation of neoglycoconjugates, *Carbohydr. Res.*, 223 (1992) 303–309.
205. N. K. Kochetkov, E. M. Klimov, N. N. Malysheva, and A. V. Demchenko, Stereospecific 1,2-*cis*-glycosylation: A modified thiocyanate method, *Carbohydr. Res.*, 232 (1992) c1–c5.
206. Y. A. Knirel, N. A. Paramonov, A. S. Shashkov, N. K. Kochetkov, R. G. Yarullin, S. M. Farber, and V. I. Efremenko, Structure of the polysaccharide chains of *Pseudomonas pseudomallei* lipopolysaccharides, *Carbohydr. Res.*, 233 (1992) 185–193.
207. N. E. Nifantiev, A. S. Shashkov, G. M. Lipkind, and N. K. Kochetkov, Synthesis, NMR, and conformational studies of branched oligosaccharides. 7. Synthesis and ^{13}C NMR spectra of 2,3-di-*O*-glycosyl derivatives of methyl α-L-rhamnopyranoside and methyl α-D-mannopyranoside, *Carbohydr. Res.*, 237 (1992) 95–113.
208. P. I. Kitov, Yu. E. Tsvetkov, L. V. Backinowsky, and N. K. Kochetkov, Reactivity of 1,2-*O*-cyanoalkylidene sugar derivatives in trityl-cyanoalkylidene condensation, *Russian Chem. Bull.*, 42 (1993) 1423–1428.
209. P. I. Kitov, Yu. E. Tsvetkov, L. V. Backinowsky, and N. K. Kochetkov, Reactivity of sugar trityl ethers in trityl-cyanoalkylidene condensation, *Russian Chem. Bull.*, 42 (1993) 1909–1915.
210. Y. E. Tsvetkov, P. I. Kitov, L. V. Backinowsky, and N. K. Kochetkov, Unusual regioselective glycosylation of sugar secondary trityloxy function in the presence of primary ones, *Tetrahedron Lett.*, 34 (1993) 7977–7980.
211. N. K. Kochetkov, E. M. Klimov, N. N. Malysheva, and A. V. Demchenko, Stereospecific synthesis of 1,2-*cis*-glycosides of 2-amino sugars, *Carbohydr. Res.*, 242 (1993) c7–c10.
212. N. K. Kochetkov, V. I. Betaneli, I. A. Kryazhevskikh, and A. Y. Ott, Glycosylation by sugar 1,2-*O*-(1-cyanobenzylidene) derivatives: Influence of glycosyl-donor structure and promoter, *Carbohydr. Res.*, 244 (1993) 85–97.
213. N. A. Kocharova, Y. A. Knirel, A. S. Shashkov, N. E. Nifantiev, N. K. Kochetkov, L. D. Varbanets, N. V. Moskalenko, O. S. Brovarskaya, V. A. Muras, and J. M. Young, Studies of O-specific polysaccharide chains of *Pseudomonas solanacearum* lipopolysaccharides consisting of structurally different repeating units, *Carbohydr. Res.*, 250 (1993) 275–287.

214. N. A. Kocharova, J. E. Thomas-Oates, Y. A. Knirel, A. S. Shashkov, U. Dabrowski, N. K. Kochetkov, E. S. Stanislavsky, and E. V. Kholodkova, The structure of the O-specific polysaccharide of *Citrobacter* O16 containing glycerol phosphate, *Eur. J. Biochem.*, 219 (1994) 653–661.
215. E. V. Vinogradov, Y. A. Knirel, A. S. Shashkov, N. A. Paramonov, N. K. Kochetkov, E. S. Stanislavsky, and E. V. Kholodkova, The structure of the O-specific polysaccharide of *Salmonella arizonae* O21 (*Arizona* 22) containing *N*-acetylneuraminic acid, *Carbohydr. Res.*, 259 (1994) 59–65.
216. Y. A. Knirel, N. A. Paramonov, E. V. Vinogradov, N. K. Kochetkov, Z. Sidorczyk, and K. Zych, 2-Acetamido-4-*O*-[(*S*)-1-carboxyethyl]-2-deoxy-D-glucose, a new natural isomer of *N*-acetylmuramic acid from the O-specific polysaccharide of *Proteus penneri* 35, *Carbohydr. Res.*, 259 (1994) c1–c3.
217. A. V. Demchenko, N. N. Malysheva, and N. K. Kochetkov, Influence of experimental conditions on trityl-thiocyanate condensation, *Dokl. Akad. Nauk*, 341 (1995) 635–637.
218. Y. E. Tsvetkov, P. I. Kitov, L. V. Backinowsky, and N. K. Kochetkov, Highly regioselective glycosylation of a secondary position in sugar primary-secondary ditrityl ethers, *J. Carbohydr. Chem.*, 15 (1996) 1027–1050.
219. N. A. Kocharova, O. V. Shcherbakova, A. S. Shashkov, Y. A. Knirel, N. K. Kochetkov, E. V. Kholodkova, and E. S. Stanislavsky, Structure of the O-specific polysaccharide of the bacterium *Providencia alcalifaciens* serogroup O23 containing a novel amide of D-glucuronic acid with N^{ε}-(1-carboxyethyl)lysine, *Biochemistry (Moscow)*, 62 (1997) 501–508.
220. A. S. Shashkov, N. A. Paramonov, S. N. Veremeychenko, H. Grosskurth, G. M. Zdorovenko, Y. A. Knirel, and N. K. Kochetkov, Somatic antigens of pseudomonads: Structure of the O-specific polysaccharide of *Pseudomonas fluorescens* biovar B, strain IMV 247, *Carbohydr. Res.*, 306 (1998) 297–303.
221. A. V. Perepelov, S. N. Senchenkova, A. S. Shashkov, N. A. Komandrova, S. V. Tomshich, L. S. Shevchenko, Y. A. Knirel, and N. K. Kochetkov, First application of triflic acid for selective cleavage of glycosidic linkages in structural studies of a bacterial polysaccharide from *Pseudoalteromonas* sp. KMM 634, *J. Chem. Soc. Perkin Trans.* 1 (2000) 363–366.
222. N. P. Arbatsky, V. N. Shibaev, and N. K. Kochetkov, A new approach to localization of the *O*-glycosylation sites in glycoproteins: Mass-spectrometric analysis of *O*-glycopeptides formed upon the reductive cleavage of the yeast mannoprotein by $LiBH_4$–LiOH, *Doklady Chem.*, 387 (2002) 328–331.
223. L. V. Backinowsky, P. I. Abronina, A. S. Shashkov, A. A. Grachev, N. K. Kochetkov, S. A. Nepogodiev, and J. F. Stoddart, An efficient approach towards the convergent synthesis of fully-carbohydrate mannodendrimers, *Chem.: Eur. J.*, 8 (2002) 4412–4423.
224. N. A. Kocharova, S. N. Senchenkova, A. N. Kondakova, A. I. Gremyakov, G. V. Zatonsky, A. S. Shashkov, Y. A. Knirel, and N. K. Kochetkov, D- and L-Aspartic acids: New non-sugar components of bacterial polysaccharides, *Biochemistry (Moscow)*, 69 (2004) 103–107.
225. E. A. Ivlev, L. V. Backinowsky, P. I. Abronina, L. O. Kononov, and N. K. Kochetkov, A short and simple synthesis of branched mannooligosaccharides with [1- ^{13}C]-labelled terminal mannose units, *Polish J. Chem.*, 79 (2005) 275–286.

INTERNATIONAL CARBOHYDRATE SYMPOSIA—A HISTORY☆

By Stephen J. Angyal

School of Chemistry, University of New South Wales, Sydney, NSW 2052, Australia

I. Beginnings

The first meeting calling itself the International Symposium on Carbohydrate Chemistry was held in Birmingham, England, in 1962. It was organized by Maurice Stacey to commemorate the life and work of Sir Norman Haworth, and was held in the recently opened building named after Haworth. It brought together most of the people who were then extending the frontiers of carbohydrate chemistry; there were 280 participants, more than half of them from the

☆Dedicated to Roy L. Whistler who has done more than any other person to promote and organize the International Carbohydrate Symposia.

ISBN: 978-0-12-373920-9
DOI: 10.1016/S0065-2318(07)61002-6

United Kingdom. This gathering formed a firm base for subsequent meetings; eight of the chairmen of future symposia (Angyal, BeMiller, Courtois, Dutton, Jones, Micheel, Overend, and Whistler) were present in Birmingham. Subsequently, several carbohydrate chemists supported having such symposia at regular intervals, similar to the successful symposia on the Chemistry of Natural Products, inaugurated in Sydney in 1960. The next Symposium, in Münster in 1964, carried no number but the subsequent one in Kingston, Ontario (1967) was called the IVth Symposium. By then, probably owing to the influence of Roy Whistler, it was decided that a meeting held in Gif-sur-Yvette should be regarded as the first one.

That meeting (18–21 July 1960), chaired by J. E. Courtois, was entitled the International Colloquium on the Biochemistry of Sugars—Structure and Specificity; it was organized by the Centre National de la Recherche Scientifique (CNRS) with the help of the Société de Chimie Biologique, the occasion being the opening of Edgar Lederer's new research building. It was quite a small meeting with only 66 participants, but it presented an extensive program which included a considerable amount of fundamental chemistry besides biochemistry. Forty-three papers were delivered by speakers from many countries; they were subsequently published by the CNRS in 1961. Roy Whistler, who has contributed so much to the development and organization of the International Symposia, was a joint president. In his closing address he said "Surely it is the wish of all that programs such as this will be repeated often in the years ahead," thus foreshadowing the series of Symposia. Whistler relates that on the last evening he had a discussion with Courtois, Bengt Lindberg, and Fritz Micheel. He again suggested that there should be more international meetings on carbohydrates; most of the people in the group said that they would organize such a meeting at their universities.

The first to do so was Maurice Stacey; he had the same excuse as Courtois: the opening of the new chemistry building in Birmingham. Micheel followed two years later, and then J. K. N. Jones organized one in Canada (1967). After some delays, there was another meeting in Paris (1970), again chaired by Courtois; he is the only person who chaired two international carbohydrate meetings.

Not everyone agreed with the designation of the Gif meeting as the first Symposium. Maurice Stacey, in a letter to the present author, written on 12 December 1994, says: "We always considered that the Birmingham meeting was No. 1, for we considered that the meeting now called No. 1 was only a European meeting, for no-one from Birmingham was invited to attend." He was not quite right: S. A. Barker delivered a paper in Gif for which Stacey was listed as a co-author (Picture 1).

PICTURE 1. The participants in the symposium at Gif (1960).

In the preface to the handbook of the Münster meeting (1964), F. Micheel does not mention the Gif meeting, but cites a symposium held in King's College, London, in 1953 as the first international conference on carbohydrates. However, that was one of the regular meetings of The Chemical Society (London), attended mainly by British chemists. The listing of the preceding Symposia in the Utrecht handbook (1984) seems to be in doubt about the status of Gif: it lists "1960 1 Edinburgh/Gif-sur-Yvette." The meeting in Edinburgh was held in July 1960 on "Physical, Chemical, and Biological Methods in the Study of High Molecular Weight Carbohydrates," another one of the regular meetings of The Chemical Society. The subject, obviously, was much more restricted than that at Gif and that meeting would hardly have a claim for international status. The Gif meeting is now generally accepted as the first Symposium.

The time interval between symposia was uncertain at this stage, but was settled as being two years after the Paris meeting (1970). Originally, most of the papers dealt with the chemistry of carbohydrates. However, by the 1980s more and more biochemistry appeared and ultimately whole sessions were devoted to this subject. Instead of lengthening the title of the Symposia by adding "and Biochemistry," it was decided to drop the word "Chemistry." Hence, beginning in Vancouver (1982), the meetings became the International Carbohydrate Symposia.

II. International Steering Committee for Carbohydrate Symposia

Until the mid-1960s, the holding of a Carbohydrate Symposium was a haphazard affair that depended on the goodwill and enthusiasm of a local group. Some discussion took place in Münster (1964) about the future of the Symposia, but it was inconclusive. It was Roy L. Whistler who first suggested a systematic approach. Whistler (the only person who attended every Carbohydrate Symposium until the XXIst) has been the moving spirit behind the symposia during most of their existence. In 1965 he wrote to eight carbohydrate chemists, asking them to join a proposed Steering Committee for International Carbohydrate Symposia. The eight potential members, each of whom accepted, were S. J. Angyal (Australia), J. E. Courtois (France), J. K. N. Jones (Canada), N. K. Kochetkov (USSR), B. Lindberg (Sweden), F. Micheel (Germany), Z. Nikuni (Japan), and M. Stacey (Great Britain). Two more joined soon thereafter: V. Bauer (Czechoslovakia) and R. Bognár (Hungary). The purpose of the steering committee was mainly to plan a regular succession of future symposia, every second year (Picture 2).

PICTURE 2. Roy L. Whistler.

There was some delay until J. K. N. Jones decided to hold the next symposium in Kingston (Canada) in 1967. Subsequently the Hungarians put in a claim, followed by the Czechoslovaks; then an apparently stronger claim was submitted by the French. Ultimately it was decided that the symposia be held every even-numbered year, partly to coincide in year with the Symposia on the Chemistry of Natural Products, enabling people from other continents to attend both. This is no longer taken into consideration. The next symposium was to be held in Paris in 1970. It was also decided that the symposia be held alternately in Europe and away from Europe; hence the Czech claim for 1972 was postponed until 1974, and the 1972 symposium was held in Madison, USA. (The only exception to this procedure was made for the 1982 meeting in Vancouver.) The Hungarians ultimately gave up and held their own symposium in Debrecen.

The first actual meeting of the Steering Committee took place in Paris (1970) on the Sunday preceding the symposium; it was followed by a dinner. This has become a tradition, taken over later by the International Carbohydrate Organization. Its members have generally met on the Sunday before the Symposia; the dinner, to which their spouses are invited, is usually a distinguished affair. This is the members' only reward for carrying out their (not too onerous) duties.

At their meetings on 16 and 18 August 1970, in the Pharmacy School in Paris, the Committee established a constitution and passed a number of resolutions. The 11 members who had served until then were made national representatives of their countries. It was pointed out by Whistler that they did not, at that time, officially represent their countries, and arrangements were made to have them accredited by some authoritative body in each country. That body is usually the Chemical Society or the Academy of Science. It was also decided to create a class of Co-operative Members, these being the Head of the European Carbohydrate Committee, the chairmen of the Divisions of Carbohydrate Chemistry and of Wood, Cellulose, and Fiber Chemistry of the American Chemical Society, the chairman of the British Discussion Group for Carbohydrate Chemistry, and all past presidents of the International Steering Committee for Carbohydrate Meetings. It was further decided that the national representative of the country which hosts a symposium will then become president for the next two years, and the one of the following symposium, president-elect for the same two years. Hence Courtois was elected president and Whistler president-elect for 1970–1972. The position of secretary was established for four-year terms; W. G. Overend was elected secretary for 1972–1975, and Whistler was asked to continue acting as secretary till the end of 1971. J. N. BeMiller was appointed subsequently as secretary for 1975–1979 and G. N. Richards for 1979–1982.

At the Bratislava meeting (1974), Argentina and Norway were accepted as members. In 1976 Poland, New Zealand, Spain, and Israel joined the Committee, followed in 1978 by Denmark and the Netherlands.

III. The International Carbohydrate Organization

The Steering Committee had been an informal organization. However, in 1980 it was proposed that it should be established as a formal body. The particular reason for this proposal was the inauguration of the Whistler Award (see Section VII). Good, but anonymous, friends of Whistler proposed to establish a fund for a biennial award to a distinguished carbohydrate chemist; investment and handling of money required a responsible, established body. Such a body was created and named the International Carbohydrate Organization (ICO) to take over from the Steering Committee. A constitution was drawn up and amended several times over the years. The only current functions of the Organization are the regulating of the biennial Symposia and the awarding of the biennial award; but other possible functions were foreseen, such as distribution of research grants, involvement in nomenclature problems, and other items. A logo was designed and adopted (see Section VIII).

The ICO issued guidelines, originally compiled by James N. BeMiller in 1989, for the organization of international carbohydrate symposia and these have been of considerable assistance to future organizers. The locations for future symposia are approved tentatively for six years, and definitively for four years, in advance. There has been difficulty only once, with the meeting scheduled for 2006. This date

PICTURE 3. Past-Presidents: J. Thiem, S. J. Angyal, J. N. BeMiller, A. Misaki, J. F. G. Vliegenthart, H. Baer, P. J. Garegg, H. J. Jennings, R. L. Whistler (Hamburg 2000).

was allocated to Jerusalem, but the political unrest there forced ICO to withdraw this assignment. Canada stepped in at short notice and the meeting was held at the Whistler resort north of Vancouver; the organizers had less than the usual time for preparations, but the symposium was well organized and conducted.

There is no lack of nations willing to hold the symposia. The next meeting will be in Oslo (Norway) in 2008. The tentative location for 2010 is Japan and for 2012 Spain (Madrid). The most likely location for 2014 is India, for 2016 USA, for 2018 Portugal, and for 2020 Brazil (Picture 3).

Because the number of national representatives had grown considerably, the Organization became somewhat unwieldy, and in Vancouver (1982) the previous constitution was modified by omitting the co-opted members of the American, British, and French carbohydrate groups; only national representatives, past presidents and other co-opted members remained.

The ICO has been an Associated Organization of the International Union of Pure and Applied Chemistry since 1970 and an Interest Group of the International Union of Biochemistry since 1982. The plenary lectures of the Carbohydrate Symposia between 1972 and 1998 were printed in Pure and Applied Chemistry.

The ICO had originally no income; the secretaries covered incidental expenses from other sources. In 1996, however, each Symposium was allocated $1000 from its funds to the secretariat.

New nations continued to be admitted to the Organization. In 1980 South Africa joined, in 1984 Italy, in 1988 Finland and Switzerland, in 1990 India, in 1996 Austria, Cuba, and Portugal, in 1998 Ireland, in 2000 Poland and Japan, and in 2002 Brazil, China, and Taiwan. Following the separation of Slovakia from the Czech Republic, the latter was left without a representative on the ICO until 2000, when it was admitted. Since the constitution specified that national members must resign when they are no longer active in chemistry, new representatives have gradually been appointed for most nations. The present list of representatives is shown in Appendix 2. The secretaries of the Organization were: K. Wilkie for 1983–1987, G. G. S. Dutton for 1987–1991, R. Gigg for 1991–1995, B. Smestad Paulsen for 1995–2001, Elizabeth Hounsell for 2001–2006, and Johannis P. Kamerling was elected for 2007–2012. The secretary's term was extended to six years in 1994.

With rapid advances in carbohydrate chemistry, the interest in carbohydrate meetings has increased and biennial meetings appeared to be insufficient to meet the demand. A new series of meetings, restricted to Europe, started in Vienna in 1981 under the name of Eurocarb, was followed by one in Budapest in 1983. These meetings take place biennially in those years in which there is no International Carbohydrate Symposium; in 2007 the location is Lübeck (Germany). They are modeled on the International Symposia and are controlled by the European Carbohydrate Organization, representing countries in Europe in which carbohydrate research is carried out. The participants are not exclusively European, for example, in Lisbon (2001) 21% of the plenary lecturers, 11% of the invited lecturers, and 24% of the session chairmen came from countries outside Europe.

IV. Format

The format of the symposia has become fairly constant but took some time to evolve. At the Gif, Birmingham, and Münster meetings there were no plenary lectures: all talks were of equal duration and every lecturer was invited. In Birmingham, however, there were two levels of lectures: less than half were delivered to the whole audience whereas the others were given in two parallel sessions. An incident should be mentioned here. Two of the invited lecturers did not turn up: J. Staněk from Prague and L. Vargha from Budapest; they were not given exit permits by their respective governments. Their places were filled by two Russians who were not originally registered for the Symposium. Similar events occurred at some of the other early symposia. The Russians were unreliable; some who had registered did not appear while others arrived at short notice.

In Kingston (1967) and in Paris (1970) there were only four plenary lectures; at later symposia there were approximately 10; two on each day. Shorter,

30-minute invited lectures were first included into the program of the Sydney meeting (1980); later such invited lectures have been delivered in parallel sessions. Posters were included at the London symposium (1978), and with the enormous number of contributions offered in later symposia, they have had to cover the bulk of the presentations. In Ottawa (1994) it was considered necessary to limit the number of posters to one per author—a decision that has not drawn favorable response. In 1989 it was decided that all papers given orally in concurrent sessions (that is, other than plenary and invited lectures) been to have presented as posters. With the large increase in the number of posters in later years, this decision was enforced for a short time only. Current practice, introduced in Ottawa (1994), is that all contributions are submitted as posters, and the organizing committee then selects some of them for oral presentation.

The presentation of some posters warrants improvement; many contain far too much material in small typeface. Better instructions on the standards for posters in the form of a sample layout could be issued by the organizers.

Posters were originally open for inspection only during scheduled hours, but the later preferred practice has been have them freely accessible and for more than one day, but with the authors present only at specified times. This was done in Sydney (2000) and at the Whistler meeting (2006).

Parallel lecture sessions were introduced at the Birmingham (1962) meeting but at Kingston (1967) and Paris (1970) there were none. All but the plenary lectures are now given in such sessions and in Madison (1972) even plenary lectures were given in parallel. In San Diego (1998) there were at times five parallel sessions; this is probably excessive. In Sydney (1980) a lecture session was run in parallel with the poster session; it was argued that this is less restrictive than two parallel sessions, since the posters could be seen any time. This system was not adopted at later meetings.

In order to avoid two lectures on similar topics being delivered simultaneously in parallel sessions, it has become customary to divide carbohydrate chemistry into four sections: structure and analysis (A), synthesis (B), biochemistry and medicinal chemistry (C), and biotechnology and industrial applications (D). The section on synthesis has usually been the largest. This division was first used in Kyoto (1976). The practice has not been applied uniformly: in London (1978) there were seven sections and in San Diego (1998) five. In Cairns (2002) there were eight sections. In Glasgow (2004) there was no division into sections on the basis of subject. In Whistler (2006) the sections were synthesis, structure, function and engineering, glycobiology and glycomics, and therapeutics.

The symposia have generally started with a mixer on the first night to permit the renewal of old acquaintances. There have frequently been conflicts for the ICO members attending a dinner reception scheduled at the same time.

The early symposia usually finished with a symposium dinner on Friday night, but the dinner was moved to Thursday evening, beginning with the Madison meeting (1972) as the Symposia now generally terminate on Fridays after lunch. Wednesday afternoons are usually free of chemical activities, to give opportunities for sightseeing. In Madison (1972) the whole day of Wednesday was taken up by an excursion to Wisconsin Dells; this proved to be too long (until almost midnight) and tiring in the prevailing heat for many delegates. Only the Sydney meeting (1980) was organized differently. This was the only Symposium arranged in winter (though it was not the coldest; that honor goes to the London meeting in April 1978). For most participants this was their first opportunity to see Sydney and because the days were short and the evenings not very suitable for outside activities, three afternoons (Monday–Wednesday) were left free of chemical program; lectures and a poster session were held in the evenings on those days. The afternoon excursions were regarded as part of the program and tickets were given to all participants. The Tuesday cruise on Sydney Harbour also served as an efficient mixer.

The schedule in Glasgow (2004) was different: the meeting started on Friday evening and continued through the weekend without a break, closing on Tuesday, one day shorter than usual, supposedly reducing the cost of the symposium. There were three sessions running continuously, except for the plenary lectures; the two poster sessions, in the evenings, had to cater to 491 posters. This format did not prove popular. The customary half-day break in the middle of the symposia, giving a rest from the constant sequence of lectures and providing an opportunity for mixing with other participants was not scheduled, and excursions proposed to take part after the meeting did not prove popular. The next meeting, Whistler (2006) again started on Sunday evening.

Special local events were highlights of certain symposia. In Birmingham (1962) there was a visit to the Royal Shakespeare Theatre at Stratford-on-Avon for a performance of *Cymbeline*; in Sydney (1980) there was a concert in the famous Opera House, while in Milan (1996) the participants attended a moving concert in the Basilica San Marco. In Stockholm (1988) the banquet took place in the City Hall, the site of the Nobel Prize ceremonies, and there was also an organ concert, given by Bert Fraser-Reid, one of the symposium participants. The banquet in Madison (1972) included entertainment by a barbershop quartet. In Paris (1970) the banquet was held in the Berlioz and Opera Salon of the famous Grand Hotel, while in Cairns (2002) there was, of course, an excursion to the Great Barrier Reef.

In Paris (1970) there were nine roundtable discussions on selected topics. They did not prove very useful, and after several attempts at other symposia, they were abandoned. They were resurrected in Milan (1996) as a thematic plenary session, a discussion on perspectives in glycosciences, by a small panel of four (B. Ernst, R. Lane, Y. Nagai, and J. F. G. Vliegenthart).

The language used at the symposia is English; although this has not always been so. In Münster (1964) many of the lectures were given in German, and one was in French (even the program was printed in German); in Paris (1970) many of the lectures were delivered in French. After that, English became the exclusive language.

A nice touch at San Diego (1998) was the holding of special sessions in memory of two recently deceased well-known carbohydrate chemists, Guy Dutton and Margaret Clarke, with their respective spouses in attendance. In Hamburg (2000) a short commemorative session was held on the occasion of the death of Ray Lemieux; his widow was not present.

In Paris (1970) a supplementary one-day meeting was held, devoted to only one subject, trehalose. In Australia (1980) two pre-symposium and one post-symposium meetings were organized at locations different from that of the Symposium. Such satellite (or affiliated) meetings were held in connection with several other symposia and the ICO encouraged them at their meeting in 1988. Some acquired an independent existence; for example, a meeting on Conformational Studies of Carbohydrates was held in 1990, 1992, 1994, and 1996, with an average attendance of 70. The International Symposium on Cereal and Other Plant Carbohydrates, first arranged by Bruce Stone in Melbourne before the Sydney Symposium (1980), was again held in 1982 and then every second year until 1992 before each International Carbohydrate Symposium. However, in Milan (1996) the ICO felt that such events should be held as part of the Symposium. It was not in favor of independent satellite meetings, as they could detract from the main meeting. None were held in conjunction with the meetings in San Diego (1998) and Hamburg (2000).

Programs for accompanying members (originally termed ladies' programs) have been part of the symposia ever since the Birmingham meeting, allowing them to take part in the mixer and in similar social events. In Münster (1964) no program was arranged for them. Optional tours have usually been provided for accompanying members; in Sydney (1980) they were extensive, but in Glasgow (2004) there were none, and there were very few in Whistler (2006). Some meetings have provided a pleasant room for accompanying guests, as made available in Madison (1972), in Hamburg (2000), in Cairns (2002), and in Whistler (2006) and there was none in Glasgow (2004).

The number of accompanying members has usually been about 15% of that of the participants, varying according to the distance of the location from population centers and the tourist attraction of the location.

V. Numbers

The number of participants at the symposia has shown a steady increase. In Birmingham (1962) there were 280 participants; such a number could easily be handled by the University. In Kingston, 275 participants and 57 accompanying members were mostly accommodated in university residences. In Paris (1970) there were 296 participants. In Bratislava, about 450 delegates were housed mostly in commercial accommodations, but not far apart. In London (1978) there was a steep increase to 750 participants (from 37 countries), but the university system handled most of them. Vancouver had a smaller number (about 440), partly because of its distance from Europe. After that, the number stabilized for a while at 600–700. American universities could handle such numbers in campus locations (as in Cornell, with 670 participants), but in most other countries hotels had to be used extensively (Stockholm—620, Yokohama—670, Utrecht—750 participants), which led to a considerable reduction of contact and mixing between delegates. The worst example of this was the symposium in Paris (1992)—with a record number of 950 participants and about 200 accompanying members—where the hotels at the symposium center were far too expensive for most budgets. Participants were distributed widely over the Paris area and had to travel long distances to the meeting, so that there was little contact between them once the lectures were over. Paris is the only location in which two international carbohydrate symposia were held.

A number not much smaller (about 720) was well handled in Ottawa, where the symposium was held in a very large hotel that accommodated a substantial proportion of the participants, and the university accommodation was quite close by. Similar numbers (765 registrants, plus about 100 accompanying members) were rather inadequately catered for in Milan (1996). An important feature of that meeting was the large number (about 200) of students, who benefited from a lower registration fee, a policy much appreciated by the ICO. By 1998 (San Diego), this number increased to 230 and then to 290 in Hamburg. A large number (238) of students attended in Glasgow (2004), thanks to special inexpensive accommodation arrangements for them and lower registration fees.

In San Diego a record number of 1010 participants from 40 countries were almost all accommodated in colleges of the very large and spacious campus of the University of California; although some of them had to walk considerable distances to lectures and meals. A similar number (1020) of participants came to Hamburg (2000) from 41 countries; maybe an attendance of about 1000 will become a regular feature of the symposia. Again accommodation was scattered over a wide area, with only a few hotels within walking distance. Cairns (2002), a small town, provided sufficient accommodation close to the symposium; but because of the remoteness of the location, there were only 620 participants. In Glasgow (2004) the number of participants was, surprisingly, only 764, and again, the accommodations were widely scattered.

In Whistler there were 793 delegates from 46 countries.

The host nation has usually provided the largest number of participants. Thus in San Diego (1998) there were 302 registrants from the USA, followed by Japan (124) and Germany (88), with smaller numbers from Canada (57), the United Kingdom (54), and Sweden (36). By contrast, the largest group in Hamburg (2000) was the Germans with 291 representatives; and the second largest group was again the Japanese (147), followed by USA (69), with Russia, the United Kingdom, and Sweden coming far behind (37 each). Seven countries provided only one representative each. Since there are comparatively few carbohydrate chemists in Australia, it is not surprising that in Cairns (2002) the Japanese formed the largest group; their country is much closer to Cairns than Europe or America. (It is noteworthy that there was not a single Japanese participant in Birmingham in 1962.) In Glasgow (2004) there were 764 participants; the largest number (161) coming from the United Kingdom, with the Japanese being second (110); the number from the USA (55) was surprisingly small, and equal to that from Germany. In Whistler (2006), the Japanese again formed the largest group.

Attendance at the symposium in Sydney (1980) was quite small (about 220), owing to the travel distance and to the curtailment of travel grants funded by the US National Science Foundation for that year. It was possible to have the symposium almost completely residential at the university, with most meals taken together, and delegates browsing the posters before breakfast and after dinner. Unfortunately, with much larger number of participants, this ideal arrangement now becomes almost impossible, although it was closely approximated in San Diego (1998).

The number of communications (lectures and posters) has increased even more than the number of participants. In Kingston (1967), the first occasion on which people could submit communications, there were 46, in Paris (1970) only 32, in Madison (1972) 94, in London (1978) 253, the latter showing the effect of introducing posters. In Utrecht there were 519 presentations and in Paris 687, while the subsequent meeting at Ottawa attracted 570 presentations. In San Diego (1998) there were a record 813 and in Hamburg (2000) 739. In Glasgow (2004) there were 99 oral presentations and 475 posters, and in Whistler, 205 oral presentations and 463 posters, from 46 countries.

In Hamburg, the beautifully produced Book of Abstracts, with two abstracts on each page, weighed 1.25 kg; while in Paris (1992) each abstract was given a (somewhat smaller) page and the book weighed 1.7 kg. The number of oral presentations was fewer in Hamburg than in San Diego (146 versus 177) but that is reasonable in view of the fact that the bulk of information is now conveyed by posters. In Hamburg, participants were also given a disk containing the abstracts; and sufficient computers were provided for reading them; this is essential if the abstracts are not printed. The organizing committee for the Cairns meeting decided not to print the abstracts at all; and only disks and computers were provided. The abstracts appeared on the Internet one month before the meeting and hence were available for inspection well in advance. In Glasgow (2004), printed abstracts were available but had to be paid for.

The greatly increased number of participants and of presentations has caused difficulties. At Whistler the main lecture theater was a combination of two adjacent halls; and the speaker was visible only from one of them. The symposium in Hamburg was held in the buildings of the Chemistry Department (except for the opening session) and this was hardly adequate for the task; space is needed where people can sit and discuss problems between (and also during) the sessions. The posters present the main difficulties: in Hamburg there was much crowding. In San Diego the posters were displayed in a gymnasium, which was quite adequate, but distant. However, even when there is adequate space for viewing the posters, their large number still creates a problem. In San Diego and in Hamburg there were about 600 posters displayed, constituting about 80% of the presentations. No one can adequately inspect 600 posters in a few hours, nor is there enough time between the receipt of the abstracts and the beginning of the sessions, to peruse 600 abstracts. It is important for the posters to be accessible at all times, and this has not been the case in several recent symposia.

The key to solving the poster problem is in the proper use of the subject index of the abstracts. Such an index—in addition to an index of authors—was first provided in Utrecht (1984) and was subsequently included in the books of abstracts of most, but not all, symposia. The index is now compiled from keywords provided by the contributors. In San Diego (1998) keywords were requested, but ultimately an Index was not published. Such an index enables participants to select posters of direct interest to them, and it is recommended that such an index of keywords be published at every symposium. The book of abstracts issued in London (1978), of 515 pages, had no index at all and it was only possible to find any author or subject by ascertaining the serial number of the communication in the program.

With the increase in number of participants and the complexity of some of the procedures, the cost of the symposia has also increased. It may now come as a surprise that the registration fee at the Birmingham symposium (1962) was five pounds (about 5 dollars), and half that for members of The Chemical Society. It was also stated that "Each fellow may introduce a lady guest without payment." Soon the fee rose, not only because of inflation but much more from the fact that the large numbers of participants could no longer be handled by inexpensive university facilities but required the hiring of expensive locations. In London (1982) the registration fee was £50 (£25 for students). Some comparisons are noted between fees in the local currency: Vancouver (1982) $Can 150 (50 for students) and Ottawa (1994) $Can 380 (180 for students); the former symposium was held in the University of British Columbia, the latter (much larger one) in a convention center. At Cornell (1986) the fee was US $100 ($50 for students) and at San Diego (1998) $375 (students $225); both were held at campuses but the latter one was much larger. To add to the cost, in Glasgow (2004) and Whistler (2006) the luncheons had to be paid for. The Sydney meeting (1980) was wholly accommodated in the University of New South Wales but the symposium in Cairns was held in a convention center; at the former, the fee was $Aust 100 (students $30), the latter cost $Aust 800 (students 400). The increase of cost may ultimately affect the number of participants. In Glasgow (2004) the dinner was poorly attended, probably because of its high cost (£75).

There is only one person who attended all the symposia (not counting the one in Gif): Derek Horton. Regrettably, Roy Whistler could not attend the symposium held in the Whistler resort in British Columbia.

VI. Subjects

There is significant current demand for the carbohydrate symposia as shown by the fact that 1000 people would attend, many of them at great expense. In the first half of the 20th century there were very few scientific symposia. With the great increase in quantity and complexity of research in the second half of the 20th century, researchers seek direct communication with leaders in the field and seek a chance to exchange information with others. The plenary lecturers at these symposia are chosen from those in the forefront of research and hence the subjects of their talks give a good historical record of the development of carbohydrate chemistry during the past forty years.

Of the 146 plenary lecturers so far, only three (Lemieux, Montreuil, and Sinaÿ) gave three lectures. If the Whistler Award lecture is counted as a plenary lecture, then the names of Bock and Wong can be added to this list. There were 19 who delivered two plenary lectures each. One third (53) of the plenary lecturers came from the USA, followed by 22 from the UK and 15 from Canada. Altogether, 22 nations have provided plenary lecturers.

At the first two symposia where plenary lectures were held, R. U. Lemieux gave talks on NMR in carbohydrate chemistry (Kingston 1967) and conformational analysis (Paris 1970). Lemieux was the outstanding carbohydrate chemist of the second half of the century, and these were the subjects where he made his most important contributions. In his talk on NMR, although it was on a minor aspect (allylic couplings) he underlined the great importance of this tool in carbohydrate chemistry. In Paris he emphasized that very few compounds actually assume the ideal chair (or boat or skew) form but show intermediate conformations depending on internal interactions, solvation, and association with other molecules. In his third plenary lecture (London 1978) he introduced the importance of different oligosaccharides to biological functions; in this case, the human blood-group glycoproteins that are responsible for the A, B, and O(H) specificities.

Further developments in NMR were presented in Sydney (1980): on high-resolution NMR by J. F. G. Vliegenthart, and on two-dimensional NMR by L. D. Hall. More applications of conformational analysis were discussed by K. Bock (Vancouver 1982) on the conformations of oligosaccharides, by H. J. Jennings (Utrecht 1984), on the conformations of some polysaccharides, by I. Tvaroska (Stockholm 1988), on computational methods for determining conformations, and by Y. Kishi (Paris 1992) on the conformations of *C*-glycosyl compounds.

The assumed boundaries between organic and carbohydrate chemistry were first breached in Sydney where D. H. R. Barton, a "non-carbohydrate" chemist, was invited to talk on new synthetic methods applicable to sugars. This trend was continued in Vancouver where K. B. Sharpless, another non-carbohydrate chemist, spoke on asymmetric epoxidation, a reaction important for carbohydrate syntheses. Then the boundary disappeared altogether. In Stockholm (1988) S. J. Danishefsky spoke on synthetic methods and in Yokohama (1990) K. C. Nicolaou also discussed new methods and synthetic strategies applicable to carbohydrates. Both of these eminent speakers are organic chemists whose work includes many reactions with carbohydrates.

Carbohydrates are used increasingly for the synthesis of antibiotics and their analogues. This subject was discussed by J. G. Moffat and S. Umezawa in London (1978) and by C.-H. Wong in Hamburg (2000). Enzymes came into more extensive use: S. David discussed the use of immobilized enzymes in preparative carbohydrate chemistry (Ithaca 1986) and C.-H. Wong further developed this subject in Paris (1992). The mechanism of the action of enzymes was discussed by S. G. Withers in his Whistler Award lecture in Cairns (2002).

By this time the synthesis of specific oligosaccharides became the most important subject in carbohydrate chemistry. This was a problem too difficult to handle in earlier times. The physical and biological properties of oligosaccharides were discussed in Paris (1992) by J. P. Carver and W. Saenger. Methods for their synthesis became one of the most important subjects for plenary lectures: P. J. Garegg and J. H. van Boom discussed this in Milan (1996), K. Bock in San Diego (1998), D. R. Bundle in Cairns (2002), H. Overkleeft in Glasgow (2004). Also in Cairns, P. H. Seeberger presented an extensive survey of oligosaccharide synthesis, including starting materials, synthetic methods, automated synthesis, testing, and applications. Biological effects on and with carbohydrates became increasingly important: the program in Glasgow (2004) could have formed the basis of a biochemical symposium. This may also be said about the 2006 program at Whistler.

Apart from attending lectures, the principal aim of the symposia is to provide an opportunity for exchanges of views and discussion of problems between participants. This is only practical if adequate seating space is provided close to the lecture theaters. The best provisions for that were at San Diego (1998) with plenty of seats, indoors, and outdoors, where refreshments were also available. The worst case was in Paris (1992): there was nowhere to sit and exchange ideas. A small center with adequate accommodation close to the meeting site is the most favorable for a Symposium; this made Cairns (2002) an ideal location. Also, a small center would not induce the participants to disappear into town after the sessions. For this reason, Whistler (2006) was also a good location, but relatively few accompanying persons (52) chose to attend.

VII. The Whistler Award

The Roy L. Whistler International Carbohydrate Award was established in 1982 by an anonymous donation; the donors stipulated that it be named after

Professor Whistler. The award consists of a plaque and a prize, and it was first awarded in Utrecht (1984).

It was agreed from the beginning that the award should not duplicate the Haworth or Hudson Awards in recognizing the most outstanding carbohydrate chemists; rather, it should serve as encouragement to promising young chemists. A formal age limit was considered too restrictive, and the wording finally adopted for selection "The purpose of the award shall be the stimulation of carbohydrate chemistry internationally. The awardee shall be an active worker in carbohydrate research, distinguished with contribution of excellence and with promise of continued productivity." (The last phrase means that the award should not be given to a person about to retire.) The prize is substantial ($10,000) and serves not only as a reward but also as a contribution to future costs of research.

The selection committee for the Award consists of the immediate past president of the ICO (who takes the chair), the president, the president-elect and the secretary, and two other persons co-opted by them. The decision whom to co-opt is, however, not taken until the list of candidates has been perused, and they are chosen so as to ensure that there is at least one person on the committee knowledgeable about the research area of each nominee.

The Award is presented at the beginning of each Symposium after the opening ceremony. The awardee then delivers a lecture on his work, scheduled as the first lecture of each symposium; with the president in the chair. It is regarded as a special lecture rather than a plenary one, and is delivered directly after the presentation of the award. Thus, the Whistler lecturer is different from the plenary lecturers; being chosen on a different basis by a special body, and not by the organizing committee.

The list of Whistler Lecturers, and their titles, follows:

1984—*Tomoya Ogawa* (Japan): Synthetic approaches to glycan chains

1986—*Klaus Bock* (Denmark): Carbohydrate–protein interactions. The substrate specificity of enzymes used in the degradation of oligosaccharides related to starch and cellulose

1988—*David R. Bundle* (Canada): Oligosaccharide–protein interactions. Studies of antibody-combining sites by competitive bonding, sequence analysis and X-ray crystallography

1990—*Johannis P. Kamerling* (the Netherlands): Playing with complex carbohydrate chains. A world of fascinating developments

1992—*Andrea Vasella* (Switzerland): A new approach to the synthesis of glycosides

1994—*Chi-Huey Wong* (USA): Enzymatic and chemo-enzymatic synthesis of carbohydrates
1996—Made jointly to two scientists: *Constant A. A. van Boeckel* (Belgium) and *Maurice Petitou* (France): Heparin: from original 'soup' to well designed heparin mimetics (presented by Petitou)
1998—*Ole Hindsgaul* (Canada): Synthetic toys for glycobiology
2000—*Anne Dell* (United Kingdom): Mass spectrometry: a powerful tool for high-sensitivity covalent structure analysis
2002—*Stephen G. Withers* (Canada): Enzymatic cleavage and formation of glycosidic bonds: from glycosidases and lyases to transferases and glycosynthases
2004—Made jointly to *Anne Imberty* (France) and *Thomas Peters* (Germany): "Forty ways to bind your sugar"
2006—*Gideon J. Davies* (United Kingdom): Carbohydrate-active enzymes: how Nature makes and breaks glycosides

Sharing the Lectureship may cause problems. The two speakers in 2004 had insufficient time, in their half-an-hour, to expand their subject. The first selection committee, in 1984, had difficulty deciding between two candidates, Tomoya Ogawa and David Bundle. Finally, Ogawa was selected because he was the older of the two. On the next occasion (1986) a different selection committee nominated Bundle for the Lectureship.

Before the establishment of the Whistler lectures, at the London meeting (1978) the Haworth and the Tate and Lyle lectures, sponsored by The Chemical Society, were included in the program as plenary lectures; the Haworth lecture was delivered by Ray Lemieux. In Sydney (1980) a lecture under the auspices of the Royal Australian Chemical Institute was presented as a plenary lecture.

At the Whistler meeting a new award was introduced: the Carbohydrate Research Award for Creativity in Carbohydrate Chemistry, sponsored by Elsevier; it was awarded to Peter H. Seeberger.

VIII. Logos

Most of the logos of the symposia are a combination of carbohydrate symbols with some local features; as illustrated here. The first logo appeared at the Kyoto meeting (1976); it showed the word KYOTO within a pyranose chair form. The next one was a simple combination of a six- and a five-membered ring with the word LONDON. After that, the logos became more sophisticated. The one for Sydney (1980) has an interesting story. Stephen Angyal asked an artist at the

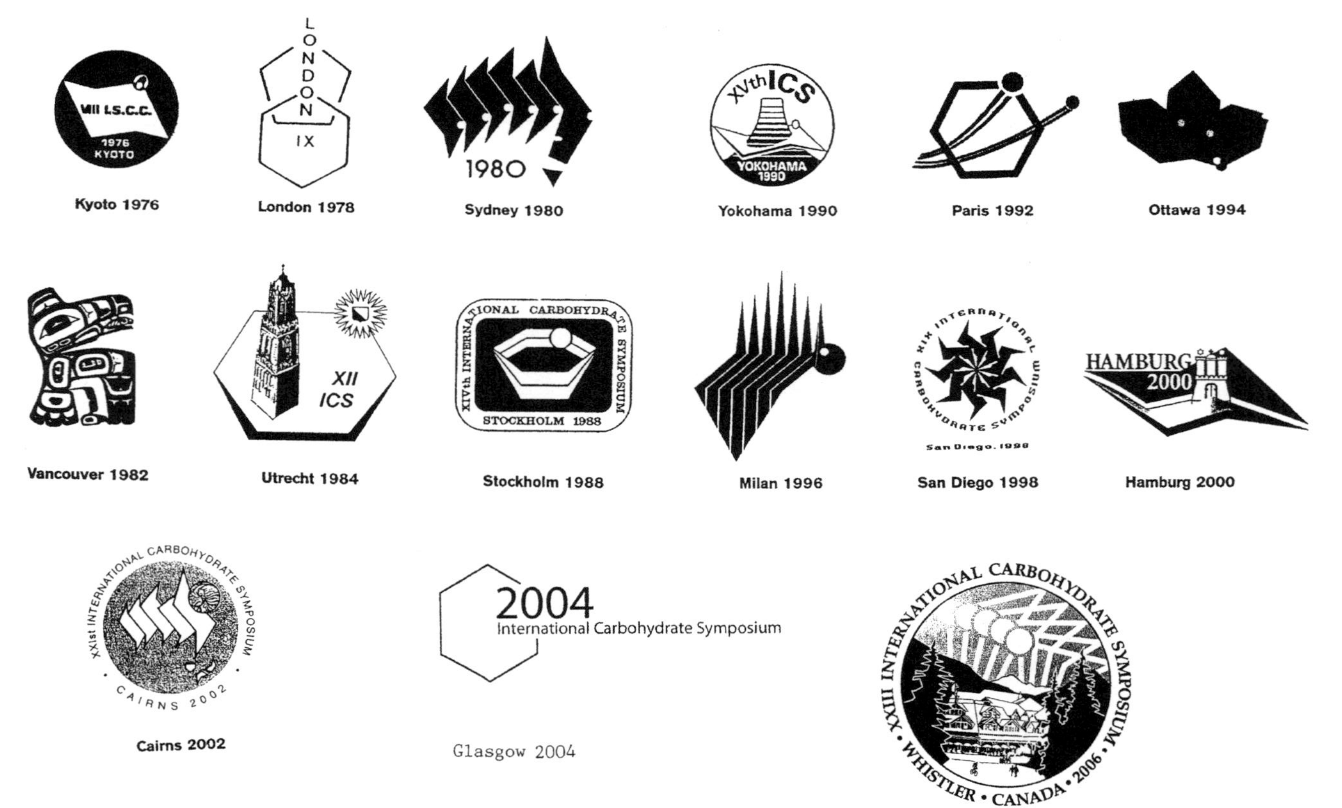

PICTURE 4. Logos of the International Carbohydrate Symposia.

PICTURE 5. Logo of the International Carbohydrate Organization.

University of New South Wales to make up a logo by combining the pyranose chair form with the map of Australia. The somewhat unexpected outcome was a map of Australia constructed from six pyranose rings. However, the southeastern end of Australia did not have the right shape, some of Victoria being amiss. It was then suggested that, since John Stevens in Sydney was working on the septanose forms of sugars and was going to present a talk at the symposium, it would be permissible to replace one of the pyranose rings with a septanose; thus the shape of Australia was more closely approached (Picture 4).

The creation of the logo of the ICO is told by this author: "In 1983 I was the president of the Whistler Award selection committee and spent quite a bit of time on the logistics of the Award. It occurred to me that the plaque should have a logo on it. I approached an artist (another one, Neil Fallows) to create something to represent progress, ambition, or endeavour by using pyranose chair forms. The result was the three figures floating (or diving?) in space. I looked at it and said that it appeared to be basically good but somewhat unfinished. Could we not put a circle around them? I said, and then my wife, who was with us, said: If a circle, why not the globe? And that is how the logo came about. It proved popular and was later adopted as the logo of the ICO" (Picture 5).

ACKNOWLEDGMENTS

The author wishes to express his gratitude to many colleagues, too numerous to be listed here, for photos, reminiscences and data on earlier meetings. He is especially grateful to Jacques Defaye and David Manners who identified the participants shown in the group-photo taken in Gif.

This chapter is a revised and updated version of a pamphlet which was published by Commonwealth Scientific and Industrial Research Organization (CSIRO) (Australia) and distributed to all participants of the 2002 Symposium in Cairns.

Appendices

Appendix 1: List of International Carbohydrate Symposia

Number	Year	Location	Chairman†
I	1960	Gif-sur-Yvette, France	Jean E. Courtois
II	1962	Birmingham, UK	Maurice Stacey
III	1964	Münster, Germany	Fritz Micheel
IV	1967	Kingston, Canada	J. K. N. Jones
V	1970	Paris, France	Jean E. Courtois
VI	1972	Madison, USA	Roy L. Whistler
VII	1974	Bratislava, Czechoslovakia	Stefan Bauer
VIII	1976	Kyoto, Japan	Konoshin Onodera
IX	1978	London, UK	W. G. Overend
X	1980	Sydney, Australia	Stephen J. Angyal
XI	1982	Vancouver, Canada	Guy S. Dutton†
XII	1984	Utrecht, the Netherlands	Johannes F. G. Vliegenthart
XIII	1986	Ithaca, USA	James N. BeMiller
XIV	1988	Stockholm, Sweden	Per J. Garegg
XV	1990	Yokohama, Japan	Akira Misaki
XVI	1992	Paris, France	Pierre Sinaÿ
XVII	1994	Ottawa, Canada	Harold J. Jennings†
XVIII	1996	Milan, Italy	Benito Casu
XIX	1998	San Diego, USA	David A. Brant†
XX	2000	Hamburg, Germany	Joachim Thiem
XXI	2002	Cairns, Australia	Robert V. Stick
XXII	2004	Glasgow, UK	Elizabeth Hounsell
XXIII	2006	Whistler, Canada	Mario Pinto

†The chairman is usually the national representative of his country on the ICO and becomes the next president after the meeting. Those two instances where this was not the case are marked by an dagger; the national representatives (Baer and BeMiller, respectively) were not available.

Appendix 2: Members of the International Carbohydrate Organization (ICO) (2006)

Argentina: Oscar Varela	Italy: Franceso Nicotra
Australia: Robert V. Stick	Japan: Yukishige Ito
Austria: Paul Kosma	New Zealand: Richard Furneaux
Brazil: Philip A. J. Gorin	Norway: Berit Smestad Paulsen
Canada: Mario Pinto	Poland: Slawomir Jarosz
Cuba: Vicente Verez Bencomo	Portugal: A. P. Rauter
China: Biao Yu	Russia: Yuriy A. Knirel
Czech Republic: Vladimir Kren	Slovakia: Jan Hirsch
Denmark: Inge Lundt	South Africa: David W. Gammon
Finland: Tapani Vuorinen	Spain: Manuel Martin Lomas
France: Serge Perez	Sweden: Stefan Oscarson
Germany: Joachim Thiem	Switzerland: Beat Ernst
Hungary: Lajos Szente	The Netherlands: J. P. Kamerling
India: Mukund K. Gurjar	Taiwan: Albert Wu
Ireland: Angela Savage	United Kingdom: Elizabeth Hounsell
Israel: Uri Zehavi	USA: Zbigniew Witczak

Past Presidents: S. J. Angyal, Hans H. Baer, Benito Casu, Per J. Garegg, H. J. Jennings, A. Misaki, W. G. Overend, Pierre Sinaÿ, J. F. G. Vliegenthart, R. L.Whistler.
Secretary: Berit Smestad Paulsen.

Appendix 3: Plenary Lectures

From many years, the plenary lectures were published in *Pure and Applied Chemistry*; all of the lectures given in Gif were published by CNRS. There were no plenary lectures at the first three Symposia.

IV. Kingston
R. U. Lemieux: Application of nuclear magnetic resonance to problems of structure, configuration, and conformation in carbohydrate chemistry
J. S. Brimacombe: Synthesis of some naturally occurring allose derivatives and some associated chemistry
J. L. Strominger: Structure and biosynthesis of bacterial cell walls in relation to the mechanism of action of antibiotics
G. Westphal: Some newer results of investigations on enterobacterial polysaccharides

V. Paris
R. U. Lemieux: Newer developments in the conformational analysis of carbohydrates
J. Montreuil: Procédés chimiques et enzymatiques d'exploration de la structure des isoglucannes
L. Mester: Structure et rôle dans la coagulation du sang des glucides liées aux protéines
K. Schmid: Characterization and structure of plasma glycoproteins

VI. Madison
R. C. Liebenow: Carbohydrates in our modern society
S. J. Angyal: Complex formation between sugars and metal ions
L. Rodén: Biosynthesis of connective tissue polysaccharides
F. Shafizadeh: Thermal reactions of cellulosic materials
A. G. Holstein: Changes in carbohydrate marketing as affected by research
A. B. Foster: Synthesis and biological activity of fluorinated carbohydrates

VII. Bratislava
N. K. Kochetkov: Pathways of synthesis of specific polysaccharides
W. G. Overend: Some approaches to the synthesis of amino sugars and branched chain sugars
C. V. N. Rao: Immunochemical approaches to the structural chemistry of polysaccharides
R. H. Marchessault: Influence of acetate groups on carbohydrate and polysaccharide conformations in the crystalline state
Ch. U. Usmanov: Physico-chemical investigations on the structure of cellulose fibers
J. Baddiley: Mechanism and control of the cell wall synthesis in bacteria
J. Montreuil: Recent data on the structure of the carbohydrate moiety of glycoconjugates. Metabolic and biological implications
U. S. Ovodov: Structural chemistry of plant glycuronoglycans
R. Kohn: Ion binding on polyuronates—alginate and pectin
D. Horton: Thio sugars. Stereochemical questions and synthesis of antimetabolites

VIII. Kyoto
L. Hough: Recent aspects of the chemistry of disaccharides
B. Lindberg: Structural studies of some bacterial polysaccharides
S. Hakomori: Status of blood group H—carbohydrate chain in ontogeny and oncology

I. J. Goldstein: Lectins as carbohydrate-binding proteins
G. O. Aspinall: The selective degradaton of carbohydrate polymers
E. D. T. Atkins: Conformational and molecular structure of polysaccharides
B. Coxon: Fourier transform NMR spectroscopy
H. Paulsen: Synthesis of amino- and branched chain mono- and oligosaccharides
L. Szabó: Structural features and biological activity of the *Bordella pertussis* endotoxin
S. Hanessian: Approaches to the total synthesis of natural products from carbohydrates

IX. London
F. W. Lichtenthaler: Sugar enolones: synthesis, reactions of preparative interest and γ-pyrone formation
P. Sinaÿ: Recent advances in glycosidation reactions
R. S. Shallenberger: The intrinsic chemistry of fructose
R. U. Lemieux: Human blood groups and carbohydrate chemistry
J. Fraser Stoddart: From carbohydrates to enzyme analogues
S. Umezawa: Recent advances in the synthesis of aminoglycoside antibiotics
C. Pedersen: Synthesis of bromodeoxy compounds from hexoses, polyols and aldonic acids
V. Shibaev: Biosynthesis of salmonella O-antigenic polysaccharide: specificity of glycosyl transferases
J. G. Moffat: Chiral synthesis of the antibiotics anisomycin and pentenomycin
A. S. Perlin: Aspects of the chemistry of glycosides

X. Sydney
D. A. Rees: Recent advances in understanding polysaccharide shapes and interactions
S. Hanessian: Recent studies in synthetic carbohydrate chemistry
Sir Derek Barton: New and selective reactions and reagents
K. -E. Eriksson: Fungal degradation of wood components
J. F. G. Vliegenthart: The applicability of high-resolution ^{1}H-NMR spectroscopy to the structure determination of carbohydrates
P. Albersheim: Structure and function of complex carbohydrates active in regulating plant–microbe interactions
V. Derevitskaya: Structure of carbohydrate chains of blood group glycoproteins
C. E. Ballou: The specific interaction of mycobacterial polymethylsaccharides with long-chain fatty acids
J. Yoshimura: Stereoselective synthesis of branched-chain sugar derivatives
L. D. Hall: Two-D or not two-D, that is the question

XI. Vancouver
R. J. Ferrier: Prostaglandins from sugars
T. J. Painter: Algal polysaccharides
E. De Clercq: Antiviral activity of 5-substituted pyrimidine nucleoside analogues
K. Bock: The preferred conformation of oligosaccharides in solution inferred from high resolution NMR data and hard-sphere exo-anomeric calculations
F. Shafizadeh: Saccharification of lignocellulosic materials
K. B. Sharpless: Application of asymmetric epoxidation in saccharide synthesis
O. Samuelson: Suppression of undesirable reactions during oxygen bleaching of wood pulp
S. Stirm: Degradation of bacterial surface carbohydrates by virus-associated enzymes: methodology and applications
N. Sharon: Oligomannoside units of membrane glycoproteins as receptors for bacteria
R. Fletterick: The structure of maltoheptaose on phosphorylase by X–ray diffraction and the nature of saccharide-protein interaction
B. A. Dmitriev: Microbial O-antigenic hexosaminoglycans

XII. Utrecht
H. Egge: Structural analysis of blood group ABH, I, i, Lewis and related glycosphingolipids. Application of high resolution NMR and FAB mass spectrometry
P. J. Garegg: Some aspects of regio-, stereo-, and chemoselective reactions in carbohydrate chemistry
H. J. Jennings: Structure, conformation and immunology of the polysaccharide capsules of human pathogenic bacteria
R. A. Khan: Chemistry and new uses of sucrose: how important?
A. Kobata: The sugar chains of γ-glutamyltranspeptidase
N. K. Kochetkov: Synthesis of fragments of bacterial polysaccharides and their application for the preparation of synthetic antigens
J. Montreuil: Spatial structures of glycan chains of glycoproteins inrelation to metabolism and function
J. C. Paulson: Selection of influenza virus variants based on sialyloligosaccharide receptor specificity
P. A. Sandford: Microbiopolysaccharides: new products and their commercial application
R. Schauer: The anti-recognition function of sialic acids: studies with erythrocytes and macrophages

XIII. Cornell
D. M. Carlson: Gene expression: from carbohydrates to cloning
S. J. Angyal: Topics in carbohydrate chemistry

K. -A. Karlsson: Glycolipids as receptors on animal cells for bacteria, bacterial toxins and viruses
G. Legler: Basic monosaccharide derivatives: tools for exploring the active site of glycohydrolases and for studies in glycoprotein biosynthesis
T. Suami: Synthetic ventures in pseudo-sugar chemistry
S. David: Immobilized enzymes in preparative carbohydrate chemistry
J. P. Carver: Site-directed processing of N-linked oligosaccharides: the role of three-dimensional structure
G. G. Stewart: The biochemistry and genetics of carbohydrate utilization by industrial yeast strains

XIV. Stockholm
L. Kenne: Computer-assisted structural analysis of regular polysaccharides
K. Nakanishi: Microscale structure determination of oligosaccharides: a chirooptical method
I. Tvaroska: Computational methods for studying oligo- and polysaccharide conformations
H. H. Baer: Recent synthetic studies in nitrogen-containing and deoxygenated mono- and di-saccharides
S. J. Danishefsky: Synthetic studies in the carbohydrate area
B. Fraser-Reid: Novel carbohydrate transformations discovered en route to natural products
R. R. Schmidt: Recent developments in the synthesis of glycoconjugates

XV. Yokohama
S. Hakomori: Specific interaction between glycolipids as a basis for specific cell recognition
E. Cabib: Carbohydrates as structural constituents of yeast cell wall and septum
H. Egge: The use of FAB-MS for the study of oligosaccharides
Y. C. Lee: Topography of binding site of animal lectins
A. Vasella: New reactions and intermediates involving the anomeric centre
P. Sinaÿ: Recent advances in glycosylation reactions
I. C. P. Smith: Structure and dynamics of carbohydrate residues on the surface of membranes as seen by ^{2}H NMR
J. Preiss: Regulatory mechanisms involved in the biosynthesis of starch
S. Suzuki: Glycosaminoglycan chains of proteoglycans: approaches to the study of their structure and function
K. C. Nicolaou: New strategies and methods for the synthesis of complex oligosaccharides
W. M. Watkins: Chemical structure, biosynthesis and genetic regulation of carbohydrate antigens: retrospect and prospect

XVI. Paris
J. C. Paulson: Bioactive sialosides
J. P. Carver: Oligosaccharides: how can flexible molecules act as signals?
Y. Kishi: Preferred solution conformation of C-glycosides and their parent glycosides
H. D. Conrad: Dissection of heparin: past and future
T. Ogawa: Recent aspects of glycoconjugate synthesis
C. -H. Wong: Practical synthesis of carbohydrates based on aldolases and glycosyltransferases
W. Saenger: Crystal structures of linear and cyclic oligosaccharides
A. Eschenmoser: Hexose nucleic acids

XVII. Ottawa
H. Brade: Chlamydial lipopolysaccharide: chemical and antigenic structure, biosynthesis, and medical applications
H. Kunz: Stereoselective syntheses using carbohydrates as chiral auxiliaries
G. W. Hart: Ubiquitous and temporal glycosylation of nuclear and cytoplasmic proteins
J. W. Gillard: Development of novel antiviral nucleoside analogues
S. J. Danishefsky: Glycals in the synthesis of oligosaccharides and glycoconjugates
C. A. A. Boeckel: Glycosaminoglycans: synthetic fragments and their interaction with serine protease inhibitors
S. G. Withers: Enzymatic cleavage of glycosides: how does it happen?
P. M. Colman: Design and antiviral properties of influenza virus neuraminidase inhibitors
A. Kobata: The sugar chain structures of carcinoembrionic antigen and related normal antigens

XVIII. Milan
P. Albersheim: Structural studies of plant cell wall polysaccharides
D. A. Brant: Shapes and motions of polysaccharide chains
P. J. Garegg: Some aspects of oligosaccharide synthesis
O. Hindsgaul: Synthetic chemistry yields new tools for glycobiology
F. W. Lichtenthaler: Perspectives in the use of low-molecular weight carbohydrates as organic raw materials
U. Lindahl: Glycosaminoglycans: polyanions with built-in messages
J. H. Shaper: Glycosyltransferases: transscriptional regulation
I. W. Sutherland: Microbial exopolysaccharides—structural subtleties and their consequences

T. Tai: Differential distribution of gangliosides in the central nervous system
J. H. van Boom: Synthesis of complex oligosaccharides

XIX. San Diego
R. R. Schmidt: New aspects of glycoside bond formation
J. D. Marth: Unexpected and physiologic roles of oligosaccharides revealed by targeted gene ablation
J. B. Robins: Bacterial polysaccharide vaccines
R. M. Brown, Jr.: Biosynthesis and structure of cellulose
G. Taylor: Sialidase structures
Y. Inoue: Diversity in sialic and polysialic acid residues and related enzymes
B. Imperiali: Chemistry and biology of asparagine-linked protein glycosylation
O. Smidsrød: Structure–function relationship in marine polysaccharides
N. Taniguchi: Cell surface remodeling by glycosyltransferase genes and growth factor signalling
K. Bock: Carbohydrate chemistry: synthetic and structural challenges towards the end of the 20th century

XX. Hamburg
M. Aebi: N-Linked protein glycosylation in the endoplasmic reticulum: from the model organism *Saccharomyces cerevisiae* to human diseases
K. Drickamer: Animal lectins that mediate cell–cell recognition events
R. J. Ferrier: The chance to discover chance in discovery
G. W. J. Fleet: Amino acids derived from carbohydrates
R. J. Linhardt: Heparin: structure and biological activities
K. Sandhoff: Functions of glycosphingolipids at membrane surfaces: pathogenesis of neurodegenerative and dermal genetic disorders
B. Swensson: Starch-degrading enzymes—structure/function insights and improvements by protein engineering
K. Tatsuta: Total synthesis and development of natural products using carbohydrates
C. -H. Wong: Synthesis of carbohydrate-based antibiotics

XXI. Cairns
D. R. Bundle: The search for high-avidity ligands: constrained univalent oligosaccharides and radially arranged multivalent oligosaccharides
G. -J. Boons: Synthetic glycoconjugates and the innate and adaptive immune system
W. R. Roush: Highly stereoselective synthesis of 2-deoxy-β-glucosides. Application to the total synthesis of landomycin A

M. Kiso: Synthetic ganglioside probes: versatile tools for the elucidation of carbohydrate functions
J. J. Hopwood: Early detection and effective therapy of lysosomal storage disorders
T. Kinoshita: Enzymes required for biosynthesis of GPI-anchored proteins
H. Kunz: Synthetic glycopeptides for the development of antitumor vaccines
M. A. J. Ferguson: GPI and glycoprotein biosynthesis as targets for anti-parasite design
P. H. Seeberger: Automated solid-phase synthesis of oligosaccharides to address biomedical problems

XXII Glasgow
H. Overkleeft: Recent advances in the design and synthesis of complex oligosaccharides
J. S. Brimacombe: Anchors away: Seeking a cure for sleeping sickness
R. Dwek: Glycosylation: disease targets and therapy
B. Ernst: From carbohydrate leads to drugs: why is it so difficult?
S. -I. Nishimura: Rational design and synthesis of carbohydrate-based drugs
S. Perez: The exquisite complexity of plant cell wall polysaccharides
C. R. H. Raetz: Structure, biosynthesis and function of lipopolysaccharide endotoxins: potent bacterial activators of innate immunity
P. Sinaÿ: Three decades with carbohydrates: selected sweet extracts
Varki: Sialic acids in biology and evolution

XXIII Whistler
B. Henrissat: A journey through families of carbohydrate-active enzymes
Y. Ito: Synthesis and functional analysis of glycoprotein glycan chains
T. Peters: NMR experiments with sugars and viruses
G. W. Hart: Dynamic cycling of O-GlcNAc on nucleocytoplasmic proteins: a nutrient/stress sensor globally regulating cellular mechanism
A. Vasella: Oligoribonucleotide foldamers with a novel architecture
A. Imberty: Structural glucobiology and host recognition by pathogens
R. R. Schmidt: New aspects of bond formation: complexity as a synthetic target
H. W. Liu: Exploring nature's strategics for making unusual sugars: mechanisms, pathways and biosynthetic applications
M. von Itzstein: Carbohydrate-recognising proteins as targets for drug research
J. C. Paulson: Sialoside sweet spots.
C. R. Bertozzi: Chemistry in living systems: a new tool for probing the glycomes
Y. Ito: Synthesis and functional envelopes of glycoprotein glycan chains
T. Feizi: Structural glycobiology

MASS SPECTROMETRY OF CARBOHYDRATES: NEWER ASPECTS

By João A. Rodrigues, Adrian M. Taylor, David P. Sumpton, James C. Reynolds, Russell Pickford and Jane Thomas-Oates

Department of Chemistry, University of York, Heslington, York YO10 5DD, UK

ISBN: 978-0-12-373920-9
DOI: 10.1016/S0065-2318(07)61003-8

I. Introduction

1. History

In 1987, Anne Dell's now classic chapter entitled 'FAB-Mass Spectrometry of Carbohydrates' appeared in this series.[1] At the time, it provided a much needed introduction to and summary of the capabilities of fast atom bombardment (FAB)[2] mass spectrometry (MS), its practicalities, and its potential applications for carbohydrate analysis. The chapter immediately became the authoritative work, and in the intervening almost 20 years it has been neither replaced nor updated.

FAB was a ground-breaking technique, that made it possible for the first time to generate, relatively easily, long-lived high-quality spectra from large, polar, thermally labile biopolymers, directly from solution without the need for derivatization. Particularly in combination with high-field magnetic sector instrumentation, it became broadly adopted and for a decade and a half was the method of choice for carbohydrate analysis, making possible enormous progress in carbohydrate research. Interestingly, in the final paragraph of the 1987 chapter, Anne Dell predicted important MS developments would take place, giving improvements in both sensitivity and mass range. This, she suggested, would allow detection of molecular species from carbohydrates such as intact rhamnogalacturonan II (RG-II), the plant cell-wall polysaccharide, and predicted that this would 'necessitate the design of new mass spectrometers that have a better high mass performance'; the first mass spectra of intact RG-II, acquired on a matrix-assisted laser desorption/ionization time-of-flight (MALDI-TOF) instrument, were published in 1996.[3,4]

The major instrumental developments that have contributed to making that prediction come true have done just what was foreseen in 1987; they have improved both sensitivity and mass range, thus making MS even more indispensable in carbohydrate analysis. They have included the introduction of the ionization techniques of electrospray (ESI)[5] and MALDI.[6,7] These have been complemented by the commercialization of a range of new mass analyzer designs, which are ideally suited to use with ESI and MALDI, including delayed extraction (DE) reflectron TOF instruments, quadrupole orthogonal acceleration TOFs, TOF/TOFs, and a range of ion trapping instruments. The combined effect has been to make MS even more user-friendly and broadly adopted, not just by the analytical community but by carbohydrate chemists and biochemists in general.

2. Aims

The brief given to us by the editor when we were approached to write the current chapter was to update the landmark 1987 article and thus to contribute a new article on the subject that reflects the rapid subsequent developments in the field of biological mass spectrometry. We have therefore set out to introduce and describe the most commonly available modern mass-spectrometric instrumentation that is used in carbohydrate studies, and to highlight, using a small selection of relevant examples, how it is now typically being used for the analysis of carbohydrates.

II. Instrumentation

Modern mass spectrometers make it possible for substances in the femtomole range (10^{-15} M) to be analyzed, and for high molecular masses of complex molecules well over 100 kDa to be obtained. Mass spectrometers allow molecular mass determination, and the generation of fragmentation data to allow sequencing and structure elucidation, generally in tandem mass-spectrometric experiments.

The coupling of MS to separation techniques, especially high-performance liquid chromatography (HPLC) and capillary electrophoresis (CE) has been the focus of intense research and development, leading to a significant increase in the range of instrumentation available.

Generically, MS is accomplished using the five-component system shown in Fig. 1.

Samples are introduced into the mass spectrometer in the gaseous, liquid, or solid phase, following which they suffer ionization in order to generate gas-phase charged species that are then transferred into the mass analyzer where they are separated according to their mass-to-charge ratios (m/z). The detection

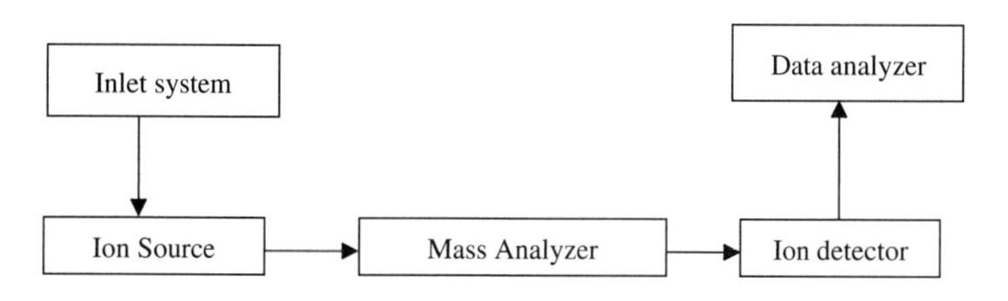

Fig. 1. Schematic of a modern mass spectrometer.

of ions is generally accomplished by using electron multiplier systems that enable m/z and abundance to be measured and displayed by means of a signal that is perceived normally by a computer, which also has the function of controlling the instrument.

1. Ionization

FAB, ESI, and MALDI all promote gentle desorption/ionization which makes possible the analysis of large, fragile biomolecules such as carbohydrates. The development of these techniques, described as 'soft' ionization methods, in which intact protonated or sodiated molecules are generally obtained on ionization allows the determination of molecular weight and makes possible the generation of structural information using collision-induced dissociation (CID) that induces the fragmentation of the intact ionized molecules on collision with an inert gas.

a. Electrospray (ESI).—Electrospray ionization sources for use with mass spectrometers were introduced by Fenn and co-workers in 1988.[5,8] This breakthrough had a tremendous impact on the practice of MS and was recognized with the 2002 Nobel Prize award in Chemistry to John Fenn for his contributions to the development of electrospray as a method to analyze biomolecules. Earlier experiments described by Dole and co-workers in 1968 had shown for the first time the ionization of macromolecules and their transfer to the gas phase at atmospheric pressure using electrospray.[9]

Electrospray allows desorption of large, non-volatile intact analytes directly from solution. Generally, an electrospray is produced by spraying a sample solution through a capillary into a strong electric field in the presence of a flow of nitrogen, transferring the ions into the gas phase (Fig. 2). As a result of the applied electric field, the surface of the liquid emerging from the capillary is highly charged and assumes a conical shape known as the Taylor cone. When the Rayleigh limit is reached (namely the point at which the surface tension of the liquid is exceeded by the repulsive forces between the charges) the surface of the cone breaks into droplets that further fragment into smaller and smaller drops, eventually producing ions in the gas phase.[10]

Two main mechanisms have been proposed for how the resulting droplets yield desolvated ions. Dole proposed the charge residue or solvent evaporation (emission) model in which ion formation is the result of an ion-desolvation process.[9,11,12] The droplets, produced by electrostatic dispersion in the liquid at the capillary tip, lose solvent molecules (aided by the curtain or nebulizer gas, usually nitrogen), and eventually produce individual ions (Fig. 3).

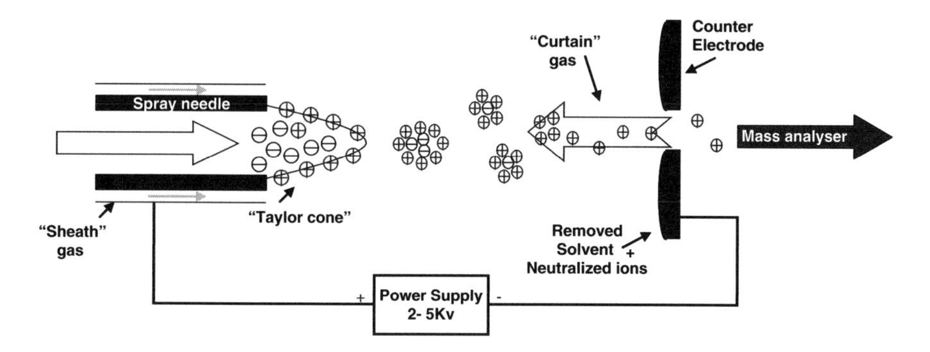

FIG. 2. Schematic of an electrospray source.

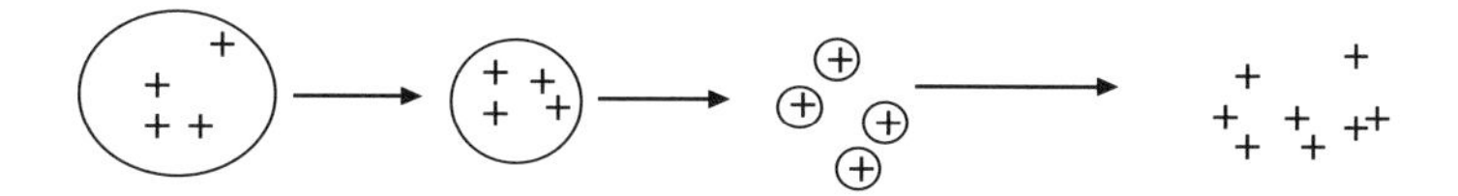

FIG. 3. Charge residue model.

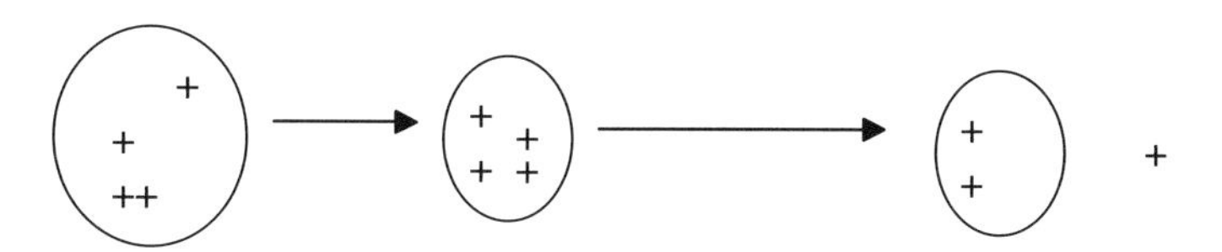

FIG. 4. Ion evaporation model.

The alternative mechanism, the ion-evaporation model, was proposed initially by Iribarne and Thomson[13] (Fig. 4) and involves desolvation of the droplets, producing an increase in charge density over the droplet surface that causes coulombic explosion and eventually leads to ejection of individual ions.

Several modifications of the ESI technique have been introduced, principally micro-electrospray[14] and nano-electrospray[15] that have the advantage of using much lower flow rates, reducing the amount of analyte needed for a mass-spectrometric analysis; this is performed using adapted probed tips. When performing nanospray, the sample is loaded into a fine hollow needle with a conductive coating. When a potential is applied the solution flows due to the electrospray process itself, generating very fine droplets at nL/min flows. This technique enables lower limits of detection to be achieved than with conventional electrospray, mainly due to the higher tolerance to salts and a more efficient production of protonated/cationized molecules in the source due to the smaller droplets produced.[16]

It is perhaps worth noting that the source settings required to obtain successful, high-quality ESI spectra of carbohydrates are rather different from the settings, often set as instrument defaults, that are required for peptide ionization. In particular, the voltage on the counter electrode (often known as the skimmer, cone, or orifice) generally needs to be set significantly higher than it is when acquiring peptide spectra. Typical solvents are 50:50 acetonitrile–water or 50:50 methanol–water.

b. Matrix-Assisted Laser Desorption/Ionization (MALDI).—MALDI has several unique advantages for the analysis of such biomolecules as carbohydrates and proteins. It is widely accepted that MALDI is generally less sensitive to contaminants such as salts from biological buffers than other ionization methods, including ESI and the process lends itself to automation and high-throughput analysis using robotic sample handling.

MALDI evolved from a progression of similar ionization techniques developed from the late 1960s, coinciding with the introduction of laser technology.[17] Unfortunately, such laser desorption (LD) techniques were limited to an upper mass range of approximately *m*/*z* 2000 and because of the high laser fluences required for LD, extensive thermal degradation of the analyte was often observed.

A method to overcome this molecular mass limitation was developed by two groups independently, led by Tanaka and Hillenkamp. The two groups essentially applied the same principle which was to combine the analyte with a matrix to facilitate desorption/ionization without inducing fragmentation.

Tanaka's approach used glycerol mixed with a fine cobalt powder as a matrix.[7] Ionization was demonstrated using the protein lysozyme (molecular weight (MW) 14,306) and Koichi Tanaka received a share in the Nobel Prize for Chemistry in 2002 for being the first to ionize large proteins using this novel ionization approach. Hillenkamp used an organic compound as the matrix, the first of which was nicotinic acid. Using nicotinic acid as a matrix, Hillenkamp successfully ionized several proteins, including bovine albumin (approximately MW 67,000).[6]

The process of modern-day MALDI analysis is remarkably similar to the early experiments of Hillenkamp and Karas; it can be split into distinct stages (Fig. 5). The first is sample preparation; the analyte is suspended within a matrix

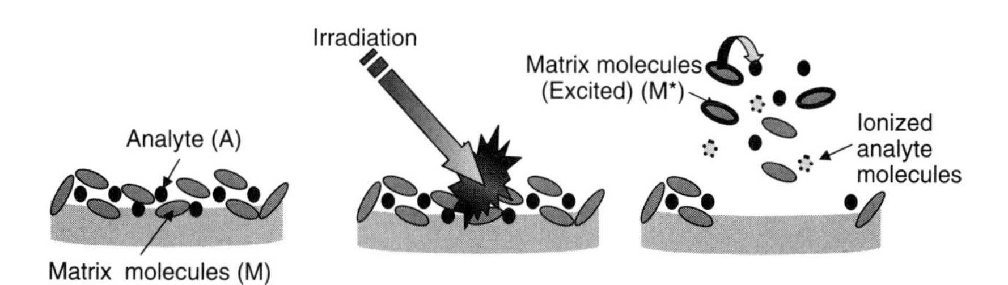

FIG. 5. Three stages of MALDI.

with which it co-crystallizes, normally on a stainless-steel MALDI target. Typically the analyte (~1–10 μM) is mixed with an equal volume of matrix solution (typical matrix concentrations are close to saturation, 10 g/L) and ~0.5 μL of this mixture is deposited upon the MALDI plate and left to dry. This methodology is often referred to as the dried droplet method and is the original[6] and preferred method for MALDI sample preparation today (details of alternative, less common approaches are conveniently summarized by Sigma-Aldrich in its 2001 newsletter AnalytiX 6 on MALDI mass spectrometry, which is available via their website).

Once the sample spot is dry the target plate is placed within the MALDI source, which is normally held under vacuum. The desorption/ionization process is initiated with a laser beam, commonly from nitrogen or Nd:YAG lasers, which emit radiation in the ultraviolet (UV) range. The matrix absorbs the laser energy which produces a plume of analyte molecules and excited matrix molecules in the source. Ionization occurs in the gas phase within this excited plume, generally by the attachment of hydrogen or alkali metal ions such as sodium or potassium. Ions produced using MALDI are almost always singly charged. Many varied chemical and physical pathways have been suggested for MALDI ion formation[18] and this is still an area of active research. Discussion of the details of the various ion formation pathways are beyond the scope of this chapter (for further insight, reviews discuss the role of desorption[19] and in-plume processes[20] as well as other models[21–23]).

For successful analysis using MALDI, homogeneous sample preparation is critical. Many variables influence the preparation of a good MALDI sample, from the concentrations of the matrix and analyte, to the choice of matrix and the amount of internal energy the matrix imparts to the analyte during ionization, but may also include the hydrophobicity/hydrophilicity of the analyte, and the presence of contaminants such as detergents.

Today a wide range of matrices is available to facilitate ionization (Table I). To produce strong ion signals from carbohydrates, the choice of the correct matrix is essential. The benchmark and most commonly used matrix for the analysis of carbohydrates is 2,5-dihydroxybenzoic acid (DHB) introduced by Strupat, Karas, and Hillenkamp in 1991, originally for the analysis of proteins.[24] DHB on crystallization forms characteristic long needle-shaped crystals, usually along the outer edge of the sample spot reaching in towards the center. Sodiated analyte ($M+Na^+$) molecules are the major species observed using DHB, accompanied by less intense corresponding potassiated ($M+K^+$) molecules.

Since the introduction of DHB much work has been undertaken in an attempt to improve the ionization properties of this matrix. This has been achieved by the addition of other compounds to the matrix solution prior to crystallization with the analyte. It is believed that matrix additives can improve ionization either by improving the co-crystallization of the analyte and matrix or by effectively sequestering salts and other contaminants away from the analyte. Karas introduced a successful combination of DHB with 10% 2-hydroxy-5-methoxybenzoic acid, referred to as 'Super DHB',[25] which improved the detection of glycopeptide signals within a mixture compared to normal DHB. Other dopants include 1% 1-hydroxyisoquinoline (HIQ), which improved carbohydrate ionization[26] and α-L-fucose[27] for better resolution and reproducibility. DHB and spermine[28] have been used for improved negative-mode analysis of sialylated glycans. For further information, Harvey[29,30] covered types of matrix and matrix selection for carbohydrate research in detail. Table I summarizes the main matrices and their application for carbohydrate research.

Recent research has concentrated on the use of new materials as possible matrices for MALDI. Two promising areas of research are the use of ionic liquids and of carbon nanotubes. The use of a liquid matrix is not novel; Tanaka's matrix in his pioneering work was liquid based (glycerol), but ionic liquids are different and have proved successful in a large range of chemical applications. An ionic liquid is a salt, which usually contains an organic cation (usually containing a nitrogen or phosphorus) and a large inorganic or organic anion, and which have melting points generally below 100 °C. Armstrong and co-workers[42] evaluated 38 ionic liquids for use as matrices for MALDI, 20 of which were effective for more than one type of analyte. Ionic liquid matrices (ILMs) afforded greater vacuum stability, higher sample homogeneity, lower detection limits, and overall better shot-to-shot reproducibility than conventional matrix systems. Ionic liquid salts of more traditional solid matrices have also

TABLE I
Matrices Used for MALDI-MS of Glycoconjugates

Matrix	Comments
2,5-Dihydroxybenzoic acid (DHB)	Benchmark matrix for the analysis of carbohydrates including glycopeptides and glycoproteins (most effective <5000 Da), free sugars and glycolipids.[24] A number of additives have been used to alter the matrix's properties (see text)
3-Amino-4-hydroxybenzoic acid	First matrix to be developed specifically for carbohydrates (free glycans) found to be useful for the analysis of glycoprotein *N*-glycans[31]
α-Cyano-4-hydroxycinnamic acid (HCCA)	More commonly used for the analysis of proteins and peptides;[32] more effective than DHB for the analysis of glycopeptides and glycoproteins. Hotter matrix than DHB (i.e. transfers more internal energy to the analyte on ionization). Neutral glycosphingolipids[33]
Sinapinic acid[34]	Analysis of glycopeptides and glycoproteins (<5000 Da)
2-(*p*-Hydroxyphenylazo)-benzoic acid (HABA)	Glycopeptides and glycoproteins, free glycans, gangliosides;[35] Hotter matrix than DHB
2,6-Dihydroxacephenone (DHAP)	Mixed with diammonium hydrogen citrate to limit cation adduction, offers improved ionization of sialylated glycopeptides with limited in-source fragmentation compared to DHB and HCCA[36]
2,4,6-Trihydroxyacetophenone (THAP)	THAP can be used for the analysis of oligosaccharides and can offer improved sensitivity for detection of acidic glycopeptides over HCCA.[37] Ionization with THAP can be improved with the addition of di-ammonium hydrogencitrate to limit cation adduction
5-Chloro-2-mercapto-benzothiazole (CMBT)[38]	CMBT gives a higher signal-to-background for *N*-linked glycans than DHB, thus offering better sensitivity for glycan analysis. CMBT is a successful matrix for the ionization of gangliosides[38]
Esculetin (6,7-dihydroxycoumarin)	Provides strong signals for neutral glycosphingolipids[33]
Arabinose phenylosazone	Arabinose phenylosazone has proven to provide better resolution and sensitivity than DHB for neutral free glycans,[39] also useful for the analysis of sialylated and sulfated glycans[40,41] particularly in negative ion mode

been investigated[43] and have been successfully applied in the MALDI analysis of a range of analytes, including highly sulfated oligosaccharides.[44]

Carbon nanotubes are another new material being exploited as novel matrices for MALDI, particularly in the analysis of small molecules offering excellent sensitivity and reproducibility, as few matrix background ions are observed.[45–47]

2. Mass Analyzers

'Mass analyzer' is the term given to the part of the mass spectrometer that discriminates between ions on the basis of their mass-to-charge ratios (m/z). The most common instruments currently in use with MALDI and ESI sources for the analysis of carbohydrates are quadrupole-based instruments, TOF mass spectrometers, and ion-trapping instruments (ion cyclotron resonance and quadrupole ion-trap designs).

a. Quadrupoles.— *(i) Linear Quadrupole Mass Analyzers.* Quadrupole mass analyzers are one of the cheapest analyzers currently available and are capable of unit resolution when dealing with ions of masses up to a few thousand daltons. The analyzer uses a dynamic electric field which only allows ions of a specific m/z value to be transmitted.[48] To obtain a mass spectrum, a scan of the electric field has to be performed. These devices therefore require a continuous beam of ions, which makes them ideal for coupling with ESI sources.

Linear quadrupole mass analyzers consist of four parallel metal rods aligned along the instrument's z-axis (Fig. 6).

The two pairs of opposing rods are connected to a voltage made up of a direct (DC) potential, U, and alternating (RF) potential, $V \cos \omega t$. One pair of rods has a positive DC potential applied, and the other a negative potential. This leaves their net charges (over time) positive and negative, respectively. Ions are introduced into the quadrupoles at low kinetic energy (< 10 eV) and pass through the analyzer at constant velocity where they are subjected to an electric field in the x–y plane consisting of the alternating field superimposed on the constant field resulting from the potentials applied to the two sets of rods (Eq. (1)).

$$\Phi_0 = +(U - V \cos \omega t) \quad \text{and} \quad -\Phi_0 = -(U - V \cos \omega t) \tag{1}$$

where Φ_0 is the potential applied to the rods (V), ω the angular frequency (rad/s) ($= 2\pi f$, where f is the frequency of the RF field), U the direct potential, and V the RF voltage.

The motion of ions in a quadrupolar field is described mathematically by a second-order linear differential equation (Eq. (2)), the solutions to which were

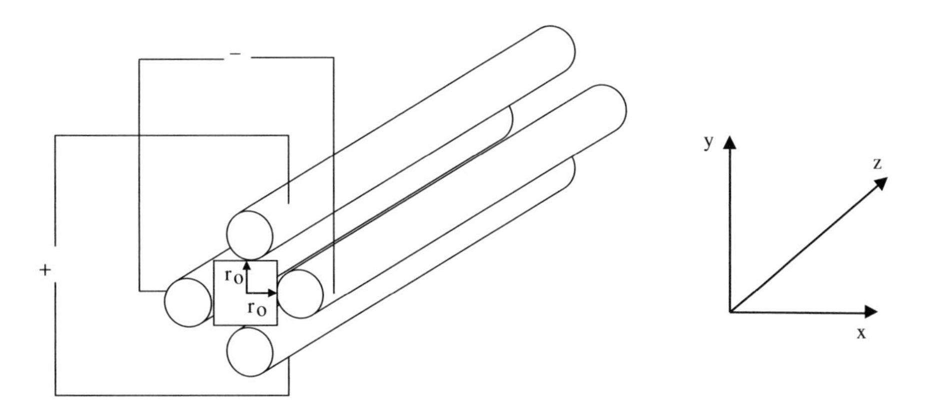

FIG. 6. The arrangement of rods in a quadrupole device.

described by E. Mathieu in 1868, originally describing vibrations across a stretched membrane. The results of the Mathieu equations can be used to define what instrumental settings are required for an ion of a particular m/z to have a stable trajectory and therefore be transmitted through the quadrupole mass analyzer.

$$\frac{\delta^2 u}{\delta \xi^2} + (a_u - 2q_u \cos 2\xi)u = 0 \tag{2}$$

Before this equation can be applied to describe the movement of ions in a quadrupolar field the parameters a, q, and ξ must be defined. This is possible using classical mechanics. The force in the x-direction, F_x, experienced by an ion of mass m and charge e may be expressed as

$$F_x = ma = m\frac{\mathrm{d}^2 x}{\mathrm{d}t^2} = -Z\frac{\mathrm{d}\Phi}{\mathrm{d}x} \tag{3}$$

where m is the mass of the ion, a the acceleration in the x-direction, Z the charge on the ion (ze, charge state multiplied by the charge on an electron), and $\mathrm{d}\Phi/\mathrm{d}x$ the differential of the electric field ($\Phi = (U - V\cos\omega t)$) in the x-direction.

Further mathematical operations allow the equation of motion of an ion in the x-plane of a quadrupole field (Eq. (6)) to be determined as follows.

If we define a_x and q_x as

$$a_x = \frac{8\lambda eU}{mr_0^2\omega^2} \quad \text{and} \quad q_x = \frac{4\lambda eV}{mr_0^2\omega^2} \tag{4}$$

Considering forces in the x-direction, $\lambda = 1$, and for forces in the y–direction, $\lambda = -1$.

To allow the Mathieu equation to be used, the factor ξ is introduced, where

$$\xi = \frac{\omega t}{2} \tag{5}$$

resulting in the equation of motion for ions in a quadrupole field in the x-direction:

$$\frac{d^2x}{d\xi^2} + (a_x - 2q_x \cos 2\xi)x = 0 \tag{6}$$

and in any direction (u):

$$\frac{d^2u}{d\xi^2} + (a_u - 2q_u \cos 2\xi)u = 0 \tag{7}$$

where

$$a_u = \frac{8\lambda eU}{mr_0^2\omega^2} \quad \text{and} \quad q_u = \frac{4\lambda eV}{mr_0^2\omega^2} \tag{8}$$

The equation of motion (Eq. (6)) allows calculation of trajectories for an ion of mass m at DC voltage U and RF voltage V. The operating values for the DC voltage and the RF voltage required to obtain a stable trajectory for ions of a particular m/z value can therefore be obtained. Mass selection is obtained by setting the operating voltages so that the trajectory of the ion at the desired m/z value is stable and the trajectories of ions at other m/z values unstable. This results in transmission of only ions with that selected m/z value.

As ions move away from the center of the quadrupole, the force experienced increases, driving them back towards the center. Quadrupoles are therefore focusing devices as well as mass filters.

A quadrupole mass-filter stability diagram can be created. The shaded area in Fig. 7 shows regions in a–q space where the calculated ion trajectories are stable.

The stability areas for the x- and y-dimensions are both shown in Fig. 7, the x-stable areas pointing down and the y-stable pointing up. There are regions in this a–q space where ions can be x-stable but not y-stable and vice versa.

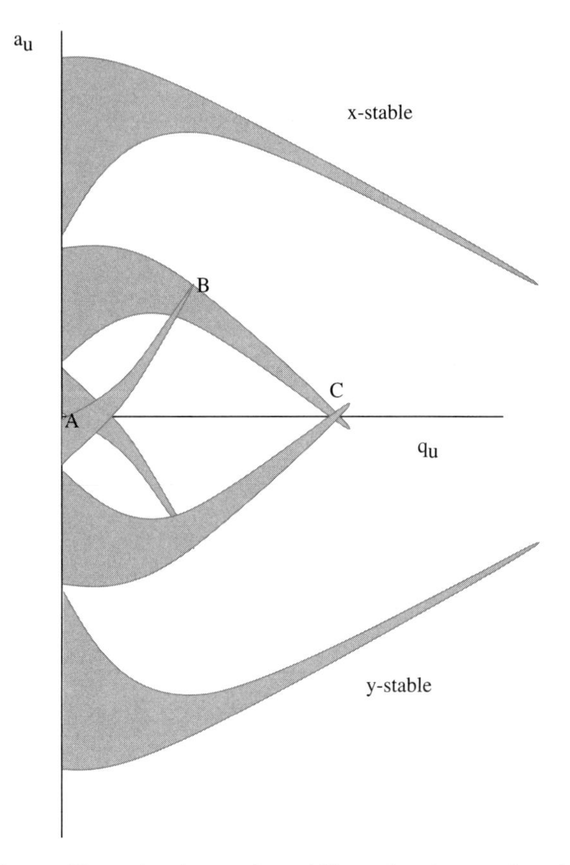

FIG. 7. Cartoon illustrating the x and y stability regions in a quadrupole mass filter.

Overlapping regions are labeled A, B, and C. These are the regions in a,q space which can be exploited to give mass discrimination in MS. Regions B and C would involve application of large potentials, as well as impractical ion trajectories. Though stable, the trajectories would be too great in the x and y-directions for containment in a simple lab-based instrument so that area A is exploited in quadrupole mass filter devices.

Mass discrimination (namely selective transmission of specific m/z ratios) can be achieved by changing the values of q and a. This is achieved by in turn varying U and V, the applied voltages. As one m/z value moves into region A and is transmitted, another leaves the stable region and is discharged into the rods.

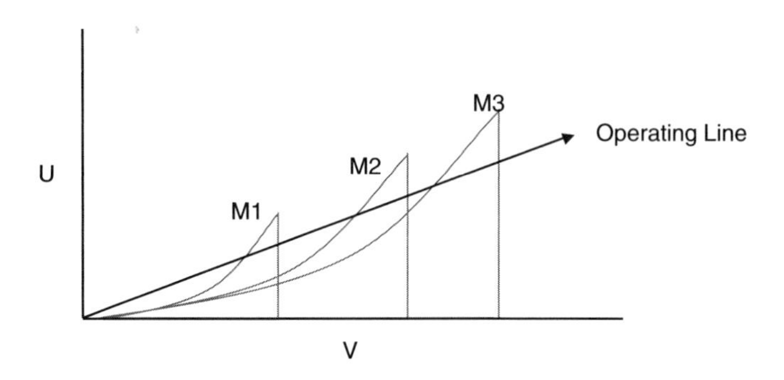

FIG. 8. 'Operating line' of a quadrupole mass analyzer and the stability regions of three different m/z values, M1, M2, and M3.

If U and V are maintained at a fixed ratio a linear operating line, or scan line, may be plotted. This line has a fixed slope of a/q. Rearrangement of Eq. (8) gives

$$\frac{a_x}{q_x} = \frac{2U}{V} \tag{9}$$

Maintaining a constant ratio of U to V also maintains the ratio of a to q, generating an operating line as shown in Fig. 8.

The lighter lines show the positive half of stability region A for ions of three different m/z ratios, M1, M2, and M3. U and V are scanned at a constant ratio and different m/z values move onto the operating line and are transmitted. As the scan proceeds they move out of the stable regions, are discharged and the next m/z value is transmitted. Each stable area is m/z specific and so ions at each m/z are transmitted individually and sequentially as the instrument scans.

The concept of resolution can be clearly demonstrated using the same stability diagram as shown earlier with additional scan lines plotted (Fig. 9).

Consider the cases of scan lines a, b, and c. Line a has a shallow gradient and does not discriminate between any of the three masses; resolution is not achieved to any extent. Line b has a steeper gradient and separation of the three species occurs—they are resolved. Line c passes through the apices of the stability regions; it represents the highest possible resolution. It should be noted that a steeper line than this may not pass through all the stability regions and the ions would therefore not be transmitted through the device.

(ii) Tandem Mass Spectrometry and Triple Quadrupoles. Modern, soft ionization methods give the analyte very little internal energy; it is transferred intact into the gas phase. To obtain detailed information concerning the analyte,

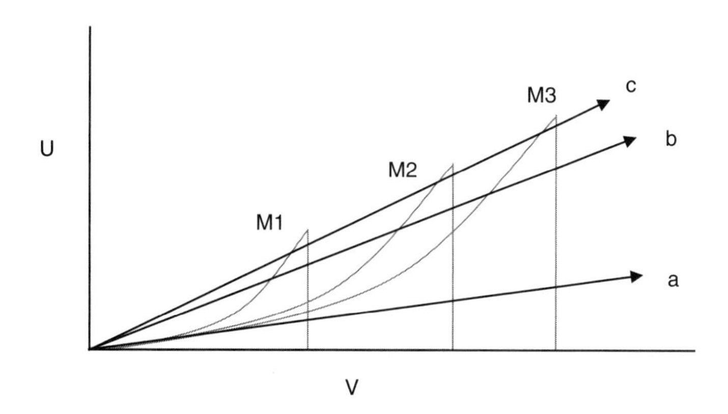

FIG. 9. Resolution in a quadruple mass analyzer.

structurally diagnostic fragment ions need to be generated. Fragmentation of ionized analytes can be promoted by absorption of laser energy, interaction with low-energy electrons or, most frequently, colliding the analyte ions with neutral gas molecules in a collision cell.[49–51] This is called collision-induced dissociation (CID) or collision-activated dissociation (CAD). In the collision process, a fraction of the ion's kinetic energy is converted into internal energy, which causes the ion to dissociate. The number and form of the product ions depends upon the total internal energy of the precursor ion. Ions dissociate in a manner which can be rationalized, although not always predicted, and so structural information can be obtained.

CID is most conveniently carried out using a tandem mass spectrometer; two stages of mass analysis are separated by a CID stage. This allows the precursor ion to be selected in the first stage of mass analysis, fragmented and then the product ions analyzed in the second stage of mass analysis. The selectivity introduced by using two stages of mass analysis allows connectivity between precursor and product ions to be established. Tandem MS is frequently referred to as MS/MS.

There are two classes of tandem instrument. The original development was CID tandem-in-space MS involving arrangement of two mass analyzers separated by a collision cell. A cartoon of a tandem-in-space instrument is shown in Fig. 10.

Common tandem-in-space instruments employ a quadrupole as the first mass analyzer, a multipole collision cell (usually hexapole) operated in RF-only mode, and then either a second quadrupole or a TOF tube as the second mass analyzer. These instruments are termed 'triple or tandem quadrupole' and 'quadrupole-time-of-flight' mass spectrometers.

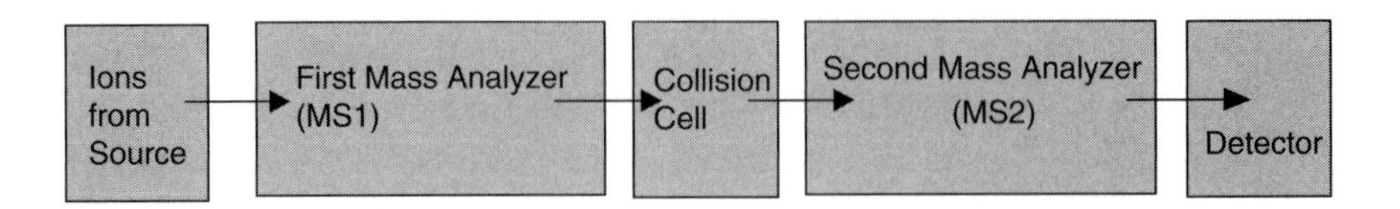

FIG. 10. Cartoon of a tandem-in-space instrument.

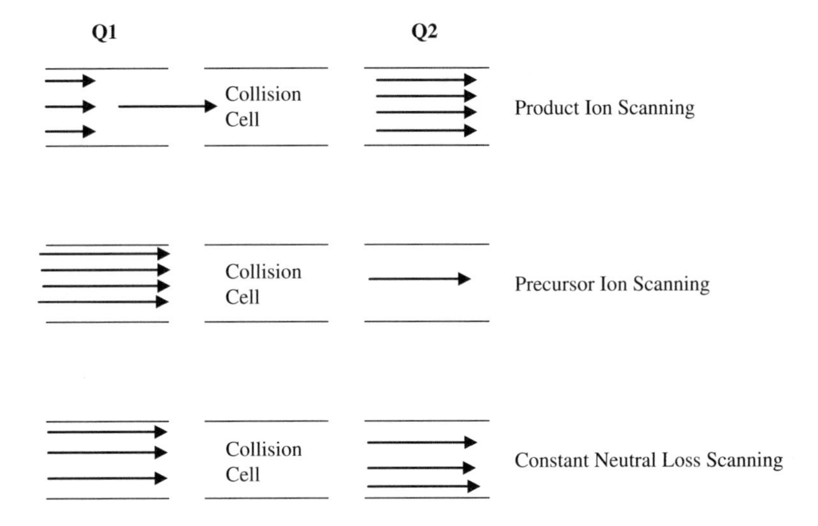

FIG. 11. The three common tandem MS experiments for carbohydrate analysis.

As instrumentation developed, tandem-in-time approaches were developed using ion trap and Fourier transform ion cyclotron resonance (FT-ICR) instruments. During tandem-in-time experiments, the sequential stages of mass selection, CID, and mass analysis are performed within the same, trapping, mass analyzer.

The amount of internal energy transferred to the precursor depends on the type of mass analyzer used, as well as user settings. The number of collisions occurring and the energy of these collisions as well as the mass of the neutral collision gas used affect the degree of fragmentation.

There are three MS/MS experiments that are of use in carbohydrate analysis. These can be depicted schematically (Fig. 11) and are described using a tandem quadrupole instrument as an example. Quadrupole-based tandem-in-space instruments operate at low CID energies (up to 100–200 eV).

The schematic illustrates the principles using a tandem quadrupole device. The first (Q1) and second (Q2) quadrupoles act as mass analyzers with a collision cell in between. The most commonly used mode is product ion scanning. A precursor ion is selected by fixing the RF and DC voltages of the first mass analyzer. The selected ion is transmitted into the collision cell. The ions produced after excitation and fragmentation of the precursor are analyzed by scanning the second quadrupole to successively transmit them. Quadrupole-time-of-flight instruments perform this experiment well due to their relatively high duty cycle for both precursor and product ions.

Precursor ion scanning involves fixing Q2 so that a product ion of fixed *m/z* is transmitted to the detector. Its precursors are determined by scanning Q1. A signal is observed only when a precursor ion selected in Q1 fragments on CID analysis to give the *m/z* value set by Q2. This is conceptually the reverse of product ion scanning. An application of this is in glycopeptide analysis where the product ion corresponding to a glycan fragment ion is monitored (such as *m/z* 163 for Hex^{+}) and all precursors producing this ion are shown in the mass spectrum. Tandem quadrupole instruments are ideal for this type of experiment because they transmit the characteristic fragment ion with very high efficiency (100% duty cycle).

Neutral loss scanning involves detecting all the fragmentations leading to the loss of a specific neutral. This is achieved by scanning both Q1 and Q2 at a fixed offset corresponding to the neutral of interest. Examples include the neutral glycan species lost from a glycopeptide.

The latter two experiments can be best performed on a tandem quadrupole instrument, although software data manipulation can allow this type of data to be extracted from experiments on other instruments.

b. Ion Traps.— *(i) Linear Ion Trap (LIT).* The linear quadrupole can be modified to be used as an ion-storage device—an ion trap—as well as an ion-transmission device. Ions are confined radially by the quadrupole field, but must also be contained axially (along the *z*-axis). To this end a potential well is created along the *z*-axis, with the center of the well sitting at the center of the axis. This well can most simply be created by the use of positively charged electrodes at the ends of the quadrupole.

LITs have two major advantages over 3D ion traps: a larger ion-storage capacity due to higher internal volume and a higher trapping efficiency. LITs can be operated as stand-alone mass analyzers, with the trapped ions being ejected axially or radially into a detector,[52] or coupled with other mass analyzers, for example with time-of-flight (LIT-TOF-MS), FT-ICR, or as part of a tandem quadrupole instrument.[53]

(ii) 3D Ion Trap. The quadrupole ion trap or Paul trap was described on the same patent application as the linear quadrupole described in the previous section.[54] Ion traps benefit from being relatively inexpensive when compared to other types of mass analyzer, and have the ability to perform multiple stages of fragmentation, which can be very useful when trying to elucidate complicated structures. Ions with *m/z* values of up to 3000 thompsons (Th) can be routinely analyzed in an ion trap, making them ideal for analysis of most oligosaccharide samples, and, by taking advantage of multiple charging with ESI, even large polysaccharides can be analyzed. Linking a chromatographic separation to an ion trap is relatively simple if a complex mixture of carbohydrates needs to be analyzed.

The ion trap is a 3D analog of the linear quadrupole. Two of the rods form the end-cap electrodes, whilst the other pair of rods is bent round to form the doughnut-shaped ring electrode (Fig. 12).

The ion trap works on a similar principle to the linear quadrupole. Oscillating RF and DC voltages are applied to the electrodes. These voltages create a quadrupolar field in the ion trap, and ions can be retained in stable trajectories in this field. The force acting upon the ions in the trap is directly proportional to the distance of the ions from the center of the ion trap. Therefore the quadrupolar field acts to store the ions as a packet at the center of the trap. For an ion to be retained in the ion trap it needs to have a stable trajectory in both the axial (z) and radial (r) directions. Whether the ion is stable in one or both of these directions is given by the solutions to the Mathieu equation previously used to describe ion motion through the linear quadrupole with the reduced Mathieu parameters:

$$a_z = -2a_r = \frac{-16eU}{m(r_0^2 + 2z_0^2)\Omega^2} \tag{10}$$

$$q_z = -2q_r = \frac{8eV}{m(r_0^2 + 2z_0^2)\Omega^2} \tag{11}$$

where U is the applied DC voltage, V the applied RF voltage, Ω the main RF drive frequency applied to the ring electrode, r_0 the internal radius of the ion trap, m the mass of the ion and e the charge on the ion.

Areas in which ions are stable in the axial and radial directions are shown in Fig. 13. Areas on this figure where the two regions of stability overlap show where ions are stable in both directions, and may thus be stored in the trap. All commercial ion traps operate using the area of stability nearest the origin of the graph.

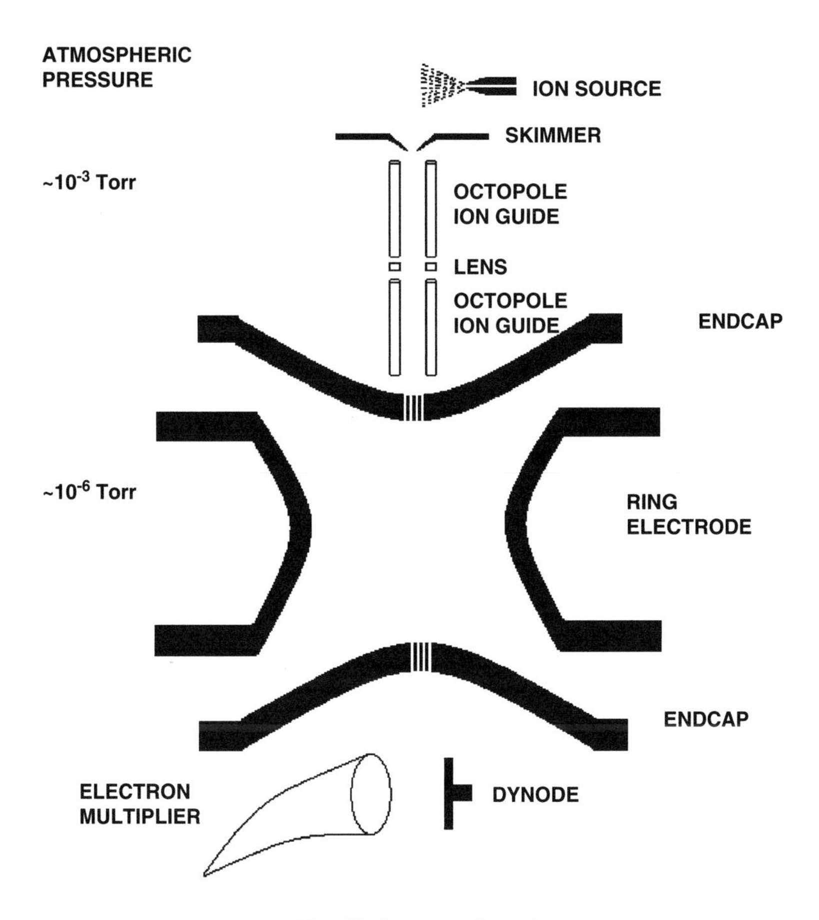

FIG. 12. Ion trap schematic.

By expanding the circled region in Fig. 13, the ion-trap stability diagram (Fig. 14) plotted in terms of the parameters a_z and q_z is obtained. These parameters are directly related to the RF (q_z) and DC (a_z) voltages applied to the ion-trap electrodes. The areas of stability have boundaries where the β_u parameters ($u = z$ or r) have values 0 and 1. β_u is a complex function of a_u and q_u and is directly related to the fundamental secular frequency of the ion (ω_u) and the main RF frequency (Ω) by the equation

$$\omega_u = \beta_u \frac{\Omega}{2} \tag{12}$$

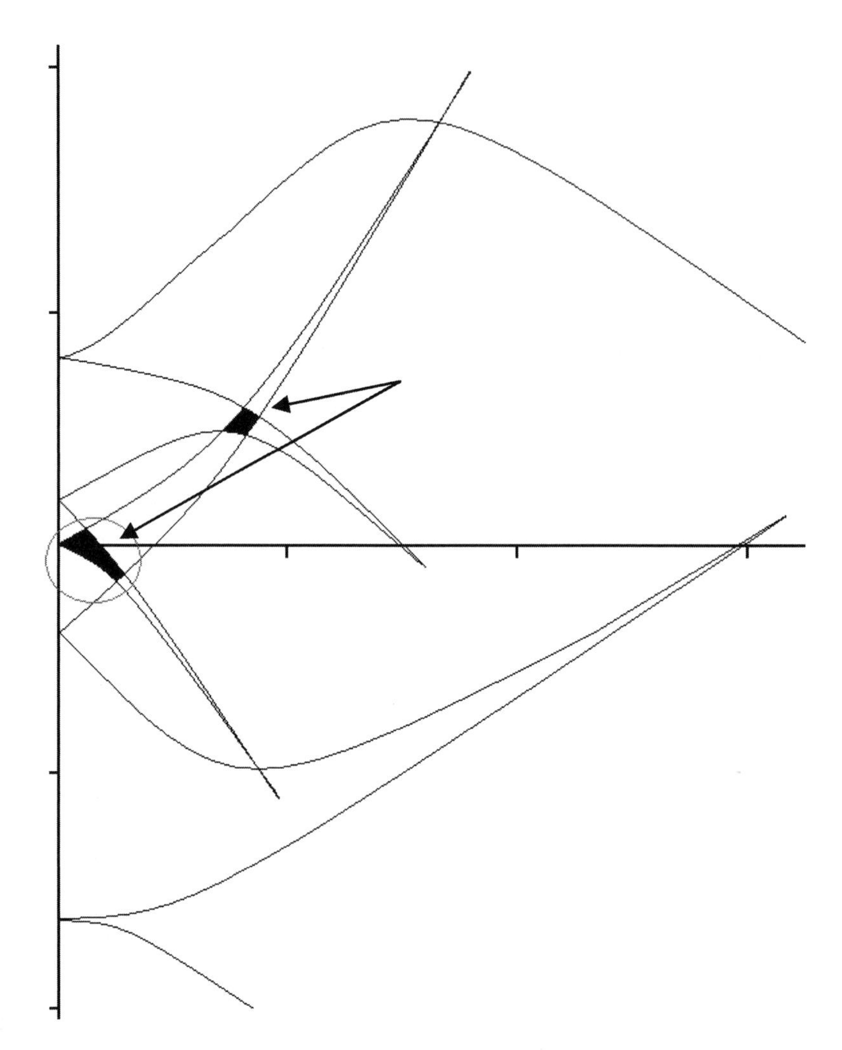

FIG. 13. Stability areas in the r and z-directions in an ion trap. The highlighted area of common stability is displayed enlarged in Fig. 14.

Usually no DC voltage is applied ($a_z = 0$) and the RF frequency and trap geometry are fixed so the expression for the reduced Mathieu parameter that determines stability has the form

$$q_z = k \frac{V}{m/e} \tag{13}$$

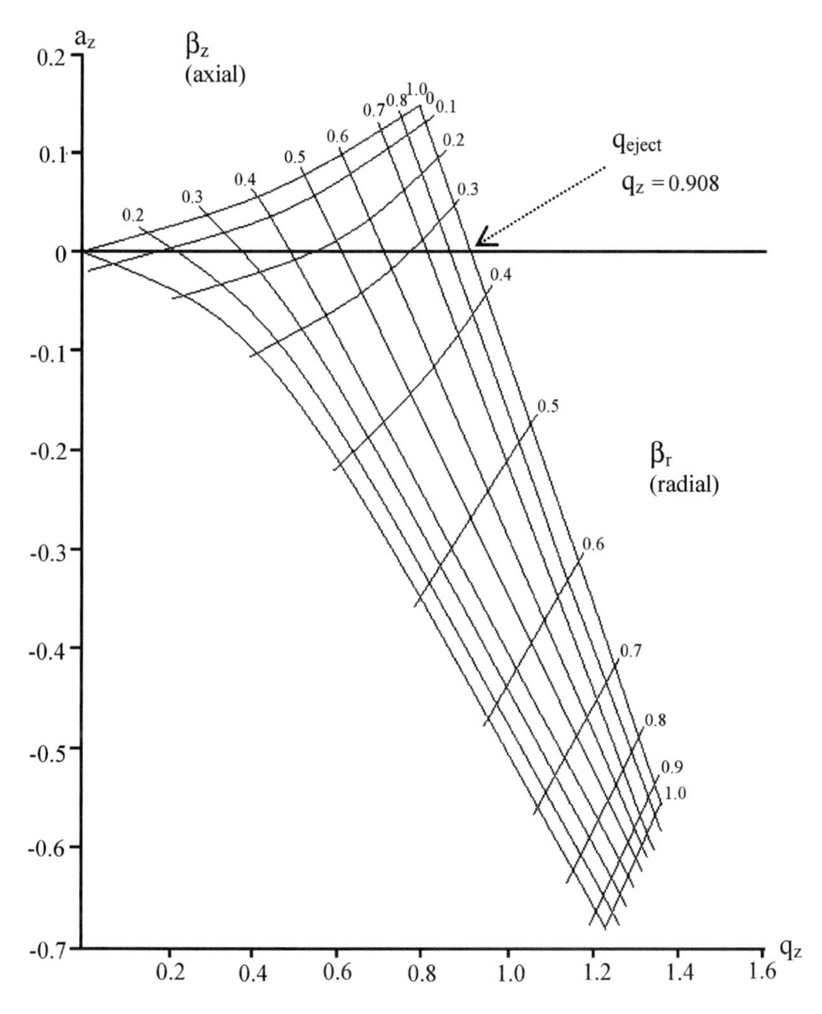

FIG. 14. Simultaneous stability diagram, showing stability in both radial and axial directions.

where k is a constant. All ions with $m/e > k\ V/0.908$ are stable and are stored within the trap.

(iii) Ion-Trap Operation. Once the ions are stored in the trap, they need to be analyzed according to their m/z values. Modern ion traps achieve this using the mass-selective instability mode of operation.[55] This works by making the ions in the trap selectively enter unstable trajectories so that they leave the trap in the order of increasing m/z and can be detected. In the mass-selective instability scan the ion

trap is operated in RF-only mode where the ring electrode RF voltage (V) is ramped. This causes the q_z values of the stored ions to increase in proportion to the applied RF voltage. When the ions reach the point on the stability diagram marked q_{eject} ($q_z = 0.908$) they become unstable. The ion's β_z value at this point is 1, therefore its secular frequency is exactly half of the main RF drive frequency. The ions pick up energy from the drive frequency and exit the ion trap through one of the end-caps. Only half of the ions pass through the correct end-cap to allow them to reach the detector.

Early on in the development of ion-trap instruments it was noticed that operating the ion trap with a background pressure of a light buffer gas such as helium, greatly increased the performance of the instrument.[56] Today, all ion traps operate with a buffer gas. The buffer gas increases sensitivity because analyte ions in the trap undergo low-energy collisions with buffer gas molecules. This has the effect of collisional cooling, the ions' motion is damped and they are forced into a tight bunch at the center of the ion trap. Sensitivity and resolution are improved, because the spatial distribution of the ions in the trap is reduced and therefore ions of a similar m/z ratio arrive at the detector over a smaller time distribution.

If the population of ions in the trap goes above a certain level (approximately 10^6 ions), then like charges on the ions repel each other. This effect, known as space charging, has the opposite effect to the buffer gas, resulting in a more diffuse cloud of ions, reducing the sensitivity and resolution of the instrument. Modern ion traps eliminate this problem by restricting the number of ions entering the trap. A short prescan is generally taken before the main analytical scan to determine the strength of the ion current and the accumulation time for the main analytical scan is adjusted to allow an optimal number of ions to be trapped; this approach is called automatic gain control (AGC).[57]

(iv) Tandem Mass Spectrometry in an Ion Trap. The ion trap is ideally suited to perform tandem MS, and is even capable of performing sequential stages of tandem MS (MS^n). A precursor ion is selected by applying a broadband waveform across the ion-trap end-cap electrodes. This waveform has a window corresponding to the secular frequency of the precursor ion. The effect of this waveform is that all ions except the required precursor ion become unstable and leave the ion trap. With the precursor ion isolated, a supplementary RF 'tickle' voltage is applied across the end-caps at the secular frequency of the precursor. The ion picks up energy from this voltage which increases the amplitude of the ions' trajectory in the trap. As a result, the ions undergo increased collisions with buffer gas molecules, imparting vibrational energy to the ions and causing them to

fragment. The fragment ions are not excited by the tickle voltage (they have different secular frequencies than the precursor) and so relax to the center of the trap. For this reason, ion-trap product ion spectra only show the lowest energy fragmentations, and so from a single stage of tandem MS give less information than other tandem instruments. However, the ion trap can be used to perform further stages of MS. A fragment ion can be isolated and fragmented in a similar fashion to that used on the original precursor ion. This process can be repeated over a number of stages provided the ion current is high enough to give reasonable fragment ion spectra. This process is called MS^n where the n term represents the number of consecutive stages of isolation and fragmentation. The advantage of this approach is that connectivity can be assigned between the different fragment ions. This can be particularly useful when trying to determine complex carbohydrate structures, as single monosaccharide residues can be removed with each stage enabling sequencing.

(v) MALDI Ion Trap. Ion traps can also be used with MALDI, thanks to the development of atmospheric pressure MALDI sources. Creaser and coworkers[58] give an example of the analysis of carbohydrates using atmospheric pressure MALDI ion-trap MS. This work analyzed native N-linked glycans, and linear glucan polymers, using the multi-stage capability of the trap to follow the fragmentation of the oligosaccharides through several stages of MS.

(vi) Hybrid Ion-Trap Instruments. Recently hybrid ion-trap instruments combining an ion trap with another type of mass analyzer have become available. Shimadzu has developed an ion-trap TOF hybrid instrument. This instrument has been used to analyze complex N-linked glycans.[59] The advantage of this instrument is that the MS^n capability of the ion trap can be used to generate nth generation fragment ions whose mass can then be accurately mass measured using the TOF mass analyzer.

c. FT-ICR.—Fourier-transform ion-cyclotron-resonance mass spectrometry (FT-ICR-MS)[60,61] was introduced by Comisarow and Marshall in 1974.[62,63] The FT-ICR is a powerful ion trapping instrument, capable of making mass measurements with higher resolution and mass accuracy than is available from any other mass spectrometer. Attention focused on the analysis of biomolecules using ESI and MALDI has facilitated coupling of these external ion sources to the FT-ICR instrument. The difficulty in using an external ion source lies in the transfer of ions from the source to the mass analyzer. This problem has been solved in modern commercial instruments using a variety of guiding optics. FT-ICR instruments analyze and detect ions using methods unique among mass spectrometers.

FT-ICR instruments have four common components. The first is a magnet, usually superconducting. The quality of mass measurement obtained from the instrument improves as the magnetic field strength increases. Field strength in production instruments available ranges between 3 and 12 T, although fields of up to 25 T have been used in experimental instruments.[64] The second component is a cell or Penning trap, where the ions are stored, mass analyzed, and detected. The cell consists of six plates, orientated in the magnetic field so that one opposing pair of plates lies orthogonal to the direction of the magnetic field and two pairs lie parallel to the field (Fig. 15). Each pair of electrodes has a function—trapping, exciting, or detecting. The third feature is an ultra-high vacuum system, the performance of an FT-ICR instrument being sensitive to pressure, and typically requiring pressures of 10^{-9} or 10^{-10} Torr when detecting ions. The final feature is a data system. Components of this include frequency synthesizers, broadband radio frequency (rf) amplifiers and preamplifiers, a fast digitizer and a computer to control and coordinate these devices, as well as to acquire and process data.

The ions are guided from the external ESI or MALDI source into the cell using optics, which generally include a multipole and several focusing elements. Ions pass through several stages of pumping as the pressure is lowered from atmospheric in the source to high vacuum in the cell. Once in the cell, the ions are trapped by a combination of magnetic and electronic forces. The ions are

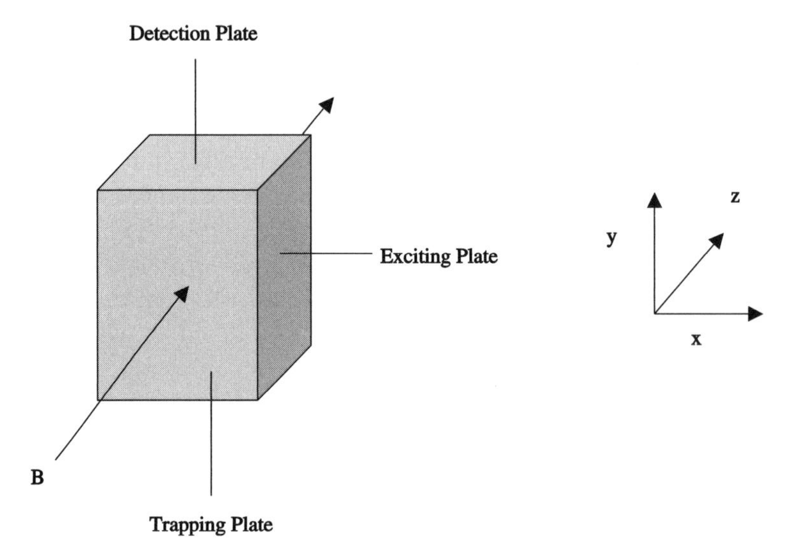

FIG. 15. The cell in an FT-ICR instrument.

trapped in the z-direction—along the axis of the magnetic field—using DC potentials applied to one pair of plates. Ion trapping in the x–y plane is achieved using the magnetic field, which causes the ions to orbit the z-axis. The vector cross product of the magnetic field, B, and the ion's velocity, v, results in a force, F, acting orthogonally to both B and v, termed the Lorentz force.

$$F = zv \times B \tag{14}$$

where z is the charge on the ion.

The combination of the Lorentz force and the ion's initial velocity upon entering the cell acts upon the ion and creates a circular trajectory—cyclotron motion (Fig. 16).

The frequency of cyclotron motion, that is, how rapidly an ion precesses about the orbit, is m/z dependent. Applying Newton's Second Law

$$zvB = \frac{mv^2}{r} \tag{15}$$

v/r is frequency (ω), and so

$$\omega = \frac{zB}{m} \tag{16}$$

Therefore by measuring the frequency of an ion's precession about the cyclotron radius, with magnetic field strength B known, it is possible to determine m/z for the ion. Frequency is the physical property which can be measured most accurately with modern electronics, resulting in high-quality mass measurements.

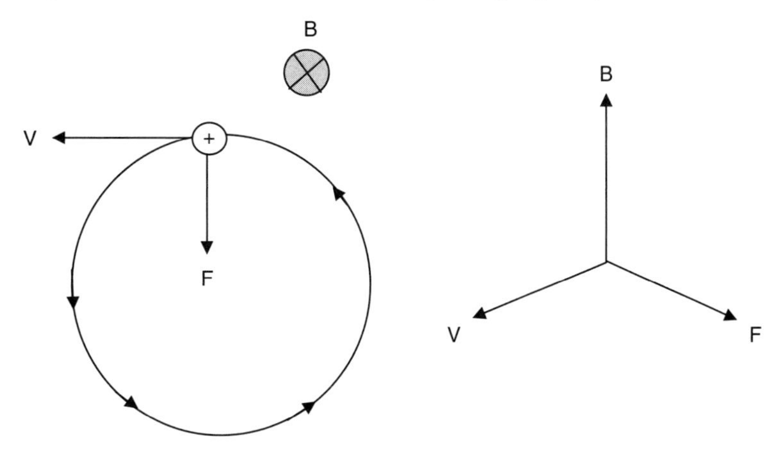

FIG. 16. Cyclotron motion within the cell of an FT-ICR instrument.

The ions are trapped in the cell, allowing thousands or millions of measurements to be made, enhancing the quality of data obtained.

Ion-cyclotron motion within the cell is normally of too small a radius and too incoherent to be detectable. Therefore, for detection, ions are selectively excited to a larger cyclotron radius using an rf electric field generated by the exciting plates of the cell. Ions absorb rf energy when it is at the same frequency as their cyclotron motion. The absorption of energy causes the radius of the cyclotron motion to increase (Fig. 17), allowing detection. As ions at different m/z values have different cyclotron frequencies, excitation of single, multiple, or a range of m/z values is possible. Excitation profiles containing the necessary frequencies are created by the data system and generated by the exciting plates.

Once excited, the ions remain at the larger radius until they relax due to collisional cooling. Re-excitation and re-measurement is possible and can be essential. The (excited) cyclotron radius of an ion is independent of m/z, depending on the rf energy applied and the magnetic field strength. This means that no mass discrimination is seen when exciting ions over a range of m/z values.

Ions can be singly excited or ejected (by inputting enough energy to increase the cyclotron radius beyond the size of the cell) or broadband excited/ejected depending on the desired experiment. Broadband frequencies can be generated which excite or eject most or all of the ion population. Notches can be created in these frequencies which allow, for example, the trapping of ions of three selected m/z values and the ejection of all other ions from the trap. The combination of rf frequencies necessary to perform a desired experiment are generated automatically by the data system, generally using stored-waveform inverse Fourier transform (SWIFT).

Ion detection in FT-ICR instruments is unique (Fig. 18). It is a non-destructive detection method, which means multiple measurements can be performed on the same ions.

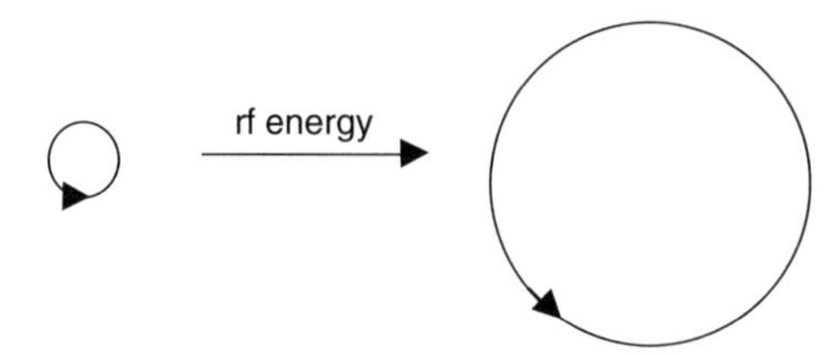

FIG. 17. The effect of rf energy on ion radius within the ICR cell.

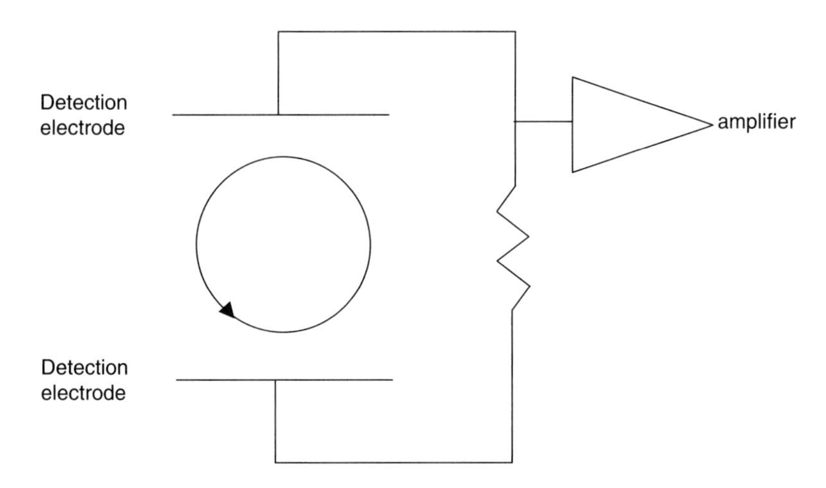

FIG. 18. Detection of ions in an ICR cell.

A positively charged ion induces an image current between the two detection electrodes as it cycles; this is measured in time and then converted into frequency by Fourier transformation. Ions generate a larger image current when they are coherent and at a larger radius, hence the need for excitation prior to detection. The signals from ions with different m/z values, which have different cyclotron frequencies, are additive and are superimposed then deconvoluted as part of the Fourier transform operation (Fig. 19). The signal is proportional to the number of ions and approximately linear.

FT-ICR instruments are also capable of performing MS^n experiments. The most popular method of ion activation is sustained off-resonance irradiation (SORI), where ions are excited to a larger cyclotron radius using rf energy, undergo collisions with a neutral gas pulsed into the cell and dissociate. Other methods are available, including infrared multiphoton dissociation (IRMPD)[65] and electron capture dissociation (ECD)[66] which is of particular value in glycopeptide analysis (Section VI.4).

d. Time of Flight Mass Analyzers. — *(i) TOF.* The principles of TOF mass analyzers are relatively simple.

In a simple linear TOF instrument (Fig. 20), ions are formed within the source region(s) in the presence of an electrical field (V). The field accelerates the ions into a field-free region (d), which the ions all ideally enter with the same kinetic energy (KE) defined by the accelerating potential. Subsequently as the ions drift

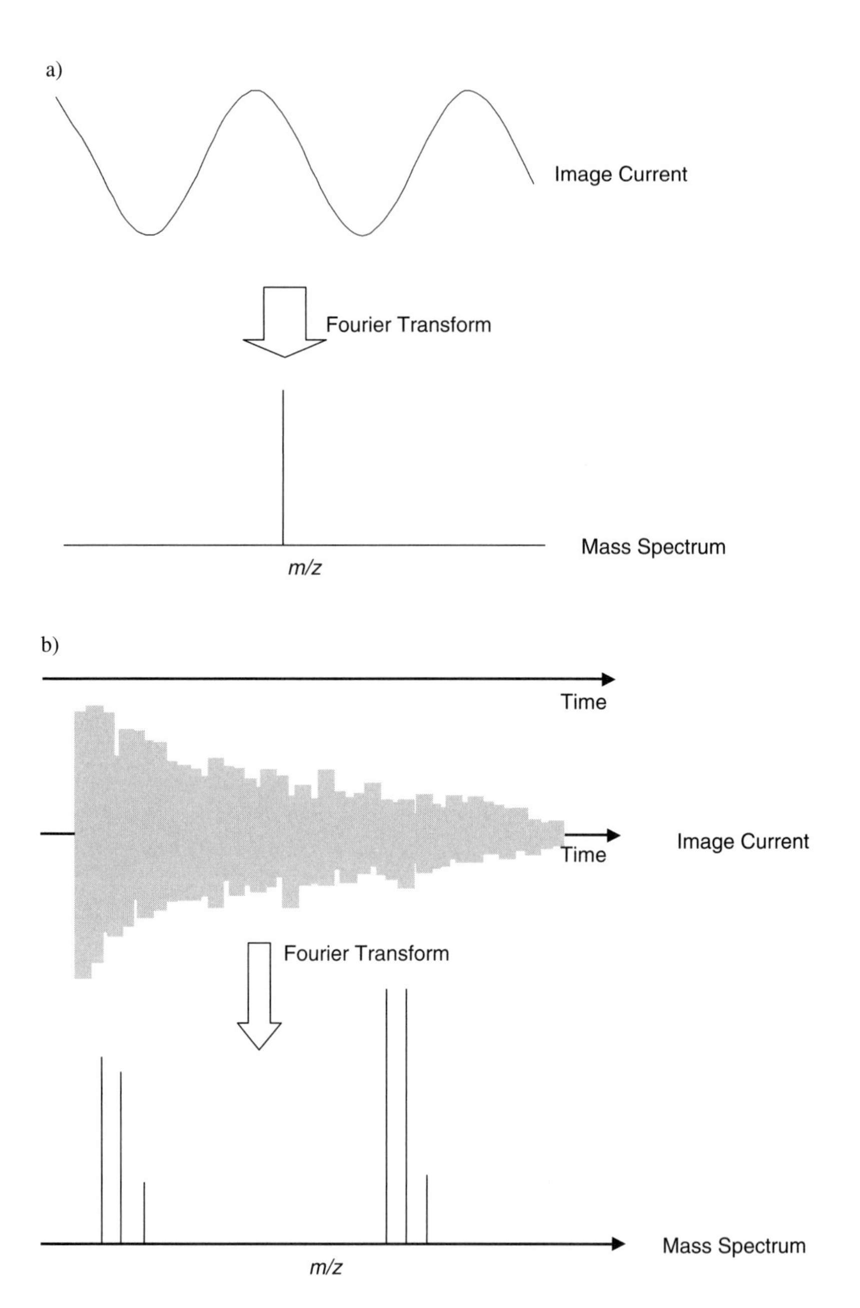

FIG. 19. Cartoon of conversion of frequency to m/z using Fourier transformation (a) single frequency and (b) multiple frequencies.

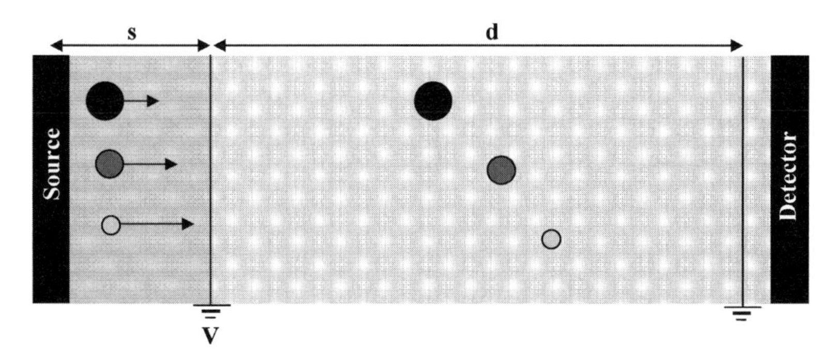

FIG. 20. A simple linear TOF analyzer.

with no external field, they separate as a function of their masses by traveling at different velocities based on their specific mass-to-charge ratios:

$$KE = zV = \frac{1}{2}mv^2$$

The KE of the ion is equal to the product of the ion's charge (z) and the applied voltage (V), or can be described with the energy–mass relationship, where m is the mass of the ion and v its velocity.

The time required to traverse the field-free region is given by:

$$t = \frac{d}{v}$$

Substituting for v and expanding,

$zV = md^2/2t^2$ and rearranging for t, $t^2 = m/z(d^2/2V)$

If the parameters in parentheses remain constant, then m/z can be calculated from the ions' flight times.

Unfortunately ions produced in the source are formed under less than ideal conditions (Fig. 21). Ions with the same m/z may have a distribution of kinetic energies; consequently, the ions have different flight times and arrive at the detector at different times (Fig. 21A). Furthermore, ions with the same m/z can be formed at different locations in the source (Fig. 21B). This spatial distribution

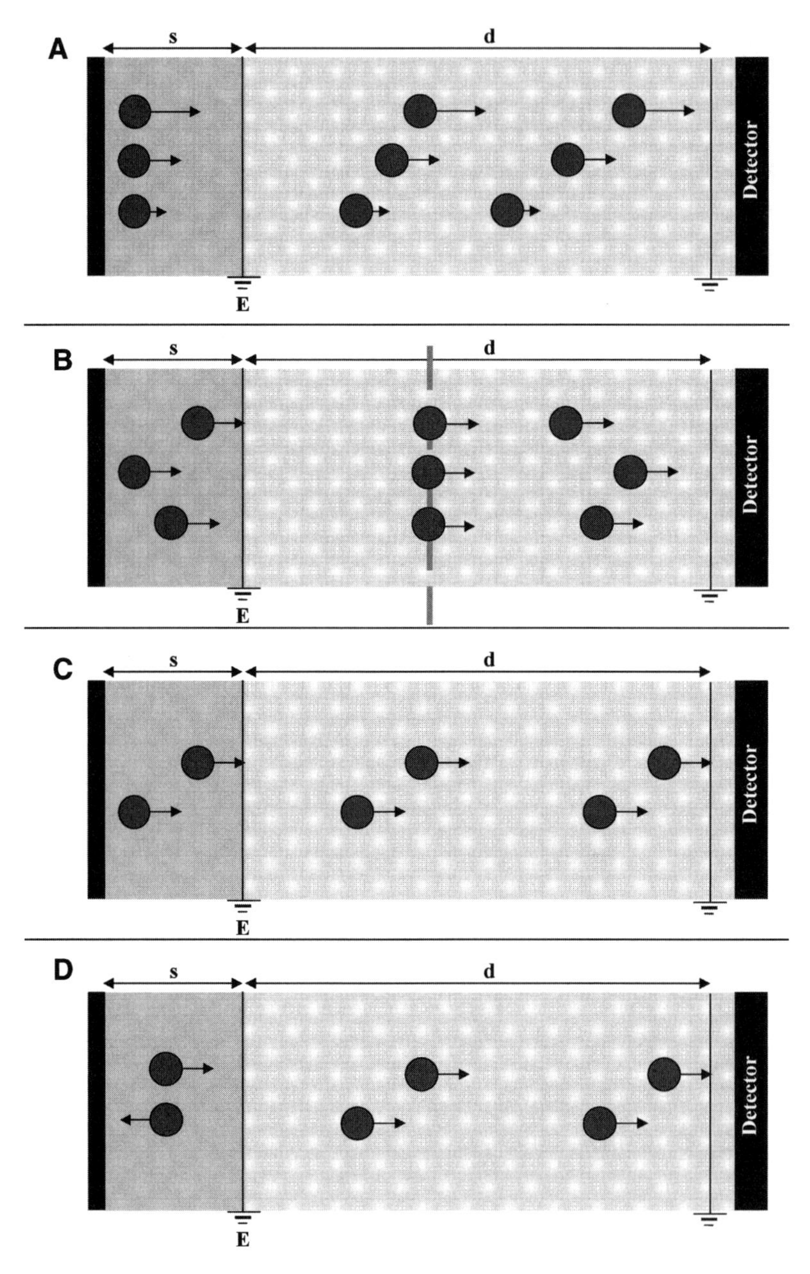

FIG. 21. Detrimental ion distributions (A) kinetic energy (B) spatial (C) time (D) turn-around time.

affects the magnitude of the accelerating potential. Ions formed closer to the extraction plate are extracted into the field-free region with lower KE than ions formed further from the extraction voltage, causing ions with the same m/z to reach the detector at different times. Ions with the same m/z formed at different times traverse the field-free region maintaining the difference in time and therefore arrive at the detector at different times (Fig. 21C). Finally, ions can be formed travelling away from the drift tube. These ions require extra time to 'turn around' and leave the source region, and so arrive at the detector later than ions with velocities in the 'forward' direction (Fig. 21D).

These distributions are the major origins of the poor resolution observed using simple linear TOF instrumentation. Resolution (R) for a TOF mass spectrometer is dependent on peak width (Δt) and the total flight time (t):

$$R = \frac{m}{\Delta m} = \frac{t}{2\Delta t} \tag{17}$$

The deleterious effects of the different ion distributions are to effectively broaden the peak and therefore reduce resolution. Modern TOF instruments have overcome these inherent limitations using a number of different strategies. A detailed description of each of these is beyond the scope of this chapter; readers are directed to the key publications and reviews.[67–70]

The first significant instrument improvement was pioneered by Wiley and McLaren in 1955 and was originally known as time-lag focusing (TLF);[71] it has now evolved into a technique referred to as delayed extraction (DE),[72,73] which is commonly used in unison with MALDI. TLF improved TOF resolution by attempting to correct for both the initial kinetic energy and spatial distributions. The new instrument design included a dual-stage extraction region coupled to gas-phase ionization using electron ionization. Fundamentally, Wiley and McLaren assumed ionization occurred in a relatively small space close to the source and they applied a time delay between ion formation and ion extraction. This time delay in effect corrects for the initial kinetic-energy distribution. During the delay after ion formation, the ions are effectively reorganized as their initial velocities cause the ions to spread out within the first extraction region (Fig. 22). This results in a large spatial dispersion which is correlated to the initial velocity distribution and can be refocused using the dual-extraction source.

There is a point along the flight tube, called the 'space focus plane', where the spatial distribution is minimized for ions of any given m/z. If the space focus plane is not at the detector then, after travelling through the space focus plane, ions with the same m/z will have different flight times to the detector (Fig. 21B).

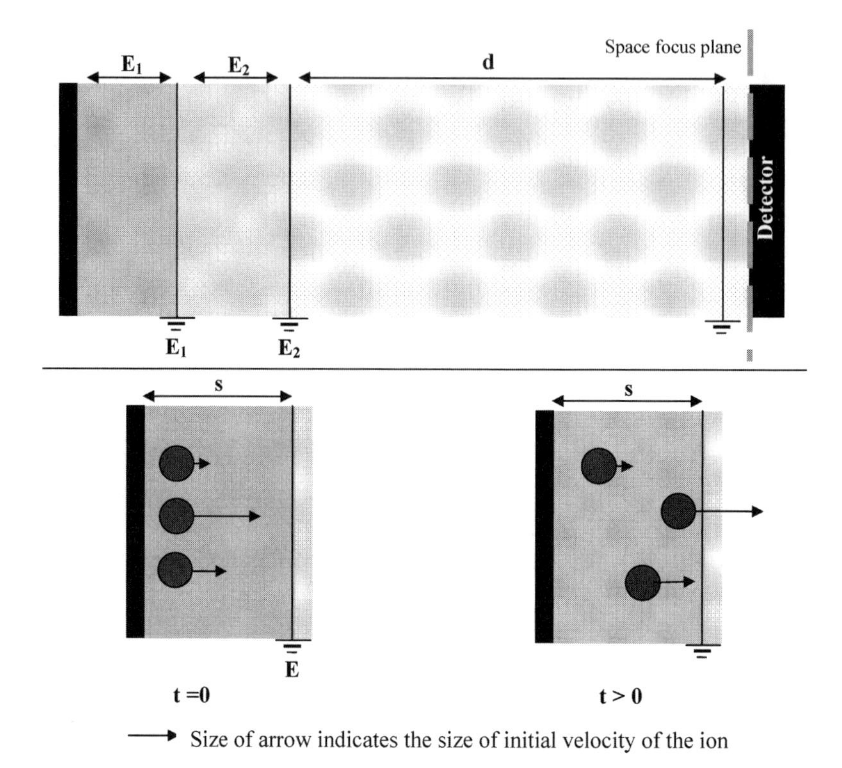

FIG. 22. Dual-extraction source.

The space focus plane is essentially due to two opposing factors affecting the spatially distributed ions during extraction and how this spatial distribution affects the ions' arrival time. Firstly, ions in the source that are more distant from the detector have further to travel to the detector, and secondly, ions formed more distant from the detector experience a larger extraction potential and therefore reach a higher velocity in the drift tube.

In a single-extraction source, the space focus plane is fixed at twice the length of the extraction region. Wiley and McLaren showed mathematically that in a dual-extraction source the space focus plane can be moved by changing the ratio of the potentials in the extraction regions (E_1/E_2) (Fig. 22). By moving the space focus plane to the detector plane by application of the correct extraction potentials in both extraction regions then the spatial distribution can be corrected at the detector. Unfortunately simultaneous kinetic and spatial focusing is difficult. The spatial focussing requires the first extraction potential to be weaker than the second, but use of an initial weak extraction potential is detrimental to KE focusing. The lag time to correct the kinetic energy is mass dependent and therefore the resolution can only be improved over a limited mass range.

DE, which is used today in MALDI sources, is based on the principles of TLF but with some notable differences. Since MALDI is a surface desorption technique, this significantly reduces the initial spatial distribution, any 'turn around' issues, as well as temporal distributions within the ion source. Therefore the major origin of resolution loss during MALDI is the distribution in the ions' KEs due to collisions within the matrix plume during ionization. Therefore simultaneous kinetic and spatial focusing is unnecessary, and DE is optimized to focus the ions' initial KE distribution. After an optimized delay between ion formation and extraction, a large potential is used to extract the ions into the drift region, thus minimizing the effect of the initial KE distribution on resolution. The time delay also significantly reduces the number of collisions within the source during ionization, further improving the resolution. DE is only applicable over a limited mass range but during MALDI the initial velocity of desorbed ions is nearly independent of mass. Therefore the initial KE of the ion is proportional to mass; this enables predictable changes in the pulsed extraction to optimally focus ions with different mass-to-charge ratios.

In addition, generally all modern-day TOF instruments, regardless of the method of ionization, have a device called a reflectron to improve resolution. The reflectron was introduced by Mamyrin and co-workers in 1973[74,75] to correct for ions' kinetic-energy distributions. The device consists of a series of rings or grids; each ring has a higher voltage applied than the preceding one (Fig. 23). The voltages increase linearly (in the simplest design) up to a voltage slightly

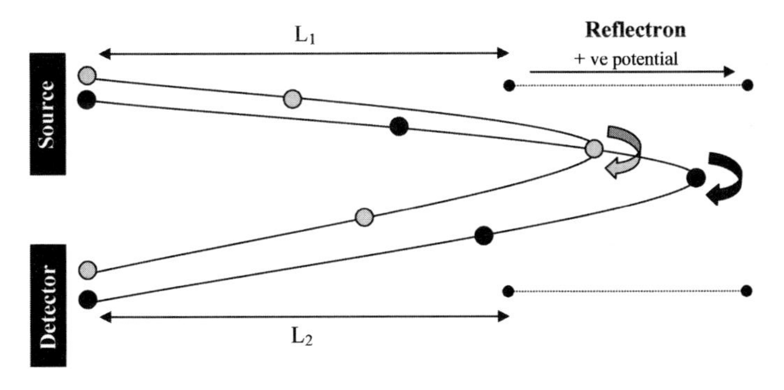

FIG. 23. Cartoon of an ion's flight within a reflectron.

higher than the extraction voltage of the source. The reflectron is positioned after the field-free drift region and acts as an ion mirror by the use of the retarding field produced between the rings/grids. Ions leave the reflectron with the same velocities with which they entered, but traveling in the opposite direction. Ions with greater energy penetrate further and therefore take longer to be reflected. Using the correct voltages brings ions of the same mass-to-charge ratio but different energies to the detector at the same time. In effect, this focusing can be considered to be similar to TLF or DE; ions leaving the reflectron are brought to a space–time focus plane positioned at the detector. The other key advantage of the reflectron is that it dramatically increases the flight time of the ion without enlarging the instrument significantly, which leads to further increases in resolution. The key disadvantage of the reflectron is that it is unable to focus ions with a wide range of kinetic energies and furthermore the reflectron cannot correct for temporal distributions in ion formation.

(ii) TOF/TOFs. Most modern tandem mass spectrometers operate at low collision energies, so that fragmentation is dominated by low-energy pathways. However, TOF mass analyzers are ideal for transmission and analysis of high-energy ions (>1 keV). This led to the design of a new instrument, the MALDI time-of-flight/time-of-flight tandem mass spectrometer (MALDI TOF/TOF).

In order for the MALDI TOF/TOF design to be successful, one major obstacle had to be overcome. To record a product ion spectrum, ions with a broad range of KEs must be focused prior to detection, which is impossible directly using a classic reflectron design. This problem arises as, when an ion fragments in a field-free region, the resultant fragment ions maintain the same velocity as the precursor ion but have a wide range of KEs. The kinetic energy (Ef) is proportional to the fragment ion's mass, such that it is equal to the product of the ratio of its mass (Mf) to the precursor's mass (Mp), and the precursor ion's original kinetic energy (Ep).

$$Ef = \frac{Mf}{Mp} Ep \tag{18}$$

Several commercial instrument designs have effectively overcome this inherent problem and can record high-energy CID product ion spectra. The Applied Biosystems 4700 Proteomics Analyzer was one of the two originally available commercial TOF/TOF designs[76,77] (Fig. 24). When the instrument is operated to obtain a mass spectrum, it acts as a standard inline reflectron TOF instrument. In product ion experiments the precursor ion is isolated at the end of the first flight tube by the timed-ion selectors (TIS). The MALDI source uses DE to correct for

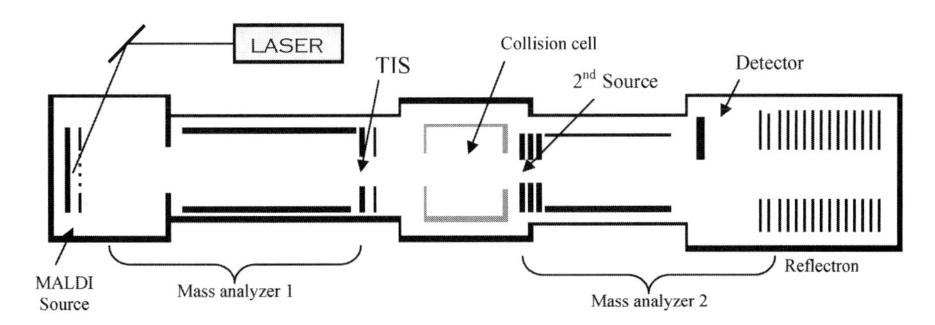

FIG. 24. Cartoon demonstrating the effect of the reflectron (ion mirror) on the ions' flight path in a time of flight mass analyser.

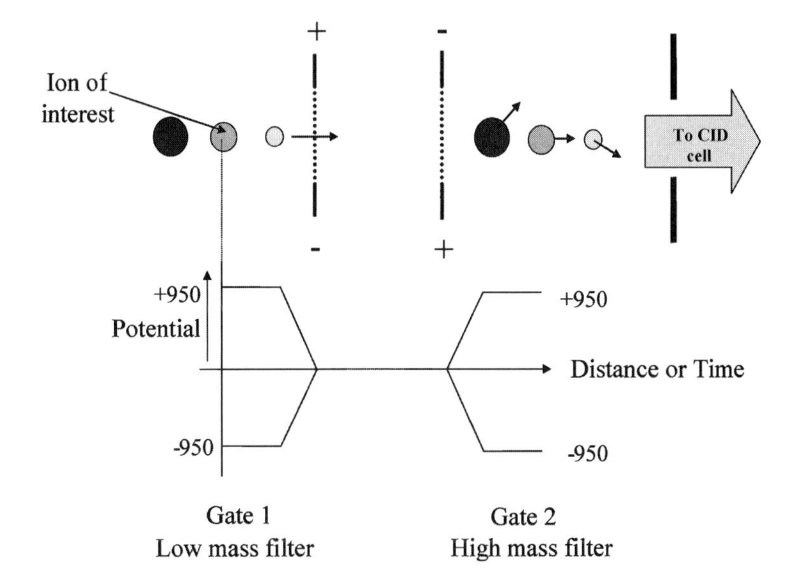

FIG. 25. Time-ion selector gate.

initial kinetic energy spread. The space focus plane created by the DE is centered on the TIS gates to improve the resolution for ion selection.

The mechanism of precursor selection using the TIS relies on changes in potential applied to two gates (Fig. 25). Gate 1 acts as a low-mass filter and is normally 'closed' with the application of $+/-950$ V potential. The second gate (Gate 2) acts as the high-mass filter and is normally in the 'open' position, potential free. The time of flight of the selected precursor to the entrance of the

TIS can be calculated. Gate 1 via the applied potential is set to deflect any ion entering the TIS sufficiently to prevent transmission into the collision cell. At the arrival time of the precursor, the potential on Gate 1 is rapidly switched from 'closed' to 'open' (0 V) allowing the transmission of the precursor, after deflecting ions of a lower m/z which arrived before the precursor. Once the precursor has cleared Gate 2, Gate 2 is pulsed from 'open' to 'closed' by the application of $+/-950$ V potential across the electrodes. Gate 2 now deflects any ions with a higher m/z than the precursor away from the collision cell.

The collision energy within the collision cell is defined by the difference in potential between the acceleration voltage in the source (collision offset) and the potential of the collision cell. Normally ions are slowed down with a retarding lens before entering the collision cell, typically lowering the collision energy down to 1–3 keV. Once the ions exit the collision cell they enter an ion acceleration region referred to as the second source. The parent ion and its fragments essentially leave the collision cell with the same velocity and therefore reach the second source at the same time. The acceleration pulse is timed to coincide with the ion's entry into the region. This reacceleration pulse provides the start time for the TOF measurement in the second analyzer, as well as making it possible for the reflectron to transmit fragments across the full m/z range with the same voltages (the reflectron is only able to focus ions in a narrow energy range for a given set of voltages).

The second commercial design, the Ultraflex TOF/TOF introduced by Bruker Daltonics uses a different approach to recording TOF/TOF spectra and has differing TOF ion optics[78] (Fig. 26). The most notable difference is that the collision cell is placed before both the first and second mass analyzers.

Like the previous TOF/TOF design, when operated to obtain a mass spectrum it acts as a standard inline TOF instrument with DE. Precursor ion selection is

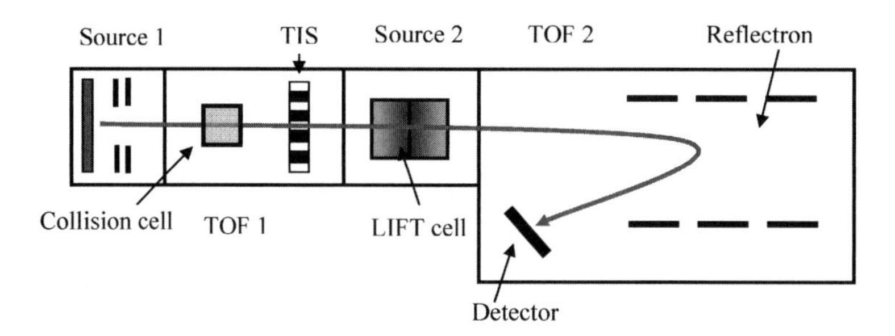

FIG. 26. Bruker Ultraflex TOF/TOF.

achieved using a TIS gate, but takes place *after* CID in the collision cell. This is possible as fragment ions maintain the same velocity as the precursor ion, so that both the precursor and its fragments are isolated for mass analysis in the second drift region, at the end of the first field-free region.

Once the selected precursor and its fragment ions leave the TIS gate they enter the LIFT cell. The LIFT cell comprises four grids arranged into three stages; within the first LIFT stage a high voltage is rapidly applied while the second and third stages act in unison conceptually as a DE source, time-focusing the ions onto the detector. The LIFT cell allows a complete product ion spectrum to be obtained in a single analysis; the rapid potential increase is the crucial step. By increasing the ions' kinetic energy all the ions become focusable in the reflectron. A reflectron is able to successfully focus and reflect a fragment ion with its precursor ion if the fragment's KE is at least 60% of the precursor's KE; obviously lower m/z fragment ions do not meet this criterion. In this TOF/TOF design, the selected precursor and its fragment ions enter the LIFT cell together and with the same velocity and at an energy of, for example, 8 keV. A LIFT potential of 12 keV is then rapidly applied. The precursor ion will therefore leave the LIFT cell with $8+12=20$ keV. The smallest fragment will leave with a energy slightly higher than the LIFT potential (12 keV). Since 12 is 60% of 20, all ions will have a range of KEs focusable by a single set of reflectron voltages.

Both instrument designs have been successfully used for carbohydrate structural analyses.[79–81]

(iii) Orthogonal Acceleration TOF. TOF analyzers are conventionally employed with a pulsed ion source such as MALDI and have several advantages over quadrupoles including their unlimited mass range and high-duty cycle (scanning is not involved). With the resolution of modern instruments, mass accuracy and response times are also high. Orthogonal injection also allows TOFs to be used with ions generated as continuous beams,[82,83] such as by electrospray, as well as with MALDI (Fig. 27). Since TOF analyzers can tolerate a spread of ions in the plane perpendicular to the flight path, orthogonal injection, which introduces minimal spread in the direction of the flight path, allows high resolution.

The collimated (focused) ion beam leaves the hexapole and enters the ion gating system—the push and pull electrodes (Fig. 27). It is this system which provides the pulse of ions necessary for TOF analysis. When the region in front of the push electrode is filled by the ion beam, the ions are pushed towards the extraction electrodes. As this packet of ions leaves the gate it is reopened and another packet is admitted. At the extraction electrodes, ions are accelerated

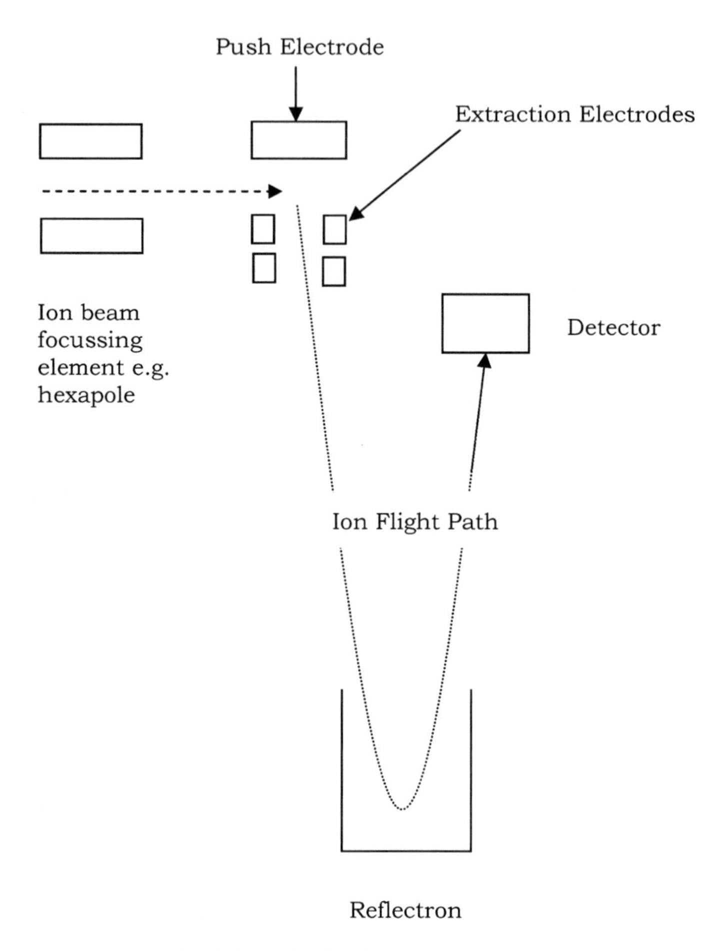

FIG. 27. Schematic of orthogonal TOF mass analyzer.

into the TOF tube with an acceleration potential. The ions then pass via the reflectron to the detector. The extraction pulse acts as the start time for the TOF measurement, analogous to the DE pulse.

The main factor affecting orthogonal TOF is the duty cycle. This is how often ions can be pulsed into the TOF analyzer. When waiting between extraction pulses, ions are being wasted (they pass through the extraction region and are lost) and so the ideal is obviously to have this process as near to continuous as possible. The factor limiting the duty cycle is the time taken by the slowest (heaviest) ion to fly down the tube to the detector. Modern instruments operate at a frequency of a few kilohertz, meaning the maximum flight time is of the order of a few hundred microseconds.

III. Hyphenation

To obtain a comprehensive analysis of carbohydrate mixtures from biological matrices that display a high degree of compositional and structural heterogeneity, a multidimensional approach is required. The combination into a so-called "hyphenated system," of liquid chromatography (LC) or CE with MS provides the advantage of selective and efficient separation with the mass specificity, sensitivity, and structural information gained from MS.

1. LC-MS

Liquid chromatography (LC) is a commonly employed separation technique that can handle large numbers of highly complex samples with minimal pre-processing, is highly robust and compatible with most labeling procedures, and is therefore a valuable method for carbohydrate analysis. In coupling LC with MS, real-time results are produced, whilst the separation efficiency obtained from the LC is maintained, enabling successful MS analysis. Coupling the two analytical techniques is therefore advantageous. A number of obstacles, however, had to be overcome to allow the original on-line coupling of MS with HPLC, involving the differences in operational pressures of the two systems and the incompatibility of typical LC mobile phases with MS. The advent of the atmospheric-pressure ionization technologies of ESI[84] and atmospheric pressure chemical ionization (APCI) addresses both of these incompatibilities and so facilitated the widespread use of HPLC-MS. APCI is more suitable to the analysis of thermally stable, non-polar, small molecules, and so is less appropriate to the analysis of carbohydrates than is ESI. However, HPLC coupled on-line with ESI-MS/MS has become a powerful tool for the characterization of complex glycan samples.[85–93]

During the past few years, downscaling to capillary-scale columns of 20–100 μm i.d. and the use of nanoelectrospray technology[15,94] to introduce the eluent into the mass spectrometer with flow rates ranging from nL/min to a few μL/min, has meant that capillary LC-MS analysis has become popular, requiring only very small volumes of sample to be injected.[83,84] Schulz *et al.*[95] and Kawasaki *et al.*[96] demonstrated the gains in sensitivity of capillary LC-MS over conventional LC/ESI-MS in the structural analysis of oligosaccharides after reductive β-elimination and exoglycosidase digestion, respectively, of gel-separated glycoproteins.

Today, commonly used electrospray interfaces involve pumping eluent through a capillary with flow rates of 0.1–10 μL/min and applying a 2–5 kV potential. If the LC flow rate is compatible with the ESI flow rate, a transfer capillary transfers the eluent directly from the LC column to the ESI needle. If the LC flow rate is too high for the ESI interface it has to be split. Fused-silica capillary LC columns are typically packed with small diameter spherical particles (1.5–3 μm) under high pressure. Separation efficiency is increased with decrease in particle size but with an increased backpressure. A compromise between high backpressure, length of capillary column, and complete separation of complex mixtures has to be made. A number of stationary phases have been coupled on-line with MS. These include reversed phase (RP), normal phase (NP), and graphitized carbon columns (GCCs), as well as ion-exchange chromatography columns.

a. Reversed Phase (RP).—Oligosaccharides generally do not exhibit strong retention on C_{18}-RP-LC columns and therefore require derivatization with a hydrophobic reagent to allow efficient separation. A number of reducing-terminal derivatizations compatible with both RP separation and on-line MS detection include 1-phenyl-3-methyl-5-pyrazolone (PMP)[97] and phenylhydrazine labeling[98] as well as reductive amination with 2-amino-5-bromopyridine (ABP).[99] Derivatization with 8-aminonaphthalene-1,3,6-trisulfonic acid (ANTS) has been used for RP-LC-MS analysis with negative-ion mode MS detection but is also compatible with NP-LC-MS and CE-MS. Both reducing-terminal labeled and permethylated oligosaccharides have also been analyzed with improved sensitivity using RP-nano-LC-MS over RP-LC-MS.

b. Normal Phase (NP).—Analysis of glycans by NP-HPLC is normally performed after the carbohydrates are labeled with either anthranilic acid (2-AA; 2-aminobenzoic acid)[100,101] or 2-aminobenzamide (2-AB)[102] to enable fluorescence detection. On-line NP-LC-MS of 2-AB labeled glycans led to later miniaturization with NP-nano-LC-MS analysis of similarly derivatized glycans using a 75 μm i.d. column and flow rates of 0.3 μL/min, achieving low femtomole sensitivity.[103] Wuhrer *et al.*[103] have demonstrated an on-line NP-nano-LC-MS analysis of underivatized complex oligosaccharides, allowing a sensitive and rapid analysis of glycosylation patterns and individual oligosaccharides.

c. High-Performance Anion-Exchange Chromatography (HPAEC).—Using HPAEC, separation of carbohydrate analytes can be achieved not only according to charge but also by size, monosaccharide composition and linkage, and is a widely used technique for carbohydrate analysis (see Section VI.4). Typically mass spectrometric carbohydrate analysis is complicated by the presence of salt in the samples, compromising MS sensitivity. This is especially magnified when analyzing fractions obtained from ion-exchange chromatography, often suppressing the signal completely. A common approach to enable coupling of HPAEC with MS is to employ a membrane desalter prior to the MS analysis.[104] An alternative approach to remove salt is to derivatize the hydroxyl groups via acid-catalyzed peracetylation using trifluoroacetic anhydride—glacial acetic acid to partition the carbohydrates into dichloromethane on water—dichloromethane partition.[105,106] These methods are used in off-line coupling to MS, however on-line membrane desalters can also be used.[107–109]

d. Graphitized Carbon Columns (GCCs).—The use of GCCs in HPLC has produced successful separations of oligosaccharides, mainly as their reduced forms.[110] This stationary phase has also been used by Kawasaki *et al.* in LC-MS analyses of unreduced carbohydrates.[88] Good resolution and separation of isomeric glycans is achieved and the columns are stable over a wide pH range. GCC-LC-MS provides a sensitive technique for analysis of both native and derivatized glycans in both negative and positive-ion modes. It has been shown to be successful in revealing carbohydrate heterogeneity with regard to sialylation, acetylation, and sulfation patterns and distinguishing between closely related carbohydrate structures.[111] Kawasaki *et al.* used a combination of RP-LC-MS(/MS) and GCC-LC-MS(/MS) for the mapping of both peptide/glycopeptides and oligosaccharides, respectively, determining the differences in the carbohydrate heterogeneity among glycosylation sites.[89] Later the same group showed the successful use of microbore GCC-LC-MS for the sensitive analysis of high mannose-type, hybrid-type, and complex-type oligosaccharides.[92] Although the mobile phases used with GCCs are more compatible with MS coupling than the mobile phase used in HPAEC, Cohen *et al.*[112] showed that the two chromatographic methods are complementary and neither should replace the other. A combination of the two techniques is advantageous. With this in mind, an LC-MS method using a porous graphitized carbon (PGC) column (Hypercarb) has been used to successfully characterize the peaks in an oligosaccharide profile from an HPAEC with pulsed amperometric detection (PAD) analysis, without further manipulation of the collected fractions.[113] The PGC column was demonstrated to not only provide efficient desalting before MS but also provided further fractionation of oligosaccharides that coeluted on HPAEC.

2. LC-MALDI

Disadvantages of on-line LC-ESI-MS include (a) the generation of multiply charged ions that complicate spectra; (b) complete sample consumption during the run; and (c) short analysis times which limit the chance of performing multiple mass-spectrometric experiments on individual ions. MALDI on the other hand typically produces singly charged species, provides the opportunity to preserve an aliquot of sample after the LC step for performing further detailed analysis, and is more tolerant towards contaminants from LC buffers and salts. MALDI-MS has been shown to be a useful technique to analyze large carbohydrates with little sample preparation.[114] Several interfaces for LC-MALDI coupling have been developed to enable the matrix-sample co-crystallization necessary for MALDI analysis. Methods for off-line coupling by continuous sample deposition from the LC to a target plate include the use of heated capillary nebulizers[97,115] or non-continuous as a series of discrete spots using either micro dispensers,[116,117] or a pulsed electric field for automated non-contact deposition.[118,119] On-line LC-MALDI-MS methods include continuous flow, in which the LC eluent and the matrix solution are mixed in a mixing tee and directed to the flow probe[120] that in one example employs continuous analyte/matrix co-crystallization into a porous frit installed at a capillary end which is used as the target for MALDI.[121] Aerosol[122] and rotating-ball inlet[123] interfaces have also been developed. Although interest in LC-MALDI coupling has been mainly confined to peptide analysis, coupling nano-flow HPLC, a Probot microfraction collector, and MALDI-TOF-MS (nano-LC-MALDI-TOF-MS) has been used in analyzing glycoproteins.[124] This allowed localization and characterization of peptides and carbohydrate structures.

3. CE-MS

a. On-Line Hyphenation.—CE, in which separation depends on analyte charge and size, is efficient and more rapid than LC and offers high resolving power. CE is conducted in a fused-silica capillary, the ends of which are placed into buffer reservoirs connected to high voltages (10–40 kV). Separations result from differences in the electrophoretic mobilities of analytes and occur superimposed upon a bulk electro-osmotic flow (EOF) towards the cathode. This EOF originates from the migration of ions in the electrical double layer at the capillary–buffer interface and moves all analytes in the same direction.

Carbohydrates are very difficult compounds to analyze using CE because of their inherent structural complexity, frequent lack of a charge and lack of a UV or fluorescent chromophore. They thus require derivatization to facilitate ion formation and detection.

The complexities involved in carbohydrate CE analysis have resulted in very few examples of on-line CE-MS coupling for carbohydrate analysis. Almost all examples use ESI to couple CE to MS. In conventional CE the cathode is immersed in a buffer reservoir, which needs to be replaced by the MS on hyphenation. The earliest and commonest interface for CE/ESI-MS uses a coaxial sheath-liquid interface.[125–127] The sheath liquid provides the solvent flow necessary for achieving stable ionization by maintaining electrical contact for the ESI voltage and also acts as the CE cathode. The solvent used as the sheath liquid, appropriate for ESI, picks up the analyte migrating out of the CE capillary and the resulting mixture is sprayed into the MS. A major disadvantage of this set up, however, is the diluting effect the sheath-liquid has on the CE buffer which compromises the sensitivity of analysis.

A sheath-liquid CE-MS system has been used to analyse ANTS[128] and 1-aminopyrene-3,6,8-trisulfonate (APTS)-labeled carbohydrates. The chromophore imparts a negative charge, enhancing electrophoretic separation and improving electrospray ionization. A variation on this sheathflow interface uses a double coaxial stainless-steel tubing assembly surrounding the fused-silica separation capillary that enables both addition of liquid make-up and the delivery of a sheathflow gas. The addition of the gas assists in the formation and stability of the spray. This interface was used in CE-MS identification of heparin oligosaccharides.[129]

An alternative to the sheathflow interface is the high-sensitivity sheathless configuration, where coating a conductive layer directly on the sprayer tip provides the electrical contact for both CE and ESI.[130–133] In combination with micro- or nano-ESI sources, the flow into the MS is compatible with the EOFs generated on CE and the sample is sprayed directly from the tip of the CE capillary into the MS. The resulting higher analyte concentrations improve sensitivity. However, in spite of experimenting with different coating materials (metal,[130,133] fairy dust[131] or graphite[132]) it is not always possible to obtain a stable spray, the technique lacks robustness and reproducibility and is not commercially available.

Despite the number of sheathless CE-ESI interfaces described, few have been used for the analysis of complex carbohydrate mixtures. Zamfir *et al.* described a sheathless on-line CE/ESI-MS interface consisting of a CE capillary with the terminus shaped into a copper-coated microsprayer.[134,135] This was described as being convenient, sensitive, and reliable, and exhibiting a long-lasting, stable ESI spray for the analysis of neutral and acidic carbohydrates.

b. Off-Line Hyphenation.—Off-line CE-MS is attractive since the experimental conditions for CE and MS are decoupled. A number of MS ion sources can be used. Off-line CE separation, combined with fraction collection of CE-UV peaks, and subsequent MALDI-TOF-MS structural characterization of APTS-derivatized oligosaccharides has been shown.[136] To overcome the dilution effect on fraction collection, a chip-based microfluidic fraction-collection approach improved throughput and sensitivity.[137] Chip-based MS provided efficient ionization of carbohydrates arising from the CE. MS/MS analysis of species in the fractions collected from CE was possible because of the long-lasting and stable spray achieved through the chip-based system.

c. Multiple Hyphenations.—Triple hyphenation of separation methods can allow the identification of a large number of oligosaccharides or glycoforms found in heterogeneous biological samples. This is achieved through the comparison and then integration of different degrees of orthogonal separation. To illustrate this, Udiaver *et al.*[138] first fractionated a sample by LC followed by CE and then carried out MALDI-TOF-MS.

IV. Derivatization

1. Permethylation and Peracetylation

The use of derivatization prior to mass-spectrometric analysis of glycoconjugates can provide significant advantages in both simplifying fragmentation and improving sensitivity of detection by increasing the analytes' surface activity. The two most common procedures that produce such results are permethylation[139] and peracetylation.[140] These procedures (producing methyl ethers and acetyl esters, respectively) can be routinely performed in high yield.[141]

Peracetylation provides a quick and reversible derivatization that is effective in improving the sensitivity of mass-spectrometric detection of carbohydrates for measuring the mass of intact derivatized molecules over their underivatized native counterparts.[142–144]

Permethylated oligosaccharides also provide increased ion signals over their underivatized counterparts, allowing the sensitive analysis of minor components.[142,145,146] This method provides a smaller mass increase and a greater volatility than the peracetylation method,[147] however the use of peracetylation prior to methylation can aid the removal of undesired salts and other impurities; the peracetylation procedure therefore acts as a sample clean-up step prior to subsequent methylation. Such a reaction in which the acetyl groups are replaced by methyl groups has been shown to be successful[143,148] using Hakomori methylation[149] which is catalyzed by the methylsulfinyl carbanion in dimethyl sulfoxide.

In 1984, a procedure for permethylation of carbohydrates using iodomethane, catalyzed by sodium hydroxide[139] was introduced that was shown to result in higher yields of permethylated products in a shorter reaction time than the Hakomori permethylation. It is this method of permethylation that has become routinely used.

The choice between peracetylation and permethlyation is dependent on the experiment being performed. When dealing with samples in complex, salty matrices, acid-catalyzed peracetylation proves to be the right choice[107,145] while permethylation tends not to be as effective under such conditions.[143]

2. Reducing-Terminal Labeling

The reducing-terminal aldehyde group of a carbohydrate generally reacts readily with excess alkylamine. The reaction proceeds through the formation of a Schiff's base with the aldehyde, which may be reduced. The product of this reaction is a secondary amine, which is much more stable than the unreduced glycosylamine intermediate. The use of this kind of chemistry increases the hydrophobicity of the carbohydrate and increases sensitivity for MS detection using both ESI and MALDI. Another advantage of this derivative is that it leads to the introduction of a chromophore for additional chromatographic detection. Several different amine groups have been used for this type of reaction and have been thoroughly reviewed.[29,150]

V. Mass Spectral Interpretation

1. Carbohydrate Fragmentation

For structural determination, it is necessary to observe mass-spectrometric fragment ions that are characteristic of the structure of the ion from which they were formed. Whether and how fragmentation occurs is dependent on the ion's internal energy. Although considered a soft ionization technique, FAB imparts more internal energy than do MALDI and ESI. Although MALDI, particularly with DHB as the matrix, may introduce sufficient internal energy to cause some fragmentation (either in-source[151,152] or post-source[153,154]), CID is generally carried out to promote reliable fragmentation and generate structurally diagnostic fragment ions which are most conveniently recorded using tandem mass spectrometers. However, so-called cone voltage fragmentation[155,86] can be induced in the ESI source by increasing the acceleration of ions through the high-pressure region towards the counter electrode and can generate fragment ions, although without the possibility of precursor mass selection, sacrificing precursor–product connectivity.

The soft ionization techniques generate predominantly even-electron ions (whether $[M+H]^+$, $[M+Na]^+$, $[M\text{-}H]^-$, etc.) which fragment to generate even-electron fragments. The fragmentation of MALDI- and ESI-generated ions is therefore very similar to that of ions produced using FAB, and summarized by Dell.[1] The range of different fragment ions that can be generated from soft ionization-generated large glycoconjugates was recognized by Domon and Costello,[156] who proposed a formalized nomenclature now adopted as standard. The fragmentation and nomenclature are broadly applicable to ions generated in both positive and negative modes, to all classes of carbohydrate and glycoconjugate, and to all instrumental and experimental set ups.

The weakest bond in a glycoconjugate and thus generally the easiest to fragment mass-spectrometrically is the glycosidic bond. Cleavage of the glycosidic bond generates ions from the reducing/aglycone-containing (Y_j) or non-reducing (B_i) termini (Fig. 28); since these bonds require the lowest energy input to fragment, they tend to be observed even if the energy input is too low to observe any other type of fragment ion. Less readily cleaved is the bond on the other side of the glycosidic oxygen, which generates reducing/aglycone-containing (Z_j) or non-reducing (C_i) fragments. The least labile bonds in a carbohydrate are the bonds in the monosaccharide ring, in particular the C–C bonds, so that the non-reducing A_i and reducing/aglycone-containing X_j fragments require the most energy.

Most of the positive-ion fragments can be rationalized as having similar structures whether the charge-bearing species is a proton or an alkali metal cation, most commonly a sodium ion. However, while B ions derived from protonated molecules can be considered to have the structure of an oxonium ion (Fig. 29a), B ions from sodiated molecules are proposed to be produced by an elimination reaction which generates a C-1–C-2 double bond, with the sodium ion imparting the charge[157] (Fig. 29b).

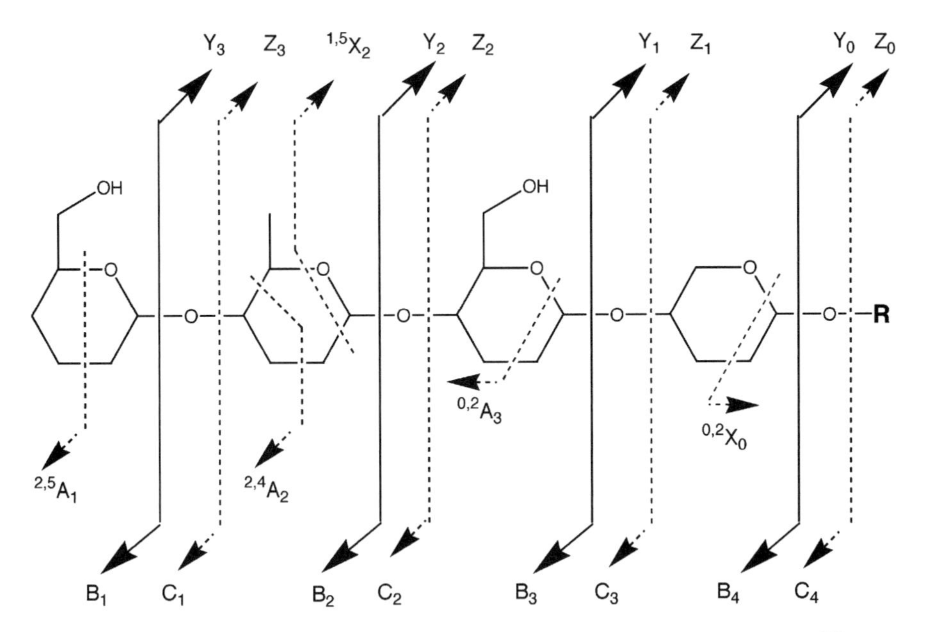

FIG. 28. Carbohydrate fragmentation showing nomenclature of Domon and Costello.[156] Superscripts indicate which ring bonds are broken to give cross-ring cleavage A and X ions; subscripts indicate which glycosidic bond is broken counting from the non-reducing terminus or from the reducing terminus or aglycone (bond to the aglycone is numbered zero). In the positive-ion mode B and Z ions arise by simple-bond cleavage, while C and Y ions involve protonation and a transfer onto the fragment of a single H. In the negative-ion mode, C and Y ions involve simple-bond cleavage, while B and Z ions require deprotonation and transfer away from the fragment of a single H.

It is worth mentioning that it is very rare to observe B_1 ions from sodiated molecules, while this is not the case for B_1 ions generated from protonated molecules. This can be rationalized on consideration of the site of localization of the charge-bearing species. The charge-bearing proton is generally considered to be located on the glycosidic oxygen and thus to drive fragmentation of the glycosidic bond. In contrast, based on empirical observations and modeling approaches, the sodium ion has been proposed to interact with the ring oxygen and the glycosidic oxygen.[158,159] Since the B_1 ion does not contain a glycosidic oxygen, presumably the interaction with sodium is destabilized, making a sodiated B_1 ion less favored.

Multiple cleavages may occur, generating ions corresponding to structures that contain neither the original reducing terminus/aglycone nor the original non-reducing terminus of the precursor. Such ions have been termed double or

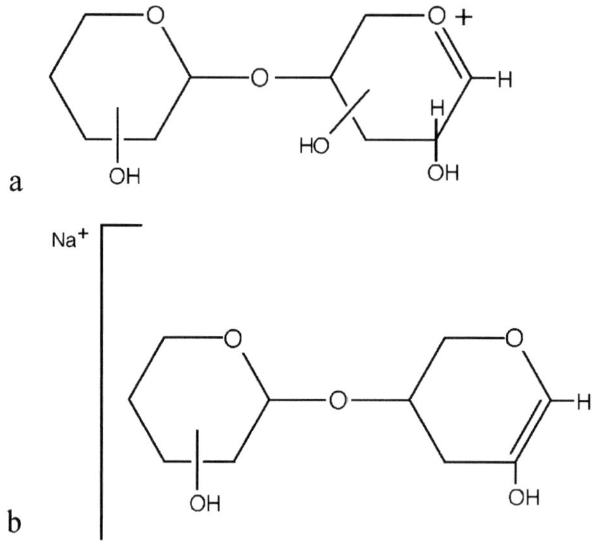

FIG. 29. Structure of positive-ion mode *B* ions deriving from (a) protonated and (b) sodiated molecules.

multiple cleavage ions by Dell and colleagues[1] and internal fragments by Harvey.[160] Such ions are difficult to discern in underivatized samples, although permethylation or peracetylation permits their identification.[1]

One of the main challenges of carbohydrate structural analysis is the differentiation of isomers, whether these are linkage isomers, epimers, or anomers. A good deal of effort has been invested in trying to identify differences in the mass-spectrometric fragmentation behavior of carbohydrate isomers that might allow their differentiation. While A and X ions allow linkages to be differentiated, especially in permethylated derivatives (for instance Reinhold and co-workers,[161] among many others), it has proven much more difficult to demonstrate robust, generally applicable approaches to differentiation of epimers and anomers on the basis of mass-spectrometric fragmentation. While there are some specific cases in which it has been possible, using model compounds differing only in the epimer (such as Ohashi *et al.*[162]) or the anomer (for example Mulroney *et al.*[163]), to observe differences in product ion intensities that correlate with the different isomeric forms, these differences are generally restricted to the spectra of the model species and frequently do not translate to the spectra of 'real' samples.

In addition, it should be noted that, inconveniently, the bond through which a glycan is attached to the peptide in a glycopeptide (whether N- or O-linked) is even weaker than the interglycosidic bonds, and so is generally cleaved even more readily than the bonds in the glycan (and the peptide) backbone. There are a few heroic, and perhaps rather fortuitous examples where low-energy spectra have been obtained in which low-intensity ions are observed for peptide backbone fragments which retain the carbohydrate moiety, allowing the site of glycan attachment to be deduced.[164,165] However, such ions are not routinely observed, and while, if the amounts of sample permit, it may be worth determining whether such ions can be detected, the survival of such fragments cannot be relied upon.

2. Rearrangement Ions

Knowing how carbohydrates fragment and how the major ions relate to the structure of the precursor makes MS extremely useful for determining monosaccharide sequence and linkage, as well as branching and substitution patterns in glycoconjugates. However, this means that fragment ions that are indistinguishable from the predictable and structurally useful B_i and Y_j glycosidic cleavage ions and yet arise via different routes can undermine the utility of MS in structural studies, unless their existence is recognized and their origin understood.

During a systematic study carried out using series of closely related synthetic glycoconjugates, it became clear to us that ions that cannot be explained in terms of straightforward glycosidic-bond cleavages are very commonly generated on both FAB and ESI-CID-tandem MS.[157,166–168] The formation of these ions was shown unambiguously to arise by the loss of internal glycosyl residues[167] (that is, monosaccharide residues that are bound on both sides by glycosidic linkages) and was thus dubbed ‘internal residue loss’ (IRL). The unique masses of the different monosaccharide residues in the model compounds made it possible to demonstrate that the loss is clearly that of an internal residue. As a result of these studies it became clear that such rearrangement ions occur very commonly and do not correlate with any particular structural features; it was shown that IRL was not dependent on anomericity, linkage, or the identity of the residue itself.[168] These systematic studies, further underpinned by subsequent work of others,[169–172] have made it clear that if the mass of the internal residue makes it distinguishable from a terminal residue, then it can be shown that ions arising by IRL are generally detectable to some extent, in spectra obtained at both high and low collision energies, whether the sample is native or derivatized, and from both free glycans and glycoconjugates. It has been proposed that the only requirement for the process to occur is a mobile proton, since it was shown that sodiated, lithiated, and deprotonated precursor ions failed to produce such product ions, while protonated and ammoniated molecules did.[157,169]

Following the realization that such rearrangement ions are so common in both the standard compounds through which we first recognized the phenomenon, as well as in the spectra of many of our biological samples[173] we scrutinized literature spectra for other possible examples of such ions. Although in very many published spectra the masses of the residues make it impossible to distinguish normal *B* and *Y* ions from IRL ions, some examples were identified. Interestingly, in 1983 the loss of an internal glycosyl residue from a per-*O*-alkylated oligosaccharide alditol was described on chemical ionization.[174] A similar ion was reported in a conference poster which described a rearrangement ion in the tandem mass spectrum of an underivatized lipochitin oligosaccharide, which apparently lost an internal *N*-acetylglucosamine residue.[175] This loss was attributed on the poster as being 'due to an isomerization of long living ions prior to fragmentation'.

In addition, we found three reports of FAB and ESI-CID spectra of peracetylated or permethylated samples in which ions appeared at an *m*/*z* value consistent with a fully derivatized smaller oligosaccharide. The origins of these ions were not further investigated in those papers and were explained in the papers as deriving from methyl,[176] methoxy,[177] or acetyl[178] transfers (instead of the much more normal hydrogen transfers that makes possible the identification of fragment ions of such derivatives, Fig. 28). Because the analytes were peralkylated or peracylated, it was possible to clearly distinguish these ions from normal B or Y ions. These ions are also fully consistent with the IRL rearrangement.

It is important to note that, while IRL ions represent a rearrangement of the original structure, they should not present a problem in terms of using MS to determine structure unambiguously, as long as the possibility of their existence is recognized and accounted for when interpreting spectra.

VI. Applications

1. Free Glycans

a. Cyclic *β*-Glucans.—Cyclic β-glucans are produced mainly by soil bacteria where they are thought to have a role in aiding the bacteria to infect the roots of plants. They can make up a significant part of the dry weight of some bacterial species with up to 20% of their mass coming from cyclic β-glucans. All bacteria of the Rhizobiaceae[179] and certain other species of bacteria (such as *Xanthomonas campestris*[180] and *Burkholderia solanacearum*[181]) produce cyclic β-glucans. As a general rule, *Rhizobium and Agrobacterium* species produce only β-(1→2) linked cyclic glucans. *Bradyrhizobium* species produce cyclic glucans which are linked by both β-(1→3) and β-(1→6) linkages. Although the only monosaccharide residue in these structures is glucose, they can differ greatly in the number of monosaccharides. Branching residues may also be present, and they may be substituted with a range of different non-carbohydrate groups, for example, anionic moieties such as ethanolamine phosphate and glycerol phosphate.[181]

Linear glucans also exist in many species of bacteria, so identifying a cyclic species can be difficult. A cyclic glucan has no reducing (or non-reducing) terminus, therefore a native cyclic glucan gives an m/z value 18 Th lower than an equivalent native linear glucan. To confirm that a glucan is cyclic requires permethylation or peracetylation. A derivatized cyclic glucan incorporates one fewer methyl (or acetyl) group than its linear counterpart, facilitating its identification.[1]

MS is a central tool for the analysis of cyclic β-glucans, and several groups have used MS to elucidate such structures. Komaniecka *et al.*[182] used Smith degradation followed by permethylation and MALDI-TOFMS to characterize β-(1→3), β-(1→6)-linked cyclic β-glucans from *Azorhizobium caulinodans.* Smith degradation was used to determine the distribution of the (1→3) linkages, which are resistant to this reaction. The products identified were glucose polymers of up to seven monomers. The intensities of the signals for oligomers of two, three, and four glucose units were the highest suggesting that the most common structures contain two, three, or four adjacent (1→3)-linked glucoses. As well as using MS in this way to determine the linkages within a glucan, it can also be used to identify substituents. For example Jung *et al.*[183] used MALDI-TOFMS and ESI triple quadrupole MS in conjunction with nuclear magnetic resonance (NMR) spectroscopy to determine the structures of glycerophosphorylated cyclic β-glucans from *Xanthomonas campestris,* detecting cyclic glucans with zero, one, and two glycerophosphoryl moieties. The singly glycerophosphorylated form was shown to be the most common.

b. Miscellaneous.—ESI ion-trap mass spectrometry has been used to analyze oligosaccharides from human milk.[184] Negative-ion electrospray identified a complex mixture of oligosaccharides. The ion trap's MS^n capability was used to elucidate the structures of five isomeric fucosylated lacto-*N*-hexaoses, including three novel structures. A parent ion at *m*/*z* 1217 corresponded to five possible isomers. Upon fragmentation it gave a large number of fragments which suggested the presence of isomers. MS^3 analysis of these fragments confirmed the presence of five different structures consisting of a GlcNAc-Gal backbone with a branching fucose residue.

Malto-oligosaccharides are produced during fermentation via the enzyme-catalyzed breakdown of large polysaccharides and are typically chains of 3–10 glucose monomers. ESI-MS was used to determine the distribution of malto-oligosaccharides in beer.[185] Calibration used known concentrations of standard malto-oligosaccharides. Peak areas obtained on injections of diluted beer samples were then compared to the calibration, from which concentrations of malto-oligosaccharides in the beer were determined.

Cyclodextrins have been shown to be useful for improving the solubility and stability of drugs *in vivo*,[186] forming non-covalent inclusion complexes. Guo *et al.*[187] used ESI ion-trap MS to study non-covalent interactions between cyclomaltohexa-, hepta-, and octa-oses (α-, β-, and γ-cyclodextrins), and rutin and quercetin. Non-covalent complexes of the cyclodextrins and drug molecules were observed. The collision energy at which each of the complexes fragmented in the ion trap was used to give an indication of the stability of the complexes in the gas phase.

2. Glycolipids

Glycolipids are glycosyl derivatives of lipids and form a large subclass of glycoconjugates. Different types of glycolipid can be distinguished based upon their lipid moieties (Fig. 30).

Glycosphingolipids are secondary metabolites common to all eukaryotes that have also been found in several prokaryotes (reviewed by Olsen and Jantzen[188]). Numerous biological functions have been assigned to glycosphingolipids. A recent review highlights key roles of sphingolipids within plant metabolism and function.[189] In contrast, the sphingolipids of mammalian cells have been more intensively researched. Several diseases have been shown to be caused by defects in sphingolipid metabolism (extensively reviewed by Kolter and Sandhoff[190]). Glycolipids are a very difficult class of glycoconjugate to analyze mass-spectrometrically. They tend to contain very labile groups, such as sialic acids, and structural heterogeneity in both the glycan and lipid can be enormous. Glycolipids have been analyzed with a wide range of mass-spectrometric methods. Soft ionization approaches, including FAB, MALDI,[191] and ESI[192] have been used. Mass-spectrometric analysis of glycosphingolipids was extensively reviewed by

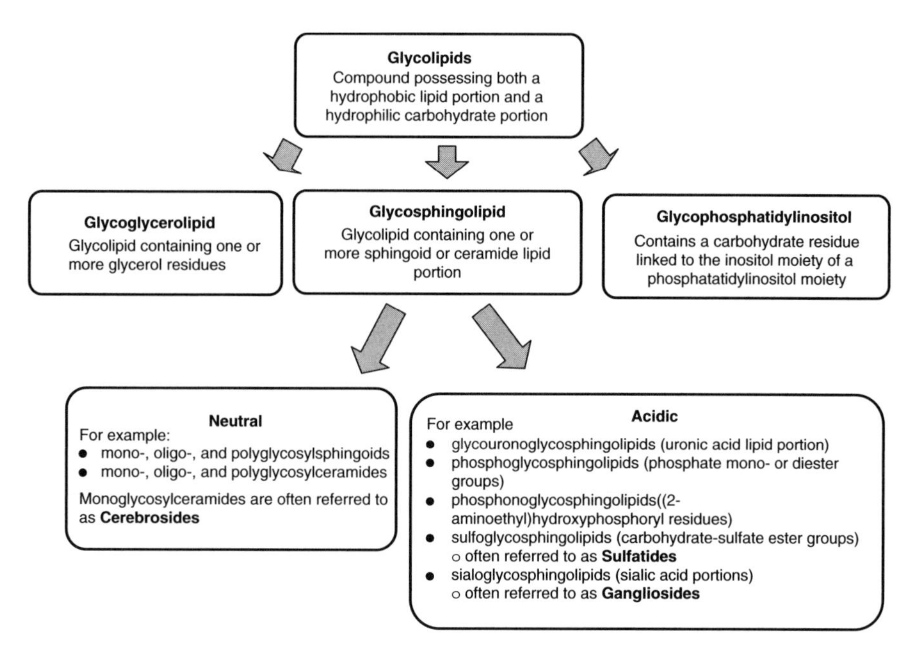

FIG. 30. Glycolipid nomenclature.

Levery.[193] Illustrated here are typical examples of the application of MALDI and ESI to glycolipid research.

Juhasz and Costello reported the first analysis of ganglioside structures by MALDI in 1992.[191] Several matrices were assessed: 2,5-DHB, 1,5-diaminonaphthalene, 4-hydrazimobenzoic acid, and 6-aza-2-thiothymine (ATT) proved to be the most efficient. Spectra could be recorded in both positive and negative mode but significantly better quality spectra were recorded in negative mode, exhibiting higher resolution and signal-to-noise and less in-source fragmentation. Permethylation increased the sensitivity by up to two orders of magnitude; resolution was also improved.

Harvey[33] in 1995 examined a range of underivatized sphingo- and glycosphingolipids using positive-ion MALDI. The nature of the matrix altered the type and abundance of the ions observed as well as the extent of fragmentation. For neutral glycosphingolipids it was found that 2,5-DHB, 4-HCCA, and 6,7-hydroxycoumarin gave the best ion signals. Similiarly to Juhasz and Costello, Harvey found that negative-ion signals from the acidic species with MALDI were much stronger and clearer; this was attributed to the decrease in alkali salt formation.

Commonly, using positive-mode MALDI, the most abundant ion signal is the sodiated molecule $[M+Na]^+$. Harvey also observed that gangliosides are much less stable than other glycolipids and fragment by the loss of CO_2, H_2O, and sialic acid. This fragmentation was much more pronounced using a 'hot' matrix such as 4-HCCA than a cooler matrix. 2,5-DHB proved to be the best matrix for limiting fragmentation during ionization in this particular study. Similarly, the 'cool' matrix ATT was also found to reduce fragmentation on ionization.[194]

Gangoliside in-source fragmentation can be reduced but not eradicated, by doping the matrix with large cations such as cesium, by permethylation,[191] conversion of the carboxyl group into a methyl ester,[195] and the use of a novel ionic–liquid matrix which has soft desorption capabilities.[144]

In-source fragmentation makes FT-ICR-MS particulary difficult, due to the long time between ion formation and detection. O'Connor and Costello have built an experimental MALDI ion source which operates with a high background pressure of bath gas during ionization. The bath gas minimizes fragmentation by providing collisional cooling to the desorbed ions.[196,197] Gangliosides with up to five sialic acid residues were ionized with minium fragmentation, showing the potential of the design.

Thin-layer chromatography (TLC) is one of the standard techniques for separation and isolation of gangliosides from complex mixtures. TLC coupled to MS for structural determination provides a powerful tool for gangolioside analysis. Guittard *et al.* coupled TLC with MALDI-TOF. Although sensitive, poor resolution and mass accuracy were obtained because of the uneven TLC plate surface.[198] However, application of MALDI-FT-ICR-MS using the vibrationally cooled source[199] combines the advantages of TLC with the high mass accuracy and resolution of MS.

ESI, like MALDI is compatible with negative- and positive-mode analyses and for analysis of native or permethylated neutral glycosphingolipids. The most abundant signal is also often the singly sodiated ion $[M+Na]^+$. In-source fragmentation can be tuned by varying the ion source temperature and skimmer voltage.

Recently a direct comparison of MALDI- and ESI-FT-ICR was undertaken to compare the two ionization methods for ganglioside analysis.[200] Both performed better in negative mode, ESI-FT-ICR proved to be more gentle than MALDI producing less fragmentation on ionization. A range of ion spieces was formed under both regimes such as $[M+Na–2H]^-$ and $[M+2Na–3H]^-$ along with losses of water $[M+Na–H_2O–2H]^-$ and sialic acid fragments: $[M–Neu5Ac–2H+Na]^-$ in MALDI and $[M–2H]^{2-}$ and $[M–Neu5Ac–H]^-$ in ESI.

Doping with alkali-metal ions can improve ionization and stop the ion current being split. Lithium adduction has proved to be most successful, developed by Ann and Adams[201,202] for the analysis of ceramides, offering improved sensitivity and improving the yield of useful fragmentation upon CID. The technique has been applied to the analysis of cerebrosides from fungi[203,204] and mammals[201,202,205] at a range of CID energies.

ESI is the driving force behind the emerging field of lipidomics, which can be simply described as the characterization of lipid constituents of cells or tissues often using multiple analytical techniques. As glycolipids are important secondary metabolites and play an active role in a number of biological processes and their metabolism has been linked to a number of diseases, lipidomics has great potential for furthering our understanding of essential life processes. The role of ESI in lipidomics has been reviewed.[206–209] The emphasis of recent studies is to provide rapid analysis with quantification.

Aebersold and co-workers[210] showed that complex profiling of ceramides was possible using ESI-MS. They validated a precursor-ion method to detect ceramides in complex mixtures. By scanning for precursors which fragment to m/z 264 (indicative of sphingosine-containing species) or m/z 266 (for dihydrosphingosine) it is possible to identify ceramides within a complex mixture. This was applied to a biological extract from Jurkat T cells. Finally as a proof of principle, by adding a non-naturally occurring ceramide into the sample matrix as an internal standard, they demonstrated relative quantification.

More recently Han and Cheng characterized and directly quantified cerebrosides using a similar ESI-based shotgun lipidomics approach.[211] An optimized protocol for screening and sequencing complex sialylated and sulfated glycosphingolipid mixtures by negative-ion ESI-FT-ICR was published recently,[212] along with a method for profiling gangliosides in animal tissues with ESI.[213]

MALDI has also been applied to quantify biological material.[214,215] This approach was first used to investigate serum glycolipids.[216] Sulfatides were quantified using MALDI-TOF which proved to be sensitive and reliable. More recently, the levels of sphingolipids from cardiac valves of patients with Fabry disease were quantified similarly.[217]

3. Glycoproteins

Glycans can be attached to the protein backbone via the amide nitrogen of asparagine (Asn) in the consensus sequence –Asn–X–Ser/Thr– (X is any amino acid except proline, *N*-glycosylation, Fig. 31A) or the hydroxyl group most

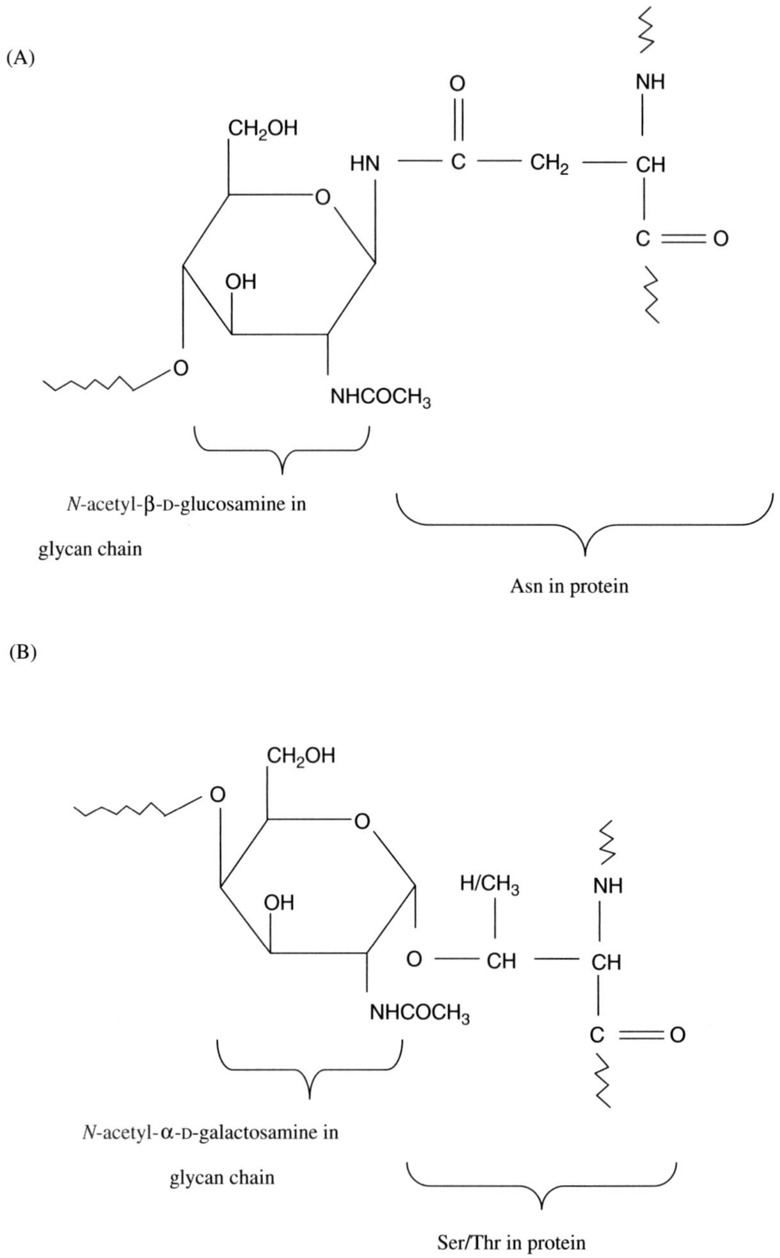

(A)

CH_2OH HN — C(=O) — CH_2 — CH(NH)(C=O) OH O NHCOCH$_3$

N-acetyl-β-D-glucosamine in glycan chain

Asn in protein

(B)

CH_2OH O OH H/CH_3 NH O — CH — CH C=O NHCOCH$_3$

N-acetyl-α-D-galactosamine in glycan chain

Ser/Thr in protein

FIG. 31. The two types of glycosidic linkages (A) N-glycosylic bond—glycan linkage via the amido nitrogen of the side chain of asparagine (B) O-glycosylic bond—linkage via the oxygen in the hydroxyl group of serine or threonine.

commonly of serine (Ser) or threonine (Thr) generally in regions of high hydroxyamino acid density (O-glycosylation, Fig. 31B).

Glycoprotein analysis must determine the extent to which each potential N- and O-glycosylation site is occupied, and then determine the different oligosaccharide structures at each occupied site. Glycoprotein analysis is made challenging by the sheer structural diversity of glycans attached to proteins and because each glycosylated polypeptide is generally associated with a population of different glycan structures frequently attached at more than one site, possibly via more than one different type of chemical linkage. Consequently, detailed structural analysis of one or more glycans is impractical while the glycans are still attached to the polypeptide; release of the glycan is therefore required, using either chemical or enzymatic approaches.

a. Methods for the Release of N-Linked Glycans.—N-Linked oligosaccharides are most frequently released using peptide-N^4-(N-acetyl-β-glucosaminyl)asparagine aminidase enzymes (usually PNGase F), or less frequently endoglycosidase H (Endo H). PNGase F has a broad specificity, cleaving most N-linked oligosaccharides at the N-glycosylic bond yielding intact oligosaccharides and a protein modified by converting Asn to Asp at the site of glycosylation.[218]

Endo H cleaves within the chitobiose core of the N-linked glycan leaving behind an *N*-acetylglucosamine (GlcNAc) residue attached to the Asn and has a strict substrate specificity for α-Man-(1→3)-α-Man-(1→6)-β-Man-(1→4)-GlcNAc, therefore cleaving only oligomannose and hybrid-type N-linked oligosaccharides.

b. Methods for Release of O-linked Glycans.—Analysis of O-glycosylation is more challenging than N-glycosylation because no distinctive universal consensus sequence has been described for O-glycosylation, and no enzyme equivalent to PNGase F has been identified with broad specificity for the release of intact O-linked glycans. Monosaccharides can be removed sequentially from glycoprotein glycans, using a series of exoglycosidases and then removal of the O-glycan core using endo-O-glycosidase but with no modification of the serine or threonine residues, which prevents assignment of the site of O-glycan attachment. Identification of the attachment site is also complicated by the fact that many potential sites of attachment are often found close together in the peptide backbone, all of which may or may not be glycosylated.

The lack of a pan-specific or broad specificity O-glycanase makes chemical methods the most appropriate approach for the non-specific release of intact O-linked oligosaccharides from glycoproteins for site analysis.[219] The most common chemical methods employed for the release of O-linked glycans are based on β-elimination. The methods were originally developed to release the glycans for analysis, and were later adapted for site analysis. There are therefore distinct approaches used depending upon whether the glycan is being released for glycan analysis or for attachment site identification.

(i) Releasing the Glycan Using Classical Reductive β-Elimination. Alkaline reductive cleavage of the alkali-labile bond linking the glycan via the Ser or Thr side-chain hydroxyl group, was first used by workers in the 1960s when studying mucins[220] and blood group substances.[221–223] The glycoproteins were treated with alkaline borohydride solution, releasing reduced oligosaccharides.

Anderson *et al.*[224] first described β-elimination from an O-substituted Ser with the creation of a double bond and a reduction in amounts of recoverable Ser, when investigating the Ser-linked glycan of chondroitin sulfate. Later work by the same group on glycosaminoglycans and glycoproteins described β-carbonyl elimination of O-substituted Ser and Thr derivatives.[225] The alkaline-reductive mechanism by which O-linked glycan structures are released involves breaking the glycosidic linkage between the glycan and the amino acid, and was rationalized as a β-elimination of the side chains[226] (Fig. 32).

In the strongly alkaline conditions used, 'peeling reactions' may occur in which degradation of alkali-labile linkages in the released *O*-glycan takes place by successively dissociating monosaccharide units, starting from the reducing terminus. To prevent this, alkaline β-elimination was performed in the presence of excess reducing agent.[226] The peeling reaction is intercepted on sodium borohydride reduction of the reducing-terminal monosaccharide. This results in the recovery of oligosaccharide alditols (Fig. 33).

Alditol formation may be disadvantageous, depending on the strategies to be followed for subsequent structure determination, as it prevents further reducing-terminal labeling. However, protocols for analysis of released alditols have been devised. For example, glycans are separated from the (glyco)protein using a reversed phase C_{18} cartridge on which released glycans are not retained.[141] Recovery of the glycans from the unretained fraction may be achieved following acid-catalyzed peracetylation of the dried unretained fraction and extraction of the acetylated glycans into dichloromethane for MS analysis.[141]

(ii) Identification of Glycan Attachment-Site in Proteins Using Reductive β-Elimination. In 1964, Tanaka and co-workers[227,228] showed that treatment with $NaBH_4$ in alkaline conditions caused a decrease in both recoverable Ser and Thr and later produced results that indicated that in alkaline $NaBH_4$-treated mucin, much Thr is converted to 2-aminobutanoic acid and Ser to alanine (Fig. 34). However, it was almost another 30 years before Rademaker and

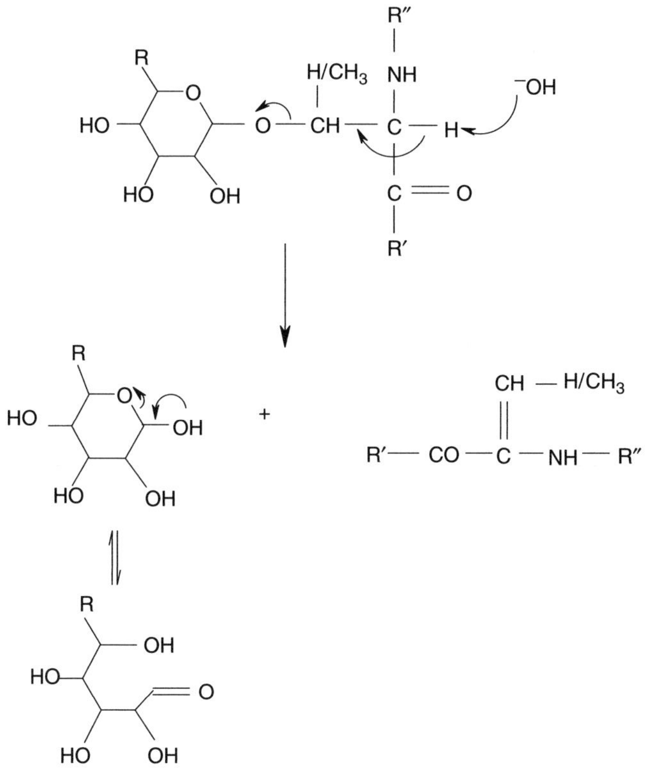

Fig. 32. General mechanism for the alkaline β-elimination used for the release of a glycan from an O-linked glycopeptide.

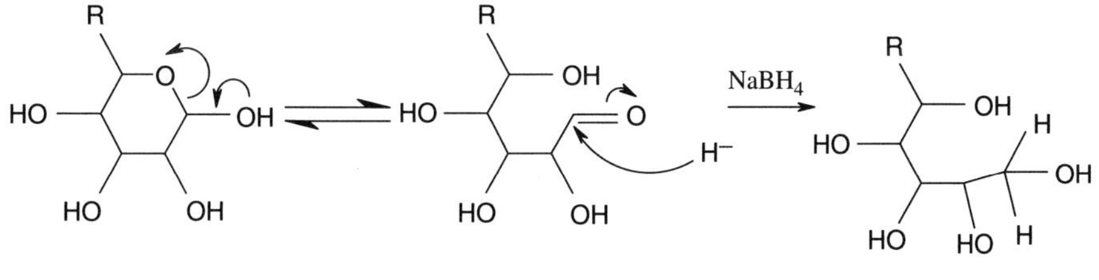

Fig. 33. Formation of alditol in presence of reducing agent.

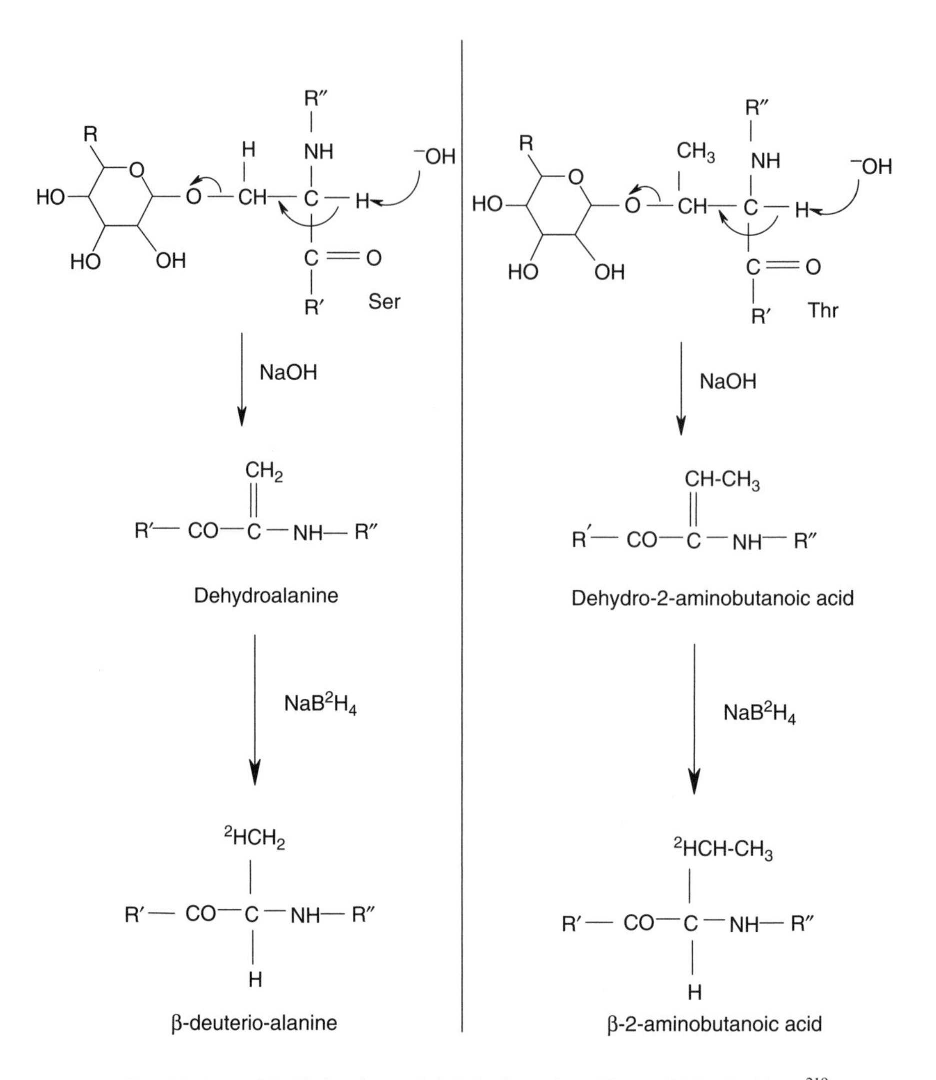

FIG. 34. Use of β-elimination to label *O*-glycosylated Ser and Thr Residues.[219]

co-workers demonstrated that the polypeptide backbone remains intact on alkaline borohydride treatment.[219]

The popular β-elimination protocols were adapted by Rademaker *et al.*[219] for mass-spectrometric analysis, taking advantage of the fact that the originally glycosylated Ser and Thr residues can be easily recognized in tandem MS by the mass shift arising from chemical modification of those residues by the reagent. Thr has a residue mass of 101 Da, whereas its dehydration product after β-elimination 'dehydro-2-aminobutanoic acid' has a residue mass of 83 Da. $NaBD_4$ causes the 'dehydro-2-aminobutanoic acid' to be reduced to 2-aminobutanoic acid (mass 86 Da), which Rademaker and co-workers used to recognize the presence of the glycosylated Thr. To recognize glycosylated Ser, the use of the deuterated reductant is critical because chemically generated β-deuterioalanine (mass 72 Da), produced on reductive β-elimination processing of Ser (mass 87 Da) may be distinguished from endogenous alanine (mass 71 Da).

Separation of the modified peptide from the released glycan may be accomplished using a C_{18} cartridge and the glycan then isolated from the unretained fraction by acid-catalyzed acetylation and organic/aqueous partition. Tandem mass-spectrometric analysis of the recovered peptide yields sequence information allowing location of the modified amino acid.

(iii) Identification of Glycan Attachment-Site in Protein Using Non-Reductive β-Elimination and Volatile Reagents. On treatment with NaOH, O-glycosylated Thr is transformed into 'dehydro-2-aminobutanoic acid' and O-glycosylated Ser to dehydroalanine (Fig. 34).[219] However, the use of NaOH introduces Na, which requires subsequent removal. More recently, Rademaker *et al.*[229] demonstrated that volatile NH_4OH can be used instead of NaOH, obviating the need for further sample clean up. β-Elimination using NH_4OH allows the site of glycosylation to be identified by incorporating the elements of NH_3 into the amino acid residue across the β-elimination-generated double bond by Michael type addition (Fig. 35).[229] The modified amino acid residue then has a distinct mass corresponding to incorporation of the 17 Da increment. Tandem mass spectra of the peptide show where the unique amino acid is located and the site of glycosylation can therefore be deduced.

To increase the mass increment for easier identification of the site of glycosylation by MS, workers have experimented with alkylamine labeling instead of ammonia. Mirgorodskaya *et al.*[230] used methylamine to produce stable methylamine derivatives and Hanisch *et al.*[231] similarly used ethylamine.

In none of these reports was it attempted to recover the released glycans for further analysis, the purpose of the studies being to develop methods for identification of the glycosylation sites.

(iv) β-Elimination to Yield Free Reducing-Terminal Glycans. Novotny and co-workers[232] have further adapted the ammonia-catalyzed glycan release protocol to yield the glycan with the reducing residue unmodified by including ammonium carbonate in the aqueous ammonium hydroxide elimination

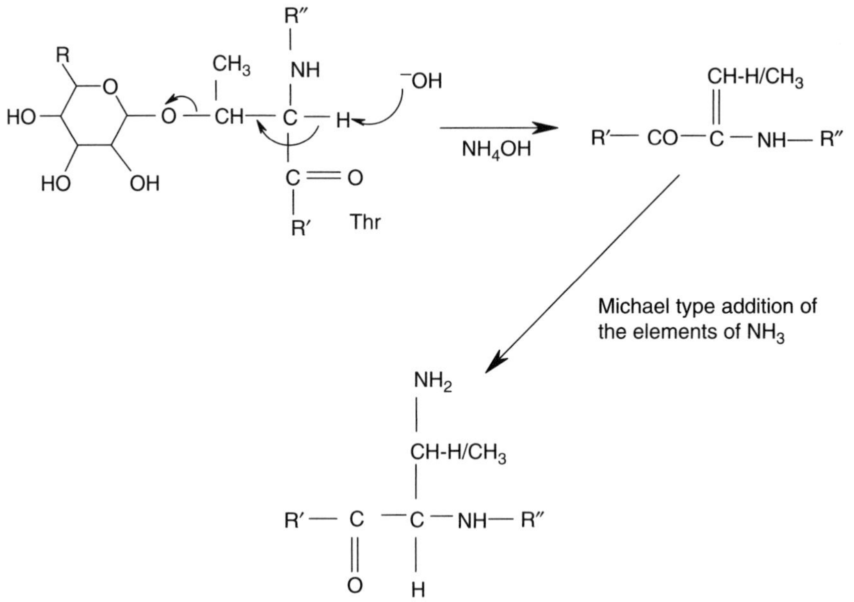

FIG. 35. Ammonia-catalyzed β-elimination for the identification of O-glycosylation sites.

medium. These reagents release alkali-stable glycan intermediates by blocking the reducing terminus of released glycans and so avoiding alkaline peeling. Boric acid treatment then generates a free reducing terminal to allow coupling to a chromophore or fluorophore through reductive amination of the reducing terminus.[102] Interestingly, this method has also been shown to release N-linked as well as O-linked oligosaccharides, yielding intact reducing N- and O-linked glycans in high yields.[232]

(v) Hydrazinolysis. Hydrazinolysis has been used extensively to cleave the *N*-glycosylic bond.[233] It is an approach that may be advantageous because it has been reported to yield both N- and O-linked reducing glycans.[234] The approach, however, is tedious, dangerous, and can generate many by-products, as well as destroying the peptide backbone. It also results in chemical modification of the original glycan, including the loss of *N*-acetyl and *N*-glycolyl groups from amino sugar residues and the introduction of significant artifacts.[233] Such losses did not occur using β-elimination.[232] Hydrazinolysis is no longer widely used for glycan release.

c. Release of Glycans from Glycoproteins in Gels.—In a proteomics context, mixtures of proteins and glycoproteins are routinely separated using SDS-PAGE.

Methods have therefore been developed to enable the profiling of glycans directly from glycoproteins on gels. Many research groups have described blotting glycoproteins from gels onto polyvinylidene fluoride (PVDF) membranes prior to either PNGase F release of N-linked glycans[235–237] or reductive β-eliminative release of O-linked glycans [238,239] and one paper has described the success of directly performing in-gel protein de-N-glycosylation using PNGase F.[240] Kilz and co-workers[241] have described an analogous in-gel HF-pyridine de-O-glycosylation; in-gel de-*O*-glycosylation/ethylamination was reported for glycosylation site identification by Hanisch *et al.*,[231] and Taylor and colleagues reported in-gel reductive and non-reductive β-eliminative release and analysis of *O*-glycans together with subsequent proteomic analysis of the deglycosylated polypeptide in a study of glycoproteins from the capsule of *Mycobacterium avium*.[242]

4. Analysis of Released Glycans

Determining the structure of released glycans is not possible using only a single technique, so that a combination is required that may include MS, NMR spectroscopy, monosaccharide composition analysis, and methylation linkage-analysis using gas chromatography-mass spectrometry (GC-MS), LC, and electrophoresis, possibly in combination with derivatization methods. Formation of methyl ethers (permethylation) or alternatively acetyl esters (peracetylation) is a common derivatization procedure before mass-spectrometric analysis of glycan structures. Derivatization for the mass-spectrometric analysis of glycoconjugates can provide advantages in improving the sensitivity of detection, directing fragmentation, and facilitating interpretation of fragment ions.[142,243,145,146]

a. LC-MS.—LC-MS is a powerful technique for studying both the proteome and glycoproteome, bridging the gap between the analysis of proteins and the need to identify not only the polypeptide but also its post-translational modifications. The combination of LC separation with MS(/MS) for the detection and structural analysis of released glycans provides detailed information. Significant simplification of the sample processing can be achieved using LC-MS methods over GC-MS, as carbohydrates are commonly analyzed in their native form. The development of robust ESI methods for carbohydrate analysis has facilitated the use of LC-MS carbohydrate analysis. Released glycans can be analyzed by LC-MS in their reducing form, as alditols, after permethylation or after reducing-terminal labeling (refer to Section IV). LC coupled on-line with electrospray MS(/MS) is an increasingly common approach to the analysis of oligosaccharide derivatives.[87–92]

A number of stationary phases are suitable for coupling on-line with MS, including RP, NP, and graphitized carbon (refer to Section III).

b. CE-MS.—The unique separation selectivity and high resolving power of CE relative to LC has encouraged development of carbohydrate CE separation methodologies coupled with MS that are applicable to glycans released from proteins. Methods, described above, that release both O- and N-linked complex glycans yielding free reducing-terminal oligosaccharides, presents a site for derivatization via a simple reductive amination reaction using, for example, APTS. This technique provides charge for electrophoretic migration and facilitates mass-spectrometric analysis (refer to Section III).

c. High-Performance Anion-Exchange Chromatography with Pulsed Amperometric Detection (HPAEC-PAD).—HPAEC-PAD is commonly used to separate released glycoprotein glycans.[244,245] It is carried out at high pH; the quaternary amine-bonded stationary phase provides high selectivity for carbohydrates, as reducing sugars, oligosaccharides, and alditols. Since neutral oligosaccharides have p*K*a values between 12 and 14, they are considered weak acids. Therefore at high pH (12–14) they are either partially or completely ionized and can be separated by anion-exchange mechanisms. Depending upon the composition of the eluent, resolution of the analytes can be achieved not only according to charge but by size, monosaccharide composition, and linkage. The mobile phase typically contains sodium hydroxide and sodium acetate necessary for the chromatographic separation of mixtures of neutral and acidic sugars. Incorporating pulsed amperometric detection provides a sensitive and selective detection method for analyzing carbohydrates, which lack a chromophore or easily detectable functional group. The detection method, which does not require derivatization for improving carbohydrate detection, uses a repeating sequence of three applied potentials in a cyclic voltammetry process using a gold electrode. Oligosaccharides can then be characterized by comparison of retention times with standards before and/or after exoglycosidase treatments. However, for more convincing identification and for structural analysis, oligosaccharides are commonly collected post-column and analyzed using MS (refer to Section III).

d. Additional Strategies Used to Identify Glycosylation Sites.—Although on collisional activation, the glycan-peptide bond usually fragments very readily so that information on the site of glycan attachment is generally lost (see Section V.1), glycosylated ions can occasionally be identified with ESI-Q-TOF-MS/MS, usually using low collision energies and long acquisition times. Nano-Q-TOF-MS/MS has been used to determine glycosylation sites in *O*-glycosylated,[246–249] O-fucosylated,[250,251] as well as the robust *C*-mannosylated[251] peptides.

Huddleston *et al.*[86] designed an on-line HPLC method that allowed the selective detection of glycopeptides in protein digests by collisional fragmentation of glycopeptides using ESI microbore LC-MS/MS. Glycopeptides are usually identified by precursor-ion scanning, in which carbohydrate oxonium fragment ions at m/z 163 $[Hex]^+$, 204 $[HexNAc]^+$, and 366 $[Hex–HexNAc]^+$ produced on CID of glycopepetides are monitored. This approach was extended, by the same group, to the selective identification and differentiation of N- and O-linked oligosaccharides in glycoproteins. Detection of both N-linked and O-linked glycopeptides at the low picomole level in digests of glycoproteins was achieved using the precursor-ion experiments just described. N-Linked oligosaccharides were then differentiated from O-linked by LC-ESI-MS/MS analysis of the glycoprotein digest prior to and after selective removal of N-linked carbohydrates using PNGase F; glycopeptide signals identified in the second LC-ESI-MS/MS experiment are then due to the undigested O-glycopeptides. The method also facilitates the isolation of glycopeptides for further characterization of the carbohydrate entities.[252]

ECD[253] is a radical-site dissociation method that cleaves different and many more backbone bonds in peptides and proteins than conventional CID and IRMPD used with FT-ICR instruments, by adding energy directly to the uneven-electron ions.[254] These ions mainly fragment by cleavage of the N–Cα bonds of the peptide chain generating c and $z^{\bullet}$ ions. ECD is a soft fragmentation technique, to date uniquely commercially implemented in FT-ICR instruments, and does not cause the loss of labile post-translational modifications such as glycosylation, but also carboxylation, sulfation, and phosphorylation.[255–259] FT-ICR-MS with ECD is therefore a very useful technique that allows unambiguous assignment of peptide glycosylation sites.

5. Polysaccharides

a. Bacterial Polysaccharides.—Many bacteria produce polysaccharide capsules, which are called the K-antigens (Kapsel antigen) to distinguish them from the O-antigens of lipopolysaccharides. Capsules represent the outermost layer and normally present a mucoid type appearance. More than 80 different capsule types have been described in *Escherichia coli*. The capsular polysaccharides exhibit extraordinary diversity in structure, but are frequently acidic, and may contain hexuronic acids, Kdo or a sialic acid in combination with neutral or amino sugars.

One clear example of the importance of capsular polysaccharides is in the symbiotic relationship between nitrogen-fixing rhizobial bacteria and leguminous plants, where MS has played a central role in helping elucidate structures. Here we aim to highlight, using the specific example of bacterial K-antigen studies, the central role of modern mass-spectrometric techniques in polysaccharide analysis.

Bacterial invasion of the plant cells and the formation of nitrogen-fixing nodules require Nod factors (unique glycolipids with a carbohydrate backbone of *N*-acetylglucosamine residues (GlcNAc) and a fatty acyl chain attached to the non-reducing-terminal residue) but also additional signals, including bacterial surface components.[260,261] Exopolysaccharides (EPSs), lipopolysaccharides (LPSs), cyclic β-glucans, and surface capsular polysaccharides (K-antigens) have all been shown to play roles in nodulation and nodule development. These polysaccharides probably have several functions, such as in cell-to-cell communication, shielding of the bacteria from plant defense mechanisms, or in signaling to suppress plant defense reactions. It is likely to be the overall combination of all these glycoconjugates that results in a successful nitrogen-fixation relationship.[262]

Biological studies in combination with powerful analytical tools, especially modern MS, for the structural characterization of low-abundance complex molecules, has enabled the phenotyping of numerous examples of these structures.

Reports describing the structures of KPS from several soybean symbionts have shown that there are clear differences among them. All of these studies rely on the use of a combination of analytical instruments including modern mass spectrometers with electrospray and MALDI ionization sources. In many of the structures identified, a conserved structural motif becomes evident since the repeating unit is formed of a variable hexosyl residue linked to a Kdo-type residue (Kdo or a Kdo-related residue Kdx). This conserved structural motif (Hex-Kdx) appears in the KPS of three out of four *Sinorhizobium meliloti* strains studied. The exception produces a homopolymeric KPS composed only of Kdo. In that study, the authors also presented for the first time evidence for the existence of a phospholipid anchor of a rhizobial KPS and present a partial structure for it. It is composed of a phosphoglycerol moiety bearing a hydroxy-octacosanoic acid.[263] This was determined using high-resolution ESI-FT-ICR-MS. The hexose-Kdx motif has also been shown to be present in the KPS of the broad host-range *Sinorhizobium* sp. NGR234 and in four *Sinorhizobium fredii* strains.[264,265] In these studies, negative-ion MALDI-TOF-MS was used to determine the nature of the capsular polysaccharide after mild acid hydrolysis to provide fragments of observable size. Ions corresponding to different degrees of polymerization were observed as $[M+H]^+$ or $[M+Na]^+$, namely three oligosaccharides (*m/z* 1163) and four (*m/z* 1545) disaccharide repeating units.

These *S. fredii* strains all form nitrogen-fixing symbioses with Asiatic soybean cultivars but fail to nodulate American soybean cultivars. In contrast, three *S. fredii* wild-type strains that are able to establish successful symbioses with both Asiatic and American soybean cultivars, produce KPS repeating units that do not show the Hex-Kdx motif. A recent example has been added of structural characterization of a capsular polysaccharide from a fifth *S. fredii* strain,[266] an effective symbiont of Chinese soybean, using a combination of electrospray orthogonal quadrupole TOF and other techniques such as NMR. In this study, the structure of the capsular polysaccharide was found to be composed of repeating-units of [→6)-2,4-di-*O*-methyl-α-D-Gal*p*-(1→4)- β-D-Glc*p*A-(1→]. An interesting aspect of this study is that the use of CID of the native tetrasaccharide isolated after hydrolysis of the polysaccharide provided not only structurally diagnostic ions from the cleavage of the glycosidic bonds that enabled the assignment of the constituents of the polymer but also the nature of the substituents in the hexose moiety. A tetrasaccharide fragmented to yield *Y* ions indicating an alternating sequence of hexuronic acid (HexA) and a residue with mass 190 Da (consistent with a hexose residue (Hex) plus 28 Da). A 28 Da mass increment could suggest either an ethyl ether or two methyl ethers. Since the Y_3 ion loses 14 and 32 Th, consistent with β-cleavage and β-elimination, respectively, of a methyl group, and not 28 and 46 Th, it was possible to assign a dimethylated instead of an ethylated hexose.

VII. Perspectives for the Future

The past 20 years has witnessed an explosion of technical advances that have transformed the field of MS of carbohydrates; our aim has been that this chapter should describe the most important of these and highlight examples of good practice in the application of state-of-the-art MS in carbohydrate analysis. We are now entering an era in which MS instruments are becoming more affordable and a great deal more user-friendly, which is leading to a complete change of approach. MS use is no longer the preserve of specialists, and large numbers of mass spectrometers are now being purchased and used routinely by biologists to great effect. The demands of these users for higher throughput, greater ease of use, higher sensitivity, better mass accuracy, more facile interfacing with higher performance separations techniques, better quantitative approaches, and for more methods of generating fragmentation are driving the MS specialists to develop the instrumentation ever faster.

And nothing enhances development in an area more than the introduction of a new instrument. This is certainly true of MS, with a good deal of interest currently being focused on novel instrument designs, such as the recently introduced orbitrap (a linear ion-trapping device), and new methods of fragmenting ions to generate additional tandem MS data, such as electron-transfer dissociation (ETD). These new developments will in turn drive research into a better understanding of the biological roles of carbohydrates. Carbohydrate research has certainly never been as exciting or promised greater potential.

In addition, we also envisage MALDI-MS imaging becoming more mainstream in the area of carbohydrate research, offering the added dimension of spatial information on top of the specificity of the structural information that MS delivers.

The field of post-genomic science is already taking off in the much more challenging areas of glycomics and glycoproteomics, areas where MS plays an essential role. These disciplines, deriving from the onset of the post-genomic revolution at the beginning of the 21st century, desperately need a combination of instrumentation that will allow a higher throughput with increased sensitivity, hyphenated with very powerful separations technologies. The metabolomics community which seeks to study systematically the unique chemical fingerprints that specific cellular processes leave behind, namely products of cellular processes such as glycolysis, are already demanding accurate quantitation of extremely low concentrations of such carbohydrate molecules.

The future will undoubtedly see modern MS's application to carbohydrate research playing a major role in the emerging field of Systems Biology, the discipline which seeks to integrate high-throughput biological studies in an attempt at a global understanding of biological systems and how they function. By studying the relationships and interactions between various parts of a biological system (for example metabolic pathways, organelles, cells, physiological systems, organisms, etc.) the aim is that eventually an understandable model of the whole system can be achieved.

References

1. A. Dell, F.A.B.-mass spectrometry of carbohydrates, *Adv. Carbohydr. Chem. Biochem.*, 45 (1987) 19–72.
2. M. Barber, R. S. Bordoli, R. D. Sedgwick, and A. N. Tyler, Fast atom bombardment of solids (F.A.B.): A new ion source for mass spectrometry, *J. Chem. Soc. Chem. Commun.* (1981) 325–327.

3. M. A. O'Neill, D. Warrenfeltz, K. Kates, P. Pellerin, T. Doco, A. G. Darvill, and P. Albersheim, Rhamnogalacturonan-II, a pectic polysaccharide in the walls of growing plant cell, forms a dimer that is covalently cross-linked by a borate ester. *In vitro* conditions for the formation and hydrolysis of the dimer, *J. Biol. Chem.*, 271 (1996) 22923–22930.
4. P. Pellerin, T. Doco, S. Vidal, P. Williams, J.-M. Brillouet, and M. A. O'Neill, Structural characterization of red wine rhamnogalacturonan II, *Carbohydr. Res.*, 290 (1996) 183–197.
5. M. Yamashita and J. B. Fenn, Electrospray ion source. Another variation on the free-jet theme, *J. Phys. Chem.*, 88 (1984) 4451–4459.
6. M. Karas and F. Hillenkamp, Laser desorption ionisation of proteins with molecular masses exceeding 10.000 daltons, *Anal. Chem.*, 60 (1980) 2299–2301.
7. K. Tanaka, H. Waki, Y. Ido, S. Akita, Y. Yoshida, and T. Yoshida, Protein and polymer analysis up to *m/z* 100.000 by laser ionisation time-of-flight mass spectrometry, *Rapid Commun. Mass Spectrom.*, 2 (1988) 151–153.
8. J. B. Fenn, M. Mann, C. K. Meng, S. F. Wong, and C. M. Whitehouse, Electrospray ionisation for mass spectrometry of large biomolecules, *Science*, 246 (1989) 64–71.
9. M. Dole, L. L. Mach, R. L. Hines, R. C. Mobley, L. D. Ferguson, and M. B. Alice, Molecular beams of macroions, *J. Chem. Phys.*, 49 (1968) 2240–2249.
10. P. Kebarle and L. Tang, From ions in solution to ions in the gas phase, *Anal. Chem.*, 65 (1993) 972A–986A.
11. R. B. Cole, Some tenets pertaining to electrospray ionization mass spectrometry, *J. Mass Spectrom.*, 35 (2000) 763–772.
12. P. Kebarle, A brief overview of the present status of the mechanisms involved in electrospray mass spectrometry, *J. Mass Spectrom.*, 35 (2000) 804–817.
13. J. V. Iribarne and B. A. Thomson, On the evaporation of small ions from charged droplets, *J. Chem. Phys.*, 64 (1976) 2287.
14. M. R. Emmet and R. M. Caprioli, Micro-electrospray mass spectrometry: Ultra-high-sensitivity analysis of peptides and proteins, *J. Am. Soc. Mass Spectrom.*, 5 (1994) 605–613.
15. M. Wilm and M. Mann, Analytical properties of the nanoelectrospray ion source, *Nature*, 379 (1996) 466–469.
16. R. Juraschek, T. Dulcks, and M. Karas, Nanoelectrospray—more than just a minimized-flow electrospray ionization source, *J. Am. Soc. Mass Spectrom.*, 10 (1999) 300–308.
17. M. Karas and F. Hillenkamp, Matrix-assisted laser desorption/ionisation, an experience, *Int. J. Mass Spectrom.*, 200 (2000) 71–77.
18. R. Zenobi and R. Knochenmuss, Ion formation in MALDI mass spectrometry, *Mass Spectrom. Rev.*, 17 (1998) 337–366.
19. K. Dreisewerd, The desorption process in MALDI, *Chem. Rev.*, 103 (2003) 395–425.
20. R. Knochenmuss and R. Zenobi, MALDI ionization: The role of in-plume processes, *Chem. Rev.*, 103 (2003) 411–452.
21. M. Karas, M. Gluckmann, and J. Schafer, Ionization in matrix-assisted laser desorption/ionization: Singly charged molecular ions are the lucky survivors, *J. Mass Spectrom.*, 35 (2000) 1–12.
22. M. Gluckmann, A. Pfenninger, R. Kruger, M. Thierolf, M. Karas, V. Horneffer, F. Hillenkamp, and K. Strupat, Mechanisms in MALDI analysis: Surface interaction or incorporation of analytes? *Int. J. Mass Spectrom.*, 210 (2001) 121–132.
23. V. E. Frankevich, J. Zhang, S. D. Friess, M. Dashtiev, and R. Zenobi, Role of electrons in laser desorption/ionization mass spectrometry, *Anal. Chem.*, 75 (2003) 6063–6067.
24. K. Strupat, M. Karas, and F. Hillenkamp, 2,5-Dihydroxybenzoic acid—a new matrix for laser desorption ionization mass-spectrometry, *Int. J. Mass Spectrom. Ion Process.*, 111 (1991) 89–102.

25. A. Tsarbopoulos, M. Karas, K. Strupat, B. N. Pramanlk, T. L. Nagabhushan, and F. Hillenkamp, Comparative mapping of recombinant proteins and glycoproteins by plasma desorption and matrix-assisted laser desorption/ionization mass-spectrometry, *Anal. Chem.*, 66 (1994) 2062–2070.
26. M. D. Mohr, K. O. Bornsen, and H. M. Widmer, Matrix-assisted laser-desorption ionization mass-spectrometry—improved matrix for oligosaccharides, *Rapid Commun. Mass Spectrom.*, 9 (1995) 809–814.
27. A. I. Gusev, W. R. Wilkinson, A. Proctor, and D. M. Hercules, Improvement of signal reproducibility and matrix/comatrix effects in Maldi analysis, *Anal. Chem.*, 67 (1995) 1034–1041.
28. Y. Mechref and M. V. Novotny, Matrix-assisted laser desorption ionization mass spectrometry of acidic glycoconjugates facilitated by the use of spermine as a co-matrix, *J. Am. Soc. Mass Spectrom.*, 9 (1998) 1293–1302.
29. D. J. Harvey, Matrix-assisted laser desorption/ionization mass spectrometry of carbohydrates, *Mass Spectrom. Rev.*, 18 (1999) 349–450.
30. D. J. Harvey, Matrix-assisted laser desorption/ionization mass spectrometry of carbohydrates and glycoconjugates, *Int. J. Mass Spectrom.*, 226 (2003) 1–35.
31. K. K. Mock, M. Davy, and J. S. Cottrell, The analysis of underivatised oligosaccharides by matrix-assisted laser desorption mass spectrometry, *Biochem. Biophys. Res. Commun.*, 177 (1991) 644–651.
32. R. C. Beavis, T. Chaudhary, and B. T. Chait, α-Cyano-4-hydroxycinnamic acid as a matrix for matrix assisted laser desorption mass spectrometry, *Org. Mass Spectrom.*, 27 (2) (1992) 156–158.
33. D. J. Harvey, Matrix-assisted laser desorption/ionization mass spectrometry of sphingo- and glycosphingo-lipids, *J. Mass Spectrom.*, 30 (1995) 1311–1324.
34. R. C. Beavis, B. T. Chait, and H. M. Fales, Cinnamic acid derivatives as matrices for ultraviolet laser desorption mass spectrometry of proteins, *Rapid Commun. Mass Spectrom.*, 3 (1989) 432–435.
35. P. Juhasz, C. E. Costello, and K. Biemann, Matrix-assisted laser desorption ionization mass spectrometry with 2-(4-hydroxyphenylazo)benzoic acid matrix, *J. Am. Soc. Mass Spectrom.*, 4 (1993) 399–409.
36. J. J. Pitt and J. J. Gorman, Matrix-assisted laser desorption/ionization time-of-flight mass spectrometry of sialylated glycopeptides and proteins using 2,6-dihydroxyacetophenone as a matrix, *Rapid Commun. Mass Spectrom.*, 10 (1996) 1786–1788.
37. D. I. Papac, A. Wong, and A. J. S. Jones, Analysis of acidic oligosaccharides and glycopeptides by matrix-assisted laser desorption/ionization time-of-flight mass spectrometry, *Anal. Chem.*, 68 (1996) 3215–3223.
38. N. Xu, Z. H. Huang, J. T. Watson, and D. A. Gage, Mercaptobenzothiazoles: A new class of matrices for laser desorption ionization mass spectrometry, *J. Am. Soc. Mass Spectrom.*, 8 (1997) 116–124.
39. P. Chen, A. G. Baker, and M. V. Novotny, The use of osazones as matrices for the matrix-assisted laser desorption/ionization mass spectrometry of carbohydrates, *Anal. Biochem.*, 244 (1997) 144–151.
40. S. F. Wheeler and D. J. Harvey, Negative ion mass spectrometry of sialylated carbohydrates: Discrimination of *N*-acetylneuraminic acid linkages by MALDI-TOF and ESI-TOF mass spectrometry, *Anal. Chem.*, 72 (2000) 5027–5039.
41. S. F. Wheeler and D. J. Harvey, Extension of the in-gel release method for structural analysis of neutral and sialylated *N*-linked glycans to the analysis of sulfated glycans: Application to the glycans from bovine thyroid-stimulating hormone, *Anal. Biochem.*, 296 (2001) 92–100.

42. D. W. Armstrong, L. K. Zhang, L. He, and M. L. Gross, Ionic liquids as matrixes for matrix-assisted laser desorption/ionization mass spectrometry, *Anal. Chem.*, 73 (2001) 3679–3686.
43. M. Mank, B. Stahl, and G. Boehm, 2,5-Dihydroxybenzoic acid butylamine and other ionic liquid matrixes for enhanced MALDI-MS analysis of biomolecules, *Anal. Chem.*, 76 (2004) 2938–2950.
44. T. N. Laremore, S. Murugesan, T. J. Park, F. Y. Avci, D. V. Zagorevski, and R. J. Linhardt, Matrix-assisted laser desorption/ionization mass spectrometric analysis of uncomplexed highly sulfated oligosaccharides using ionic liquid matrices, *Anal. Chem.*, 78 (2006) 1774–1779.
45. S. Y. Xu, Y. F. Li, H. F. Zou, J. S. Qiu, Z. Guo, and B. C. Guo, Carbon nanotubes as assisted matrix for laser desorption/ionization time-of-flight mass spectrometry, *Anal. Chem.*, 75 (2003) 6191–6195.
46. S. F. Ren, L. Zhang, Z. H. Cheng, and Y. L. Guo, Immobilized carbon nanotubes as matrix for MALDI-TOF-MS analysis: Applications to neutral small carbohydrates, *J. Am. Soc. Mass Spectrom.*, 16 (2005) 333–339.
47. S. F. Ren and Y. L. Guo, Oxidized carbon nanotubes as matrix for matrix-assisted laser desorption/ionization time-of-flight mass spectrometric analysis of biomolecules, *Rapid Commun. Mass Spectrom.*, 19 (2005) 255–260.
48. R. E. March and R. J. Hughes, Quadrupole storage mass spectrometry, *Chemical Analysis Series*, Vol. 102, John Wiley, New York, 1989.
49. K. R. Jennings, Collision-induced decompositions of aromatic molecular ions, *Int. J. Mass Spectrom. Ion Process.*, 1 (1968) 227–235.
50. W. F. Haddon and F. W. McLafferty, Metastable ion characteristics. VII. Collision-induced metastables, *J. Am. Chem. Soc.*, 90 (1968) 4745–4746.
51. A. K. Shukla and J. H. Futrell, Tandem mass spectrometry: Dissociation of ions by collisional activation, *J. Mass Spectrom.*, 35 (2000) 1069–1090.
52. J. C. Schwartz, M. W. Senko, and J. E. P. Syka, A two-dimensional quadrupole ion trap mass spectrometer, *J. Am. Soc. Mass Spectrom.*, 13 (2002) 659–669.
53. G. Hopfgartner, E. Varesio, V. Tschappat, C. Grivet, E. Bourgogne, and L. A. Leuthold, Triple quadrupole linear ion trap mass spectrometer for the analysis of small molecules and macromolecules, *J. Mass Spectrom.*, 39 (2004) 845–855.
54. W. Paul and H. S. Steinwedel, US Patent (1960) 2 939 952.
55. G. C. Stafford, P. E. Kelley, J. E. P. Syka, W. E. Reynolds, and J. F. J. Todd, Recent improvements in and analytical applications of advanced ion trap technology, *Int. J. Mass Spectrom. Ion Process.*, 60 (1984) 85–98.
56. P. H. Dawson, Quadrupole mass analyzers—performance, design and some recent applications, *Mass Spectrom. Rev.*, 5 (1986) 1–37.
57. G. C. Stafford, D. M. Taylor, S. C. Bradshaw, and J. E. P. Syka, *Proc. 35th ASMS Conf. Mass Spectrom.Allied Topics*, Denver, Colorado, 1987.
58. C. S. Creaser, J. C. Reynolds, and D. J. Harvey, Structural analysis of oligosaccharides by atmospheric pressure matrix-assisted laser desorption/ionisation quadrupole ion trap mass spectrometry, *Rapid Commun. Mass Spectrom.*, 16 (2002) 176–194.
59. V. N. Reinhold and B. Stall, *Proc. 50th ASMS Conf. Mass Spectrom. Allied Topics*, Orlando, Florida, 2002.
60. I. J. Amster, Fourier transform mass spectrometry, *J. Mass Spectrom.*, 31 (1996) 1325–1337.
61. A. Marshall, C. L. Hendrickson, and G. S. Jackson, Fourier transform ion cyclotron resonance mass spectrometry: A primer, *Mass Spec. Rev.*, 17 (1998) 1–35.
62. M. B. Comisarow and A. G. Marshall, Fourier transform ion cyclotron resonance spectroscopy, *Chem. Phys. Lett.*, 25 (1974) 282–283.

63. M. B. Comisarow and A. G. Marshall, Frequency-sweep Fourier-transform ion-cyclotron resonance spectroscopy, *Chem. Phys. Lett.*, 26 (1974) 489–490.
64. S. D. H. Shi, J. J. Drader, C. L. Hendrickson, and A. G. Marshall, Fourier transform ion cyclotron resonance mass spectrometry in a high homogeneity 25 Tesla resistive magnet, *J. Am. Soc. Mass Spectrom.*, 10 (1999) 265–268.
65. K. Hakansson, H. J. Cooper, M. R. Emmett, C. E. Costello, A. G. Marshall, and C. L. Nilsson, Electron capture dissociation and infrared multiphoton dissociation MS/MS of an N-glycosylated tryptic peptide to yield complementary sequence information, *Anal. Chem.*, 73 (2001) 4530–4536.
66. F. W. McLafferty, D. M. Horn, K. Breuker, Y. Ge, M. A. Lewis, B. Cerda, R. A. Zubarev, and B. K. Carpenter, Electron capture dissociation of gaseous multiply charged ions by Fourier-transform ion cyclotron resonance, *J. Am. Soc. Mass Spectrom.*, 12 (2001) 245–249.
67. R. J. Cotter, Time-of-flight mass spectrometry for the structural analysis of biological molecules, *Anal. Chem.*, 64 (1992) 1027A–1039A.
68. M. Guilhaus, Special feature: Tutorial. Principles and instrumentation in time-of-flight mass spectrometry. Physical and instrumental concepts, *J. Mass Spectrom.*, 30 (1995) 1519–1532.
69. M. Guilhaus, V. Mlynski, and D. Selby, Perfect timing: Time-of-flight mass spectrometry, *Rapid Commun. Mass Spectrom.*, 11 (1997) 951–962.
70. D. Ioanoviciu, Ion-optical properties of time-of-flight mass spectrometers, *Int. J. Mass Spectrom.*, 206 (2001) 211–229.
71. W. C. Wiley and I. H. McLaren, Time-of-flight mass spectrometer with improved resolution, *Rev. Sci. Instrum.*, 26 (1955) 1150–1157.
72. M. L. Vestal, P. Juhasz, and S. A. Martin, Delayed extraction matrix-assisted laser desorption time-of-flight mass spectrometry, *Rapid Commun. Mass Spectrom.*, 9 (1995) 1044–1050.
73. S. M. Colby, T. B. King, and J. P. Reilly, Improving the resolution of matrix-assisted laser desorption/ionization time-of-flight mass-spectrometry by exploiting the correlation between ion position and velocity, *Rapid Commun. Mass Spectrom.*, 8 (1994) 865–868.
74. B. A. Mamyrin, I. Karataev, D. V. Shmikk, and A. F. Zagulin, The mass reflectron, a new nonmagnetic time-of-flight mass spectrometer with high resolution, *zh.Eksp.theor.Fiz.*, 64 (1973) 82–89.
75. B. A. Mamyrin, Laser assisted reflectron time of flight mass spectrometry, *Int. J. Mass Spectrom. Ion Process.*, 131 (1993) 1–19.
76. K. F. Medzihradszky, J. M. Campbell, M. A. Baldwin, A. M. Falick, P. Juhasz, M. L. Vestal, and A. L. Burlingame, The characteristics of peptide collision-induced dissociation using a high-performance MALDI-TOF/TOF tandem mass spectrometer, *Anal. Chem.*, 72 (2000) 552–558.
77. A. L. Yergey, J. R. Coorssen, P. S. Backlund, P. S. Blank, G. A. Humphrey, J. Zimmerberg, J. M. Campbell, and M. L. Vestal, De novo sequencing of peptides using MALDI/TOF-TOF, *J. Am. Soc. Mass Spectrom.*, 13 (2002) 784–791.
78. D. Suckau, A. Resemann, M. Schuerenberg, P. Hufnagel, J. Franzen, and A. Holle, A novel MALDI LIFT-TOF/TOF mass spectrometer for proteomics, *Anal. Bioanal. Chem.*, 376 (2003) 952–965.
79. Y. Mechref, M. Novotny, and C. Krishnan, Structural characterization of oligosaccharides using MALDI-TOF/TOF tandem mass spectrometry, *Anal. Chem.*, 75 (2003) 4895–4903.
80. E. Stephens, S. L. Maslen, L. G. Green, and D. H. Williams, Fragmentation characteristics of neutral *N*-linked glycans using a MALDI-TOF/TOF tandem mass spectrometer, *Anal. Chem.*, 76 (2004) 2343–2354.

81. W. Wuhrer, C. H. Hokke, and A. M. Deelder, Glycopeptide analysis by matrix-assisted laser desorption/ionization tandem time-of-flight mass spectrometry reveals novel features of horseradish peroxidase glycosylation, *Rapid Commun. Mass Spectrom.*, 18 (2004) 1741–1748.
82. I. V. Chernushevich, W. Ens, and K. G. Standing, Orthogonal-injection TOFMS for analyzing biomolecules, *Anal. Chem.*, 71 (1999) 452–461.
83. M. Guilhaus, D. Selbyand, and V. Mlynski, Orthogonal acceleration time-of-flight mass spectrometry, *Mass Spectrom. Rev.*, 19 (2000) 65–107.
84. C. M. Whitehouse, R. N. Dreyer, M. Yamashita, and J. B. Fenn, Electrospray interface for liquid chromatographs and mass spectrometers, *Anal. Chem.*, 57 (1985) 675–679.
85. J. Liu, K. J. Volk, E. H. Kerns, S. E. Klohr, M. S. Lee, and I. E. Rosenberg, Structural characterization of glycoprotein digests by microcolumn liquid chromatography-ionspray tandem mass spectrometry, *J. Chromatogr. A*, 632 (1993) 45–56.
86. M. J. Huddleston, M. F. Bean, and S. A. Carr, Collisional fragmentation of glycopeptides by electrospray ionization LC/MS and LC/MS/MS: Methods for selective detection of glycopeptides in protein digests, *Anal. Chem.*, 65 (1993) 877–884.
87. K. A. Thomsson, N. G. Karlsson, and G. C. Hansson, Liquid chromatography–electrospray mass spectrometry as a tool for the analysis of sulfated oligosaccharides from mucin glycoproteins, *J. Chromatogr. A*, 854 (1999) 131–139.
88. N. Kawasaki, M. Ohta, S. Hyuga, O. Hashimoto, and T. Hayakawa, Analysis of carbohydrate heterogeneity in a glycoprotein using liquid chromatography/mass spectrometry and liquid chromatography with tandem mass spectrometry, *Anal. Biochem.*, 269 (1999) 297–303.
89. N. Kawasaki, M. Ohta, S. Hyuga, M. Hyuga, and T. Hayakawa, Application of liquid chromatography/mass spectrometry and liquid chromatography with tandem mass spectrometry to the analysis of the site-specific carbohydrate heterogeneity in erythropoietin, *Anal. Biochem.*, 285 (2000) 82–91.
90. K. A. Thomsson, N. G. Karlsson, and G. C. Hansson, Sequencing of sulfated oligosaccharides from mucins by liquid chromatography and electrospray ionization tandem mass spectrometry, *Anal. Chem.*, 72 (2000) 4543–4549.
91. D. Schmid, B. Behnke, J. Metzger, and R. Kuhn, Nano-HPLC-mass spectrometry and MEKC for the analysis of oligosaccharides from human milk, *Biomed. Chromatogr.*, 16 (2002) 151–156.
92. S. Itoh, N. Kawasaki, M. Ohta, M. Hyuga, S. Hyuga, and T. Hayakawa, Simultaneous microanalysis of *N*-linked oligosaccharides in a glycoprotein using microbore graphitized carbon column liquid chromatography–mass spectrometry, *J. Chromatogr. A*, 968 (2002) 89–100.
93. L. A. Gennaro, D. J. Harvey, and P. Vouros, Reversed-phase ion-pairing liquid chromatography/ion trap mass spectrometry for the analysis of negatively charged, derivatized glycans, *Rapid Commun. Mass Spectrom.*, 17 (2003) 1528–1534.
94. M. S. Wilm and M. Mann, Electrospray and Taylor-cone theory, Dole's beam of macromolecules at last? *Int. J. Mass Spectrom. Ion Process.*, 136 (1994) 167–180.
95. B. L. Schulz, N. H. Packer, and N. G. Karlsson, Small-scale analysis of O-linked oligosaccharides from glycoproteins and mucins separated by gel electrophoresis, *Anal. Chem.*, 74 (2002) 6088–6097.
96. N. Kawasaki, S. Itoh, M. Ohta, and T. Hayakawa, Microanalysis of *N*-linked oligosaccharides in a glycoprotein by capillary liquid chromatography/mass spectrometry and liquid chromatography/tandem mass spectrometry, *Anal. Biochem.*, 316 (2003) 15–22.
97. D. S. Ashton, C. R. Beddell, D. J. Cooper, and A. C. Lines, Determination of carbohydrate heterogeneity in the humanised antibody CAMPATH 1H by liquid chromatography and matrix-assisted laser desorption ionisation mass spectrometry, *Anal. Chim. Acta*, 306 (1995) 43–48.

98. E. Lattova and H. Perreault, Profiling of *N*-linked oligosaccharides using phenylhydrazine derivatization and mass spectrometry, *J. Chromatogr. A*, 1016 (2003) 71–87.
99. M. Li and J. A. Kinzer, Structural analysis of oligosaccharides by a combination of electrospray mass spectrometry and bromine isotope tagging of reducing-end sugars with 2-amino-5-bromopyridine, *Rapid Commun. Mass Spectrom.*, 17 (2003) 1462–1466.
100. K. R. Anumula and S. T. Dhume, High resolution and high sensitivity methods for oligosaccharide mapping and characterization by normal phase high performance liquid chromatography following derivatization with highly fluorescent anthranilic acid, *Glycobiology*, 8 (1998) 685–694.
101. J. C. Bigge, T. P. Patel, J. A. Bruce, P. N. Goulding, S. M. Charles, and R. B. Parekh, Nonselective and efficient fluorescent labeling of glycans using 2-amino benzamide and anthranilic acid, *Anal. Biochem.*, 230 (1995) 229–238.
102. G. R. Guile, P. M. Rudd, D. R. Wing, S. B. Prime, and R. A. Dwek, A rapid high-resolution high-performance liquid chromatographic method for separating glycan mixtures and analyzing oligosaccharide profiles, *Anal. Biochem.*, 240 (1996) 210–226.
103. M. Wuhrer, C. A. M. Koeleman, A. M. Deelder, and C. H. Hokke, Normal-phase nanoscale liquid chromatography-mass spectrometry of underivatized oligosaccharides at low-femtomole sensitivity, *Anal. Chem.*, 76 (2004) 833–838.
104. J. J. Conboy and J. J. Henion, High-performance anion-exchange chromatography coupled with mass spectrometry for the determination of carbohydrates, *Biol. Mass Spectrom.*, 21 (1992) 397–407.
105. L. Brüll, M. Huisman, H. Schols, F. Voragen, G. Critchley, J. E. Thomas-Oates, and J. Haverkamp, Rapid molecular weight and structural determination of plant cell wall-derived oligosaccharides using off-line high-performance anion exchange chromatography-mass spectrometry, *J. Mass Spectrom.*, 33 (1998) 713–720.
106. G. J. Rademaker and J. Thomas-Oates, Analysis of glycoproteins and glycopeptides using fast atom bombardment, in J. R. Chapman (Ed.), *Protein and Peptide Analysis by Mass Spectrometry*, Human Press, Totowa, NJ, 1996.
107. R. C. Simpson, C. C. Fenselau, M. R. Hardy, R. R. Townsend, Y. C. Lee, and R. B. Cotter, Adaptation of a thermospray liquid chromatography/mass spectrometry interface for use with alkaline anion exchange liquid chromatography of carbohydrates, *Anal. Chem.*, 62 (1990) 248–252.
108. J. J. Conboy, M. W. Henion, M. W. Martin, and J. A. Zweigenbaum, Ion chromatography/mass spectrometry for the determination of organic ammonium and sulfate compounds, *Anal. Chem.*, 62 (1990) 800–807.
109. S. Richardson, A. Cohen, and L. Gorton, High-performance anion-exchange chromatography–electrospray mass spectrometry for investigation of the substituent distribution in hydroxypropylated potato amylopectin starch, *J. Chromatogr. A*, 917 (2001) 111–121.
110. M. Davies, K. D. Smith, A. M. Harbin, and E. F. Hounsell, High-performance liquid chromatography of oligosaccharide alditols and glycopeptides on a graphitized carbon column, *J. Chromatogr.*, 609 (1992) 125–131.
111. N. Kawasaki, M. Ohta, S. Itoh, M. Hyuga, S. Hyuga, and T. Hayakawa, Usefulness of sugar mapping by liquid chromatography/mass spectrometry in comparability assessments of glycoprotein products, *Biologicals*, 30 (2002) 113–123.
112. A. Cohen, H. Schagerlof, C. Nilsson, C. Melander, F. Tjerneld, and L. Gorton, Liquid chromatography–mass spectrometry analysis of enzyme-hydrolysed carboxymethylcellulose for investigation of enzyme selectivity and substituent pattern, *J. Chromatogr. A*, 1029 (2004) 87–95.

113. B. Barroso, R. Dijkstra, M. Geerts, F. Lagerwerf, P. van Veelen, and A. de Ru, On-line high-performance liquid chromatography/mass spectrometric characterization of native oligosaccharides from glycoproteins, *Rapid Commun. Mass Spectrom.*, 16 (2002) 1320–1329.
114. K. K. Mock, M. Davey, and J. S. Cottrell, The analysis of underivatised oligosaccharides by matrix-assisted laser desorption mass spectrometry, *Biochem. Biophys. Res. Commun.*, 177 (1991) 644–651.
115. D. B. Wall, S. J. Berger, J. W. Finch, S. A. Cohen, K. Richardson, R. Chapman, D. Drabble, J. Brown, and D. Gostick, Continuous sample deposition from reversed-phase liquid chromatography to tracks on a matrix-assisted laser desorption/ionization precoated target for the analysis of protein digests, *Electrophoresis*, 23 (2002) 3193–3204.
116. T. Miliotis, S. Kjellstrom, P. Önnerfjord, J. Nilsson, T. Laurell, L. E. Edholm, and G. Marko-Varga, Protein identification platform utilizing micro dispensing technology interfaced to matrix-assisted laser desorption ionization time-of-flight mass spectrometry, *J. Chromatogr. A*, 886 (2000) 99–110.
117. T. Miliotis, S. Kjellstrom, J. Nilsson, T. Laurell, L. E. Edholm, and G. Marko-Varga, Capillary liquid chromatography interfaced to matrix-assisted laser desorption/ionization time-of-flight mass spectrometry using an on-line coupled piezoelectric flow-through microdispenser, *J. Mass Spectrom.*, 35 (2000) 369–377.
118. C. Ericson, Q. T. Phung, D. M. Horn, E. C. Peters, J. R. Fitchett, S. B. Ficarro, A. R. Salomon, L. M. Brill, and A. Brock, An automated noncontact deposition interface for liquid chromatography matrix-assisted laser desorption/ionization mass spectrometry, *Anal. Chem.*, 75 (2003) 2309–2315.
119. J. B. Young and L. Liang, An impulse-driven liquid-droplet deposition interface for combining LC with MALDI MS and MS/MS, *J. Am Soc. Mass Spectrom.*, 17 (2006) 325–334.
120. D. S. Nagra and L. Li, Liquid chromatography-time-of-flight mass spectrometry with continuous-flow matrix-assisted laser desorption ionization, *J. Chromatogr. A*, 711 (1995) 235–245.
121. Q. Zhan, A. Gusev, and D. M. Hercules, A novel interface for on-line coupling of liquid capillary chromatography with matrix-assisted laser desorption/ionization detection, *Rapid Commun. Mass Spectrom.*, 13 (1999) 2278–2283.
122. K. K. Murray and D. H. Russel, Aerosol matrix-assisted laser desorption ionization mass spectrometry, *J. Am. Soc. Mass Spectrom.*, 5 (1994) 1–9.
123. H. Ørsnes, T. Graf, H. Degn, and K. K. Murray, A rotating ball inlet for on-line MALDI mass spectrometry, *Anal. Chem.*, 72 (2000) 251–254.
124. G. Lochnit and R. Geyer, An optimized protocol for nano-LC-MALDI-TOF-MS coupling for the analysis of proteolytic digests of glycoproteins, *Biomed. Chromatogr.*, 18 (2004) 841–848.
125. J. A. Olivares, N. T. Nguyen, C. R. Yonker, and R. D. Smith, On-line mass spectrometric detection for capillary zone electrophoresis, *Anal. Chem.*, 59 (1987) 1230–1232.
126. R. D. Smith, C. J. Barinaga, and H. R. Udseth, Improved electrospray ionisation interface for capillary zone electrophoresis-mass spectrometry, *Anal. Chem.*, 60 (1988) 1948–1952.
127. R. D. Smith and H. R. Udseth, Capillary zone electrophoresis-MS, *Nature*, 331 (1988) 639–640.
128. M. Larsson, R. Sundberg, and S. Folestad, On-line electrophoresis with mass spectrometry detection for the analysis of carbohydrates after derivatisation with 8-aminonaphthalene-1,3,6-trisulfonic acid, *J. Chromatogr. A*, 934 (2001) 75–85.
129. S. Duteil, P. Gareil, S. Girault, A. Mallet, C. Feve, and L. Siret, Identification of heparin oligosaccharides by direct coupling of CE/ionspray-MS, *Rapid Commun. Mass Spectrom.*, 13 (1999) 1889–1898.

130. D. Figeys, I. van Oostveen, A. Ducret, and R. Aebersold, Protein identification by capillary zone electrophoreis/microelectrospray ionisation-tandem mass spectrometry at the subfemtomole level, *Anal. Chem.*, 68 (1996) 1822–1828.
131. D. R. Barnidge, S. Nilsson, and K. E. Markides, A design for low-flow sheathless electrospray emitters, *Anal. Chem.*, 71 (1999) 4115–4118.
132. Y. Z. Chang and G. R. Her, Sheathless capillary electrophoresis/electrospray mass spectrometry using a carbon-coated fused-silica capillary, *Anal. Chem.*, 72 (2000) 626–630.
133. Y.-R. Chen and G.-R. Her, A simple method for fabrication of silver-coated sheathless electrospray emitters, *Rapid Commun. Mass Spectrom.*, 17 (2003) 437–441.
134. A. D. Zamfir and J. Peter-Katalinić, Glycoscreening by on-line sheathless capillary electrophoresis/electrospray ionisation-quadrupole time of flight mass spectrometry, *Electrophoresis*, 22 (2001) 2448–2457.
135. A. D. Zamfir, N. Dinca, E. Sisu, and J. Peter-Katalinić, Copper-coated microsprayer interface for on-line sheathless capillary electrophoresis electrospray mass spectrometry of carbohydrates, *J. Sep. Sci.*, 29 (2006) 414–422.
136. H. Suzuki, O. Muller, A. Guttman, and B. L. Karger, Analysis of 1-aminopyrene-3,6,8-trisulfonate-derivatised oligosaccharides by CE with MALDI TOF MS, *Anal. Chem.*, 69 (1997) 4554–4559.
137. L. Bindila, R. Almeida, A. Sterling, M. Allen, J. Peter-Katalinić, and A. Zamfir, Off-line capillary electrophoresis/fully automated nanoelectrospray chip quadrupole time-of-flight mass spectrometry and tandem mass spectrometry for glycoconjugate analysis, *J. Mass Spectrom.*, 39 (2004) 1190–1201.
138. S. Udiavar, A. Apffel, J. Chakel, S. Swedberg, W. S. Hancock, and E. Pungor, The use of multidimensional liquid-phase separations and mass spectrometry for the detailed characterisation of posttranslational modifications in glycoproteins, *Anal. Chem.*, 70 (1998) 3572–3578.
139. I. Ciucanu and F. Kerek, A simple and rapid method for the permethylation of carbohydrates, *Carbohydr. Res.*, 131 (1984) 209–217.
140. E. J. Bourne, M. Stacey, J. C. Tatlow, and J. M. Tedder, Studies on the trifluoroacetic acid 1. Trifluoroacetic anhydride as a promoter of ester formation between hydroxy-compounds and carboxylic acids, *J. Chem. Soc.* (1949) 2976–2979.
141. A. Dell, Preparation and desorption mass spectrometry of permethyl and peracetyl derivatives of oligosaccharide, *Meth. Enzymol.*, 193 (1990) 647–660.
142. A. Dell, H. R. Morris, H. Egge, H. von Nicolai, and G. Strecker, Fast-atom-bombardment mass spectrometry for carbohydrate structure determination, *Carbohydr. Res.*, 115 (1983) 41–52.
143. A. Dell, J. E. Thomas-Oates, M. E. Rogers, and P. R. Tiller, Novel fast atom bombardment procedures for glycoprotein analysis, *Biochimie*, 70 (1988) 1435–1444.
144. S. Aduru and B. T. Chait, Cf-252 plasma desorption mass spectrometry of oligosaccharides and glycoconjugates—control of ionization and fragmentation, *Anal. Chem.*, 63 (1991) 1621–1625.
145. A. Dell, M. E. Rogers, J. E. Oates, T. H. Huckerby, P. N. Sanderson, and I. A. Nieduszynski, Fast-atom bombardment mass spectrometric strageties for sequencing sulphated oligosaccharides, *Carbohydr. Res.*, 179 (1988) 7–19.
146. A. Dell, W. S. York, M. McNeil, A. G. Darvill, and P. Albersheim, Host-symbiont interactions 14. The cyclic structure of beta-D-(1,2)-linked D-glucans secreted by rhizobia and agrobacteria, *Carbohydr. Res.*, 117 (1983) 185–200.
147. M. McNeil, A. G. Darvill, P. Aman, L. E. Franzen, and P. Albersheim, Structural analysis of complex carbohydrates using high performance liquid chromatography, gas chromatrography and mass spectrometry, *Meth. Enzymol.*, 83 (1982) 3–45.

148. J. E. Thomas-Oates and A. Dell, Fast atom bombardment mass spectrometry strategies for analyzing glycoprotein glycans, *Biochem. Soc. Trans.*, 17 (1988) 243–245.
149. S. Hakomori, Rapid permethylation of glycolipid + polysaccharide catalyzed by methylsulfinyl carbanion in dimethyl sulfoxide, *J. Biochem.*, 55 (1964) 1386–1388.
150. S. Hase, Precolumn derivatization for chromatographic and electrophoretic analysis of carbohydrates, *J. Chromatogr. A*, 720 (1996) 173–182.
151. D. J. Harvey, T. J. P. Naven, B. Küster, R. H. Bateman, M. R. Green, and G. Critchley, Comparison of fragmentation modes for the structural determination of complex oligosaccharides ionized by matrix-assisted laser desorption/ionization mass spectrometry, *Rapid Commun. Mass Spectrom.*, 9 (1995) 1556–1561.
152. D. J. Harvey, P. M. Rudd, R. H. Bateman, R. S. Bordoli, K. Howes, J. B. Hoyes, and R. G. Vickers, Examination of complex oligosaccharides by matrix-assisted laser desorption/ionization mass spectrometry on time-of-flight and magnetic sector instruments, *Org. Mass Spectrom.*, 29 (1994) 753–766.
153. M. C. Huberty, J. E. Vath, W. Yu, and S. A. Martin, Site-specific carbohydrate identification in recombinant proteins using MALD-TOF MS, *Anal. Chem.*, 65 (1993) 2791–2800.
154. B. Spengler, D. Kirsch, R. Kaufmann, and J. Lemoine, Structure analysis of branched oligosaccharides using post-source decay in matrix-assisted laser desorption ionization mass spectrometry, *Org. Mass Spectrom.*, 29 (1994) 782–787.
155. V. Katta, S. K. Chowdhury, and B. T. Chait, Use of a single-quadrupole mass spectrometer for collision-induced dissociation studies of multiply charged peptide ions produced by electrospray ionization, *Anal. Chem.*, 63 (1991) 174–178.
156. B. Domon and C. E. Costello, A systematic nomenclature for carbohydrate fragmentations in FAB-MS/MS spectra of glycoconjugates, *Glycoconj. J.*, 5 (1988) 397–409.
157. L. P. Brüll, V. Kovácik, J. Thomas-Oates, J. Haverkamp, and W. Heerma, Sodium-cationized oligosaccharides do not appear to undergo 'internal residue loss' rearrangement processes on tandem mass spectrometry, *Rapid Commun. Mass Spectrom.*, 12 (1998) 1520–1532.
158. M. T. Cancilla, S. G. Penn, J. A. Carroll, and C. B. Lebrilla, Coordination of alkali metals to oligosaccharides dictates fragmentation behavior in matrix assisted laser desorption ionization/Fourier transform mass spectrometry, *J. Am. Chem. Soc.*, 118 (1996) 6736–6745.
159. E. Botek, J. L. Debrun, B. Hakim, and L. Morin-Allory, Attachment of alkali cations on beta-D-glucopyranose: Matrix-assisted laser desorption/ionization time-of-flight studies and *ab initio* calculations, *Rapid Commun. Mass Spectrom.*, 15 (2001) 273–276.
160. A. S. Weiskopf, P. Vouros, and D. J. Harvey, Characterization of oligosaccharide composition and structure by quadrupole ion trap mass spectrometry, *Rapid Commun. Mass Spectrom.*, 11 (1998) 1493–1504.
161. V. N. Reinhold, B. B. Reinhold, and C. E. Costello, Carbohydrate molecular weight profiling, sequence, linkage, and branching data: ES-MS and CID, *Anal. Chem.*, 67 (1995) 1772–1784.
162. Y. Ohashi, Y. Itoh, M. Kubota, K. Hamada, M. Ohashi, T. Hirano, and H. Niwa, Analysis of sugar epimers using mass spectrometry: *N*-acetyllactosamine-6,6′-disulfate and the 2′-epimer, *Eur. J. Mass Spectrom.*, 10 (2004) 269–278.
163. B. Mulroney, J. C. Traeger, and B. A. Stone, Determination of both linkage position and anomeric configuration in underivatized glucopyranosyl disaccharides by electrospray mass spectrometry, *J. Mass Spectrom.*, 30 (1995) 1277–1283.
164. S. Goletz, B. Thiede, F.-G. Hanisch, M. Schultz, J. Peter-Katalinic, S. Müiller, O. Seitz, and U. Karsten, A sequencing strategy for the localization of *O*-glycosylation sites of MUC1 tandem repeats by PSD-MALDI mass spectrometry, *Glycobiology*, 7 (1997) 881–896.

165. K. Alving, H. Paulsen, and J. Peter-Katalinic, Characterization of O-glycosylation sites in MUC2 glycopeptides by nanoelectrospray QTOF mass spectrometry, *J. Mass Spectrom.*, 34 (1999) 395–407.
166. L. Tip, *A mass spectrometric study of pentoses and pentose-containing oligosaccharides using fast atom bombardment*, PhD thesis, Utrecht University, The Netherlands, 1993.
167. V. Kovácik, J. Hirsch, P. Kovác, W. Heerma, J. Thomas-Oates, and J. Haverkamp, Oligosaccharide characterization using collision-induced dissociation fast atom bombardment mass spectrometry: Evidence for internal monosaccharide residue loss, *J. Mass Spectrom.*, 30 (1995) 949–958.
168. L. P. Brüll, W. Heerma, J. Thomas-Oates, J. Haverkamp, V. Kovácik, and P. Kovác, Loss of internal 1→6 substituted monosaccharide residues from underivatized and per-*O*-methylated trisaccharides, *J. Am. Soc. Mass Spectrom.*, 8 (1997) 43–49.
169. B. Ernst, D. R. Müller, and W. J. Richter, False sugar sequence ions in electrospray tandem mass spectrometry of underivatized sialyl-Lewis-type oligosaccharides, *Int. J. Mass Spectrom. Ion Process.*, 160 (1997) 283–290.
170. B. M. Warrack, M. E. Hail, A. Triolo, F. Animati, R. Seraglia, and P. Traldi, Observation of internal monosaccharide losses in the collisionally activated dissociation mass spectra of anthracycline aminodisaccharides, *J. Am. Soc. Mass Spectrom.*, 9 (1998) 710–715.
171. Y.-L. Ma, I. Vedernikova, H. Van den Heuvel, and M. Claeys, Internal glucose residue loss in protonated *O*-diglycosyl flavonoids upon low-energy collision-induced dissociation, *J. Am. Soc. Mass Spectrom.*, 11 (2000) 136–144.
172. D. J. Harvey, T. S. Mattu, M. R. Wormald, L. Royle, R. A. Dwek, and P. M. Rudd, "Internal residue loss": Rearrangements occurring during the fragmentation of carbohydrates derivatized at the reducing terminus, *Anal. Chem.*, 74 (2002) 734–740.
173. M. M. A. Olsthoorn, I. M. López-Lara, B. O. Petersen, K. Bock, J. Haverkamp, H. P. Spaink, and J. E. Thomas-Oates, Novel branched nod factor structure results from α-(1→3) fucosyl transferase activity: The major lipo-chitin oligosaccharides from *Mesorhizobium loti* strain NZP2213 bear an α-(1→3) fucosyl substituent on a nonterminal backbone residue, *Biochemistry*, 37 (1998) 9024–9032.
174. M. McNeil, Elimination of internal glycosyl residues during chemical ionization-mass spectrometry of per-*O*-alkylated oligosaccharide-alditols, *Carbohydr. Res.*, 123 (1983) 31–40.
175. M. Ferro, N. Demont, D. Promé, J.-C. Promé, C. Boivin, and B. Dreyfus, Detection and characterization of NOD factors by MALDI, LSIMS and tandem mass spectrometry, *13th Internat. Mass Spectrom. Conf.*, Budapest, Hungary, 1994, poster ThD3.
176. A. Dell, K. H. Khoo, M. Panico, R. McDowell, A. T. Etienne, A. J. Reason, and H. R. Morris, FAB-MS and ES-MS of glycoproteins, in M. Fukuda and A. Kobata (Eds.), *Glycobiology, A Practical Approach*, Oxford University Press, Oxford, 1993.
177. E. Yoon and R. A. Laine, *Linkage position determination in a novel set of permethylated neutral trisaccharides by collisional-induced dissociation and tandem mass spectrometry. Biol. Mass Spectrom.*, 21 (1992) 479–485.
178. W. S. York, H. van Halbeek, A. G. Darvill, and P. Albersheim, Structural analysis of xyloglucan oligosaccharides by ^{1}H-n.m.r. spectroscopy and fast-atom-bombardment mass spectrometry, *Carbohydr. Res.*, 200 (1990) 9–31.
179. M. W. Breedveld and K. J. Miller, Cyclic beta-glucans of the family *Rhizobiacea*, *Microbiol. Rev.*, 58 (1994) 145–161.
180. P. Talaga, B. Stahl, J.-M. Wieruszeski, F. Hillenkamp, S. Tsuyumu, G. Lippens, and J.-P. Bohin, Cell-associated glucans of *Burkholderia solanacearum* and *Xanthomonas campestris* pv citri: A new family of periplasmic glucans, *J. Bacteriol.*, 178 (1996) 2263–2271.

181. J. E. Schneider, V. N. Reinhold, M. K. Rumley, and E. P. Kennedy, Structural studies of the membrane-derived oligosaccharides of *Escherichea coli*, *J. Biol. Chem.*, 254 (1979) 10135–10138.
182. I. Komanieacka and A. Choma, Isolation and characterization of periplasmic cyclic β-glucans of *Azorhizobium caulinodans*, *FEMS Microbiol. Lett.*, 227 (2003) 263–269.
183. Y. Jung, H. Park, E. Cho, and A. Jung, Structural analyses of novel glycerophosphorylated α-cyclosophorohexadecaoses isolated from *X. campestris* pv. *campestris*, *Carbohydr. Res.*, 340 (2005) 673–677.
184. A. Pfenninger, M. Karas, B. Finke, and B. Stahl, Structural analysis of underivatized neutral human milk oligosaccharides in the negative ion mode by nano-electrospray MS^n (Part 2: Application to isomeric mixtures), *J. Am. Soc. Mass Spectrom.*, 13 (2002) 1341–1348.
185. P. Mauri, M. Minoggio, P. Simonetti, C. Gardana, and P. Pietta, Analysis of saccharides in beer samples by flow injection with electrospray mass spectrometry, *Rapid Commun. Mass Spectrom.*, 16 (2002) 743–748.
186. K. Uekama, F. Hirayama, and T. Irie, Cyclodextrin drug carrier systems, *Chem. Rev.*, 98 (1998) 2045–2076.
187. M. Guo, F. Song, Z. Liu, and S. Liu, Characterization of non-covalent complexes of rutin with cyclodextrins by electrospray ionization tandem mass spectrometry, *J. Mass Spectrom.*, 39 (2004) 594–599.
188. I. Olsen and E. Jantzen, Sphingolipids in bacteria and fungi, *Anaerobe*, 7 (2001) 103–112.
189. D. V. Lynch and T. M. Dunn, An introduction to plant sphingolipids and a review of recent advances in understanding their metabolism and function, *New Phytol.*, 161 (2004) 677–702.
190. T. Kolter and K. Sandhoff, Sphingolipids—their metabolic pathways and the pathobiochemistry of neurodegenerative diseases, *Angew. Chem. Int. Ed.*, 38 (1999) 1532–1568.
191. P. Juhasz and C. E. Costello, Matrix-assisted laser desorption ionization time-of-flight mass-spectrometry of underivatized and permethylated gangliosides, *J. Am. Soc. Mass Spectrom.*, 3 (1992) 785–796.
192. G. Siuzdak, Y. Ichikawa, T. G. Caulfield, B. Munoz, C. H. Wong, and K. C. Nicolaou, Evidence of calcium (2+)-dependent carbohydrate association through ion spray mass spectrometry, *J. Am. Chem. Soc.*, 115 (1993) 2877–2881.
193. S. B. Levery, Glycosphingolipid structural analysis and glycosphingolipidomics mass spectrometry: Modified proteins and glycoconjugates, *Meth. Enzymol.*, 405 (2005) 300–369.
194. D. J. Harvey, R. H. Bateman, R. S. Bordoli, and R. Tyldesley, Ionisation and fragmentation of complex glycans with a quadrupole time-of-flight mass spectrometer fitted with a matrix-assisted laser desorption/ionisation ion source, *Rapid Commun. Mass Spectrom.*, 14 (2000) 2135–2142.
195. A. K. Powell and D. J. Harvey, Stabilization of sialic acids in *N*-linked oligosaccharides and gangliosides for analysis by positive ion matrix-assisted laser desorption ionization mass spectrometry, *Rapid Commun. Mass Spectrom.*, 10 (1996) 1027–1032.
196. P. B. O'Connor and C. E. Costello, A high pressure matrix-assisted laser desorption/ionization Fourier transform mass spectrometry ion source for thermal stabilization of labile biomolecules, *Rapid Commun. Mass Spectrom.*, 15 (2001) 1862–1868.
197. P. B. O'Connor, E. Mirgorodskaya, and C. E. Costello, High pressure matrix-assisted laser desorption/ionization Fourier transform mass spectrometry for minimization of ganglioside fragmentation, *J. Am. Soc. Mass Spectrom.*, 13 (2002) 402–407.
198. J. Guittard, X. P. L. Hronowski, and C. E. Costello, Direct matrix-assisted laser desorption/ionization mass spectrometric analysis of glycosphingolipids on thin layer chromatographic plates and transfer membranes, *Rapid Commun. Mass Spectrom.*, 13 (1999) 1838–1849.

199. V. B. Ivleva, Y. N. Elkin, B. A. Budnik, S. C. Moyer, P. B. O'Connor, and C. E. Costello, Coupling thin-layer chromatography with vibrational cooling matrix-assisted laser desorption/ionization Fourier transform mass spectrometry for the analysis of ganglioside mixtures, *Anal. Chem.*, 76 (2004) 6484–6491.
200. S. G. Penn, M. T. Cancilla, M. K. Green, and C. B. Lebrilla, Direct comparison of matrix-assisted laser desorption/ionisation and electrospray ionisation in the analysis of gangliosides by Fourier transform mass spectrometry, *Euro. J. Mass Spectrom.*, 3 (1997) 67–79.
201. Q. H. Ann and J. Adams, Structure-specific collision-induced fragmentations of ceramides cationized with alkali-metal ions, *Anal. Chem.*, 65 (1993) 7–13.
202. Q. Ann and J. Adams, Structure determination of ceramides and neutral glycosphingolipids by collisional activation of $[M+Li]^+$ ions, *J. Am. Soc. Mass Spectrom.*, 3 (1992) 260–263.
203. R. S. Duarte, C. R. Polycarpo, R. Wait, R. Hartmann, and E. B. Bergter, Structural characterization of neutral glycosphingolipids from Fusarium species, *Biochim. Biophys. Acta-Lipids and Lipid Metabolism*, 1390 (1998) 186–196.
204. S. B. Levery, M. S. Toledo, R. L. Doong, A. H. Straus, and H. K. Takahashi, Comparative analysis of ceramide structural modification found in fungal cerebrosides by electrospray tandem mass spectrometry with low energy collision-induced dissociation of $Li+$ adduct ions, *Rapid Commun. Mass Spectrom.*, 14 (2000) 551–563.
205. A. Olling, M. E. Breimer, E. Peltomaa, B. E. Samuelsson, and S. Ghardashkhani, Electrospray ionization and collision-induced dissociation time-of-flight mass spectrometry of neutral glycosphingolipids, *Rapid Commun. Mass Spectrom.*, 12 (1998) 637–645.
206. X. L. Han and R. W. Gross, Shotgun lipidomics: Electrospray ionization mass spectrometric analysis and quantitation of cellular lipidomes directly from crude extracts of biological samples, *Mass Spectrom. Rev.*, 24 (2005) 367–412.
207. S. B. Levery, Glycosphingolipid structural analysis and glycosphingolipidomics, *Mass Spectrom.: Modified Proteins and Glycoconjugates*, 405 (2005) 300–369.
208. X. L. Han and R. W. Gross, Global analyses of cellular lipidomes directly from crude extracts of biological samples by ESI mass spectrometry: A bridge to lipidomics, *J. Lipid Res.*, 44 (2003) 1071–1079.
209. A. H. Merrill, M. C. Sullards, J. C. Allegood, S. Kelly, and E. Wang, Sphingolipidomics: High-throughput, structure-specific, and quantitative analysis of sphingolipids by liquid chromatography tandem mass spectrometry, *Methods*, 36 (2005) 207–224.
210. M. Gu, J. L. Kerwin, J. D. Watts, and R. Aebersold, Ceramide profiling of complex lipid mixtures by electrospray ionization mass spectrometry, *Anal. Biochem.*, 244 (1997) 347–356.
211. X. L. Han and H. Cheng, Characterization and direct quantitation of cerebroside molecular species from lipid extracts by shotgun lipidomics, *J. Lipid Res.*, 46 (2005) 163–175.
212. E. Vukelic, A. D. Zamfir, L. Bindila, M. Froesch, J. Peter-Katalinic, S. Usuki, and R. K. Yu, Screening and sequencing of complex sialylated and sulfated glycosphingolipid mixtures by negative ion electrospray Fourier transform ion cyclotron resonance mass spectrometry, *J. Am. Soc. Mass Spectrom.*, 16 (2005) 571–580.
213. Z. C. Tsui, Q. R. Chen, M. J. Thomas, M. Samuel, and Z. Cui, A method for profiling gangliosides in animal tissues using electrospray ionization-tandem mass spectrometry, *Anal. Biochem.*, 341 (2005) 251–258.
214. M. W. Duncan, G. Matanovic, and A. Cerpapoljak, Quantitative-analysis of low-molecular-weight compounds of biological interest by matrix-assisted laser-desorption ionization, *Rapid Commun. Mass Spectrom.*, 7 (1993) 1090–1094.
215. D. C. Muddiman, A. I. Gusev, A. Proctor, D. M. Hercules, R. Venkataraman, and W. Diven, Quantitative measurement of cyclosporin A in blood by time-of-flight mass spectrometry, *Anal. Chem.*, 66 (1994) 2362–2368.

216. E. Sugiyama, A. Hara, and K. Uemura, A quantitative analysis of serum sulfatide by matrix-assisted laser desorption ionization time-of-flight mass spectrometry with delayed ion extraction, *Anal. Biochem.*, 274 (1999) 90–97.
217. T. Fujiwaki, M. Tasaka, N. Takahashi, H. Kobayashi, Y. Murakami, T. Shimada, and S. Yamaguchi, Quantitative evaluation of sphingolipids using delayed extraction matrix-assisted laser desorption ionization time-of-flight mass spectrometry with sphingosylphosphorylcholine as an internal standard. Practical application to cardiac valves from a patient with Fabry disease, *J. Chromatogr. B*, 832 (2006) 97–102.
218. A. L. Tarentino, C. M. Gomez, and T. H. Plummer, Deglycosylation of asparagine-linked glycans by peptide: *N*-glycosidase F, *Biochem. J.*, 24 (1985) 4665–4671.
219. G. J. Rademaker, J. Haverkamp, and J. E. Thomas-Oates, Determination of glycosylation sites in *O*-linked glycopeptides: A sensitive mass spectrometric protocol, *Org. Mass Spectrom.*, 28 (1993) 1536–1541.
220. R. L. Katzman and E. H. Eylar, The isolation and characterization of a trisaccharide from porcine submaxillary glycoproteins, *Biochem. Biophys. Res. Commun.*, 23 (1966) 769–774.
221. K. O. Lloyd, E. A. Kabat, E. J. Layug, and F. Gruezo, Immunochemical studies on blood groups. XXXIV. Structures of some oligosaccharides produced by alkaline degradation of blood group A, B, and H substances, *Biochem. J.*, 5 (1966) 1489–1501.
222. G. Schiffman, E. A. Kabat, and W. Thompson, Immunochemical studies on blood groups. XXXII. Immunochemical properties of and possible partial structures for the blood group A, B, and H antigenic determinants, *Biochem. J.*, 3 (1964) 587–593.
223. G. Schiffman, E. A. Kabat, and W. Thompson, Immunochemical studies on blood groups. XXX. Cleavage of A, B, and H blood-group substances by alkali, *Biochem. J.*, 3 (1964) 113–120.
224. B. Anderson, P. Hoffman, and K. Meyer, A serine-linked peptide of chondroitin sulfate, *Biochim. Biophys. Acta*, 74 (1963) 309–311.
225. B. Anderson, N. Seno, P. Sampson, J. G. Riley, P. Hoffman, and K. Meyer, Threonine and serine linkages in mucopolysaccharides and glycoproteins, *J. Biol. Chem.*, 239 (1964) PC2716–PC2717.
226. D. M. Carlson and C. Blackwell, Structures and immunochemical properties of oligosaccharides isolated from pig submaxillary mucins, *J. Biol. Chem.*, 243 (1968) 616–626.
227. K. Tanaka, M. Bertolini, and W. Pigman, Serine and threonine glycosidic linkages in bovine submaxillary mucin, *Biochem. Biophys. Res. Commun.*, 16 (1964) 404–409.
228. K. Tanaka and W. Pigman, Improvements in hydrogenation procedure for demonstration of *O*-threonine glycosidic linkages in bovine submaxillary mucin, *J. Biol. Chem.*, 240 (1965) PC1487–PC1488.
229. G. J. Rademaker, S. A. Pergantis, L. Blok-Tip, J. I. Langridge, A. Kleen, and J. E. Thomas-Oates, Mass spectrometric determination of the sites of *O*-glycan attachment with low picomolar sensitivity, *Anal. Biochem.*, 257 (1998) 149–160.
230. E. Mirgorodskaya, H. Hassan, H. Clausen, and P. Roepstorff, Mass spectrometric determination of *O*-glycosylation sites using β-elimination and partial acid hydrolysis, *Anal. Chem.*, 73 (2001) 1263–1269.
231. F.-G. Hanisch, M. Jovanovic, and J. Peter-Katalinić, Glycoprotein identification and localization of *O*-glycosylation sites by mass spectrometric analysis of deglycosylated/alkylaminylated peptide fragments, *Anal. Biochem.*, 290 (2001) 47–59.
232. M. V. Novotny, Y. Huang, and Y. Mechref, Microscale nonreductive release of *O*-linked glycans for subsequent analysis through MALDI mass spectrometry and capillary electrophoresis, *Anal. Chem.*, 73 (2001) 6063–6069.
233. S. Takasaki, T. Mizuochi, and A. Kobata, Hydrazinolysis of asparagine-linked sugar chains to produce free oligosaccharides, *Meth. Enzymol.*, 83 (1982) 263–268.

234. T. Patel, J. Bruce, A. Merry, C. Bigge, M. Wormald, A. Jaques, and R. Parekh, Use of hydrazine to release in intact and unreduced form both *N*- and *O*-linked oligosaccharides from glycoproteins, *Biochem. J.*, 32 (1993) 679–693.
235. V. Bulone, G. J. Rademaker, S. Pergantis, T. Krogstad-Johnsen, B. Smestad-Paulsen, and J. Thomas-Oates, Characterisation of horse dander allergen glycoproteins using amino acid and glycan structure analyses, *Int. Arch. Allergy Immunol.*, 123 (2000) 220–227.
236. N. L. Wilson, B. L. Schulz, H. G. Karlsson, and N. H. Packer, Sequential analysis of *N*- and *O*-linked glycosylation of 2D-page separated glycoproteins, *J. Proteome Res.*, 1 (2002) 521–529.
237. J. Charlwood, J. M. Skehel, and P. Camilleri, Analysis of *N*-linked oligosaccharides released from glycoproteins separated by two-dimensional gel electrophoresis, *Anal. Biochem.*, 284 (2000) 49–59.
238. B. L. Schulz, D. Oxley, N. H. Packer, and H. G. Karlsson, Identification of two highly sialylated human tear-fluid DMBT1 isoforms: The major high-molecular-mass glycoproteins in human tears, *Biochem. J.*, 366 (2002) 511–520.
239. B. L. Schulz, N. H. Packer, and H. G. Karlsson, Small-scale analysis of *O*-linked oligosaccharides from glycoproteins and mucins separated by gel electrophoresis, *Anal. Chem.*, 74 (2002) 6088–6097.
240. B. Küster, S. F. Wheeler, A. P. Hunter, R. A. Dwek, and D. J. Harvey, Sequencing of *N*-linked oligosaccharides directly from protein gels: In-gel deglycosylation followed by matrix-assisted laser desorption/ionization mass spectrometry and normal-phase high-performance liquid chromatography, *Anal. Biochem.*, 250 (1997) 82–101.
241. S. Kilz, H. Budzikiewicz, and S. Waffenschmidt, In-gel deglycosylation of sodiumdodecyl sulfate polyacrylamide gel electrophoresis-separated glycoproteins for carbohydrate estimation by matrix-assisted laser desorption/ionization time-of-flight mass spectrometry, *J. Mass Spectrom.*, 37 (2002) 331–335.
242. A. M. Taylor, O. Holst, and J. Thomas-Oates, Mass spectrometric profiling of *O*-linked glycans released directly from glycoproteins in gels using in-gel reductive β-elimination, *Proteomics*, 6 (2006) 2936–2946.
243. A. Dell, J. E. Thomas-Oates, M. E. Rogers, and P. R. Tiller, Novel fast atom bombardment mass spectrometric procedures for glycoprotein analysis, *Biochimie*, 70 (1988) 1435–1444.
244. M. R. Hardy and R. R. Townsend, Separation of positional isomers of oligosaccharides and glycopeptides by high-performance anion-exchange chromatography with pulsed amperometric detection, *Proc. Natl. Acad. Sci. USA*, 85 (1988) 3289–3293.
245. P. Hermentin, R. Witzel, J. F. G. Vliegenthart, J. P. Kamerling, M. Nimtz, and H. S. Conradt, A strategy for the mapping of *N*-glycans by high-pH anion-exchange chromatography with pulsed amperometric detection, *Anal. Biochem.*, 203 (1992) 281–289.
246. F.-G. Hanisch, B. N. Green, R. Bateman, and J. Peter-Katalinić, Localization of *O*-glycosylation sites of MUC1 tandem repeats by QTOF ESI mass spectrometry, *J. Mass Spectrom.*, 33 (1998) 358–362.
247. K. Elving, H. Paulsen, and J. Peter-Katalinić, Characterization of *O*-glycosylation sites in MUC2 glycopeptides by nanoelectrospray QTOF mass spectrometry, *J. Mass Spectrom.*, 34 (1999) 395–407.
248. S. Müller, K. Alving, J. Peter-Katalinić, N. Zachara, A. A. Gooley, and F.-G. Hanisch, High density *O*-glycosylation on tandem repeat peptide from secretory MUC1 of T47D breast cancer cells, *J. Biol. Chem.*, 274 (1999) 18165–18172.
249. S. Schmitt, D. Glebe, K. Alving, T. K. Tolle, M. Linde, H. Geyer, D. Linder, J. Peter-Katalinić, W. H. Gerlich, and R. Geyer, Analysis of the pre-S2 *N*- and *O*-linked glycans of the M surface protein from human hepatitis B virus, *J. Biol. Chem.*, 274 (1999) 11945–11957.
250. B. Macek, J. Hofsteenge, and J. Peter-Katalinić, Direct determination of glycosylation sites in *O*-fucosylated glycopeptides using nano-electrospray quadrupole time-of-flight mass spectrometry, *Rapid Commun. Mass Spectrom.*, 15 (2001) 771–777.

251. A. Gonzalez de Peredo, D. Klein, B. Macek, D. Hess, J. Peter-Katalinić, and J. Hofsteenge, *C*-mannosylation and *O*-fucosylation of thrombospondin type 1 repeats, *Mol. Cell. Proteomics*, 1 (2002) 11–18.
252. S. A. Carr, M. J. Huddleston, and M. F. Bean, Selective identification and differentiation of *N*- and *O*-linked oligosaccharides in glycoproteins by liquid chromatography-mass spectrometry, *Protein Sci.*, 2 (1993) 183–196.
253. G. I. Gellene and R. F. Porter, Neutralized ion-beam spectroscopy, *Acc. Chem. Res.*, 16 (1983) 200–207.
254. R. A. Zubarev, N. L. Kelleher, and F. W. McLafferty, Electron capture dissociation of multiply charged protein cations. A nonergodic process, *J. Am. Chem. Soc.*, 120 (1998) 3265–3266.
255. N. L. Kelleher, R. A. Zubarev, K. Bush, B. Furie, B. C. Furie, F. W. McLafferty, and C. T. Walsh, Localization of labile posttranslational modifications by electron capture dissociation: The case of γ-carboxyglutamic acid, *Anal. Chem.*, 71 (1999) 4250–4253.
256. E. Mirgorodskaya, P. Roepstorff, and R. A. Zubarev, Localization of *O*-glycosylation sites in peptides by electron capture dissociation in a fourier transform mass spectrometer, *Anal. Chem.*, 71 (1999) 4431–4436.
257. A. Stensballe, O. Norregaard-Jensen, J. V. Olsen, K. F. Haselmann, and R. A. Zubarev, Electron capture dissociation of singly and multiply phosphorylated peptides, *Rapid Commun. Mass Spectrom.*, 14 (2000) 1793–1800.
258. S. D.-H. Shi, M. E. Hemling, S. A. Carr, D. M. Horn, I. Lindh, and F. W. McLafferty, Phosphopeptide/phosphoprotein mapping by electron capture dissociation mass spectrometry, *Anal. Chem.*, 73 (2001) 19–22.
259. M. Mormann, B. Maček, A. Gonzalez de Peredo, J. Hofsteenge, and J. Peter-Katalinić, Structural studies on protein *O*-fucosylation by electron capture dissociation, *Int. J. Mass Spectrom.*, 234 (2004) 11–21.
260. S. R. Long, Rhizobium legume nodulation—life together in the underground, *Cell*, 56 (1989) 203–214.
261. J. A. Leigh and G. C. Walker, Exopolysaccharides of rhizobium—synthesis, regulation and symbiotic function, *Trends Genet.*, 10 (1994) 63–67.
262. E. L. Kannenberg, B. L. Reuhs, L. S. Fosberg, and R. W. Carlson, in H. P. Spaink, A. Kondorosi, and P. J. J. Hooykaas (Eds.), *The Rhizobiaceae*, Kluwer Academic Publishers, Dordrecht, the Netherlands, 1998, pp. 119–154.
263. N. Fraysse, B. Lindner, Z. Kaczynski, L. Sharypova, O. Holst, K. Niehaus, and V. Poinsot, *Sinorhizobium meliloti* strain 1021 produces a low-molecular mass capsular polysaccharide that is a homopolymer of 3-deoxy-D-manno-oct-2-ulosonic acid harbouring a phospholipidic anchor, *Glycobiology*, 15 (2005) 101–108.
264. B. L. Reuhs, R. W. Carlson, and J. S. Kim, *Rhizobium fredii* and *Rhizobium meliloti* produce 3-deoxy-D-manno-2-octulosonic acid-containing polysaccharides that are structurally analogous to group II K antigens (capsular polysaccharides) found in *Escherichia coli*, *J. Bacteriol.*, 175 (1993) 3570–3580.
265. B. L. Reuhs, D. P. Geller, J. S. Kim, J. E. Fox, V. S. Kumar Kolli, and S. G. Pueppke, *Sinorhizobium fredii* and *Sinorhizobium meliloti* produce structurally conserved lipopolysaccharides and strain-specific K antigens, *Appl. Environ. Microbiol.*, 64 (1998) 4930–4938.
266. M. A. Rodriguez-Carvajal, J. A. Rodrigues, M. E. Soria-Diaz, P. Tejero-Mateo, A. Buendia-Claveria, R. Gutierrez, J. E. Ruiz-Sainz, J. Thomas-Oates, and A. M. Gil-Serrano, Structural analysis of the capsular polysaccharide from *Sinorhizobium fredii* HWG35, *Biomacromolecules*, 6 (2005) 1448–1456.

ADVANCES IN CARBOHYDRATE CHEMISTRY AND BIOCHEMISTRY, VOL. 61

DEOXY SUGARS: OCCURRENCE AND SYNTHESIS

BY ROSA M. DE LEDERKREMER AND CARLA MARINO

CIHIDECAR, Departamento de Química Orgánica, Facultad de Ciencias Exactas y Naturales, Universidad de Buenos Aires, Pabellón II, Ciudad Universitaria, 1428 Buenos Aires, Argentina

I. INTRODUCTION

Several deoxy sugars, notably 2-deoxy-D-*erythro*-pentose (2-deoxy-D-ribose) the sugar component of DNA, 6-deoxy-L-mannose (L-rhamnose), 6-deoxy-L-galactose (L-fucose), 6-deoxy-D-glucose (quinovose), and their derivatives, occur very widely in natural products. Also relevant are dideoxy and trideoxy sugars, such as 3,6-dideoxyhexoses, components of the antigenic determinants bacterial

ISBN: 978-0-12-373920-9
DOI: 10.1016/S0065-2318(07)61004-X

lipopolysaccharides, and the 2,6-dideoxy hexoses present in many steroidal glycosides, antibiotics, and antitumor compounds. In part, the role that they play in many physiological processes is attributed to the enhanced hydrophobicity that they display with respect to the oxygenated analogues. A previous survey on these compounds, by Hanessian, appeared in Vol. 21 of this series.[1] Chemical and biochemical aspects of deoxy sugars and deoxy sugar oligosaccharides have been reviewed.[2] The synthesis of 2-deoxyglycosides has been reviewed for the period 1988–1999.[3] Methods for the total synthesis of deoxy sugars, starting from noncarbohydrate precursors has been recently compiled,[4] and only a few examples will be given in this chapter. The naturally occurring 6-deoxyhexoses were reviewed previously in this series.[5]

Deoxy derivatives have been prepared for studies on the specificity or as inhibitors of glycosidases,[6–8] or glycosyltransferases,[9] and also to establish which hydroxyl groups are involved in the interaction with lectins.[10]

II. General Methods for Deoxygenation of Monosaccharides

In this chapter representative recent applications of general methods for the preparation of deoxy sugars are presented. It is not the scope of this survey to cover all reports for a specific technique. Deoxy sugars may be synthesized by reductive methods, using as starting materials epoxides, thio sugars, *C*-halo sugars, carboxylate esters, sulfonates, or even by direct reduction of hydroxyl groups. Alternative routes make use of standard methods of reduction of carbonyl groups in sugars (Wolff–Kishner and Clemmensen reductions, conversion into dithioacetals and subsequent reduction), but these methods are generally drastic. Unsaturated sugars are ready sources of deoxygenated derivatives, via electrophilic addition of hydrogen halides, and particularly glycals, through acid-catalyzed addition of water or alcohols.

1. Preparation from Deoxyhalo Sugars

The iodo and bromo derivatives of monosaccharides can be reduced by a variety of reducing agents to afford the corresponding deoxy sugar. Many such examples have been presented. The more stable chloro derivatives can be reduced with Raney nickel (Scheme 1). Selective reduction of a secondary chloride with respect to a primary chloride may be achieved if the reduction is performed in the presence of triethylamine.[11,12] Selective reduction of a secondary chloride has been also achieved by using organotin hydrides.[13] The radical initiator

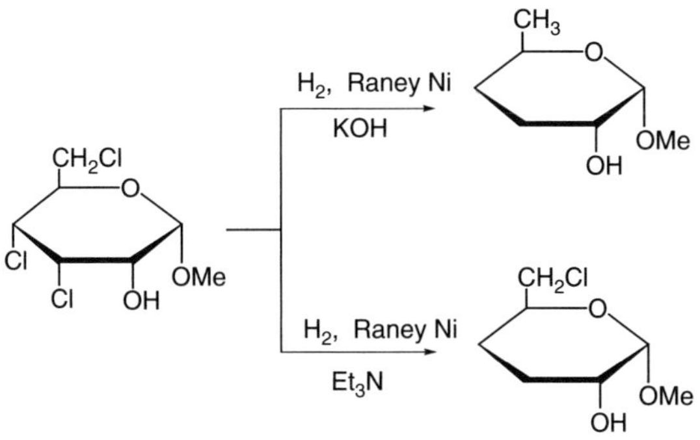

SCHEME 1.

2,2′-azobis(2-methylpropanenitrile) (azobisisobutanonitrile, AIBN) was found essential for the reduction.

2. Preparation from Sulfonates

Mesylates or tosylates may be reduced directly or via the intermediate halides or epoxides. Epoxides are involved in the reduction of tosylates by lithium aluminum hydride ($LiAlH_4$) or lithium triethyl borohydride ($LiEt_3BH$). When a good leaving group is present at the vicinal position, diaxial ring-opening governs the regioselectivity. Thus, treatment of tosylate **1** with $LiEt_3BH$ leads presumably via an epoxide to the **3**-deoxy glycoside **2** in 90% yield. When reduction of the isomeric epoxide **4** (obtained from ditosylate **3**) was performed, the 2-deoxy-α-D-*arabino*-glycoside **5** was obtained in high yield.[14]

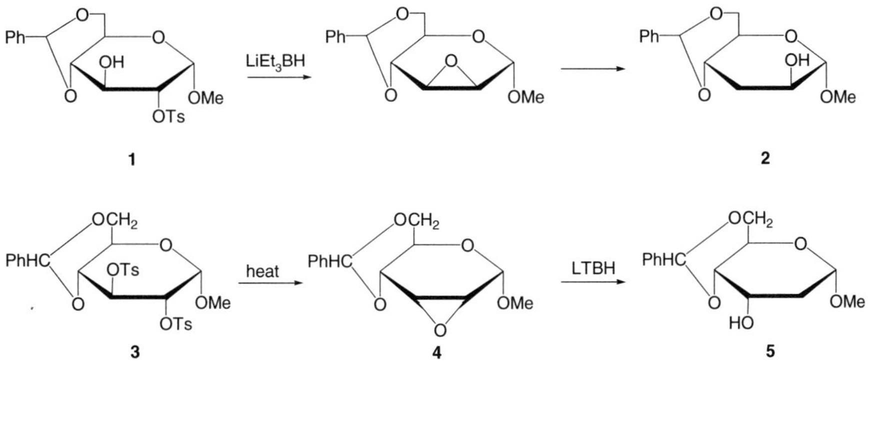

3. Radical-Mediated Deoxygenation

Deoxygenation of a hydroxyl group in sugar can be accomplished through conversion into a suitable thiocarbonyl derivative, followed by reduction with tributyltin hydride. Barton and McCombie applied this method with great success to 1,2:5,6-di-*O*-isopropylidene-α-D-glucofuranose (**6**), which was converted into the corresponding methyl xanthate **7**. Reduction with tributyltin hydride afforded the corresponding 3-deoxy derivative **8** in high yield.[15]

Bu_3SnH, AIBN

AIBN = azobisisobutanonitrile

6 R = H
7 R = C(S)-SMe

8

This reagent remains very useful, and it was recently used for an efficient synthesis of 2-deoxy-L-*erythro*-pentose (**9**), starting from L-arabinose (Scheme 2).[16]

3-Deoxy and 4-deoxy-β-D-hexopyranosides were prepared from a galactoside precursor by the same strategy, but by employing imidazol-1-ylthiocarbonyl

a R = C(S)SMe
b R = C(S)NHPh
c R = C(S)Im
d R = C(S)OPh

9

SCHEME 2. 2,2-dimethoxypropane, *p*-TsOH, ii. (a) NaH, CS_2, MeI, THF; (b) NaH, PhNCS, THF; (c) $C(S)Im_2$, THF, reflux; (d) PhOC(S)Cl, Py, CH_2Cl_2, iii. $n Bu_3SnH$, AIBN, toluene, reflux, iv. 4% TFA, 40 °C.

derivatives and subsequent reduction with tributyltin hydride.[10] The tributyltin hydride-mediated radical reduction of thiocarbonyl derivatives has also been used for the preparation of 2′-, 3′-, and 4′-deoxy derivatives of phenyl β-lactoside.[9] Triphenyltin hydride can also be employed,[17] and evidence for a radical mechanism was published.[18]

Although tributyltin and triphenyltin hydrides are excellent reagents, they are toxic, and the byproducts of the reaction are difficult to remove. The tin reagent has been replaced with advantage by silane derivatives. The first one to be used was $(SiMe_3)_3SiH$ [tris(trimethylsilyl)silane].[19] Reviews on the use of organosilanes as radical-based reducing agents have been published.[20] The $(SiMe)_3SiH$ reagent has been replaced by other less expensive silanes such as diphenylsilane.[21] Barton *et al.* have also used phenylsilane[22] and triphenylsilane for deoxygenations.[23] Trialkylsilanes have been also used for hydrogen delivery.[24]

Reagents based on the cleavage of a P–H bond have been used with success for radical deoxygenation. Hypophosphorous acid and its salts proved to be excellent reducing agents.[25] It has the advantage of nontoxicity, low price, and ease of removal from the reaction mixture.

Benzylidene acetals may undergo a thiol-catalyzed radical-chain redox rearrangement, resulting in deoxygenation at one of the HO groups involved in the acetal, and formation of a benzoate ester at the other one. For example, by using silanethiols ($ButO_3SiSH$, TBST) and $i Pr_3SiSH$, in conjunction with di-*tert*-butyl peroxide (Bu^tOOBu^t) as initiator, converted acetal **10** into deoxy derivative **11**. The regioselectivity is determined by the greater stability of the radical at C-2, in comparison with radical at C-1.[26]

BzO BzO BzO O O O Ph **10**
TBST / Bu^tOOBu^t
BzO BzO BzO O OBz **11**

3,4-*O*-Benzylidene derivatives led to comparable amounts of 3- and 4-deoxy derivatives,[26] whereas 4,6-*O*-benzylidene derivatives of glucose suffer deoxygenation at C-6 via the (less stabilized) primary radical, while the analogous galactose derivative led mainly to the 4-deoxy derivatives. The different regioselectivity has been explained considering the relative strain in the fused bicyclic structures.[27]

III. Monodeoxy Sugars

1. 2-Deoxy Sugars

a. General Considerations.—The most important 2-deoxy sugar is 2-deoxy-D-*erythro*-pentose (2-deoxyribose), the sugar constituent of deoxyribonucleic acids, but it is not considered further in this chapter. 2-Deoxyhexoses have been intensively used as inhibitors of biological processes.[28] In particular, 2-deoxy-D-*arabino*-hexose (2-deoxyglucose) has been described as an energy restriction mimetic agent.[29]

b. Synthesis.—Several methodologies have been developed for the preparation of 2-deoxy sugars, as they are important building blocks in the synthesis of bioactive carbohydrate-based structures. 2-Deoxy sugars and their glycosides are quite difficult to handle, as they are highly sensitive to hydrolysis and 1,2-elimination.

(i) From Glycals. Fischer's glycal method is one of the most widely used procedures for the synthesis of 2-deoxyhexopyranoses or pyranosides. It involves the conversion of an acylated glycosyl halide into a glycal, for example by treatment with zinc dust in acetic acid, and the subsequent addition of water, alcohol or a carboxylic acid to the double bond, to give 2-deoxy-aldoses, -aldosides, and -aldosyl esters (Scheme 3).[30]

The synthetic value of this reaction has been enhanced by the development of a wide range of methodologies for glycal synthesis from *O*-acylated glycosyl halides[1] or from noncarbohydrate precursors. By use of the Lewis acid-catalyzed diene–aldehyde cyclocondensation reaction, glycals of natural and non-natural sugars can be obtained (Scheme 4).[31]

The drawback of this method arises from the rearrangement of glycals, under the acidic conditions of the reaction, to the 2,3-unsaturated enoses (Ferrier reaction). The other problem is the incompatibility of some acid-labile protective groups with the relatively strong reaction conditions. These difficulties led to the development of different methodologies with the aim of minimizing the

OAc O OAc AcO Br OAc — Zn, HAcO → OAc O OAc AcO — ROH / H+ → OAc O OAc AcO OR + OAc O AcO OR

Scheme 3.

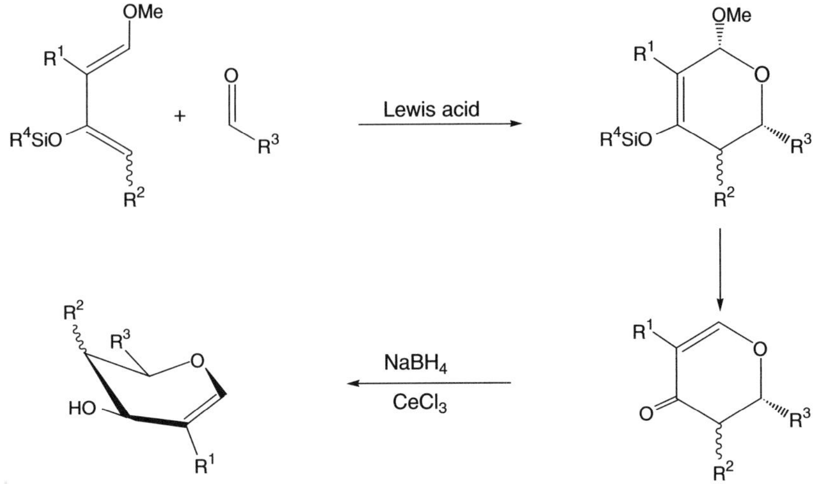

SCHEME 4. Synthesis of glycals by Lewis acid-catalyzed diene–aldehyde cyclocondensation reaction (β-R^2: gluco series; α-R^2: galacto series).

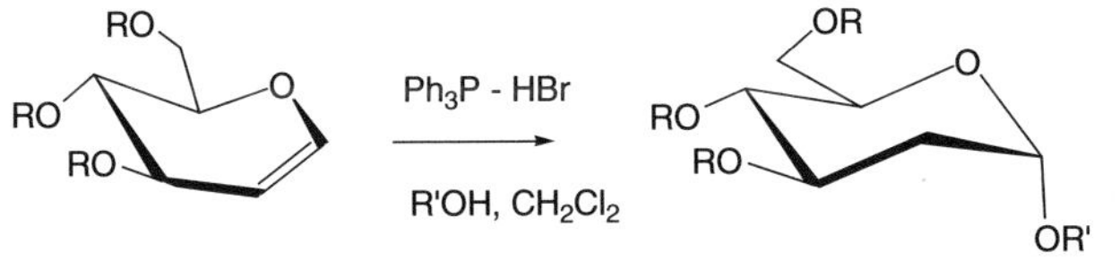

SCHEME 5.

secondary products formed under acidic conditions. For example, by using triphenylphosphine–hydrogen bromide (Ph_3P–HBr) in catalytic amount, formation of the Ferrier product was avoided and α-stereoselectivity was observed (Scheme 5). The method has been applied to galactal, glucal, L-fucal,[32] and it was shown that the stereochemistry of the addition was *cis*.[33]

A sulfonic acid resin (AG–50W X8) in the presence of lithium bromide was used under anhydrous conditions to promote the acidic addition of alcohols to glycals (Scheme 6). α-Stereoselectivity was observed for D-glucal, L-glucal, and L-rhamnal.[34]

Other mild reagents have been introduced, among them BCl_3, BBr_3,[35] $BF_3.Et_2O$,[36] a rhenium(V)-oxo complex,[37] and palladium complexes.[38] Hydroxymercuration with mercury(II) acetate in aqueous tetrahydrofuran, followed

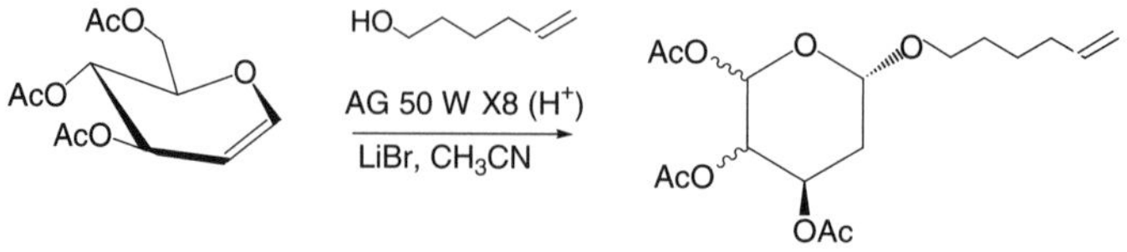

SCHEME 6.

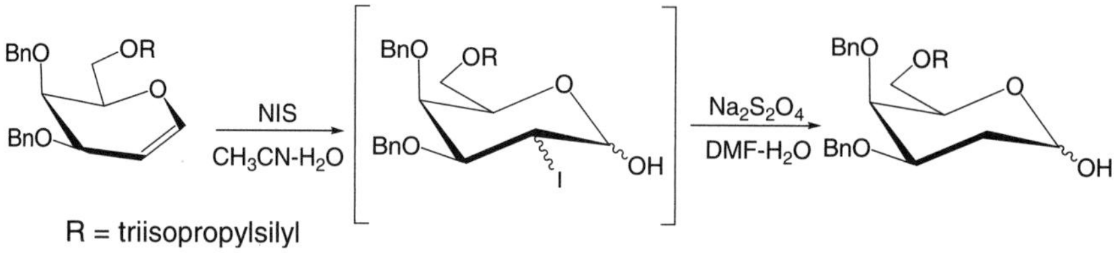

SCHEME 7.

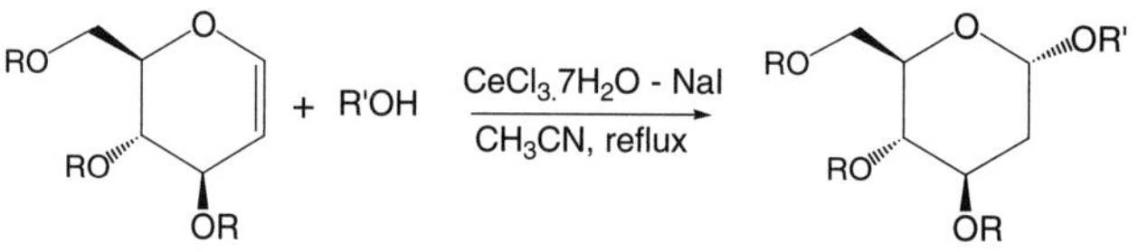

SCHEME 8.

by cleavage of the C-2–Hg bond by use of sodium borohydride,[39] was also described as a mild method for the preparation of 2-deoxy sugars, but it is not recommended for environmental considerations. A two-step, one-pot procedure was developed, using *N*-iodosuccinimide in aqueous acetonitrile, followed by reductive removal of the C-2 iodo group with $Na_2S_2O_4$ in aqueous DMF. This method is sufficiently mild to tolerate the presence of protecting groups, such as silyl or trityl ethers (Scheme 7).[40]

Lanthanide salts, acting as Lewis acids, have also been used to promote glycal addition reactions under mild conditions. Thus, $CeCl_3 \cdot 7H_2O$–NaI was used for the addition of several alcohols to such glycals as 3,4,6-tri-*O*-acetyl-D-glucal or -D-galactal, or 3,4-di-*O*-acetyl-D-xylal (Scheme 8).[41] The reaction was shown to be efficient in terms of yield and stereoselectivity, leading to the corresponding 2-deoxy-α-glycosides as a result of the thermodynamic anomeric effect. In the absence of NaI, the 2,3-unsaturated product was obtained. The reagent

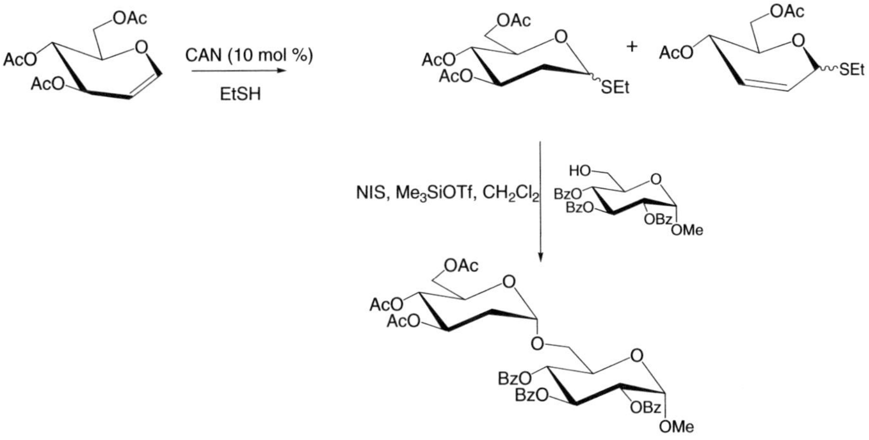

SCHEME 9.

$LaCl_3 \cdot 7H_2O$–NaI–benzyl alcohol was also reported for the same reaction, and it was used for the synthesis of 1,6-dideoxynojirimycin.[42]

From glycals, 2-deoxy-1-thioglycosides were obtained as mixtures of α- and β-anomers by using cerium ammonium nitrate (CAN) as catalyst, although the 2,3-unsaturated enose was also formed, especially from the glucal configuration (Scheme 9). The resulting 2-deoxy-1-thioglycosides were used as glycosyl donors, leading exclusively to the 2-deoxy-1-glycosides.[43]

(ii) Deoxygenation Procedures. Reductive removal of groups at C-2 has been accomplished,[44] among others, by the procedure of Barton and McCombie (see Section II.3).

A general method for the synthesis of 2-deoxyaldoses utilizes a reaction sequence involving the formation and subsequent reduction of ketene dithioacetal intermediates (Scheme 10). Reduction of ketene diethyl dithioacetal **12** with lithium aluminum hydride proceeds via the alkoxyaluminum hydride salt involving the 3-hydroxyl group. Several deoxy hexoses and pentoses were prepared by this method, and also their 2-deuterio analogues.[45]

(iii) Reduction of Glycosyl Halides with Radical Rearrangement. The reduction of acetylated or benzoylated halides or selenides with a low concentration of tributylstannane leads to 2-deoxy sugar derivatives (Scheme 11). The driving force of this radical reaction is the 1,2-*cis*-selective migration of an ester group because of the stabilization of the radical at C-2.[46]

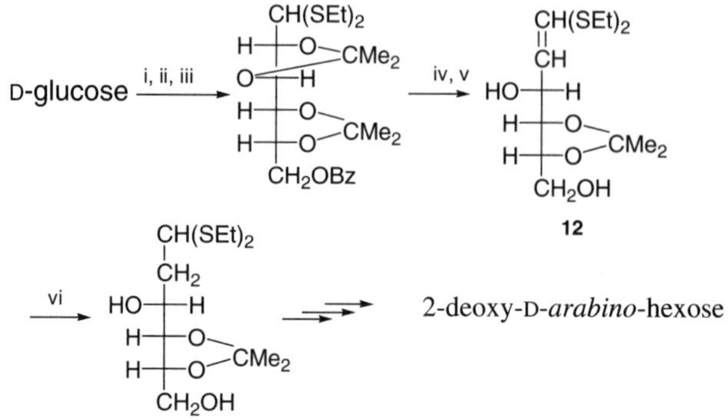

SCHEME 10. i. EtSH; ii. BzCl, Py; iii. Acetone, HCl; iv. NaOMe, MeOH; v. KOBu^t; vi. $LiAlH_4$.

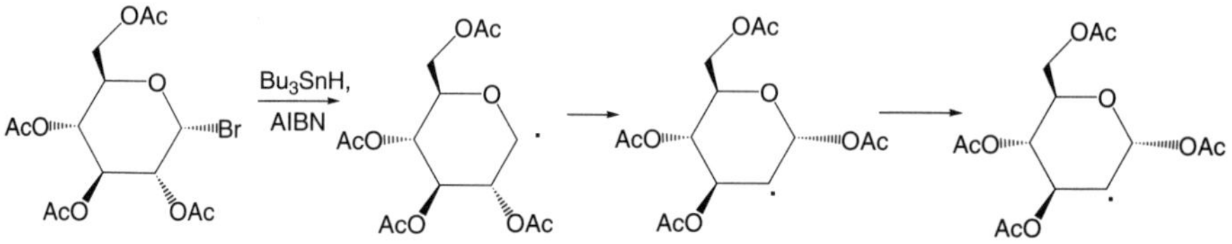

SCHEME 11.

A similar mechanism for the 2→1 migration of a phosphate group explained the formation of 2-deoxy-1-*O*-diphenylphosphoryl glycosides (Scheme 12).[47] Rearrangements that led to 2-deoxy derivatives from the glycosyl bromides by photochemical techniques have been reported.[48]

(iv) From Aldonolactones. Aldonolactones are important precursors of 2-deoxy sugars, as they offer the advantage of differential reactivity of the 2-OH group because of the inductive α-carbonyl effect. Thus, the 2-OH group can be selectively substituted and reduced by various methodologies. In our laboratory, D-galactono-1,4-lactone was chosen as the precursor of 2-deoxy-D-*lyxo*-hexofuranose (Scheme 13).[8] The α-deoxygenation step was performed by two different methods. A classical alternative involved reductive cleavage of a 2-*O*-tosyl derivative **13** in the presence of sodium iodide, to afford the 2-deoxy derivative **14**. The other method was based on deoxygenation via a photoinduced electron-transfer mechanism of the 3-(trifluoromethyl)benzoyl derivatives **15**.[49] This photochemical procedure was first used for deoxygenation of ribonucleosides[50] and for the

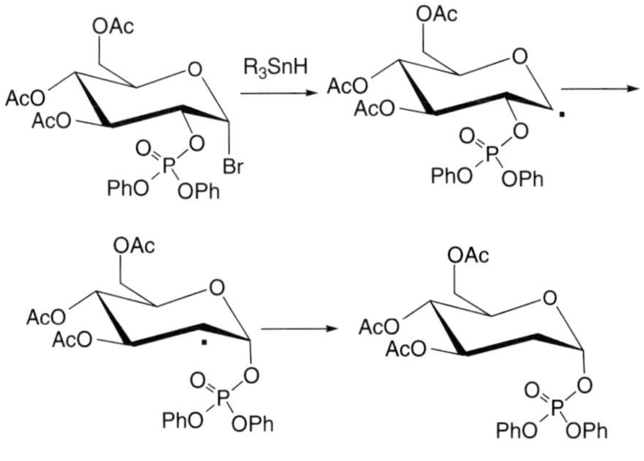

SCHEME 12.

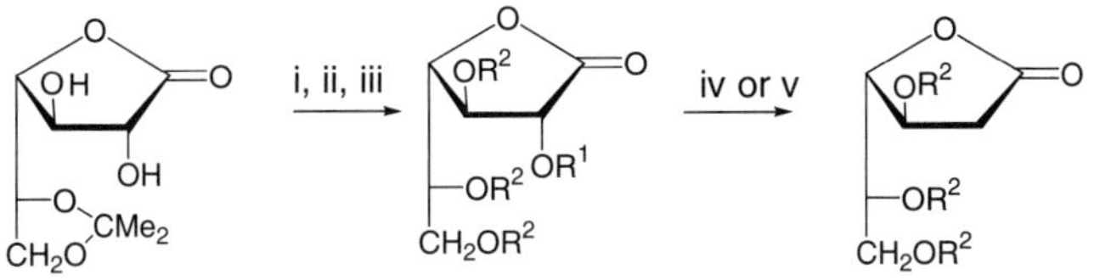

13 R^1 = Ts, R^2 = Bz or Ac **14**
15 R^1 = 3-CF_3Bz, R^2 = Bz or H

i, R^1Cl, Py; ii, AcOH-H_2O, 50-60 °C; iii, R^2Cl, Py; iv, NaI, TFA, acetone; v, 9-methylcarbazole, 9:1 ^{n}PrOH-H_2O, hν

SCHEME 13.

deoxygenation of a disaccharide analogue of moenomycin,[51] but stronger conditions were required. The carbonyl lactone allows not only the selective substitution of the 2-OH group, but also facilitates photochemical reduction at the vicinal position. The 2-deoxylactones obtained were selectively reduced to the corresponding sugars with diisoamylborane.[8]

(v) 2-Deoxy Glycosides and Oligosaccharides. Once deoxygenation at C-2 has been performed, a subsequent challenge for the synthesis of glycosides or oligosaccharides is the stereocontrolled construction of the glycosidic bond. The chemistry of 2-deoxy sugars is particularly significant, as the rate of hydrolysis of their glycosides is high compared to the hydroxylated analogues.

The absence of a substituent at C-2, usually able to influence the course of the glycosylation reaction, impedes control of the α:β ratio. Marzabadi and Franck have compiled extensive data on the relevant glycosylation methods for 2-deoxy sugars.[3]

The 2-deoxy-2-iodo derivatives have been directly used as glycosyl donors for the synthesis of glycosides and oligosaccharides, or transformed into more reactive donors, as discussed later. A glycosylation protocol has been developed involving the 2-iodo glycosyl derivatives as intermediates, using polymer-bound reagents.[52]

Several methods were reported for the selective synthesis of α-glycosides in the absence of a controlling group at C-2. The 2-deoxy sugar is first converted into a glycosyl donor by formation of phosphites,[53] phophonodithioates,[54] or phosphinothioates,[55] which react with different acceptors and promoters.

Selective β-glycosylation in the absence of a participating group at C-2 is still a major challenge. There are relatively few methods for synthesis of this important glycosidic linkage directly from 2-deoxyglycopyranosyl precursors. The reported methods lack generality and often do not proceed with high selectivity.[56–59] For example, in the case of 2-deoxy-D-*lyxo*-hexofuranose, glycosylation was not easy to achieve, and the $BF_3 \cdot$ MeOH-promoted glycosylation led to both methyl glycosides without stereoselectivity.[8]

The most extensively developed strategy for the synthesis of 2-deoxy-β-D-glycosides makes use of the temporary introduction of a group at C-2 (such as -Br,[59] -SR,[60] -SePh,[60–62] -OAc,[63]-NHCHO,[63] or 1,2-epoxy[64]), which is reductively removed after glycosylation. For example, the 2-deoxy-2-iodo-glucopyranosyl acetates **16**, obtained either from the corresponding glycal or from the iodo anhydro sugar **17**[65,66] led to several β-glucosides by the action of Me_3Si–OTf or Bu_3Si–OTf.[67] Application of this methods will be described under the 2,6-dideoxy sugars section (see Section III.1).

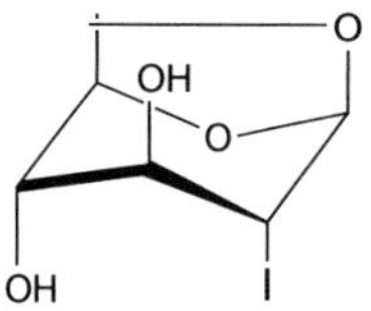

17

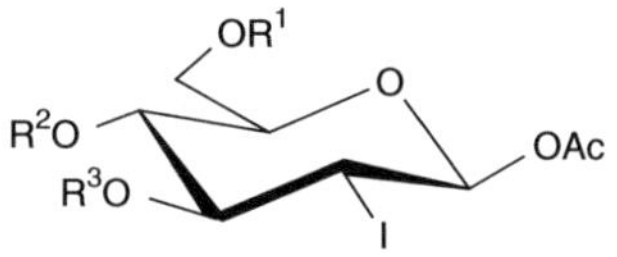

16 R^1=H, Br, OAc
R^2=Bu_3Si, Bn, Bz
R^3=Bu_3Si, Et_3Si, Bn

The 2-deoxy-2-X-glycopyranosyl acetates (X = I, Br, SePh, and SPh) were activated through the formation of the more active imidate derivatives **18**. These derivatives undergo highly stereoselective glycosylation reactions with a range of monosaccharide acceptors to give β-glycosides, precursors of 2-deoxyglycosides.[68–71]

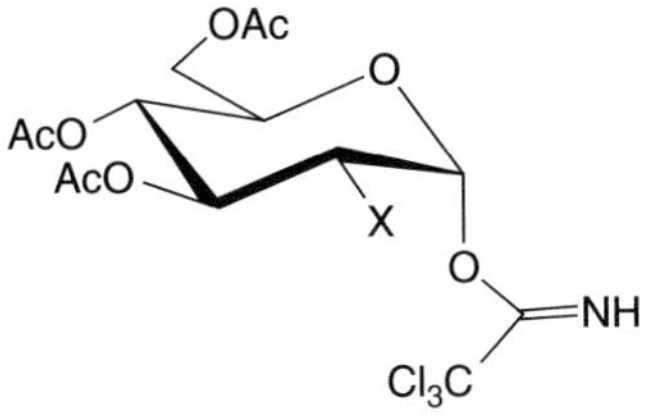

18 X = I
X = Br
X = SPh
X = SePh

Glycosidase enzymes can also be used to catalyze the addition of alcohols, or unprotected sugar derivatives, to glycals for the preparation of disaccharides with a nonreducing 2-deoxy sugar.[72] Using a β-glycosidase from *Sulfolobus solfataricus*, D-glucal as the donor, and methyl α-D-glucopyranoside as acceptor, a mixture of disaccharides **19**, and **20** was obtained, whereas phenyl 1-thio-D-glucopyranoside gave disaccharide **21**. This regioselectivity agrees with that observed for the transfer of glucose towards pyranosidic acceptors.[73]

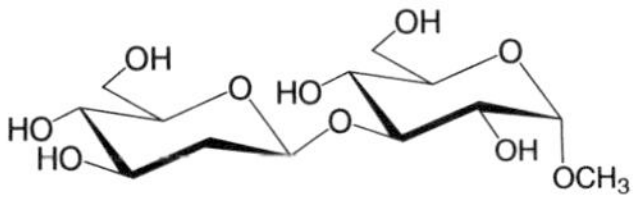

19 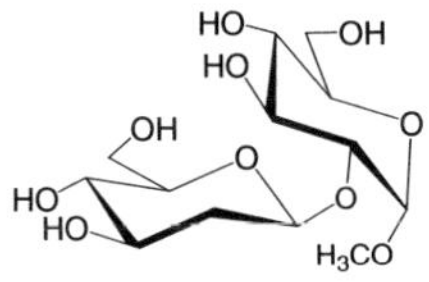 **20** 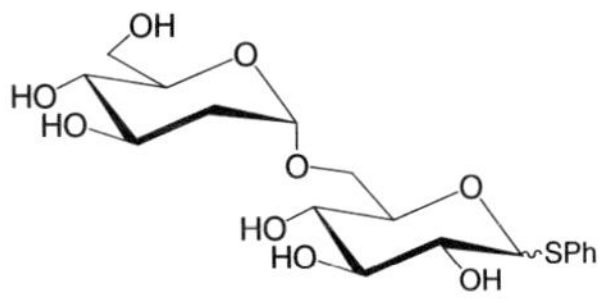 **21**

2. 3-Deoxy Sugars

a. Occurrence.—The only 3-deoxy sugar isolated from a natural source is 3-deoxy-D-*erythro*-pentose (cordycepose), a constituent of the antibiotic cordycepin (**22**), an adenine nucleoside isolated from *Cordyceps* species. A revision of the structure of cordycepin can be found in the previous chapter on deoxy sugars in this series.[1]

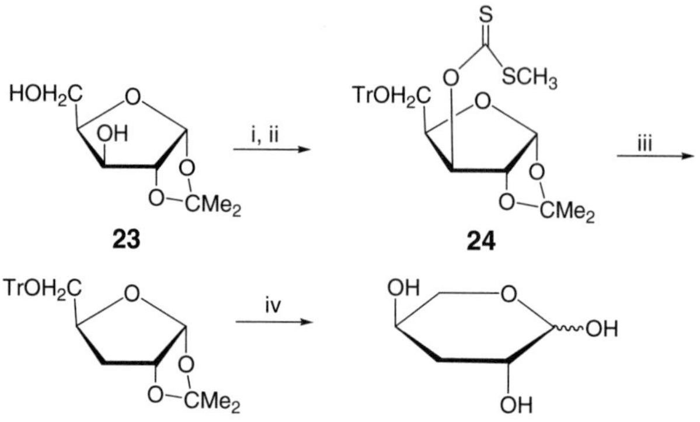

i, TrCl, Py, ii, CS_2, Me_2SO, NaOH; iii, Bu_3SnH; iv, HOAc

Scheme 14.

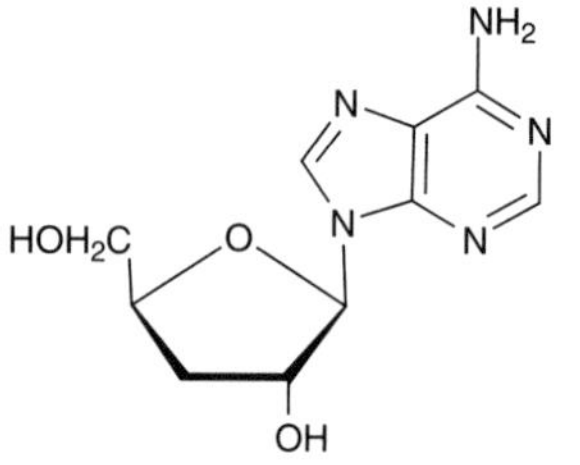

22

b. Synthesis.—General methods (see Section II) have been applied for the synthesis of 3-deoxy sugars and were described in Hanessian's article.[1] For example, the natural 3-deoxy-D-*erythro*-pentose was prepared from 1,2-*O*-isopropylidene-α-D-*xylo*-furanose (**23**) according to Scheme 14. Deoxygenation of C-3 was performed by reduction with tributyltin hydride of the thiocarbonyl derivative **24**.[74]

The 3-OH group of lactose or *N*-acetyl-lactosamine is the site of glycosylation in some biologically important glycoprotein and glycolipids.[75] For instance, L-fucose is linked to O-3 of *N*-acetyl-lactosamine in the Lewis^X trisaccharide determinant (**25**), and sialic acid is linked to O-3 of the galactose residue in sialyl

Lewis^x (SLe^X, **26**).[76] The enzyme trans-sialidase transfers sialic acid to O-3 of the galactose unit.[77] Thus, the 3′-deoxy derivatives of the disaccharides would be useful for studying biological recognition processes.

Lewis^X trisaccharide **(25)**

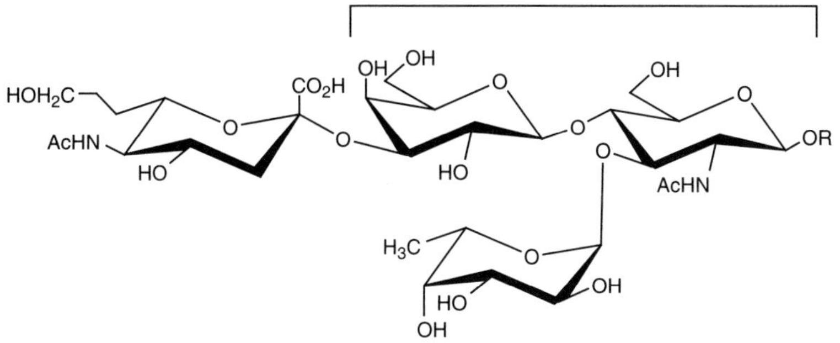

Sialyl Lewis^X **(26)**

For the preparation of a 3′-deoxy Le^X analogue, an acylated 3-deoxy-D-*xylo*-hexopyranose was used. The latter was prepared either from phenyl 1-thio-D-galactopyranoside[78] or from 1,2:5,6-di-*O*-isopropylidene-α-D-galactofuranose,[79] via the Barton and McCombie deoxygenation method.

A convenient approach for the preparation of 3-deoxy-D-*xylo*-hexopyranose (**27**) starts from 1,2:5,6-di-*O*-isopropylidene-α-D-glucofuranose and proceeds via the 3,4-unsaturated derivative **28** (Scheme 15).[80]

Facile syntheses of 3-deoxy sugars can be achieved by starting from the configurationally related aldonolactone. Thus, 3-deoxy-D-*arabino*-hexose was synthesized in four steps from the commercially available D-glucono-1,5-lactone (Scheme 16). The synthesis is based on the β-elimination reaction that takes place on benzoylation of the lactone in an excess of pyridine.[81] The enonolactone **29** obtained in 97% yield was selectively hydrogenated over palladium–charcoal to give the 3-deoxy-D-*arabino*-hexono-1,5-lactone. Diisoamylborane reduction of the lactone carbonyl group, followed by debenzoylation, afforded crystalline 3-deoxy-D-*arabino*-hexose (**30**) in 70% overall yield.[82]

The elimination of benzoylated 1,4-aldonolactones in pyridine is difficult to control, and further elimination reactions take place, decreasing the yield of monounsaturated derivative and consequently of the 3-deoxy sugar.[83]

A high yield of a 3-deoxy-1,4-lactone can be obtained by hydrogenation of a peracylated aldonolactone under pressure in the presence of triethylamine.[84] This

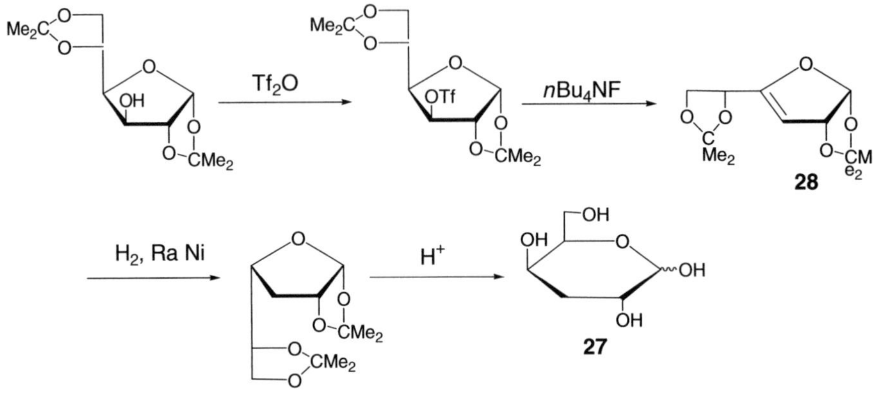

SCHEME 15.

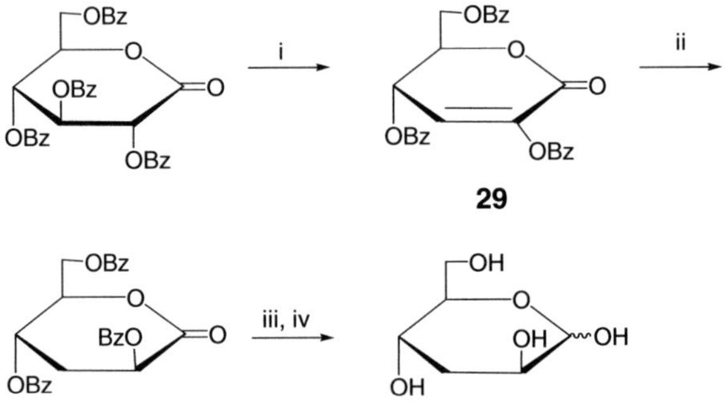

SCHEME 16. i. BzCl, py, 16 h, rt; ii. H_2, Pd/C; iii. Diisoamylborane, THF; vi. NaOMe–MeOH.

method was used for the preparation of crystalline 3-deoxy-D-*gluco*-heptose, starting from perbenzoylated D-*glycero*-D-*gulo*-heptono-1,4-lactone. The tautomeric equilibrium of 3-deoxy-D-glucoheptose studied by ^{13}C NMR spectroscopy was in close agreement with that observed for 3-deoxy-D-*ribo*-hexose, showing 20% of furanose forms.[85]

The benzoylated 3-deoxy sugars obtained from 1,4-lactones are useful for the preparation of 3-deoxy-glycofuranosides. In this way 4-nitrophenyl 3-deoxy-β-D-*xylo*-hexofuranoside was prepared from D-galactono-1,4-lactone (Scheme 17) for studies on the specificity of β-D-galactofuranosidase.[86]

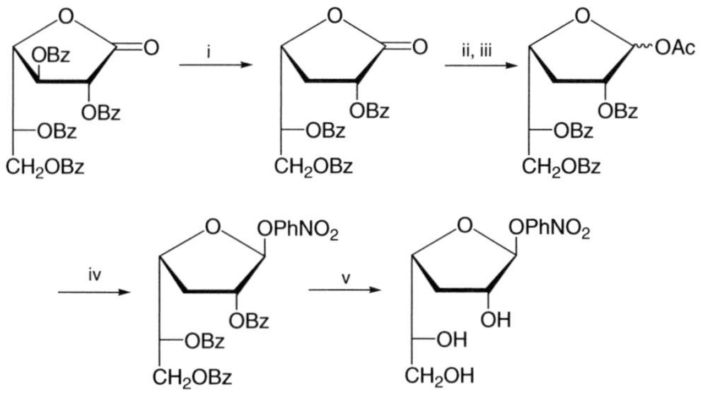

SCHEME 17. i. H_2, Pd/C, Et_3N; ii. Diisoamylborane, THF; iii. Ac_2O, Py; iv. 4-$NO_2C_6H_4OH$, $SnCl_4$, CH_2Cl_2; v. NaOMe–MeOH.

3. 4-Deoxy Sugars

a. Occurrence.—In the previous chapter in this series,[1] this class of sugars was described as not being found in Nature. More recently, 4-deoxy-D-*arabino*-hexose has been described in the O-specific polysaccharide of the lipopolysaccharide (LPS) of *Citrobacter braakii* PCM 1531.[87] The repeating unit (**31**) is a tetrasaccharide comprised of D-fucose, L-rhamnose, 4-deoxy-D-*arabino*-hexose, and *O*-acetyl groups in 2:1:1:1 molar ratio. In other *Citrobacter* strains[88,89] a homopolymer (**32**) has been characterized, and also a polysaccharide[90] having a branched trisaccharide repeating unit (**33**) in which the deoxy sugar is β-linked as a side chain to GlcNAc.

```
                        OAc
                         ↓
                         2
→3)-α-D-Fucp-(1→3)-β-L-Rhap-(1→3)-α-D-Fucp-(1→        31
                         4
                         ↑
                         1
                   α-D-Ara4dHexp

→2)-β-D-Ara4dHexp-(1→                                 32

→4)-α-D-GalpNAc-(1→6)-α-D-GlcpNAc-(1→                 33
                           3
                           ↑
                           1
                    β-D-Ara4dHexp
```

Although 4-deoxy-D-*arabino*-hexose is the immunodominant sugar in antigens **31** and **32**, these antigens are serologically unrelated, which can be accounted for by the different anomeric configuration of the 4-deoxy sugar.

b. Synthesis.—4-Deoxy hexoses have been prepared by general deoxygenation methods, starting from convenient derivatives having a free 4-OH group. These partially protected derivatives are readily obtained by reduction of 4,6-*O*-benzylidene-hexopyranosides with cyanoborohydride in an anhydrous acid medium. Benzoyl, benzyl, and *N*-acetyl protecting groups are not affected.[91]

Methyl 4-deoxy-D-*xylo*-hexopyranoside (**35**) was obtained via the 4-tosyl derivative **34**. Reaction with sodium iodide, followed by catalytic hydrogenation and debenzoylation, afforded **35** in ~50% yield.[92]

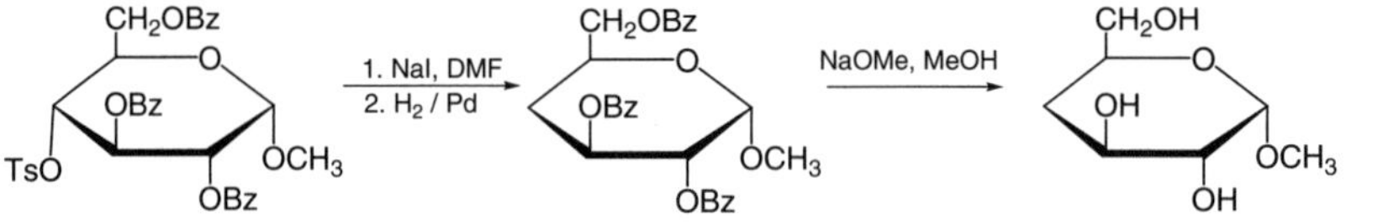

A triflate group undergoes ready reduction by sodium borohydride. The reaction has been applied to the preparation of methyl 2,3,6-tri-*O*-benzyl-4-deoxy-α-D-*xylo*-hexopyranoside. By using sodium borodeuteride, the 4-deoxy-α-D-(4- ^{2}H)-galactopyranoside was obtained.[93]

Deoxygenation of the partially substituted lactoside derivative **36**[94] was accomplished by the Barton reduction method to afford the 4′-deoxy-β-lactoside **37**.[6] The 4′-deoxy derivative was not a substrate for the β-D-galactosidase (E.C. 3.2.1.23) of *Escherichia coli.*

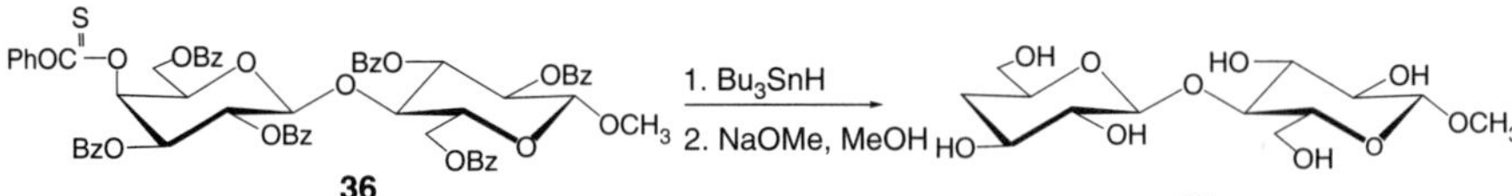

A synthetic route to 4-deoxy-D- and L-hexoses from noncarbohydrate starting materials has been reported.[95] Derivatives of 4-deoxy-D-*lyxo*-hexose, erroneously referred as 5-deoxy-mannose derivatives, have been synthesized by a hetero Diels–Alder reaction.[96]

4. 5-Deoxy Sugars

This class of carbohydrates has not been found in Nature. The absence of 5-OH group restricts the sugar to the furanose ring form. The first synthetic 5-deoxyhexose was the glucose analogue, 5-deoxy-D-*xylo*-hexose (**40**), and the different methods reported have been reviewed by Overend.[97] One of the obvious approaches was the reduction of compounds conveniently derivatized at C-5. For example, treatment of the 5-tolylsulfonyl derivative **38**, with lithium aluminum hydride gave compound **39**, which upon deprotection led to **40**.

The isomeric 5-deoxy-D-*ribo*-hexofuranoside (**42**) was obtained either by reduction of the tosylate **41**[98] or by configurational inversion of the mesylate derivative of **39**.[99]

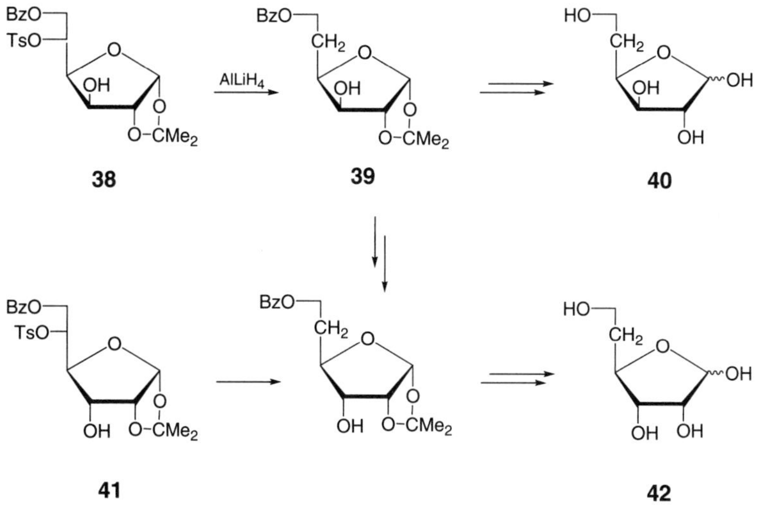

Another strategy for the synthesis of 5-deoxyhexoses involves the anti-Markovnikov hydration of a 5,6-alkene derivative, as first developed by Wolfrom.[97,100] Since that report, the same approach has been followed by several authors, and conditions for the key step have been improved. For example, starting from tosylate **43**, treatment with sodium iodide resulted in the alkene **44**. Addition of iodine trifluoracetate (produced *in situ* by the reaction of silver trifluoroacetate and iodine) to **44**, followed by hydrogenation over

Raney nickel catalyst in ethanol containing triethylamine, led to the 5-deoxy derivative **45**.

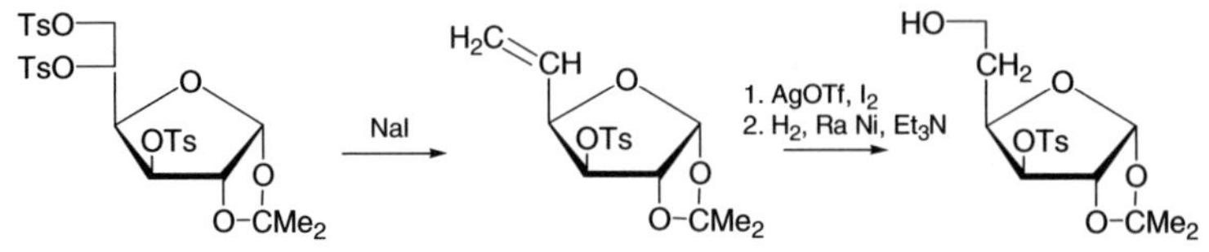

Photochemical desulfonylation, and acid treatment for removal of the isopropylidene group, afforded the free sugar **40**. Similarly, the 5-deoxygenated analogue from D-galactose, 5-deoxy-L-*arabino*-hexofuranose (**46**),[101] and 5-deoxy-D-*ribo*-hexofuranose (**42**) were prepared. By invertion of the configuration at C-2 in **42**, 5-deoxy-D-*arabino*-hexofuranose (**47**) was obtained. Nucleosides of the 5-deoxyhexoses have also been described.[102]

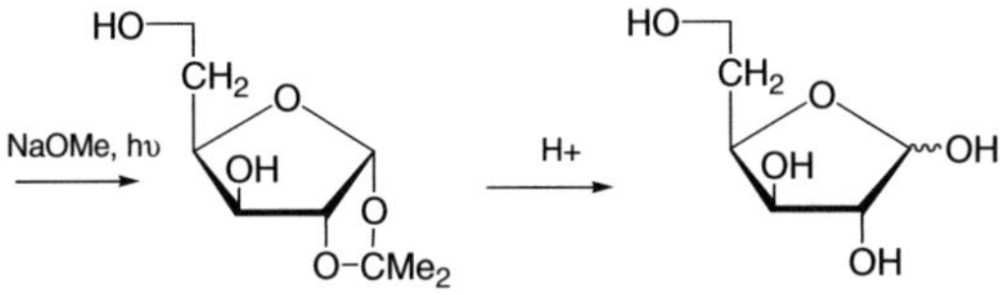

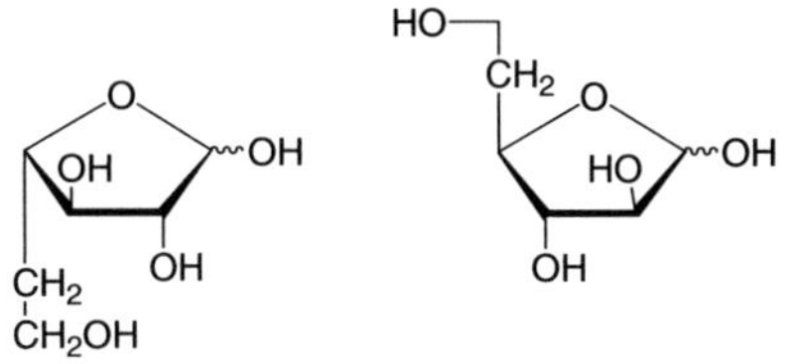

The 5,6-*O*-benzylidene derivative **48** led to **49** and **50** in 4.9:1 ratio by a ring-opening redox rearrangement of the benzylidene acetal catalyzed by tri-*tert*-butoxysilanethiol (Bu^tO_3SiSH).[103]

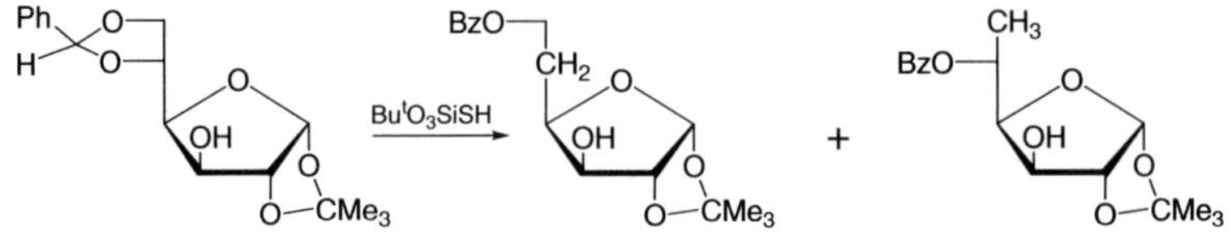

A microbial glucose isomerase (EC 5.3.1.5) that catalyzes the conversion of D-glucose into D-fructose was able to convert 5-deoxy-D-*xylo*-hexose (**40**) into the 2-hexulopyranose **51**.[104]

51

IV. Dideoxy Sugars

Deoxygenation at C-2, C-3, and C-4 of 6-deoxyhexoses and substitution of one of the remaining OH-groups by amino, di-*N*-methyl, *O*-methyl, and *C*-methyl groups creates significant structural diversity. In secondary metabolites, such as the polyketide antibiotics, 2,6-dideoxy- and 2,3,6-trideoxy-hexoses, contribute to the antibiotic and antitumor activity.

1. 2,6-Dideoxy Sugars

a. Structures and Occurrence.—All diastereomers of the 2,6-dideoxyhexoses are found in Nature in compounds with biological activities. 2,6-Dideoxy-3-*O*-methyl-hexopyranoses and 2,6-dideoxy-3-amino-hexopyranoses have been also described as components of natural products with medicinal properties.

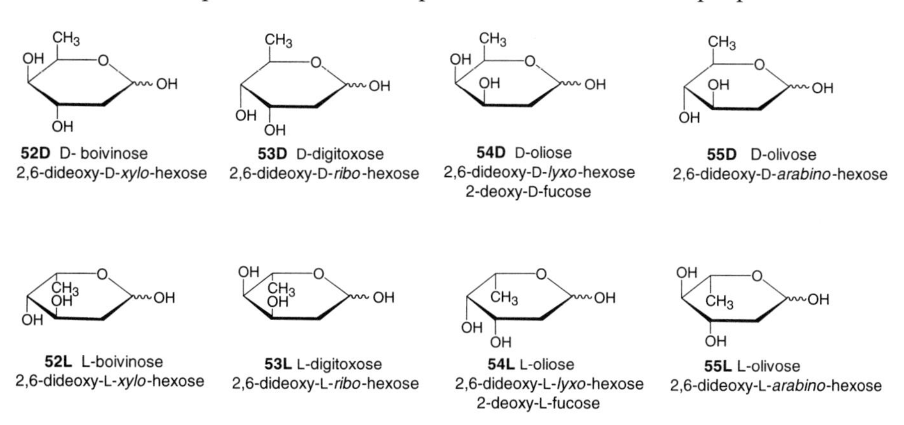

52D D- boivinose
2,6-dideoxy-D-*xylo*-hexose

53D D-digitoxose
2,6-dideoxy-D-*ribo*-hexose

54D D-oliose
2,6-dideoxy-D-*lyxo*-hexose
2-deoxy-D-fucose

55D D-olivose
2,6-dideoxy-D-*arabino*-hexose

52L L-boivinose
2,6-dideoxy-L-*xylo*-hexose

53L L-digitoxose
2,6-dideoxy-L-*ribo*-hexose

54L L-oliose
2,6-dideoxy-L-*lyxo*-hexose
2-deoxy-L-fucose

55L L-olivose
2,6-dideoxy-L-*arabino*-hexose

D-Boivinose (2,6-dideoxy-D-*xylo*-hexose, **42D**) was described as component of a cardenolide glycoside isolated from the seeds of *Corchorus olitorius* L. The structure was established as cannogenol 3-*O*-β-D-glucopyranosyl-*O*-β-D-boivinopyranoside.[105]

From the leaves of *Rhamnella inaequilatera* three apigenin *C,O*-bisglycosides, A (**56**), B (**57**), and C (**58**) were isolated, containing D-oliose, D-boivinose, and D-olivose, as the *C*-linked sugars, respectively.[106]

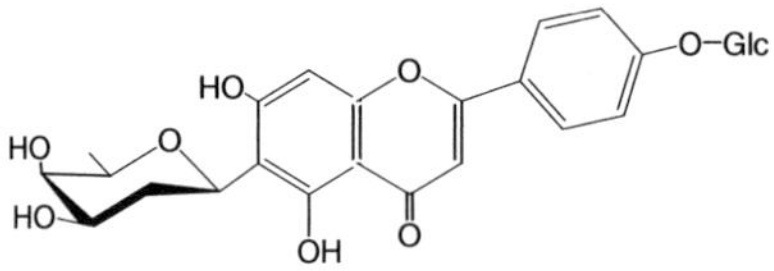

56

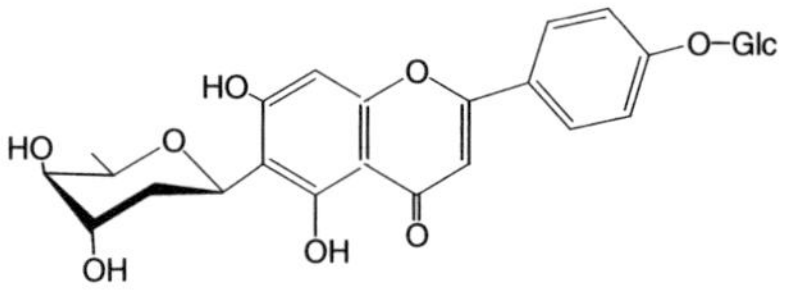

57

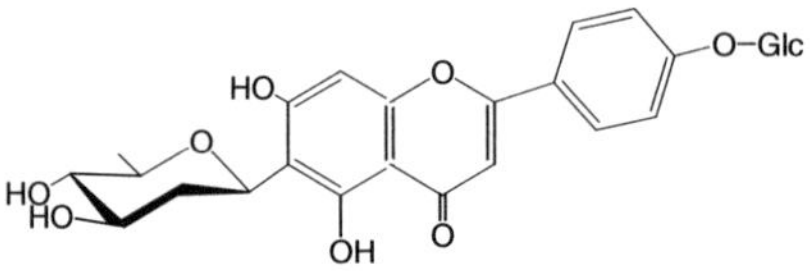

58

L-Boivinose (2,6-dideoxy-L-*xylo*-hexose, **52L**), the less abundant enantiomer of boivinose, was found as the flavone *C*-glycosyl compounds, 6-*C*-β-L-boivinopyranosyl chrysoeriol-7-β-D-glucopyranoside (**59**) and the 6-*C*-β-L-boivinopyranosyl glycoside **60** in corn (*Zea mays*).[107] L-Boivinoside **60** has been previously isolated from alligator weed (*Alternanthera philoxeroides*).[108]

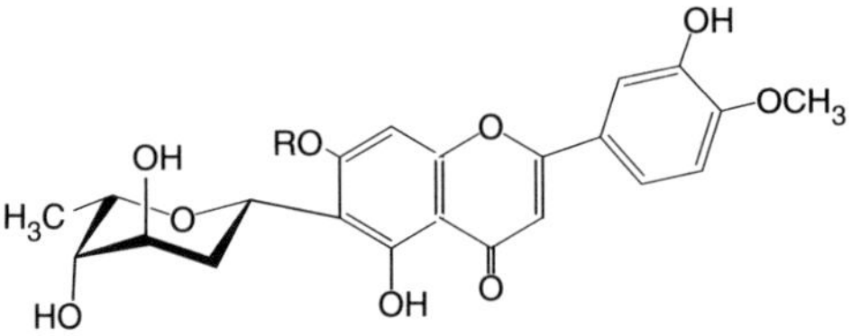

59 R = Glc
60 R = H

D-Digitoxose (2,6-dideoxy-D-*ribo*-hexose, **53D**) is well known as a constituent of plant cardiac and other steroidal glycosides.[109] Digoxin (**61**), extracted from the leaves of the foxglove *Digitalis lanata*, is composed of the aglycone digoxigenin linked to a trisaccharide of D-digitoxose. The conformational behavior of digoxin has been studied by NMR spectroscopy and molecular-dynamics calculations.[110]

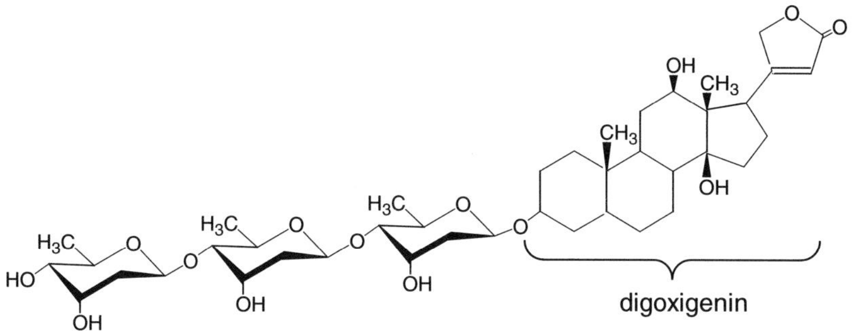

digoxigenin

61

Digoxin-like immunoreactive factors (DLIF) are present in serum and tissues of humans and animals.[111] They were identified as genin glycosides, with one or two D-digitoxose units. Subtle structural differences were found in all three important portions of the molecule, the sugar, the lactone ring, and the sterol nucleus.[112]

Several pregnan glycosides containing D-digitoxose in their oligosaccharide moieties have been isolated from plants of the *Asclepiadaceae* family. From *Telosma procumbens*, sweet-tasting glycosides have been characterized (Scheme 18). One of them, telosmoside A_{15}, was 1000 times sweeter than sucrose. D-Digitoxose, β-linked to the aglycone, is a constituent of the oligosaccharides along with cymarose (3-*O*-methyl-D-digitoxose, Cym), oleandrose (3-*O*-methyl-D-olivose, Ole), and thevetose (3-*O*-methyl-6-deoxy-D-glucose, The), as other 2,6-dideoxy sugars. 6-Deoxy-3-*O*-methyl-D-allose (Alm) was also detected in one of the telesmosides. The presence of a digitoxose unit attached directly to the aglycone was common to all the sweet glycosides.[113]

Steroidal glycosides containing 1–3 D-digitoxose residues in the oligosaccharides were isolated from other *Asclepias* spp.[114,115]

Eight glycosides, all containing 3-*O*-methyl-D-digitoxose (cymarose) β-linked to the steroid 8,14-*seco*-pregnane, were isolated from the plant *Cynanchum aphyllum*. Three of them also contained D-digitoxose linked to cymarose in the

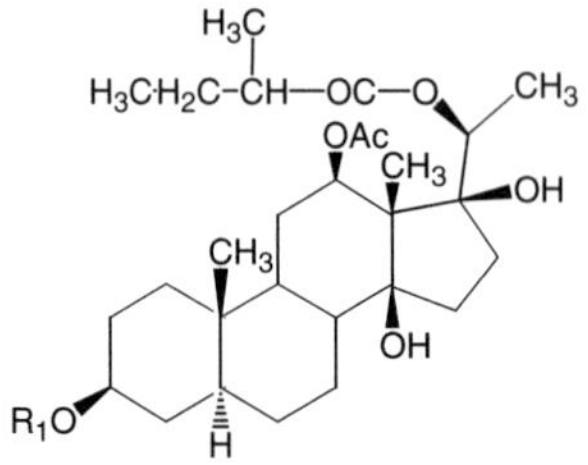

	R_1	Taste
A_1	-Cym^4-Ole^4-The	N
A_2	-Cym^4-Ole^4-The^4-Glc	B
A_3	-Cym^4-Ole^4-The^4-Glc^4-Glc	N
A_4	-Cym^4-Cym^4-Ole	N
A_5	- Cym^4-Cym^4-Ole^4-Glc^4-Glc	N
A_6	-Dig^4-Cym^4-Ole	N
A_7	-Dig^4-Cym^4-Ole^4-Ole	N
A_8	-Dig^4-Cym^4-Ole^4-Glc	S
A_9	-Dig^4-Ole^4-The^4-Glc	S
A_{10}	-Dig^4-Cym^4-Ole^4-Ole^4-Ole	S
A_{11}	-Dig^4-Cym^4-Ole^4-Ole^4-The	S
A_{12}	-Dig^4-Cym^4-Ole^4-Ole^4-Glc	S
A_{13}	-Dig^4-Dig^4-Ole^4-Ole^4-The	S
A_{14}	-Dig^4-Cym^4-Ole^4-Ole^4-Ole^4-Glc	S
A_{15}	-Dig^4-Cym^4-Ole^4-Ole^4-The^4-Glc	S
A_{16}	-Dig^4-Cym^4-Ole^4-Ole^4-Glc^4-Glc	S
A_{17}	-Dig^4-Cym^4-Ole^4-Ole^4-Alm^4-Glc	S
A_{18}	-Dig^4-Cym^4-Ole^4-The^4-Glc^4-Glc	S

SCHEME 18. S: sweet, N: tasteless, B: bitter.

oligosaccharide.[116] The methylated sugars cymarose, D-oleandrose, and D-thevetose were identified in the glycosides of *Asclepias tuberosa*.[117] From *Sarcostemma* species, pregnane glycosides were isolated. Tri- or tetra-saccharides containing D-digitoxose and D- and L-cymarose were described as comprising the glycone moiety.[118]

L-Digitoxose (2,6-dideoxy-L-*ribo*-hexose, **53L**) has been only described in products from actinomycetes. It is present in a unique family of antibiotics, the jadomycins. Glycosylation with L-digitoxose distinguishes jadomycin B (**62**) from jadomycin A. Jadomycin B exists in a dynamic equilibrium of the 3a*S* and the 3a*R* diastereomers.[119,120] The dideoxy sugar is considered important for the bioactivity, as only **62** displayed anti-yeast activity.[121] They are also active against Gram-positive and Gram-negative bacteria. Jadomycin B is the main product of fermentation of *Streptomyces venezuelae* ISP 5230 under stress conditions.[122]

The genes for the construction of L-digitoxose from D-glucose 1-phosphate and its transfer to the angucycline aglycone have been identified.[121]

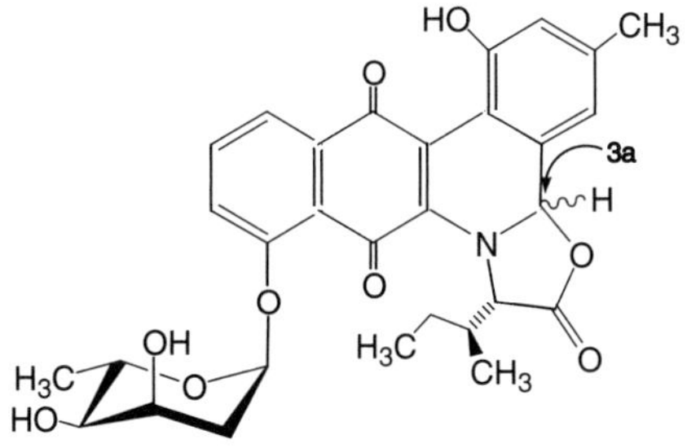

62

L-Digitoxose (**53L**) and its 4-*O*-acetyl derivative are present in a tetrasaccharide glycosidically linked in tetrocarcin A, an antibiotic inhibitor of the anti-poptotic Bcl-2 protein family.[123] Other tetrocarcins have been isolated from the culture of *Micromonospora chalcea* KY 11091.[124] From another strain of *Micromonospora*, analogues of tetrocarcin A, called arisostatins A and B, containing isobutanoyldigitoxose instead of acetyldigitoxose were isolated.[125]

D-Oliose (2,6-dideoxy-D-*lyxo*-hexose or 2-deoxy-D-fucose, **54D**) is present as the 6-*C*-β-D-olioside apigenin (**63**) in torosaflavone A, a constituent of the leaves of *Cassia torosa* Cav I.[126]

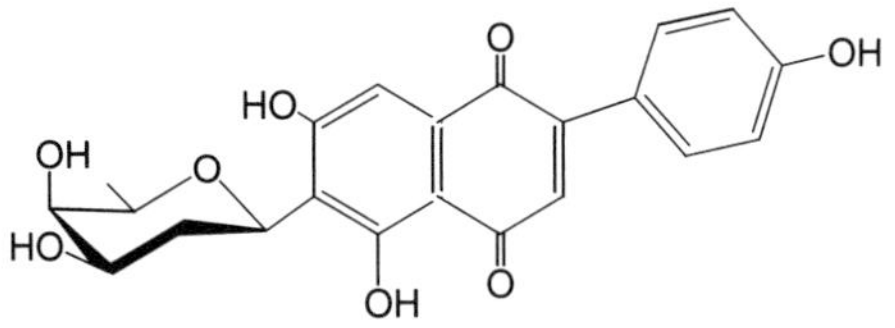

63

D-Oliose, 4-*O*-acetyl-D-oliose, 4-*O*-methyl-D-oliose, together with D-olivose (2,6-dideoxy-D-*arabino*-hexose, **55D**) are components of the oligosaccharides in the aureolic acid family of antitumor drugs[127–129] which comprise mithramycin, UCH 9[130] with a tricyclic polyketide as aglycone, and chromocyclomycin with a tetracyclic polyketide. The more-recently isolated durhamycin A is a potent inhibitor of HIV-1 Tat transactivation.[131] All structures are glycosylated at two opposite sides of the aglycon by saccharides of different lengths, all of them containing 2,6-dideoxy hexoses, D-olivose being the main sugar (Scheme 19). They are produced by different *Streptomyces* strains. The members of this family inhibit growth and

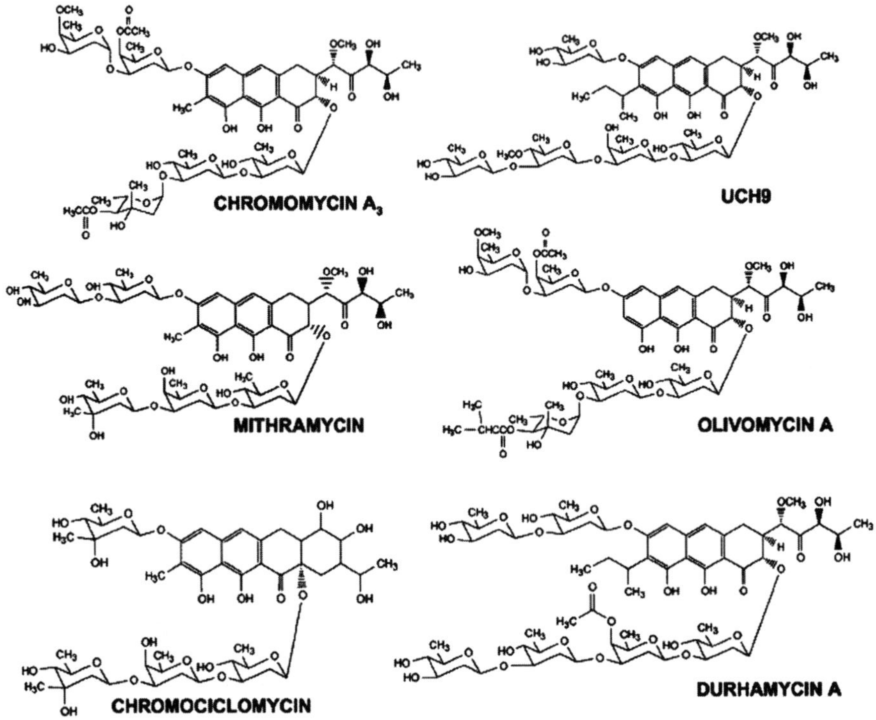

SCHEME 19. Structure of members of the aureolic acids family of antitumor drugs.

multiplication of several tumor-cell lines. Mithramycin is clinically used for the treatment of some testicular cancers and other diseases.[132] The carbohydrate moieties are major structural contributors to the biological activities. In particular, the acetyl groups in the saccharides of chromomycin, with a similar structure (Scheme 19), contribute with additional H-bonding to DNA complex formation.[133,134] The glycosylation sequence of the mithramycin biosynthesis has been studied.[135]

Three mithramycins having antitumor properties, but differing in the aglycone, have been produced by *Streptomyces argillaceus*, through combinatorial biosynthesis.[136] The biosynthetic gene cluster for chromomycin A_3 has been characterized, and the genes involved in deoxy sugar biosynthesis and transfer were identified. The organization of the cluster differs from that of the closely related mithramycin.[137]

L-Oliose (2,6-dideoxy-L-*lyxo*-hexose, **54L**) is the internal unit in the trisaccharide chain glycosidically linked to the anthracycline Aclacinomycin

A (aclarubicin, **64**).[138] The sugar moiety linked to the 7-OH group of the aglycone greatly affects the biological activity. This anthracycline drug is produced by *Streptomyces galilaeus* and is used in Japan and France for the treatment of acute leukemias and non-Hodgkin's lymphomas.[139,140] Crystallographic analysis of anthracycline DNA complexes indicate that the oligosaccharide linked to C-7 is critical for the interaction. Therefore, new anthracycline like antitumor compounds have been synthesized, by changing the sugar moiety.[141,142] The L-oliose disaccharide-containing MEN 10755 (**65**) is in clinical trials.[143] The anthracycline glycosyltransferase *AknK* catalyzes the addition of L-oliose, using dTDP-L-oliose and rhodosaminyl aklavinone. Interestingly, *AknK* also catalyzes the addition of a second L-oliose to the natural disaccharide chain to produce a variant of the natural aclacinomycin.

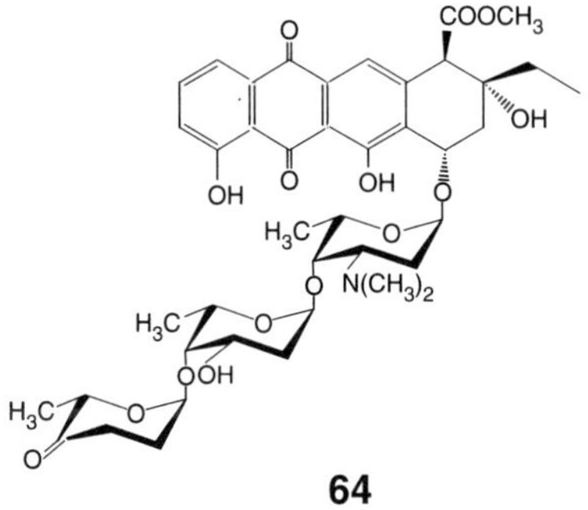

64 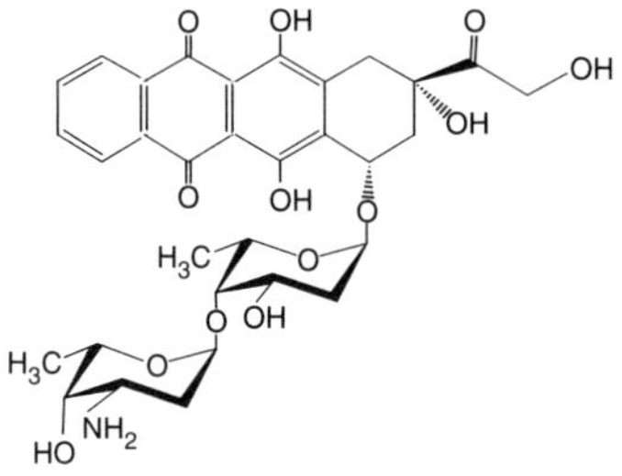 **65**

D-Olivose (2,6-dideoxy-D-*arabino*-hexose, **55D**) is part of the glycosidically linked hexasaccharide in Landomycin A (**66**),[144] a member of the angucycline antibiotic family.[145]

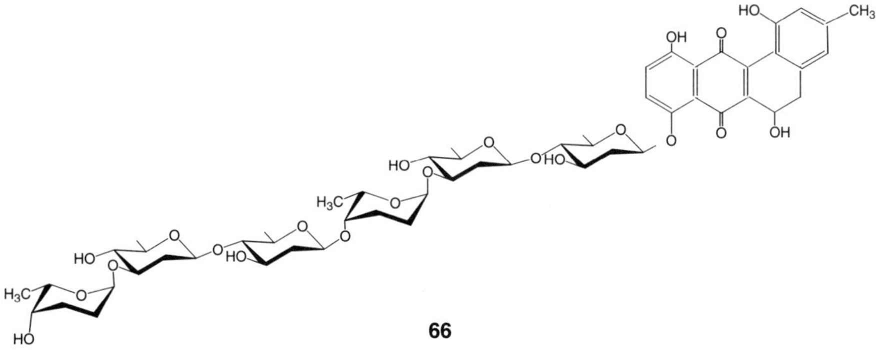

66

D-Olivose is C-glycosylically linked in aquayamycin (**67**)[146] which is an intermediate in several biosynthetic pathways.[147]

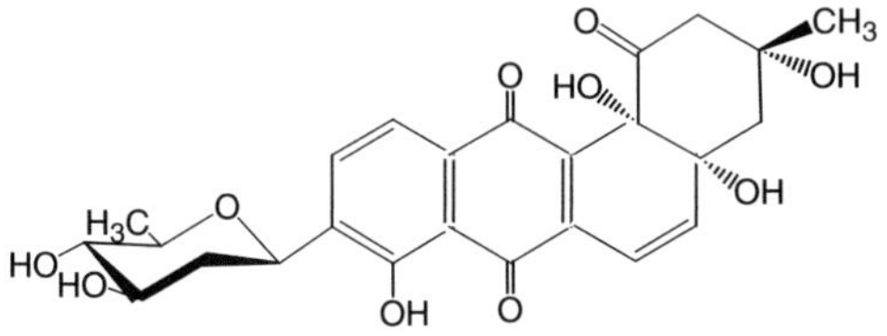

67 Aquayamycin

The derivatives 3-*O*-methyl-D-olivose (D-oleandrose, 3-*O*-methyl-2,6-dideoxy-D-*arabino*-hexose), and 3-*O*-methyl-D-oliose (3-*O*-methyl-2,6-dideoxy-D-*lyxo*-hexose) are components of steroidal glycosides obtained from *Cynanchum othophyllum* Schneid, a traditional Chinese medicine.[148] A derivative 4-*O*-methyl-D-olivose has been identified as the terminal sugar of the tetrasaccharide moiety in the phenolic glycolipid of *Mycobacterium kansasii.*[149]

L-Olivose (2,6-dideoxy-L-*arabino*-hexose, **55L**) is not common in Nature. The 3-*O*-methyl derivative, L-oleandrose has been found in oleandomycin, a macrolide antibiotic produced by *Streptomyces antibioticus*. Interestingly, it was proved that methylation occurs in the L-oliviosyl glycoside once L-olivose is transferred to the aglycone.[150]

D-Oleandrose and L-olivomycose (3-*C*-methyl-L-olivose) together with 6-deoxy-4-*O*-methyl-L-glucose are constituents of the oligosaccharide in apoptolidin, isolated from *Nocardiopsis* sp., which was the subject of intense synthetic and biological investigation because of its potential for the treatment of cancer.[151–155]

The genes involved in the biosynthesis of 2,6-dideoxy sugars have been the object of several studies.[156,157]

b. Synthesis.—Total enantioselective syntheses of 2,6-dideoxyhexoses from noncarbohydrate starting materials have been described. Thus, D-olivose (**55D**), D-digitoxose (**53D**), D-oliose (**54D**), D-cymarose (3-*O*-methyl-D-digitoxose), and racemic boivinose (**52**) have been obtained in four to seven steps and moderate yields.[158,159] Stereospecific synthesis of 2,6-dideoxy sugars of the L-series was achieved by addition of diallylzinc to α,β-dialkoxy chiral carbonyl compounds.[160] All 2,6-dideoxyhexoses have been synthesized from benzylidene-protected dihydroxyacetone (1,3-dihydroxy-2-propanone), and enantioselectivity was achieved with convenient chiral reagents.[161]

Starting from simple precursors via stereodefined alkenic esters, and ring-closing metathesis using Grubbs' catalyst produced β,γ-unsaturated lactones, which could be converted into 2,6-dideoxyhexopyranoses.[162]

D-Boivinose (**52D**) has been obtained by a tellurium method and sharpless asymmetric epoxidation.[163]

The preparation of D-digitoxose (**53D**) from methyl 4,6-*O*-benzylidene-2-deoxy-α-D-*ribo*-hexopyranoside (**68**) in an overall yield of 58% was described in detail (Scheme 20).[164] From the same precursor, via methyl 3,4-di-*O*-benzoyl-6-bromo-2,6-dideoxy-α-D-*ribo*-hexopyranoside, L-oliose (**54L**) was prepared in good yield.[165]

L-Digitoxose (**53L**) and L-cymarose (2,6-dideoxy-3-*O*-methyl-L-*ribo*-hexopyranose were prepared, starting from methyl 2,3-*O*-benzylidene-α-L-rhamnopyranoside, according to Scheme 21.[166]

L-Digitoxose (**53L**) and L-oliose (**54L**) have been obtained as methyl glycosides starting from methyl 2,3,6-trideoxy-α-L-*erythro*-hex-2-enopyranoside (**69**) and its α-L-*threo* analogue **70**, respectively (Scheme 22).[167]

D-Oliose (**54D**) was prepared from D-galactose in eight steps, with 25% overall yield (Scheme 23).[168]

A derivative of L-oliose (**54L**) suitable for glycosylation was obtained from L-fucose[169] using the Barton deoxygenation method for introduction of the 2-deoxy function. Another approach for the synthesis of L-oleosyl synthons started from the 3,4-di-*O*-acetyl L-fucal (**71**), readily converted into the corresponding chloride (Scheme 24).[138] This method was used for the synthesis of dTDP-β-L-oliose, a glycosyl donor in the construction of aclacinomycin A (**72**).

Scheme 20.

BuLi, THF

NaBH$_4$

1. NaH, CH$_3$I
2. HAcO

L-cymarose

HAcO

L-digitoxose (**53L**)

MEM: 2-methoxyethoxymethyl

SCHEME 21.

69

I(collidine)$_2$ClO$_4$

Ba(OH)$_2$

53L

diethylazodicarboxylate
Ph$_3$P, PhCOOH

70

I(collidine)$_2$ClO$_4$

Ba(OH)$_2$

54L

SCHEME 22.

D-galactose

1. BnOH, HCl
2. 2,-dimethoxypropane, p-TSA

PhSSPh, *n*Bu$_3$P

Ra-Ni, EtOH

NaH, CS$_2$, CH$_3$I, imidazole

H$_3$PO$_2$, Et$_3$N, AIBN

1. Ca, NH$_3$(l)
2. Dowex 50X (H$^+$)

SCHEME 23.

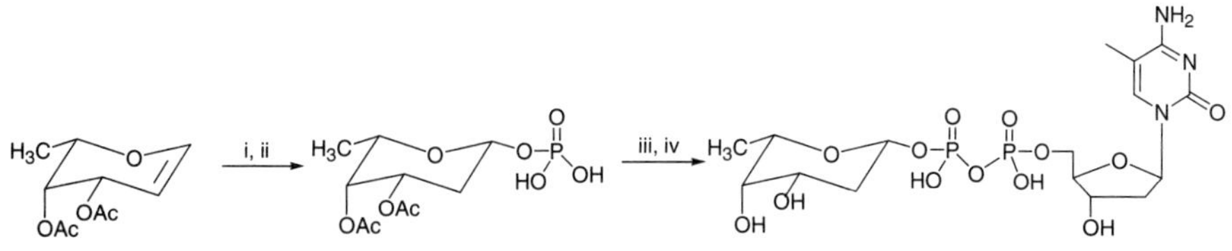

71

SCHEME 24. i. HCl (g), benzene; ii. $Bu_4NH_2PO_4$, ethyldiisopropylamine, CH_2Cl_2; iii. ribosylthymine monophosphate morpholidate, tetrazole, Py; iv. Et_3N, MeOH, H_2O.

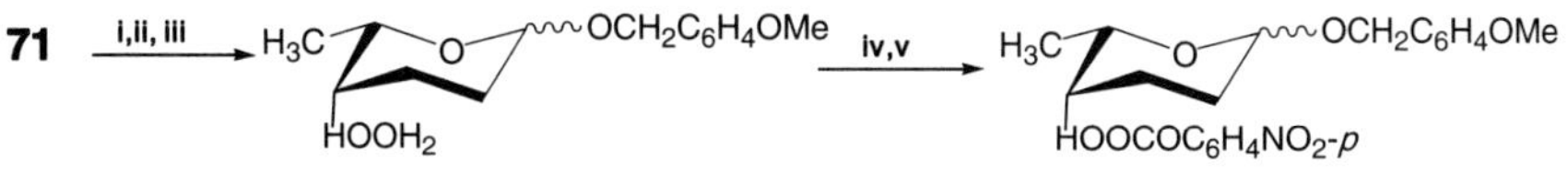

73

SCHEME 25. i. $(p\text{-MeO})C_6H_4CH_2OH$, NIS; ii. Bu_3SnH, AIBN; iii. NaOMe–MeOH; iv. Bu_2SnO; v. $(p\text{-NO}_2)BzCl$, Et_3N.

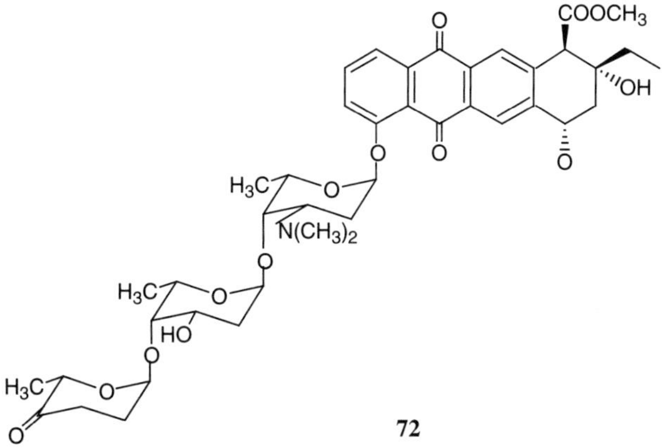

72

The fucal derivative **71** was also used for the synthesis of an L-oleosyl derivative (73) with a free 4-OH group, suitable for (1→4) glycosylation (Scheme 25).[141]

L-Olivose (**55L**) could be readily obtained from L-rhamnose via 3,4-di-*O*-acetyl-L-rhamnal[170] or from L-rhamnono-1,4-lactone via the 2-bromo-2-deoxy-L-rhamnono-1,4-lactone (Scheme 26). Starting from the enantiomeric D-rhamnono-1,4-lactone, D-olivose (**55D**) was obtained.[171,172]

Octyl and dodecyl glycosides (**75,76**) of L-olivose were prepared by the addition of octanol or dodecanol to the double bond of the deoxyglycal **74**. The antibacterial

SCHEME 26.

and antifungal activities of these surface-active glycosides were evaluated. The dodecyl glycoside alone was quite active against some *Bacillus* species.[173]

74

75 $n = 7$
76 $n = 11$

The chemoenzymatic synthesis of dTDP-β-L-olivose and dTDP-α-L-olivose, donor substrates for the biosynthesis of polyketides and other drugs, has been described. Starting from 2-deoxy-D-*arabino*-hexose 6-phosphate, dTDP D-oliose was also synthesized.[174]

Some examples of the synthesis of oligosaccharides of natural products containing 2,6-dideoxy sugars have been recorded. The synthesis of the aureolic acids and anthracycline oligosaccharides containing 2,6-dideoxyhexoses and derivatives thereof has been reviewed.[175] Subsequently, Thiem and Köpper published the sequential synthesis of the tetrasaccharide component of kijanimicin, consisting entirely of L-digitoxose residues.[176] The α-linked trisaccharide moiety, uniformly composed of L-digitoxose (**53L**) residues, was prepared in a one-pot procedure.[177] The *N*-iodosuccinimide procedure was used for self-condensation of the benzylated glycal **77**, and a final glycosylation with benzyl alcohol was included. The iodo trisaccharide **78** was obtained in 30% yield from the monomers, and hydrogenation under basic conditions afforded the substituted trisaccharide **79** (Scheme 27).

The iodinative coupling of glycals was also used by Danishefsky and coworkers for synthesis of the trisaccharide component **84** of ciclamycin 0 and its attachment to the aglycone. The L-fucal derivative **80** was used as precursor of the L-oliose units and I(*sym*-collidine)$_2$ClO$_4$ as the reagent for coupling with the 3,6-dideoxy-L-glycal derivative **81**. The synthesis of the trisaccharide glycal **84** for coupling to the aglycon was achieved according to Scheme 28.[178]

Another approach was used by Kahne *et al.* for synthesis of the L-oliose-containing trisaccharide of ciclamycin 0. The sulfoxide glycosylation reaction

OH O
H_3C
BnO
OH
H_3C O
BnO
X
H_3C O
BnO
O
NIS, BnOH
77
X
H_3C O
BnO
OBn
X
H_2, Pd, Et_3N **78** X = I (30%)
79 X = H

SCHEME 27.

H_3C O
OH OBz **80**
H_3C O
BnO
81
1. I(*sym*-collidine)$_2$ClO$_4$
2. Ph_3SnH
3. $LiAlH_4$
4. $tBuMe_2SiCl$
H_3C O
OSiBu^{t}Me$_2$
O
H_3C O
BnO
H
82
H_3C O
OH OBz **80**
I(*sym*-collidine)$_2$ClO$_4$
H_3C O
OBz
O
H_3C O
OSiBu^{t}Me$_2$
O
H_3C O
BnO
H
83
H_3C O
OSiMe$_3$
O
H_3C O
OSiMe$_3$
O
H_3C O
O
H
84

SCHEME 28.

was employed in a polymerization reaction from the three L-oliose derivatives **85, 86,** and **87**.[179]

SPh
H_3C O
HO OBn
85
+
O
SC$_6$H$_4$OMe-*p*
H_3C O
Me$_3$SiO OBn
86
+
O
SPh
H_3C O
O
87
TfOH
SPh
H_3C O
OBn
O
H_3C O
OBn
O
H_3C O
O
88

However, the trisaccharide sulfoxide obtained from **88** was not convenient for glycosylation of the aglycone to obtain ciclamycin 0. In a later paper the

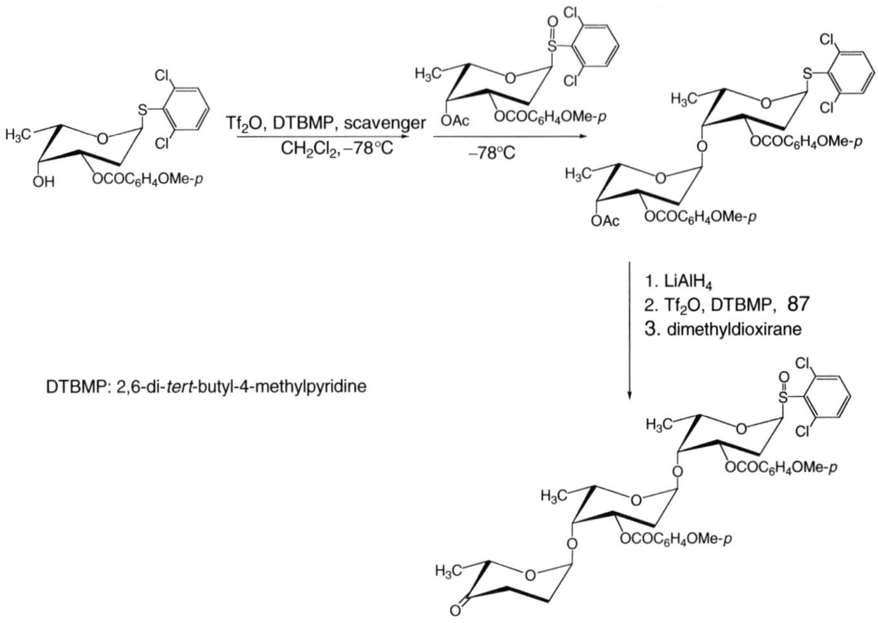
Tf2O, DTBMP, scavenger
CH2Cl2, −78°C
−78°C
1. LiAlH4
2. Tf2O, DTBMP, 87
3. dimethyldioxirane
DTBMP: 2,6-di-tert-butyl-4-methylpyridine
OH
OAc
OCOC6H4OMe-p

89

SCHEME 29.

protecting groups of the L-oliose monomers were modified for the preparation of the trisaccharide **89**, used successfully for coupling to the aglycone (Scheme 29).[180]

An L-oliose trisaccharide glycal (**90** and its acetate **91**) suitable for further condensation was synthesized via iterative alkynol cycloisomerization and acid-catalyzed glycosylation.[181] Glycosylation of oligosaccharides bearing the glycal of L-fucose as the reducing unit favors formation of the axial (α) anomeric 2,6-dideoxyglycosides.

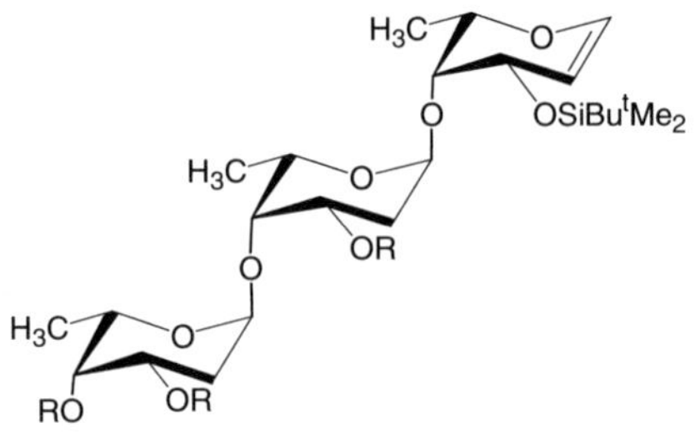
OSiBu^{t}Me2
OR
RO

90 R = H
91 R = Ac

A highly stereoselective synthesis of a functionalized precursor (**92**) of the tetrasaccharide unit in durhamycin A (Scheme 19) was described.[182]

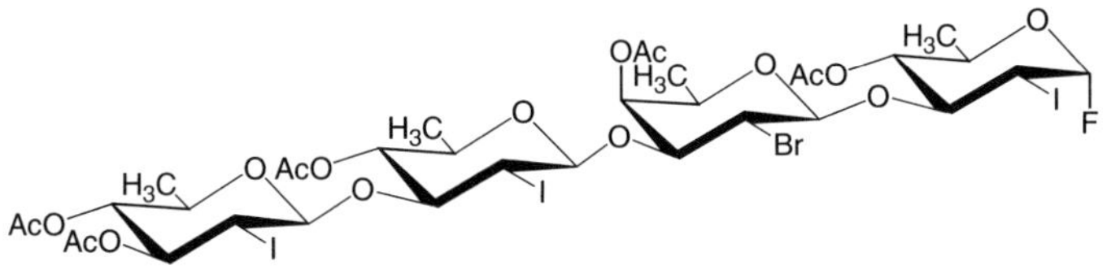

92

2. 3,6-Dideoxy Sugars

The 3,6-dideoxyhexose class of sugars was identified and extensively characterized in lipopolysaccharides (LPS) of Gram-negative bacteria.[183,184] These sugars occur as terminal nonreducing units and are the immunodominant monosaccharides in the O-antigens. There are numerous examples describing the immunogenicity of *Salmonella* oligosaccharide protein conjugates.[185–188] Antibodies induced by the vaccine conjugate were largely directed against the particular 3,6-dideoxy hexose.[188]

The 3,6-dideoxy sugars are readily released by mild hydrolysis and are resistant to periodate oxidation when glycosidically linked. They are known by the common names abequose (**93**, 3,6-dideoxy-D-*xylo*-hexose), ascarylose (**94**, 3,6-dideoxy-L-*arabino*-hexose), colitose (**95**, 3,6-dideoxy-L-*xylo*-hexose), and tyvelose (**96**, 3,6-dideoxy-D-*arabino*-hexose). These sugars are generally found as terminal α-pyranosyl groups. On the other hand, the other natural 3,6-dideoxy sugar, paratose (**97**, 3,6-dideoxy-D-*ribo*-hexose), exists as a terminal β-furanosyl (**97a**) group in the LPS from *Yersinia pseudotuberculosis* type 1B, and as terminal β-pyranosyl groups (**97b**) in the LPS from type III of the same species.[183]

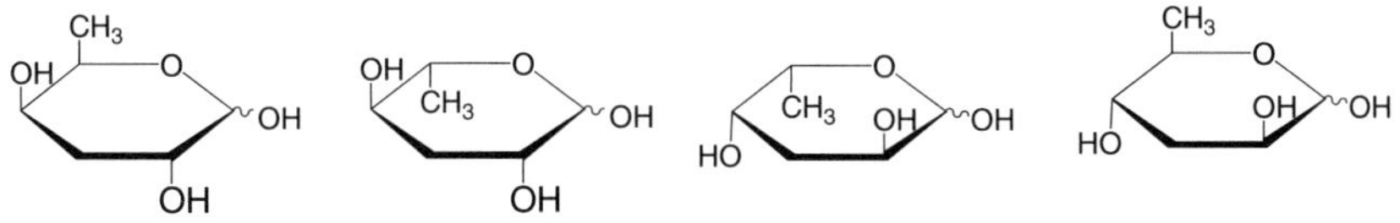

93 **94** **95** **96**

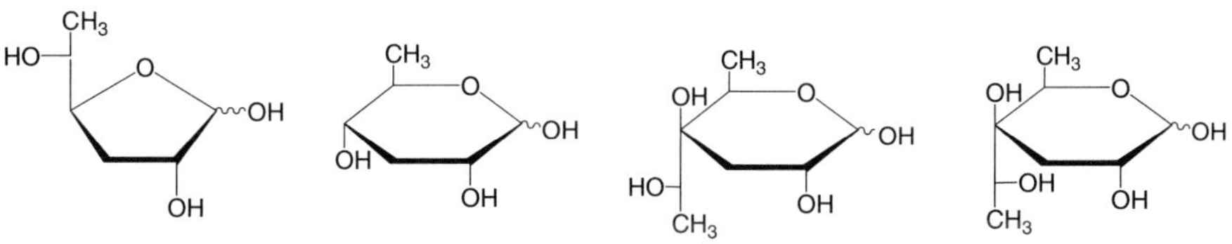

97a **97b** **98** **99**

Branched 3,6-dideoxyhexoses, yersiniose A (**98**, 3,6-dideoxy-4-*C*-(L-*glycero*-4^1-hydroxyethyl)-D-*xylo*-hexose),[189] yersiniose B (**99**, 3,6-dideoxy-4-*C*-(D-*glycero*-4^1-hydroxyethyl)-D-*xylo*-hexose), and other C-4 alkyl chain derivatives have been described.[190]

Colitose is biosynthesized from GDP-D-mannose,[191] while the other 3,6-dideoxy hexoses are made from CDP-D-glucose.[192]

Besides being found in LPS of bacteria, ascarylose was first discovered in the eggs of *Ascaris* nematodes,[193] and tyvelose was identified in larval glycoproteins of *Trichinella spiralis*.[194]

a. Abequose. — *(i) Occurrence.* Abequose (3,6-dideoxy-D-*xylo*-hexose, **93**) is α linked to the 3-position of D-mannopyranose in the lipopolysaccharide O-antigen of serogroup B *Salmonella*. Following cases of *Salmonella* food poisoning, sera contain antibodies specific for the LPS O-antigen of the infective bacteria. It was shown that the monoclonal antibody Se 155-4 binds to the branched trisaccharide α-D-Gal*p*-(1→2)[α-D-Abe*p*-(1→3)]-α-D-Man*p* present in *Salmonella*.[195]

(ii) Synthesis. Synthesis of 3,6-dideoxyhexoses usually starts from a derivative of the 6-deoxy analogue, when commercially available. Starting from methyl D-fucofuranoside (100), abequoside 104, and (3*S*)[3-^{2}H] abequoside (3*S*)104 may be obtained (Scheme 30). (3*R*)[3-^{2}H]Abequoside was also synthesized.[196,197]

The disaccharide α-Abe-(1→3)-Man was synthesized as the 4-pentenyl glycoside using the di-*O*-benzyl thioabequoside **107** as donor (Scheme 31).[198] The abequoside diacetate **106** was previously prepared from propenyl 4,6-*O*-benzylidene-α-D-galactopyranoside **105**.[199]

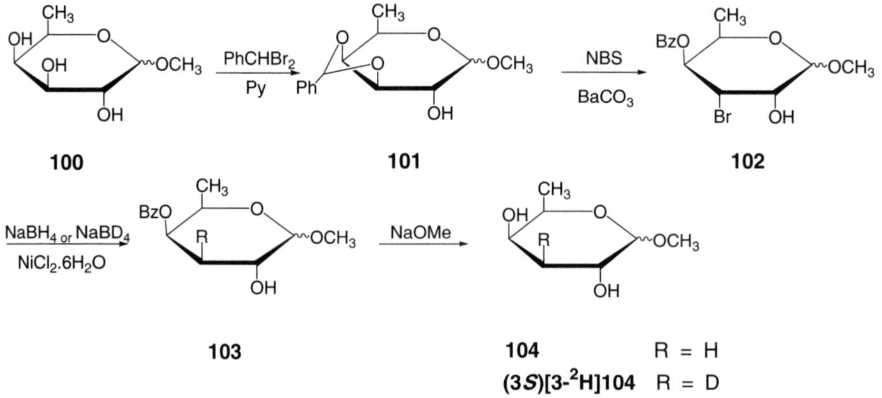

SCHEME 30.

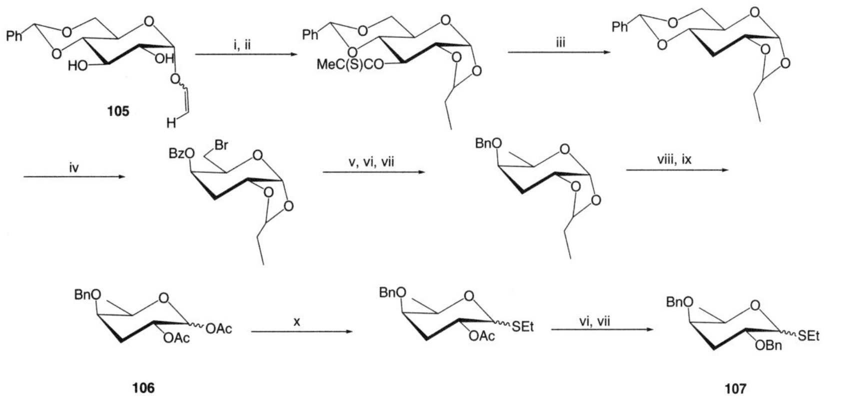

SCHEME 31. i. Camphor-10-sulfonic acid, MS 4 Å; ii. CS_2, NaH, imidazole; iii. Bu_3SnH, $NCC_6H_{10}N = NC_6H_{10}CN$; iv. NBS, $BaCO_3$; v. H_2.Pd-C, $NaHCO_3$; vi. NaOMe–MeOH; vii. BnBr, NaH, DMF; viii. 10% H_2SO_4, THF; ix. Ac_2O, Py; x. EtSH, $BF_3.Et_2O$.

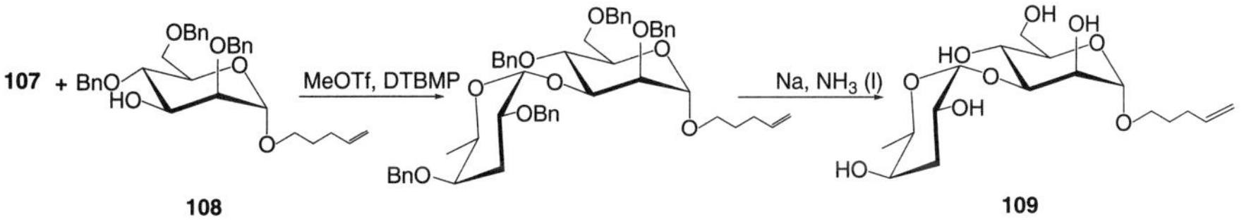

DTBMP: 2,6-di-*tert*-butyl-4-methylpyridine

SCHEME 32.

Glycosylation of the abequose donor **107** with the convenient mannopyranosyl acceptor **108** afforded 4-pentenyl 3,6-dideoxy-α-D-*xylo*-hexopyranosyl-(1→3)-α-D-mannopyranoside **109** which was further functionalized with an ω amino-containing aglycone (Scheme 32).

The allyl glycoside of α-D-Abe-(1→3)-α-D-Man **115** was prepared by a different approach.[200] The ethyl 1-thio-D-abequopyranoside donor **113** was obtained from methyl β-D-galactopyranoside derivative **110** according to Scheme 33. The cyclic sulfate intermediate **111** was the precursor for the stereoselective reduction with tetrabutylammonium borohydride to **112** which was further derivatized to the thioglycosyl donor **113**. Donor **113** was reacted with acceptor **114** to give **115** after deprotection.

SCHEME 33. i. $LiAlH_4$, THF; ii. 50 % HOAc; iii. (a) $SOCl_2$, Py, (b) $RuCl_3$, $NaIO_4$; iv. $n Bu_4NBH_4$, THF; v. BzCl, Py; vi. $EtSSiMe_3$, ZnI_2, $n Bu_4NI$.

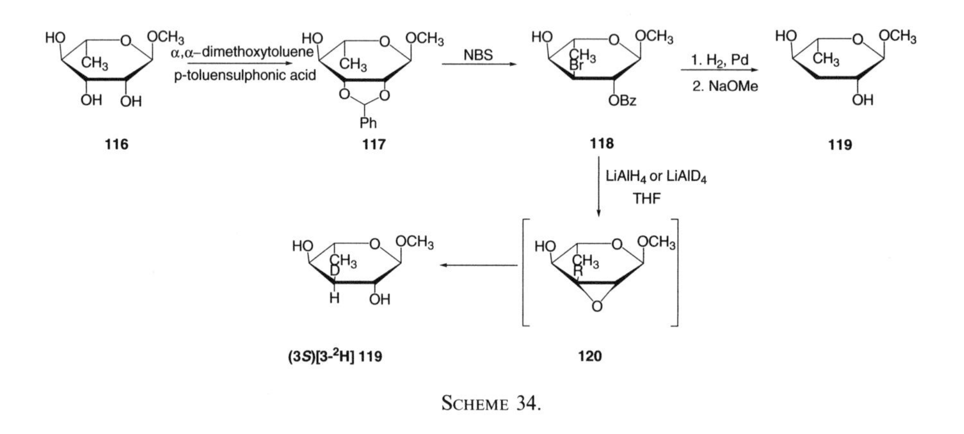

SCHEME 34.

b. Ascarylose. — *(i) Occurrence.* Ascarylose (3,6-dideoxy-L-*arabino*-hexopyranose, **94**) is a constituent of the glycolipid ascarosides A, B, and C isolated from the egg membrane of the tapeworm *Parascaris equorum*, and has also been characterized in the LPS of *Yersinia pseudotuberculosis*.[183]

(ii) Synthesis. Methyl 3,6-dideoxy-α-L-*arabino*-hexopyranoside (**119**) was obtained in four steps starting from methyl α-L-rhamnopyranoside (**116**, Scheme 34). The intermediate 2,3-*O*-benzylidene acetal **117** yielded the 3-bromo derivative **118** in more than 80% yield by treatment with *N*-bromosuccinimide. Catalytic hydrogenation with palladium, followed by debenzoylation afforded methyl 3,6-dideoxy-α-L-*arabino*-hexopyranoside (**119**).[201]

DSB: diisoamylborane

SCHEME 35.

A convenient synthesis of ascarylose in four steps starting from L-rhamnono-1,5-lactone (**121**) was reported.[202] The deoxy group in C-3 was introduced through stereoselective hydrogenation of the 2-enonolactone **122** as reported for the preparation of 3-deoxy sugars (see Section III.2). A crystalline furanose derivative of ascarylose, 2,5-di-*O*-benzoyl-3,6-dideoxy-α-L-*arabino*-hexofuranose (**125**) was also obtained (Scheme 35).[203]

When catalytic hydrogenation of the enonolactone **122** was performed with deuterium, stereospecific labeling took place at C-3 and at the 2*R* position of ascarylose.[197]

Ascarylose (**94**) and the (3*S*) deuterium-labeled ascarylose were prepared starting from methyl α-L-rhamnopyranoside via the 2,3-anhydro sugar **120** (Scheme 34).[204] Opening of the epoxide ring with lithium aluminum hydride in THF led selectively to the 3-deoxy derivative. If reduction was performed with lithium aluminum deuteride, the methyl (3*S*)-[3^2H]ascaryloside (3*S*)-[3^2H]**119** was obtained.

Photochemical deoxygenation of the 3-*O*-pivaloyl derivative **126** was also used for the preparation of the methyl ascaryloside (**119**) (Scheme 36).[205]

c. **Colitose.** — *(i) Occurrence*. Colitose (3,6-dideoxy-L-*xylo*-hexose, **95**) has been found in the O-specific polysaccharides of *Aeromonas trota*,[206] *Pseudoalteromonas tetraodonis* IAM 14160,[207] and *Vibrio cholerae* O139.[208] These

HO OCH3 CH3 OH OH **116** — PivCl, Py, - 20 °C → HO OCH3 CH3 OPiv OH **126** — hυ, HMPT/H_2O, 48 h → HO OCH3 CH3 OH **119**

PivCl: pivaloyl chloride
HMPT: hexamethylphosphorous triamide

SCHEME 36.

polysaccharides have in common the tetrasaccharide fragment **127** with two colitoses as part of a hexasaccharide repeating unit. Interestingly, tetrasaccharide **127** is a colitose (3-deoxy-L-fucose) analogue of the Lewis^b blood group antigenic determinant.

α-Col*p*-(1→ 2)-β-D-Gal*p*-(1→ 3)-[α-Col*p*-(1→ 4)]-β-D-Glc*pN*Ac

127

In *V. Cholerae*, the Gal*p* residue carries a 4,6-cyclic phosphate. The same repeating unit was found in the capsular polysaccharide. Multidimentional heteronuclear NMR and molecular-modeling studies showed that the O139 tetrasaccharide adopts a compact and tightly folded conformation that is relatively rigid and similar to the Le^b conformation. The cyclic phosphate on the β-galactopyranoside residue is in contact with the colitose residue linked to β-GlcNAc.[209] The capsular polysaccharide has been proposed as the basis for vaccine development.[210]

The trisaccharide α-Col*p*-(1→2)-β-D-Gal*p*-(1→3)-β-D-GlcpNAc, a colitose analogue of the Lewis^d (precursor) blood group antigen is part of the O-chain in the LPSs of *Salmonella enterica* serogroup O50.[211–213]

(ii) Synthesis. Methyl 3,6-dideoxy-α-L-*xylo*-hexopyranoside was synthesized from methyl α-L-fucopyranoside (**100**) by a route analogous to that reported for ascarylose.[201]

A trisaccharide fragment of *E. coli* O111 O-antigen has been synthesized as the 8-methoxycarbonyloctyl glycoside **128**. The two colitose residues were introduced via the chloride **129** prepared according to Scheme 37. The trisaccharide **128** inhibits binding of antibody raised against *E. coli* O111 or *S. adelaide*, to the respective cell wall LPS.[214]

Colitose (**95**) was synthesized as the ethyl 1-thioglycoside **130**, a convenient glycosyldonor for the introduction of terminal colitose.[215] The ethyl

128 Col = colitose

SCHEME 37.

SCHEME 38. i. α,α- Dimethoxytoluene, *p*-TsOH; ii. BnBr, NaH; iii. $NaCNBH_3$, HCl; iv. Im_2CS; v. Bu_3SnH, 2,2′-azobisisobutanonitrile.

1-thio-β-L-fucopyranoside used as precursor and deoxygenation of the 3-OH group was accomplished through formation of the thiocarbonylimidazole carbamate and reduction with tributyltin hydride (Scheme 38).

d. Tyvelose. — *(i) Occurrence.* Tyvelose (3,6-dideoxy-D-*arabino*-hexose, **96**), as abequose (**93**) and paratose (**97**), is α linked to the 3-position of mannose in the O-antigen of *Salmonella* serotype D1 LPS.[183] Interestingly tyvelose has been found β-linked as the terminal unit in N- and O-linked glycans in glycoproteins of the parasite *Trichinella spiralis*.[194,216,217]

Antibodies specific for tyvelose containing epitopes account for a large proportion of the immune response in *Trichinella*-infected patients.[218]

(ii) Synthesis. Methyl 3,6-dideoxy-α-D-*arabino*-hexopyranoside was prepared from the readily available methyl 4,6-O-benzylidene-α-D-glucopyranoside via the 2,3-anhydro derivative.[219,220] When the reaction was applied to the β-anomer **131** both 2,3-anhydro compounds **132** and **133** were produced

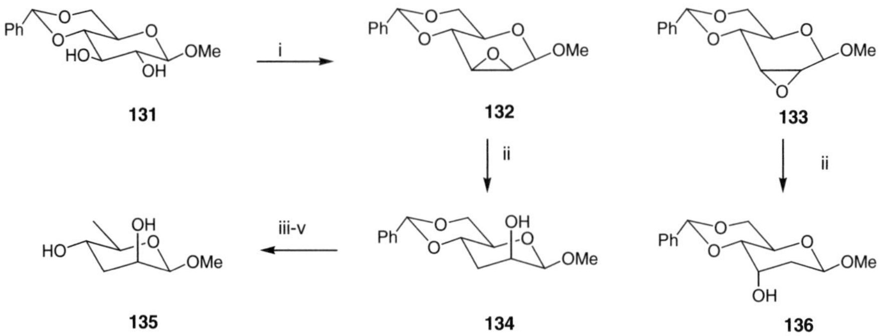

Scheme 39. i. NaH, DMF, *N*-tosylimidazole; ii. $LiAlH_4$, Et_2O; iii. NBS, CCl_4, $BaCO_3$; iv. H_2, Pd/C, Et_3N; v. NaOMe-MeOH.

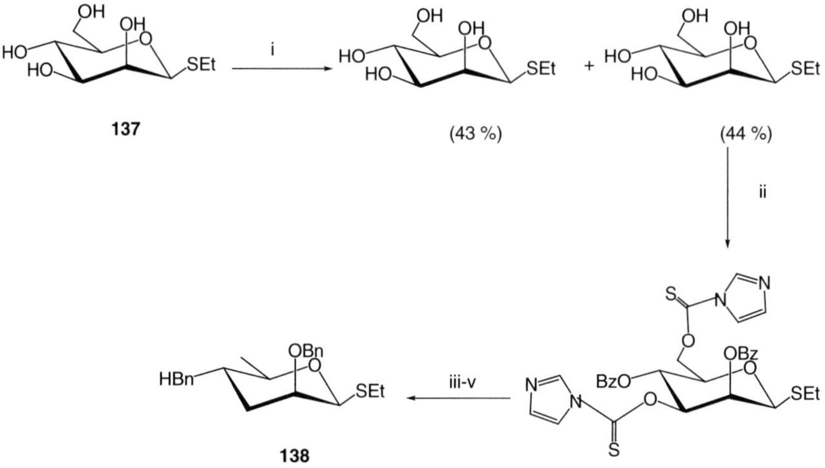

Scheme 40. i. (1)$PhC(OMe)_3$, CSA, $CHCl_3$, (2) 90 % $AcOH/H_2O$; ii. 1,1′-thiocarbonyldiimidazole, toluene, reflux; iii. Bu_3SnH, 2,2′-azobisisobutanonitrile, toluene reflux; iv. NaOMe, MeOH; v. NaH, BnBr, DMF.

(Scheme 39). Reduction of **132** led to methyl β-tyveloside **135**. The 2-deoxy-D-*ribo*-hexopyranoside **136** was obtained from epoxide **134**.[221]

The disaccharide epitope α-Tyv-(1→3)-α-D-Man present in the D1 *Salmonella* serotype was synthesized as the pentenyl glycoside. The tyvelose donor **138** was prepared from ethyl 1-thio-β-D-mannopyranoside **137** employing a single dideoxygenation step (Scheme 40). Glycosylation of **138** with the mannopyranosyl acceptor **108** was performed as for the abequose epitope **109** (Scheme 32) and the disaccharide was functionalized for the construction of self-assembled monolayers (SAMs).[198]

Glycosides of the disaccharide β-Tyv-(1→3)-β-D-GalNAc (**139**)[221] and the trisaccharide epitope **140** carrying the challenging β-1,2-*cis*-linked tyvelose of the *Trichinella* antigen were synthesized. The ω-amino tether **141** and the fluorescein conjugate **142** of the trisaccharide were described.[222]

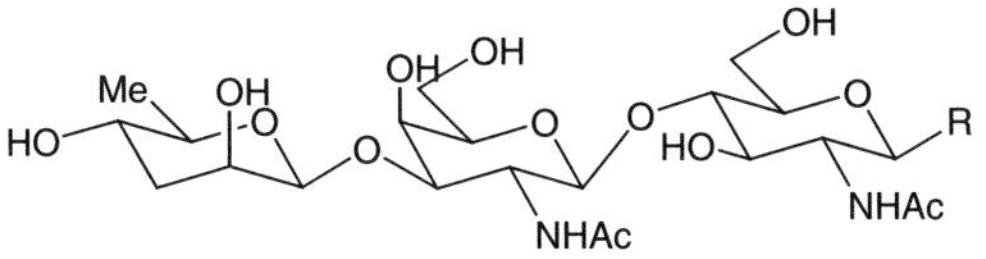

140 R = OCH_3
141 R = $OC_6H_{12}NH_2$
142 R = $OC_6H_{12}NHCSF$

In the first attempt to prepare disaccharide **139**, the dibenzyl derivative **143**[221] with a nonparticipating C-2 benzyloxi group was selected as donor. However, a mixture of both disaccharides was obtained in low yield, with predominance of the α-anomer.

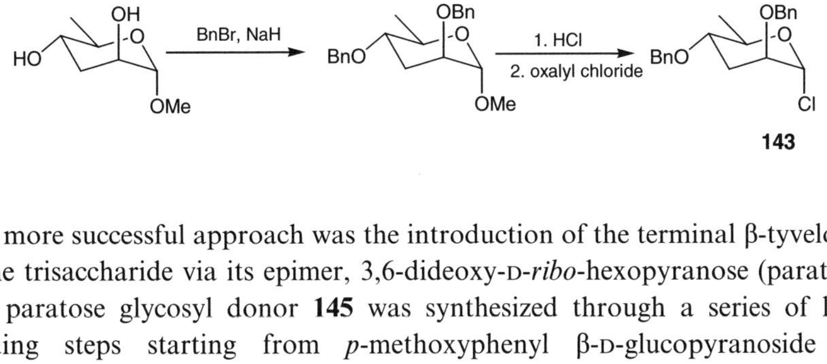

BnBr, NaH

1. HCl
2. oxalyl chloride

143

A more successful approach was the introduction of the terminal β-tyveloside in the trisaccharide via its epimer, 3,6-dideoxy-D-*ribo*-hexopyranose (paratose). The paratose glycosyl donor **145** was synthesized through a series of high-yielding steps starting from *p*-methoxyphenyl β-D-glucopyranoside **144** (Scheme 41).

The phenyl 1-thioparatoside **145** was activated with *N*-iodosuccinimide and silver triflate and reacted with a convenient derivative of the disaccharide β-D-Gal*p*NAc-(1→4)-β-D-Glc*p*NAc. After removal of the pivaloyl-protecting group with sodium methoxide, isomerization of paratose to tyvelose was performed in a one-pot reaction by oxidation with dimethyl sulfoxide and acetic anhydride, followed by reduction with L-Selectride. Selectivity of the reduction was better

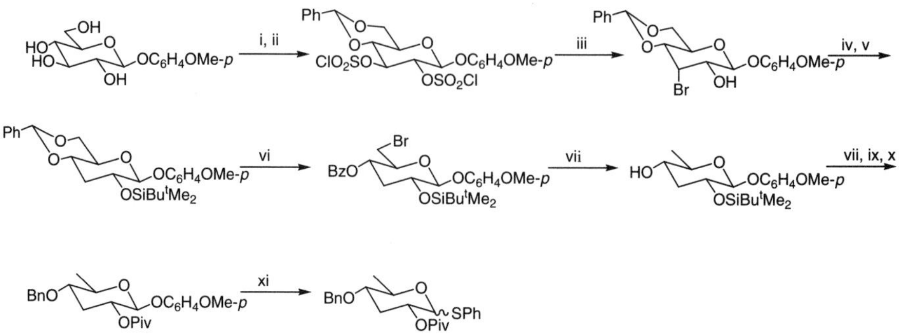

SCHEME 41. Synthesis of a paratose glycosyl donor: i. $PhCH(OMe)_2$, p-toluenesulfonic acid, CH_3CN; ii. $SOCl_2$, Py; iii. (a) Et_4NBr, CH_2Cl_2, (b) KI, $KHCO_3$, $MeOH$-H_2O; iv. Bu^tMe_2SiCl, imidazole, DMF; v. H_2, Pd/C; vi. NBS, $BaCO_3$, CCl_4; vii. H_2, Pd/C, $KHCO_3$, EtOH; viii. BnBr, NaH; ix. Bu_4NF; x. PivCl, Py; xi. PhSH, $F_3B.Et_2O$.

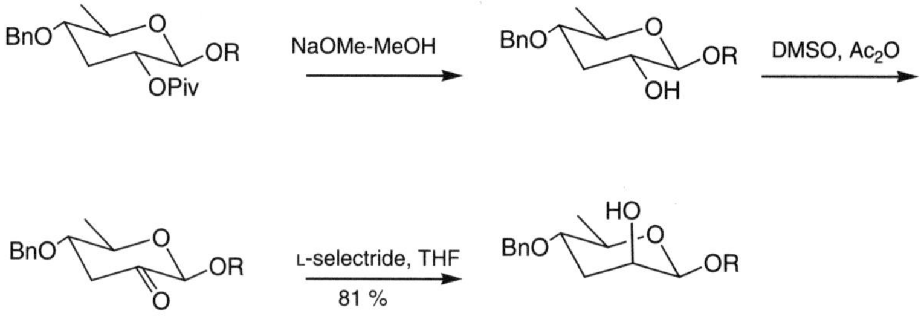

SCHEME 42.

than 95%. The protocol used for conversion of the β-paratoside to the β-tyveloside is summarized in Scheme 42.

Conversion of the 1,2-*trans* to the 1,2-*cis* configuration by oxidation and reduction of the C-2 in the paratoside trisaccharide derivative, according to Scheme 42, followed by deprotection reactions, afforded the corresponding tyveloside **140**.[222]

e. Paratose.—Paratose (3,6-dideoxy-D-*ribo*-hexose, **97**), α–(1-→3)-linked to mannose, is the antigenic determinant in the LPS of *Salmonella* serotype B. It was obtained by selective reduction of dibromide **146** obtained from methyl α-D-glucopyranoside (Scheme 43).[223]

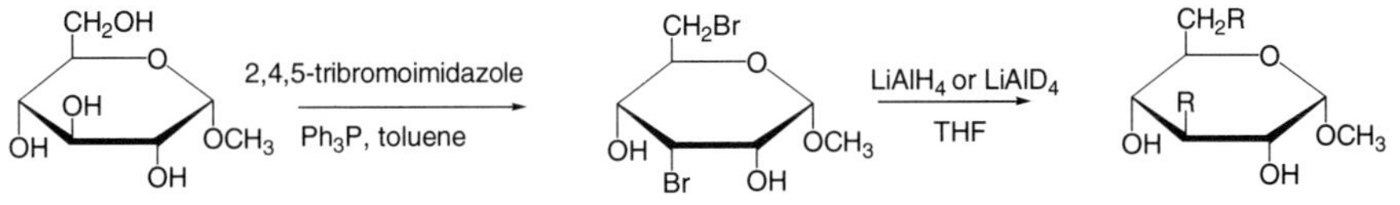

SCHEME 43.

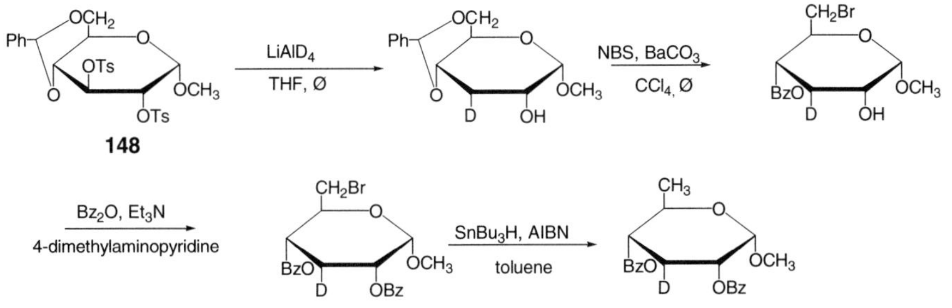

Ts: *p*-tolylsulfonyl
AIBN:2,2'-azobisisobutanonitrile

SCHEME 44.

Subsequent reduction proceeds with inversion and if $LiAlD_4$ is used (3*S*)-[3,6-$^2H_2$] paratoside **147** was obtained. A milder reduction of **146**, first with $LiAlH_4$ (THF, room temperature), which reduces preferentially the C-6 Br group, and then with $LiAlD_4$ under reflux afforded the (3*S*)-[3-^{2}H] paratoside. Compound **146** could be reduced by Barton-McCombie radical deoxygenation conditions[15] to afford **147** in 98% yield.[198] The (3*R*)-[3-^{2}H]paratoside **149** could be also obtained starting from the ditosyl glucosyl derivative **148** (Scheme 44).[197]

Another approach to prepare the paratose derivative **150**, with different subtituents at O-2 and O-4, starts from 4,6-*O*-benzylidene-1,2-*O*-propylidene-α-D-glucopyranose (Scheme 45).[199]

f. Yersinioses A and B. — *(i) Occurrence*. Yersiniose A (3,6-dideoxy-4-*C*-(L-*glycero*-4^1-hydroxyethyl)-D-*xylo*-hexose, **98**) has been first found as a component of the O-specific polysaccharide of LPS produced by species of *Yersinia*[224,225]

Bu$_3$SnH, AIBN

NBS, BaCO$_3$

1. H$_2$ / Pd-C, NaHCO$_3$
2. NaOMe-MeOH
3. BnBr, NaH, DMF

1. 10 % H$_2$SO$_4$, THF
2. Ac$_2$O, Py

PhSSiMe$_3$, TMSOTf

150

AIBN:2,2'-azobisisobutanonitrile

SCHEME 45.

and *Legionella*.[226] The epimeric sugar at C-4^1, yersiniose B [3,6-dideoxy-4-*C*-(D-*glycero*-4^1-hydroxyethyl)-D-*xylo*-hexose **99**], has been isolated from the LPS of *Y. enterocolitica* serovar O:4,32.[227] Recently, yersiniose A was found in a LPS from *Burkholderia brasiliensis*, a Gram-negative, nitrogen-fixing bacterium. For the O-polysaccharide the structure of the repeating unit consists of the branched tetrasaccharide **151**.[189] The yersiniose A residues confer hydrophobicity to the polysaccharide **151**, which may be important for the interaction between the bacterial and plant–cell surfaces.

-2)-α-D-Rha*p*-(1→3)-α-D-Rha*p*-(1→3)-α-D-Rha*p*-(1
2
↑
1
α–YerA*p* **151**

The enzyme (YerE) involved in the biosynthesis of yersiniose A has been isolated and characterized as a flavoenzyme that catalyzes the attachment of the branched chain to a 3,6-dideoxy-4-hexulose precursor, using pyruvate as the side-chain donor.[228] Substrate analogues were prepared as potential inhibitors of the biosynthesis.[229]

(ii) Synthesis. Synthesis of **98** and **99** was important to confirm their structures. They were both obtained by the same route, starting from levoglucosan. Although the synthesis was not stereospecific, the spectroscopical characterization of the eight stereoisomers obtained was important to establish the configuration of natural yersinioses A and B.[227]

3. 4,6-Dideoxy Sugars

a. Occurrence.—4,6-Dideoxy sugars occur rarely in Nature. The best known of this class of deoxy sugars is 4,6-dideoxy-3-*O*-methyl-D-*xylo*-hexose, called chalcose (**152**).

152 D - chalcose

153 4 ,6-dideoxy-D-*xylo*-hexose

154 desosamine

This sugar is one of the two sugar units found in chalcomycins (**155**), 16-membered macrolide antibiotics,[230] and in lankamycin (**156**), a 14-membered macrolide produced by *Streptomyces* sp. [231] The similar macrolide neutramycin (**157**) has been isolated with several analogues containing 4,6-dideoxy-D-*xylo*-hexose (**153**) instead of chalcose (**152**) from cultures of *S. luteoverticillatus*.[232]

155 R = CH_3
157 R = H

156

158 mycinamicin I

Desosamine (**154**), in which the 3-*O*-methyl group of chalcose is replaced by a dimethylamino group, is found in the macrolide mycinamicin I (**158**).[230,233] Genes for the synthesis of the two deoxy sugars of chalcomycin, D-chalcose, and D-mycinose were characterized in a gene cluster from *S. bikiniensis*.[230]

b. Synthesis.—Several syntheses of both enantiomers of 4,6-dideoxy-*lyxo*-hexose have been reported, in particular for studies on the role of the 4-OH group in the interaction with enzymes or in the properties of antibiotics containing an L-rhamnose unit.

One approach started from partially protected rhamnosyl derivatives. Thus, from easily obtained 2,3-*O*-isopropylidene-α-L-rhamnopyranoside, deoxygenation at C-4 by the Barton and McCombie procedure afforded the 4,6-dideoxy-α-L-*lyxo*-hexopyranoside derivative.[234]

The thioglycoside derivative of 4,6-dideoxy-L-*lyxo*-hexopyranose **160**, useful as glycosyl donor has been prepared starting from the ethyl 1-thio-α-L-rhamnopyranoside (**159**, Scheme 46).

Compound **160** was used as donor for the preparation of methyl 4,6-dideoxy-α-L-*lyxo*-hexopyranosyl-(1→3)-2-acetamido-2-deoxy-α-D-glucopyranoside, the 4′-deoxy analogue of the linkage disaccharide to the galactan chain in mycobacteria.[235]

4,6-Dideoxy-D-*lyxo*-hexose (**161**) has been obtained from methyl-α-D-mannopyranoside according to Scheme 47.[236]

4,6-Dideoxy-D-*xylo*-hexose (**163**) was obtained from methyl α-D-glucopyranoside via 2,3-di-*O*-benzoyl-4,6-di-*O*-*p*-tolylsulfonyl-α-D-glucopyranoside (**162**, Scheme 48)[237] or via methyl 4,6-dichloro-4,6-dideoxy-α-D-galactopyranoside.[238]

Syntheses of 4,6-dideoxy-D- and L-*xylo*-hexoses and the L-*lyxo* isomer starting from noncarbohydrate precursors have been also accomplished.[239]

Chalcose (4,6-dideoxy-3-*O*-methyl-D-*xylo*-hexose, **152**) was obtained from methyl 4,6-dichloro-4,6-dideoxy-α-D-galactopyranoside (**164**).[240] The dichloride **164** was readily prepared by selective chlorination of methyl α-D-glucopyranoside with

SEt
H_3C
HO
O
HO
OH
159
BzCl, Py
SEt
H_3C
HO
O
BzO
OBz
1. thiocarbonyldiimidazole
2. Bu_3SnH
SEt
H_3C
O
BzO
OBz
160

SCHEME 46.

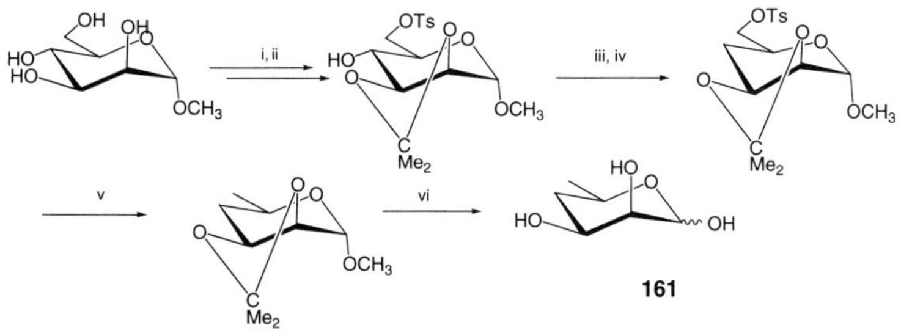

SCHEME 47. i. 2,2-Dimethoxypropane, 70 mM H_2SO_4, DMF; ii. p-toluensulfonyl chloride, Py; iii. thiocarbonyldiimidazole; iv. Bu_3SnH; v. $NaBH_4$, Me_2SO, ø; vi. Amberlite IR-120 (H^+), ø.

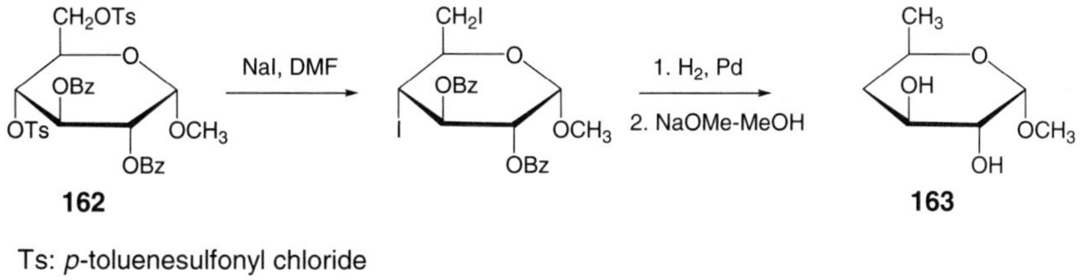

Ts: *p*-toluenesulfonyl chloride

SCHEME 48.

sulfuryl chloride, with inversion of the configuration at C-4. Dechlorination with tributyltin hydride afforded the 4,6-dideoxy sugar **165**, which was selectively benzoylated with benzoyl cyanide at O-2 and methylated at O-3. Acid hydrolysis afforded D-chalcose (Scheme 49). The sugar could be also obtained by direct methylation of **164** with diazomethane, followed by reduction.

Methods based on the cyclization of diene derivatives have been described. Thus, chalcose (**152**) has been synthesized by a cyclocondensation reaction of the silyloxy diene **166** with acetaldehyde (Scheme 50).[241]

Both, D- and L-chalcose have been synthesized from *meso*-divinylglycols.[242] Racemic D,L-chalcose and D,L-desosamine were prepared by a method based on cyclization of methyl 2-*cis*-5-hexadienoate.[243]

From methyl 4,6-dideoxy-α-D-*xylo*-hexopyranoside, by a series of substitution reactions, the isomeric D-*ribo* and D-*arabino*-hexopyranosides have been prepared.[244]

SCHEME 49.

DIBAL: diisobutylaluminium hydride

SCHEME 50.

The 4,6-dideoxy-D-fructofuranose **167** and the epimeric sorbofuranose **168** were also synthesized by a cyclization reaction.[245]

4. Other Dideoxy Sugars

The 2,3-, 2,4-, and 3,4-dideoxy sugars have not been found in Nature. Some of them have been synthesized for structure–activity relationships. 2′,

3′-Dideoxynucleosides are of interest because several members of this group exhibit potent antiviral activity against HIV, for example **169** and **170**.[246]

Enantioselective synthesis of 2,3-dideoxy-D-*erythro* and D-*threo*-hexoses has been achieved, in several steps, starting from furfural.[247]

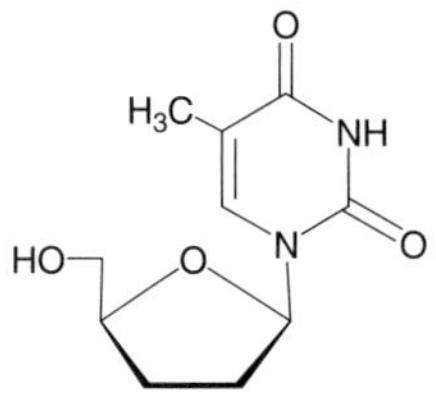

O
H_3C NH
HO O N O
O
N NH
HO O N N NH_2

169 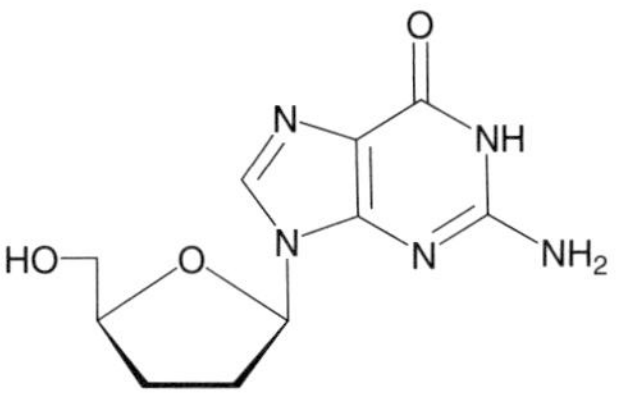 **170**

2,4-Dideoxy-D-*erythro*-hexopyranose (**172**) has been obtained from 3-deoxy-D-*erythro*-pentose (**171**, Scheme 51)[248] as an intermediate for the synthesis of the lactone moiety of inhibitors of 3-hydroxy-3-methylglutaryl-coenzyme A reductase, that lowers cholesterol levels.

Methyl 2,4-dideoxy-β-L-*erythro*-hexopyranoside has been obtained as intermediate in the synthesis of δ-hydroxy-β-lysine, from D-galacturonic acid.[249]

The syntheses of the thymine nucleosides of 2,4-dideoxy-β-D-*erythro*-hexopyranose starting from 1,2:5,6-di-*O*-isopropylidene-α-D-glucofuranose (12 steps) or from tri-*O*-acetyl-D-glucal (11 steps) have been reported.[250]

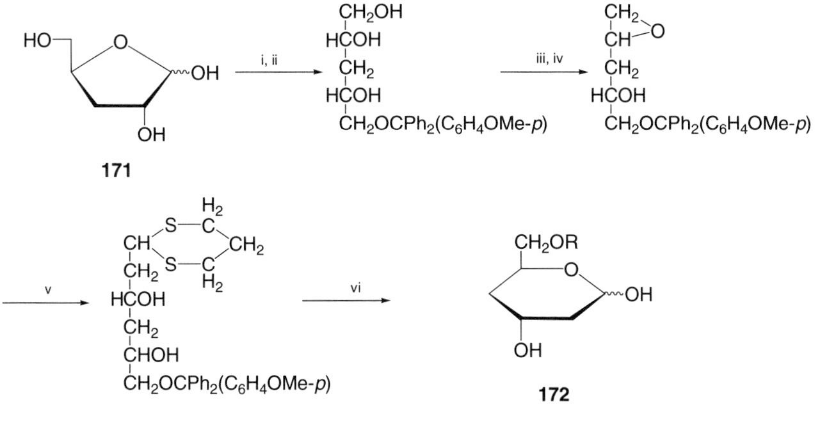

HO O OH
OH
171
i, ii
CH_2OH
HCOH
CH_2
HCOH
$CH_2OCPh_2(C_6H_4OMe$-$p)$
iii, iv
CH_2 O
CH
CH_2
HCOH
$CH_2OCPh_2(C_6H_4OMe$-$p)$
v
H_2
S C
CH CH_2
CH_2 S C
H_2
HCOH
CH_2
CHOH
$CH_2OCPh_2(C_6H_4OMe$-$p)$
vi
CH_2OR
O OH
OH
172

SCHEME 51. i. (p-$MeOC_6H_4$)Ph_2Cl, Py; ii. $NaBH_4$, EtOH; iii. TsC1, Py; iv. KOH, MeOH; v. 1,3-dithiane, BuLi; vi. NBS, 2,6-dimethylpyridine.

Modeling of several dideoxyaldohexopyranoses with the molecular mechanics algorithm MM3(92) has been described.[251]

V. Trideoxy Sugars

1. 2,3,6-Trideoxy Sugars

a. Occurrence.—The 2,3,6-trideoxy hexoses are found in Nature in important metabolites. Interestingly, amicetose (**173**) is found as the D enantiomer, whereas L-rhodinose (**174**) is the enantiomer usually found in antibiotics.

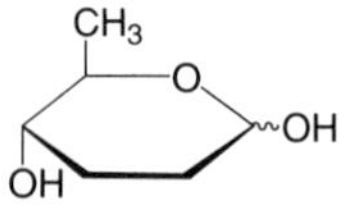

173
D-Amicetose
2,3,6-trideoxy-D-*erythro*-hexose

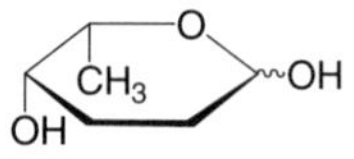

174
L-Rhodinose
2,3,6-trideoxy-D-*threo*-hexose

D-Amicetose (**173**) is a constituent of the antibiotics of the antibiotics amicetin,[252] axenomycin,[253] dutomycin,[254] and polyketomycin.[255] Amicetin is produced by *Streptomyces* spp. and the structure described in 1962.[1,256] In polyketomycin (**175**) and dutomycin, the 3-*C*-methyl branched sugar L-axenose is α-(1→4)-linked to amicetose,[257] whereas in axenomycin, a macrolide antibiotic, the two deoxy sugars are β-linked in reverse order. Mild methanolysis (0.05 M HCl, rt) of axenomycine B gave among other products the glycoside **176**.[253]

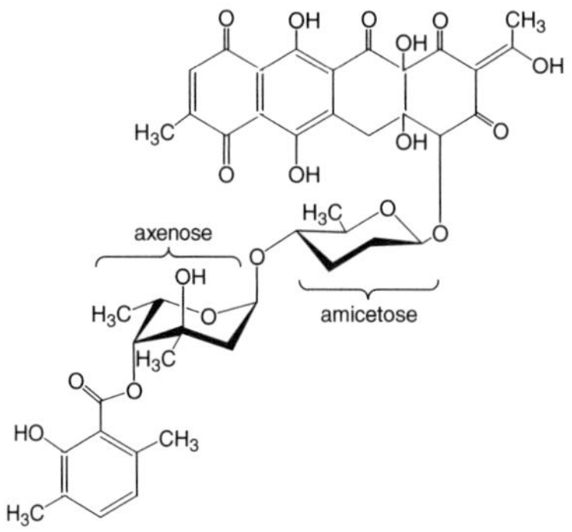

175

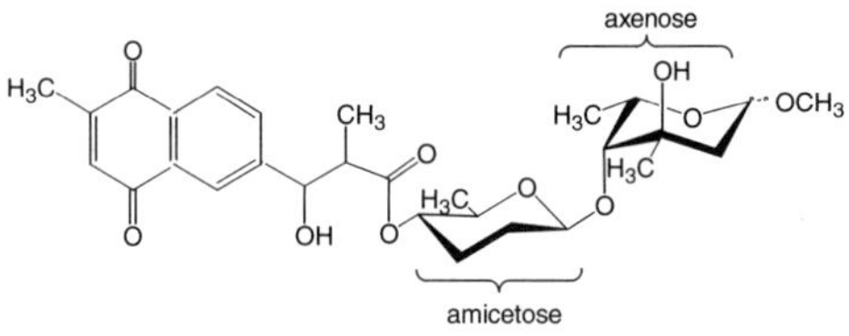

176

Polyketomycin (**175**), produced by *Streptomyces* sp. MK277-AF1, has been described as a powerful antibiotic against multidrug resistant *Staphylococcus aureus*, and as antitumor agent.[258]

Rhodinose (**174**) is the sugar linked in the antibiotic streptolydigin (**177**)[1,259] and is a component of the oligosaccharides present in the rhodomycin family of anthracycline antibiotics.[1,260]

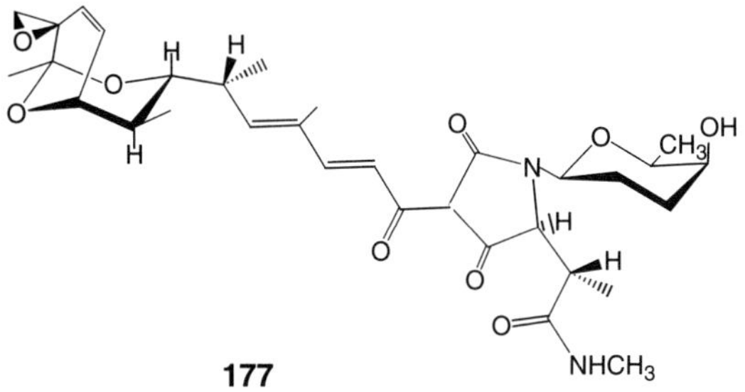

177

L-Rhodinose (**174**) is frequently found together with D-olivose in the oligosaccharide linked to an aglycone, as in Landomycin A (**66**). In Urdamycin A (**178**) an anticancer agent produced by *Streptomyces fradiae* Tü2717, L-rhodinose is the middle unit in the C-linked trisaccharide and is also O-linked as a single sugar to the aglycone. L-rhodinose was found as a di- or trisaccharide β-linked to aklavinone in aclacinomycins of mutant strain H039 of *Streptomyces galilaeus*.

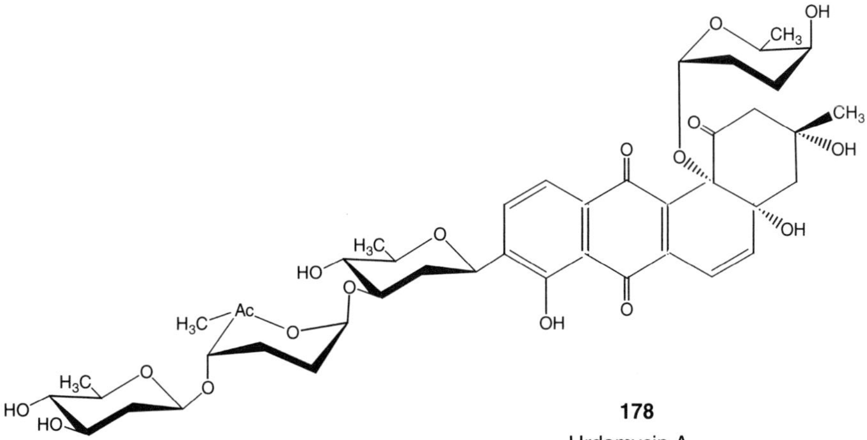

178
Urdamycin A

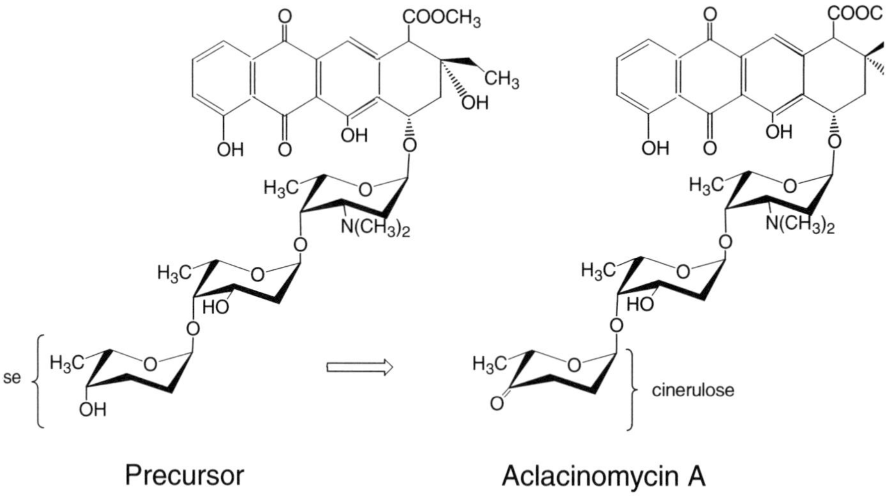

Precursor

Aclacinomycin A

SCHEME 52.

Other mutants, with rhodinose as component of the oligosaccharide, have been described.[261] For example, the aklavinone in the mutant H026 is glycosylated by a trisaccharide containing a terminal rhodinose and two other deoxy sugars. The mutant H026 is deficient in the oxidoreductase which converts Rho into cinerulose A, resulting in aclacinomycin A (aclarubicin) (Scheme 52).

b. Synthesis.—D-Amicetose (**173**) was first synthesized from D-glucal.[262] Albano and Horton synthesized methyl α-D-amicetoside from methyl α-D-glucopyranoside via the 4,6-*O*-benzylidene-2,3-dideoxy-α-D-*erythro*-hex-2-enopyranoside.[263] Since then, several other racemic and enantioselective synthesis have been described from carbohydrate and noncarbohydrate starting compounds.[264–267] More recently, methods for the synthesis of amicetose based on hydroboration of a simple enol ether were described.[268] However, attempts to perform asymmetric hydroboration of 2-isobutoxy-6-methyl-2,3-dihydropyran (**179**) with isopinocamphenylborane ($IpcBH_2$) gave the α-D-amicetoside **D–180** in only 9% yield, with 91% enantiomeric excess (ee). Racemic hydroboration gave a higher yield, and resolution of the isobutyl β-DL-amicetosides by lipase transesterification was studied. Selective acylation of the L-substrate was observed and the enriched D-amicetoside could be separated.

1. $IpcBH_2$
2. H_2O_2, NaOH
+ β-anomer (< 2%)
179
D-180 (9 %, 91 % ee)
1. borane, THF
2. H_2O_2, NaOH
D,L-181 (70 %)
lipase
D-181 + **L-182**

$IpcBH_2$: isopinocamphenylborane

The disaccharide α-L-Axe-(1→4)-D-Ami (**183**) found in polyketomycin (**175**) has been synthesized from methyl 2,3,6-trideoxy-α-D-*erythro*-hexopyranoside (**184**, methyl α-D-amicetoside, Scheme 53) and the phenyl 1-thioglycoside of 3,4-di-*O*-benzyl-2,6-dideoxy-3-*C*-methyl-L-*xylo*-hexopyranose (**186**, Scheme 54).[257] The key coupling step was performed by *N*-bromosuccinimide activation of the thioglycoside to give the (1→4)-linked disaccharide in 71% yield (Scheme 55). The methyl

1. $BF_3 \cdot OEt_2$, MeOH
2. HO^- resin
3. H_2, Pd/C
$PhCH(OCH_3)_2$, H^+
NBS, $BaCO_3$
1. separate anomers
2. nBu_3SnH or nBu_4NBH_4
3. NaOH
184

SCHEME 53.

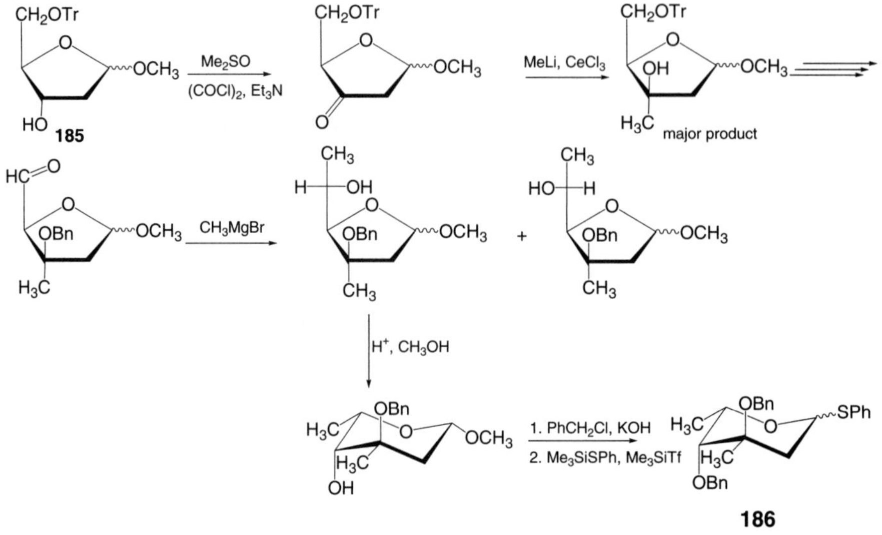

SCHEME 54.

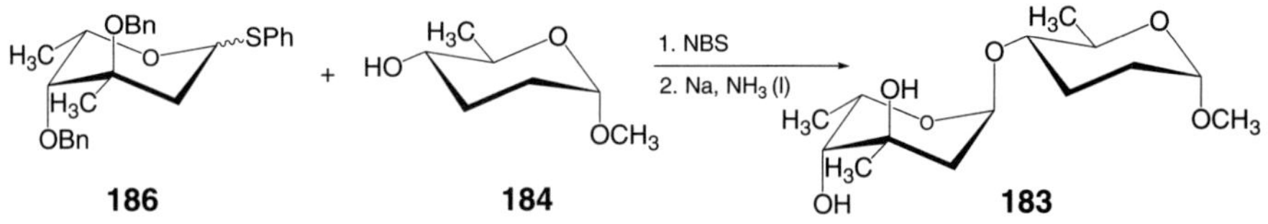

SCHEME 55.

α-D-amicetoside was prepared by a slight modification of the method described by Albano and Horton.[263]

The phenyl L-thioglycoside of the axenose donor **186** was synthesized from the methyl 2-deoxy-D-*erythro*-pentofuranoside **185** (Scheme 54).

High selectivity for α-glycosylation was achieved by NBS-activation (Scheme 55).

L-Rhodinose (**174**) was prepared from the readily available L-rhamnose.[269] The method required deoxygenation of C-2 and C-3 and inversion of configuration at C-4 (Scheme 56). Oxidation of **187** with ruthenium dioxide–IO_4^-, followed by reduction of the keto groups with lithium aluminum hydride yielded the alcohol **188**. After protection as the benzyl derivative, an alkenic linkage was

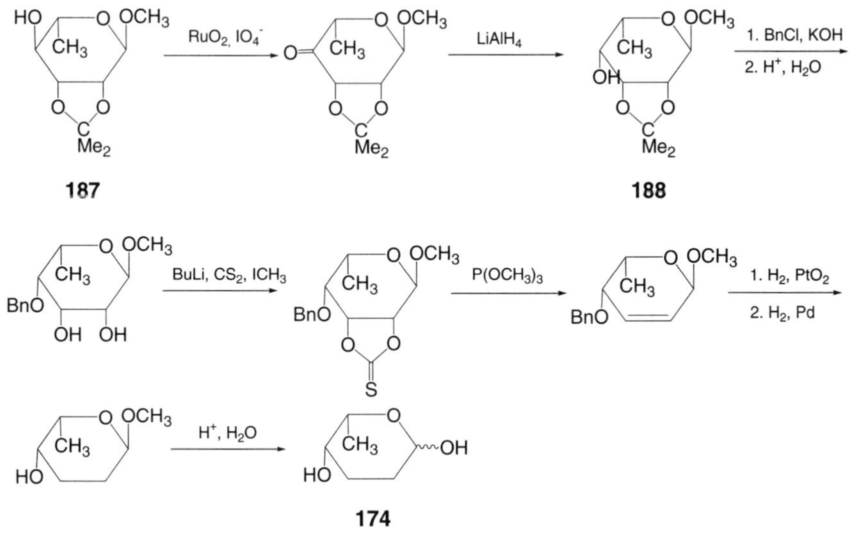

SCHEME 56.

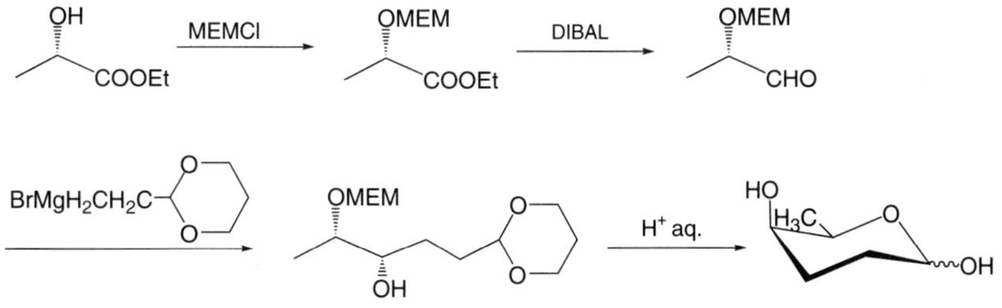

MEM: 2-methoxyethoxymethyl

DIBAL: diisobutylaluminium hydride

SCHEME 57.

introduced between C-2 and C-3 through a Corey–Winter reaction.[270] The enantiomer of natural amicetose was prepared by a similar approach, but skipping the isomerization step.

The preparation of benzyl 4-*O*-acetyl-2,3,6-trideoxy-L-*threo*-hexopyranoside, a protected form of rhodinose, starting from L-rhamnose has also been described.[271]

Starting from ethyl (*S*)-lactate, rhodinose was prepared in four steps, with 31% overall yield (Scheme 57).[272] Chelation-controlled addition of the Grignard

1. BnOH, $BF_3.OEt_2$
2. NaOMe

PPh_3, BzOH, DEAD

1. D_2, Pd/C
2. NaOMe
3. HCl aq.

DEAD: diethyl azodicarboxylate

SCHEME 58.

reagent to the chiral α-alkoxyaldehyde is the key reaction and proceeded with a 19:1 diasteroselectivity.

The chiral synthon, 2,2,5-trimethyl-L,3-dioxolane 4-carboxaldehyde, was also used for the preparation of rhodinose.[273]

The D- and L-enanatiomers of rhodinose and amicetose specifically labeled with deuterium at C-2 and C-3 were prepared from the same synthon.[274] Starting from 3,4-di-*O*-acetyl-L-rhamnal, (2*S*,3*R*)-[2,3-$^2H_2$]-L-rhodinose was synthesized as shown in Scheme 58.

2. Other Trideoxy Sugars

Disregarding nonoxygen functional groups (amino, dimethylamino, and so on) only the 2,3,6-trideoxyhexoses were isolated from natural products. However, some of the isomeric trideoxyhexoses were chemically synthesized. Thus, the methyl α-glycosides of 3,4,6-trideoxy-D-*erythro*-hexopyranose, 3,4,6-trideoxy-D-*threo*-hexopyranose and 2,4,6-trideoxy-D-*erythro*-hexopyranose were synthesized from methyl 4,6-dideoxy-α-D-*xylo*-hexopyranoside via selective substitution reactions.[275]

Starting from L-rhamnono-1,4- or 1,5-lactone, via β-elimination reactions, 3,5,6-trideoxy-α-DL-*threo*-hexose[276] or 3,4,6-trideoxy-DL-*threo*-hexose has been obtained.[277]

References

1. S. Hanessian, Deoxy sugars, *Adv. Carbohydr. Chem.*, 21 (1966) 143–207.
2. A. Kirschning, A. F. W. Bechthold, and J. Rohr, Chemical and biochemical aspects of deoxysugars and deoxysugar oligosaccharides, *Top. Curr. Chem.*, 188 (1997) 1–83; H. W. Liu and J. S. Thorson, Pathways and mechanisms in the biogenesis of novel deoxysugars by bacteria, *Annu. Rev. Microbiol.*, 48 (1994) 223–256.
3. C. Marzabadi and R. W. Franck, The synthesis of 2-deoxy-glycosides: 1988–1999, *Tetrahedron*, 56 (2000) 8385–8417.
4. A. Kirschning, M. Jesberger, and K. U. Schoning, Concepts for the total synthesis of deoxy sugars, *Synthesis* (2001) 507–540.
5. R. M. de Lederkremer and C. Gallo, Natural occurring monosaccharides: Properties and synthesis, *Adv. Carbohydr. Chem.*, 59 (2004) 9–67.
6. K. Bock and K. Adelhorst, Derivatives of methyl β-lactoside as substrates for and inhibitors of β-D-galactosidase from *E. coli*, *Carbohydr. Res.*, 202 (1990) 131–149.
7. W. Hakamata, T. Nishio, and T. Oku, Hydrolytic activity of α-galactosidases against deoxy derivatives of *p*-nitrophenyl α-D-galactopyranoside, *Carbohydr. Res.*, 324 (2000) 107–115.
8. A. Chiocconi, C. Marino, E. Otal, and R. M. de Lederkremer, Photoinduced electron transfer and chemical α-deoxygenation of D-galactono-1,4-lactone. Synthesis of 2-deoxy-D-*lyxo*-hexofuranosides, *Carbohydr. Res.*, 337 (2002) 2119–2126.
9. U. Westerlind, P. Hagback, B. Tidbäck, L. Wiik, O. Blixt, N. Razi, and T. Norberg, Synthesis of deoxy and acylamino derivatives of lactose and use of these for probing the active site of *Neisseria meningitidis* *N*-acetyltransferase, *Carbohydr. Res.*, 340 (2005) 221–233.
10. T. Lee and Y. C. Lee, Synthesis of allyl 3-deoxy- and 4-deoxy-β-D-galactopyranoside and simultaneous preparation of Gal(1→2)- and Gal(1→3)-linked disaccharide, *Carbohydr. Res.*, 251 (1994) 69–79.
11. W. A. Szarek and X. Kong, Direct halogenation of carbohydrate derivatives, in S. Hanessian (Ed.), *Preparative Carbohydrate Chemistry*, Marcel Dekker, Inc., New York, 1997, pp. 105–126.
12. W. A. Szarek, A. Zamojski, A. R. Gibson, D. M. Vyas, and J. K. N. Jones, Selective, reductive dechlorination of chlorodeoxy sugars. Structural determination of chlorodeoxy and deoxy sugars by ^{13}C nuclear magnetic resonance spectroscopy, *Can. J. Chem.*, 54 (1976) 3783–3793.
13. H. Arita, N. Ueda, and Y. Matsushima, The reduction of chlorodeoxy sugars by tributyltin hydride, *Bull. Chem. Soc. Jpn.*, 45 (1972) 567–569.
14. H. H. Baer and H. R. Hanna, Desulfonyloxylations of some secondary *p*-toluenesulfonates of glycosides by lithium triethylborohydride; a high-yielding route to 2- and 3-deoxy sugars, *Carbohydr. Res.*, 110 (1982) 19–41.
15. D. H. R. Barton and S. W. McCombie, A new method for the deoxygenation of secondary alcohols, *J. Chem. Soc. Perkin I* (1974) 1574–1585.
16. Y. Chong and C. K. Chu, Efficient synthesis of 2-deoxy-L-*erythro*-pentose (2-deoxy-L-ribose) from L-arabinose, *Carbohydr. Res.*, 337 (2002) 397–402.
17. M. Ramaiah, Radical reactions in organic synthesis, *Tetrahedron*, 43 (1987) 3541–3676.
18. D. H. R. Barton, D. O. Jang, and J. Cs. Jaszberenyi, On the mechanism of deoxygenation of secondary alcohols by tin hydride reduction of methyl xanthates and thiocarbonyl derivatives, *Tetrahedron Lett.*, 31 (1990) 3991–3994.
19. C. Chatgilialoglu, Tris(trimethylsilyl)silane—A new reducing agent, *J. Org. Chem.*, 53 (1988) 3641–3642.
20. C. Chatgilialoglu, Organosilanes as radical-based reducing agents in synthesis, *Acc. Chem. Res.*, 25 (1992) 188–194.

21. D. H. R. Barton, D. O. Jang, and J. Cs. Jaszberenyi, The invention of radical reactions. Part XXXI. Diphenylsilane: A reagent for deoxygenation of alcohols via their thiocarbonyl derivatives, deamination via isonitriles, and dehalogenation of bromo- and iodo-compounds by radical chain chemistry, *Tetrahedron*, 49 (1993) 7193–7214.
22. D. H. R. Barton, D. O. Jang, and J. Cs. Jaszberenyi, Radical deoxygenation of secondary and primary alcohols with phenylsilane, *Synlett* (1991) 435–438.
23. D. H. R. Barton, D. O. Jang, and J. Cs. Jaszberenyi, The invention of radical reactions. Part XXIX. Radical mono- and dideoxygenations with silanes, *Tetrahedron*, 49 (1993) 2793–2804.
24. J. Nicholas Kirwan, B. P. Roberts, and C. R. Willis, Deoxygenation of alcohols by the reactions of their xanthate esters with triethylsilane: An alternative to tributyltin hydride in the Barton–McCombie reaction, *Tetrahedron Lett.*, 31 (1990) 5093–5096.
25. D. H. R. Barton, D. O. Jang, and J. Cs. Jaszberenyi, The invention of radical reactions. Radical deoxygenations, dehalogenations, and deaminations with dialkyl phosphites and hypophosphorous acid as hydrogen sources, *J. Org. Chem.*, 58 (1993) 6838–6842.
26. H-.-S. Dang, B. P. Roberts, J. S. Sekhon, and T. M. Smits, Deoxygenation of carbohydrates by thiol-catalysed radical-chain redox rearrangement of the derived benzylidene acetals, *Org. Biomol. Chem.*, 1 (2003) 1330–1341.
27. B. P. Roberts and T. M. Smits, Regioselectivity in the ring opening of 2-phenyl-1,3-dioxan-2-yl radicals derived from cyclic benzylidene acetals and comparison with deoxygenation of a carbohydrate diol via its cyclic thionocarbonate, *Tetrahedron Lett.*, 42 (2001) 3663–3666.
28. R. Lagunas and E. Moreno, Inhibition of glycolysis by 2-deoxygalactose in *Saccharomyces cerevisiae*, *Yeast*, 8 (1992) 107–115; G. Kaluza, C. Scholtissek, and R. A. Rott, Inhibition of the multiplication of enveloped RNA-viruses by glucosamine and 2-deoxy-D-glucose, *J. Gen. Virol.*, 14 (1972) 251–259; A. El-Ghaouth, C. L. Wilson, and M. Wisniewski, Antifungal activity of 2-deoxy-D-glucose on *Botrytis cinerea*, *Penicillium expansum*, and *Rhizopus stolonifer*: Ultrastructural and cytochemical aspects, *Phytopatology*, 87 (1997) 772–779.
29. Z. Zhu, W. Jiang, J. N. McGinley, and H. J. Thompson, 2-Deoxyglucose as an energy restriction mimetic agent: Effects on mammary carcinogenesis and on mammary tumor cell grouth in vitro, *Cancer Res.*, 65 (2005) 7023–7030 and references therein.
30. R. J. Ferrier and J. O. Hoberg, Unsaturated sugars, *Adv. Carbohydr. Biochem.*, 58 (2003) 55–120.
31. S. Danishefsky and M. Bilodeau, Glycals in organic synthesis: The evolution of comprehensive strategies for the assembly of oligosaccharides and glycoconjugates of biological consequence, *Angew. Chem. Int. Ed. Engl.*, 35 (1996) 1380–1419.
32. V. Bolitt, C. Mioskowski, S. G. Lee, and J. R. Falck, Direct preparation of 2-deoxy-D-glucopyranosides from glucals without Ferrier rearrangement, *J. Org. Chem.*, 55 (1990) 5812–5813.
33. N. Kaila, M. Blumenstein, H. Bielawska, and R. W. Franck, Face selectivity of the protonation of glycals, *J. Org. Chem.*, 57 (1992) 4576–4578.
34. S. Sabesan and S. Neira, Synthesis of 2-deoxy sugars from glycals, *J. Org. Chem.*, 56 (1991) 5468–5472.
35. K. Toshima, H. Nagai, Y. Ushiki, and S. Matsumara, Novel glycosidations of glycals using BCl3 or BBr3 as a promoter for catalytic and stereoselective syntheses of 2-deoxy-α-glycosides, *Synlett* (1998) 1007–1009.
36. E. Wieczorek and J. Thiem, Unusual reactions of enopyranuronates with hydroxy acid esters under ferrier conditions, *Synlett* (1998) 467–468.
37. B. D. Sherry, R. N. Loy, and F. D. Toste, Rhenium(V)-catalyzed synthesis of 2-deoxy-α-glycosides, *J. Am. Chem. Soc.*, 126 (2004) 4510–4511.

38. H. Kim, H. Men, and C. Lee, Stereoselective palladium-catalyzed *O*-glycosylation using glycols, *J. Am. Chem. Soc.*, 126 (2004) 1336–1337.
39. E. Bettelli, P. Cherubini, P. D'Andrea, P. Passacantilli, and G. Piancatelli, Mercuration-reductive demercuration of glycals: A mild and convenient entry to 2-deoxy-sugars, *Tetrahedron*, 54 (1998) 6011–6018.
40. V. Costantino, C. Imperatore, E. Fattorusso, and A. Mangoni, A mild and easy one-pot procedure for the synthesis of 2-deoxysugars from glycals, *Tetrahedron Lett.*, 41 (2000) 9177–9180.
41. J. S. Yadav, B. V. S. Reddy, K. Bhaskar Reddy, and M. Satyanarayana, $CeCl_3 \cdot 7H_2O$: A novel reagent for the synthesis of 2-deoxysugars from D-glycals, *Tetrahedron Lett.*, 43 (2002) 7009–7012.
42. S. Rani, A. Agarwal, and Y. D. Vankar, $LaCl_3 \cdot 7H_2O$/NaI/benzyl alcohol: A novel reagent system for regioselective hydration of glycals: Application in the synthesis of 1,6-dideoxynojirimycin, *Tetrahedron Lett.*, 44 (2003) 5001–5004.
43. S. Paul and N. Jayaraman, Catalytic ceric ammonium nitrate mediated synthesis of 2-deoxy-1-thioglycosides, *Carbohydr. Res.*, 339 (2004) 2197–2204.
44. J. R. Rasmussen, C. J. Slinger, R. J. Kordish, and D. D. Newman-Evans, Synthesis of deoxy sugars. Deoxygenation by treatment with *N,N′*-thiocarbonyldiimidazole/tri-*n*-butylstannane, *J. Org. Chem.*, 46 (1981) 4843–4846.
45. Y. Margaret, H. Wong, and G. R. Gray, 2-deoxy-D-*arabino*-hexose, 2-deoxy-D-*lyxo*-hexose, and their (2*R*)-2-deuterio analogs, *Carbohydr. Res.*, 80 (1980) 87–98.
46. B. Giese, S. Gilges, K. S. Lambert, and T. Witzel, Synthesis of 2-deoxy sugars, *Liebigs Ann. Chem.* (1988) 615–617.
47. A. Koch, C. Lamberth, F. Wetterich, and B. Giese, Radical rearrangement of 2-*O*-(diphenylphosphoryl)glycosyl bromides. A new synthesis for 2-deoxy disaccharides and 2-deoxy ribonucleosides, *J. Org. Chem.*, 58 (1993) 1083–1089.
48. A. Alberti, M. A. Della Bona, D. Macciantelli, F. Pelizzoni, G. Sello, G. Torri, and E. Vismara, Reactivity of glucosyl radical in the presence of phenols, *Tetrahedron*, 52 (1996) 10241–10248.
49. Z. Wang, D. R. Prudhomme, J. R. Buck, M. Park, and C. J. Rizzo, Stereocontrolled syntheses of deoxyribonucleosides via photoinduced electron-transfer deoxygenation of benzoyl-protected ribo- and arabinonucleosides, *J. Org. Chem.*, 65 (2000) 5969–5985.
50. S. Riedel, A. Donnerstag, L. Hennig, P. W. Richter, K. Hobert, and D. M. van Heijenoort, Synthesis and transglycosylase-inhibiting properties of a disaccharide analogue of moenomycin A lacking substitution at C-4 of unit F, *Tetrahedron*, 55 (1999) 1921–1936.
51. I. Paterson and M. D. McLeod, Studies in macrolide synthesis: A sequential aldol/glycosylation approach to the synthesis of concanamycin A, *Tetrahedron Lett.*, 36 (1995) 9065–9068; T. Muller, R. Schneider, and R. R. Schmidt, Utility of glycosyl phosphites as glycosyl donors-fructofuranosyl and 2-deoxyhexopyranosyl phosphites in glycoside bond formation, *Tetrahedron Lett.*, 35 (1994) 4763–4766; R. Pongdee, B. Wu, and G. A. Sulikowski, One-pot synthesis of 2-deoxy-β-oligosaccharides, *Org. Lett.*, 3 (2001) 3523–3525.
52. A. Kirschning, M. Jesberger, and A. Schönberger, The first polymer-assisted solution-phase synthesis of deoxyglycosides, *Org. Lett.*, 3 (2001) 3623–3626.
53. H. Bielawska and M. Michalska, 2-Deoxyglycosyl phosphorodithioates. A novel type of glycosyl donors. Efficient synthesis of 2′-deoxydisaccharides, *J. Carbohydr. Chem.*, 10 (1991) 107–112.
54. T. Yamanoi and T. T. Inazu, A convenient 2-deoxy-α-D-glucopyranosylation reaction using dimethylphosphothionate method, *Chem. Lett.* (1990) 849–852.
55. K. Wiesner, T. Y. R. Tsai, and H. Jin, On cardioactive steroids. XVI. Stereoselective β-glycosylation of digitose: The synthesis of digitoxin, *Helv. Chim. Acta*, 68 (1985) 300–314.

56. D. Crich and T. J. Ritchie, Preparation of 2-deoxy-β-D-*lyxo*-hexosides (2-deoxy-β-D-galactosides), *Carbohydr. Res.*, 190 (1989) C3–C6.
57. D. Kahne, D. Yang, J. J. Lim, R. Miller, and E. Paguaga, The use of alkoxy-substituted anomeric radicals for the construction of beta-glycosides, *J. Am. Chem. Soc.*, 110 (1988) 8716–8717.
58. S.-i. Hashimoto, A. Sano, H. Sakamoto, M. Nakajima, Y. Yanagiya, and S. Ikegami, An attempt at the direct construction of 2-deoxy-β-glycosidic linkages capitalizing on 2-deoxyglycopyranosyl diethyl phosphites as glycosyl donors, *Synlett* (1995) 1271–1273.
59. J. Thiem and M. Gerken, Synthesis of the E-D-C trisaccharide unit of aureolic acid cytostatics, *J. Org. Chem.*, 50 (1985) 954–958 and references therein.
60. W. R. Roush, D. P. Sebesta, and R. A. James, Stereoselective synthesis of 2-deoxy-β-glycosides from glycal precursors. 2. Stereochemistry of glycosidation reactions of 2-thiophenyl- and 2-selenophenyl-α-D-*gluco*-pyranosyl donors, *Tetrahedron*, 53 (1997) 8837–8852.
61. M. Perez and J.-M. Beau, Selenium-mediated glycosidations: A selective synthesis of β-2-deoxyglycosides, *Tetrahedron Lett.*, 30 (1989) 75–78.
62. K. C. Nicolaou, H. J. Mitchell, K. C. Fylaktakidou, H. Suzuki, and R. H. Rodriguez, 1,2-Seleno migrations in carbohydrate chemistry: Solution and solid-phase synthesis of 2-deoxy glycosides, orthoesters and allyl orthoesters, *Angew. Chem. Int. Ed. Engl.*, 39 (2000) 1089–1093; K. C. Nicolaou, H. J. Mitchell, K. C. Fylaktakidou, R. H. Rodriguez, and H. Suzuki, Total synthesis of everninomicin 13,384-1—Part 2: Synthesis of the $FGHA_2$ fragment, *Chem. Eur. J.*, 6 (2000) 3116–3148; K. C. Nicolaou, H. J. Mitchell, H. Suzuki, R. H. Rodriguez, O. Bauoin, and K. C. Fylaktakidou, Total synthesis of everninomicin 13,384-1—Part 1: Synthesis of the A(1)B(A)C fragment, *Angew. Chem. Int. Ed. Engl.*, 38 (1999) 3334–3339.
63. M. Trumtel, P. Tavecchia, A. Veyrières, and P. Sinaÿ, The synthesis of 2′deoxy-β-disaccharides: Novel approaches, *Carbohydr. Res.*, 191 (1989) 29–52.
64. J. Gervay and S. Danishefsky, A stereospecific route to 2-deoxy-β-glycosides, *J. Org. Chem.*, 56 (1991) 5448–5451.
65. C. Leteux, A. Veyrières, and F. Robert, An electrophile-mediated cyclization on the 1,6-anhydro-D-glucopyranose framework, *Carbohydr. Res.*, 242 (1993) 119–130.
66. D. Tailler, J. C. Jacquinet, A. M. Noirot, and J. M. Beau, An expeditious and stereocontrolled preparation of 2-azido-2-deoxy-β-D-glucopyranose derivatives from D-glucal, *J. Chem. Soc. Perkin Trans. I* (1992) 3163–3164.
67. W. R. Roush and C. E. Bennett, A highly stereoselective synthesis of 2-deoxy-β-glycosides using 2-deoxy-2-iodo-glucopyranosyl acetate donors, *J. Am. Chem. Soc.*, 121 (1999) 3541–3542.
68. W. R. Roush, B. W. Gung, and C. E. Bennett, 2-Deoxy-2-iodo- and 2-deoxy-2-bromo-α-glucopyranosyl trichloroacetimidates: Highly reactive and stereoselective donors for the synthesis of 2-deoxy-β-glycosides, *Org. Lett.*, 1 (1999) 891–893.
69. W. R. Roush and C. E. Bennett, A highly stereoselective synthesis of 2-deoxy-β-glycosides using 2-deoxy-2-iodo-glucopyranosyl acetate donors, *J. Am. Chem. Soc.*, 121 (1999) 3541–3542.
70. W. R. Roush, D. P. Sebesta, and R. A. James, Stereoselective synthesis of 2-deoxy-β-glycosides from glycal precursors. 2. Stereochemistry of glycosidation reactions of 2-thiophenyl- and 2-selenophenyl-α-D-*gluco*-pyranosyl donors, *Tetrahedron*, 53 (1997) 8837–8852.
71. W. R. Roush, R. A. Hartz, and D. J. Gustin, Total synthesis of olivomycin A, *J. Am. Chem. Soc.*, 121 (1999) 1990–1991.
72. J.-M. Petit, F. Paquet, and J.-M. Beau, Syntheses of β-2-deoxy-D-glycosides assisted by glycosidases, *Tetrahedron Lett.*, 32 (1991) 6125–6128.
73. A. Trincone, E. Pagnotta, M. Rossi, M. Mazzone, and M. Moracci, Enzymatic synthesis of 2-deoxy-β-glucosides and stereochemistry of β-glycosidase from *Sulfolobus solfataricus* on glucal, *Tetrahedron Asymmetry*, 12 (2001) 2783–2787.

74. Z. Witczak and R. L. Whistler, A convenient synthesis of 3-deoxy-D-*erythro*-pentose, *Carbohydr. Res.*, 110 (1982) 326–329.
75. A. Varki, Biological roles of oligosaccharides: All of the theories are correct, *Glycobiology*, 3 (1993) 97–130.
76. A. Varki, Selectin ligands, *Proc. Natl. Acad. Sci.*, 91 (1994) 7390–7397.
77. A. C. C. Frash, Functional diversity in the trans-sialidase and mucin families in *Trypanosoma cruzi*, *Parasitol. Today*, 16 (2000) 282–286.
78. R. V. Stick and K. A. Stubbs, From glycoside hydrolases to thioglycoligases: The synthesis of thioglycosides, *Tetrahedron Asymmetry*, 16 (2005) 321–335.
79. C. Copeland and R. V. Stick, A synthesis of abequose (3,6-dideoxy-D-*xylo*-hexose), *Aust. J. Chem.*, 30 (1977) 1269–1273.
80. S. S. Thomas, J. Plenkiewicz, E. R. Ison, M. Bols, W. Zou, W. A. Szarek, and R. Kisilevsky, Influence of monosaccharide derivatives on liver cell glycosaminoglycan synthesis: 3-Deoxy-D-*xylo*-hexose (3-deoxy-D-galactose) and methyl(methyl 4-chloro-4-deoxy-β-D-galactopyranosid)uronate, *Biochim. Biophys. Acta*, 1272 (1995) 37–48.
81. R. M. de Lederkremer, M. I. Litter, and L. F. Sala, β-Elimination in aldonolactones: A convenient synthesis of 2,4,6-tri-*O*-benzoyl-3-deoxy-D-*arabino*-hexono-1,5-lactone, *Carbohydr. Res.*, 36 (1974) 185–187.
82. C. Du Mortier and R. M. de Lederkremer, A new synthesis of 3-deoxy-D-*arabino*-hexose and its tautomeric equilibrium, *J. Carbohydr. Chem.*, 3 (1984) 219–228.
83. R. M. de Lederkremer and O. Varela, Synthetic reactions of aldonolactones, *Adv. Carbohydr. Chem. Biochem.*, 50 (1994) 125–209.
84. K. Bock, I. Lundt, and C. Pedersen, The base catalyzed rearrangement of some 6-bromo-2,6-dideoxyaldono-1,4-lactones. Preparation of L-digitoxose, *Acta Chem. Scand. B*, 38 (1984) 555–561.
85. L. O. Jeroncic, A. Fernandez Cirelli, and R. M. de Lederkremer, Synthesis of crystalline derivatives of 3-deoxy-D-*gluco*-heptofuranose, *Carbohydr. Res.*, 167 (1987) 175–186.
86. C. Marino, A. Chiocconi, O. Varela, and R. M. de Lederkremer, The glycosyl aldonolactone approach for the synthesis of β-D-Gal*f*-(1→3)-D-Man*p* and 3-deoxy-α-D-*xylo*-hexofuranosyl-(1→3)-D-Man*p*, *Carbohydr. Res.*, 311 (1998) 183–189.
87. E. Katzenellenbogen, N. A. Kocharova, G. V. Zatonsky, D. Witkowska, M. Bogulska, A. S. Shashkov, A. Gamian, and Y. A. Knirel, Structural and serological studies on a new 4-deoxy-D-arabino-hexose-containing O-specific polysaccharide from the lipopolysaccharide of Citrobacter braakii PCM 1531 (serogroup O6), *Eur. J. Biochem.*, 270 (2003) 2732–2738.
88. E. Romanowska, A. Romanowska, J. Dabrowski, and M. Hauck, Structure determination of the O-specific polysaccharides from *Citrobacter* O4- and O27-lipopolysaccharides by methylation analysis and one- and two-dimensional ^{1}H-NMR spectroscopy, *FEBS Lett.*, 211 (1987) 175–178.
89. E. Romanowska, A. Romanowska, C. Lugowski, and E. Katzenellenbogen, Structural and serological analysis of *Citrobacter*-036-specific polysaccharide, the homopolymer of (β1–2)-linked 4-deoxy-D-*arabino*-hexopyranosyl units, *Eur. J. Biochem.*, 121 (1981) 119–123.
90. A. Gamian, E. Romanowska, A. Romanowska, C. Lugowski, J. Dabrowski, and K. Trauner, *Citrobacter* lipopolysaccharides: Structure elucidation of the O-specific polysaccharide from strain PCM 1487 by mass spectrometry, one-dimensional and two-dimensional ^{1}H NMR spectroscopy and methylation analysis, *Eur. J. Biochem.*, 146 (1985) 641–647.
91. P. J. Garegg and H. Hultberg, A novel, reductive ring-opening of carbohydrate benzylidene acetals, with unusual regioselectivity, *Carbohydr. Res.*, 93 (1981) C10–C11.
92. G. Siewert and O. Westphal, Substitution sekundärer Tosylestergruppen durch Jod Synthese von 4-Desoxy- und 4,6-Didesoxy-D-*xylo*-hexose, *Liebigs Ann. Chem.*, 720 (1988) 161–170.

93. E. P. Barrette and L. Goodman, Convenient and stereospecific synthesis of deoxy sugars. Reductive displacement of trifluoromethylsulfonates, *J. Org. Chem.*, 49 (1984) 176–178.
94. D. D. Cox, E. K. Metzner, and E. J. Reist, The synthesis of methyl 4-*O*-(4-*O*-α-D-galactopyranosyl-β-D-galactopyranosyl)-β-D-glucopyranoside: The methyl β-glycoside of the trisaccharide related to Fabry's disease, *Carbohydr. Res.*, 63 (1978) 139–147.
95. R. Caputo, M. De Nisco, P. Festa, A. Guaragna, G. Palumbo, and S. Pedatella, Synthesis of 4-deoxy-L-(and D-)hexoses from chiral noncarbohydrate building blocks, *J. Org. Chem.*, 69 (2004) 7033–7037.
96. T. Saleh and G. Rousseau, Preparation and reactions of 2-allyl-5-deoxymannose derivatives, *Tetrahedron*, 58 (2002) 2891–2897.
97. W. G. Overend, Deoxy sugars. 5-Deoxy-D-*xylo*-hexose, *Method. Carbohydr. Chem.*, 6 (1972) 173–176.
98. S. David and G. De Sennyey, Synthese des 1-(5-désoxy-β-D-*ribo*-hexofuranosyl)cytosine et 1-(2,5-didésoxy-β-D-*érythro*-hexofuranosyl)cytosine, et de leurs phosphates. Contribution à l'étude de la spécificité d'une ribonucléotide-réductase de mammifère (rat), *Carbohydr. Res.*, 77 (1979) 79–97.
99. K. J. Ryan, H. Arzoumanian, E. M. Acton, and L. Goodman, Synthesis of homoribose (5-deoxy-D-allose) and homoadenosine, *J. Org. Chem.*, 86 (1964) 2503–2508.
100. M. L. Wolfrom, K. Matsuda, F. Komitsky Jr., and T. E. Whiteley, Hydroboration in the sugar series, *J. Org. Chem.*, 228 (1963) 3551–3553.
101. W. A. Szarek, R. G. S. Ritchie, and D. M. Vyas, Synthesis of 5-deoxy-D-*xylo*-hexose and 5-deoxy-L-*arabino*-hexose, and their conversion into adenine nucleosides, *Carbohydr. Res.*, 62 (1978) 89–103.
102. M. Iwakawa, O. R. Martin, and W. A. Szarek, Synthesis of 1-(5-deoxy-β-D-*arabino*-hexofuranosyl)cytosine, *Carbohydr. Res.*, 121 (1983) 99–108.
103. B. P. Roberts and T. M. Smits, Radical-chain redox rearrangement of cyclic benzylidene acetals to benzoate esters in the presence of thiols, *Tetrahedron Lett.*, 42 (2001) 137–140.
104. M. H. Fechter and A. E. Stütz, Synthetic application of glucose isomerase: Isomerization of C-5-modified (2*R*,3*R*,4*R*)-configured hexoses into the corresponding 2-ketoses, *Carbohydr. Res.*, 319 (1999) 55–62.
105. T. Nakamura, Y. Goda, S. Sakai, K. Kondo, H. Akiyama, and M. Toyoda, Cardenolide glycosides from seeds of *Corchorus olitorius*, *Phytochemistry*, 49 (1998) 2097–2102.
106. Y. Takeda, Y. Okada, T. Masuda, E. Hirata, T. Shinzato, and H. Otsuka, C,O-bisglycosylapigenins from the leaves of *Rhamnella inaequilatera*, *Phytochemistry*, 65 (2004) 463–468.
107. R. Suzuki, Y. Okada, and T. Okuyama, Two flavone *C*-glycosides from the style of *Zea mays* with glycation inhibitory activity, *J. Nat. Prod.*, 66 (2003) 564–565.
108. Z. Bing-nan, B. Gabor, and A. C. Geoffrey, Alternanthin *C*-glycosylated flavonoid from *Alternanthera philoxeroides*, *Phytochemistry*, 27 (1988) 3633–3636.
109. T. Reichstein and E. Weiss, The sugars of the cardiac glycosides, *Adv. Carbohydr. Chem.*, 17 (1962) 65–120.
110. A. E. Aulabaugh, R. C. Crouch, G. E. Martin, A. Ragouezeos, J. P. Shockcor, T. D. Spitzer, R. Duncan Farrant, B. D. Hudson, and J. C. Lindon, The confomational behaviour of the cardiac glycoside digoxin as indicated by NMR spectroscopy and molecular dynamics calculations, *Carbohydr. Res.*, 230 (1992) 201–213.
111. I. M. Shaikh, B. W. Lau, B. A. Siegfried, and R. Valdes Jr., Isolation of digoxin-like immunoreactive factors from mammalian adrenal cortex, *J. Biol. Chem.*, 266 (1991) 13672–13678.
112. H. M. Qazzaz, S. L. Goudy, and R. Valdes Jr., Deglycosylated products of endogenous digoxin-like immunoreactive factor in mammalian tissue, *J. Biol. Chem.*, 271 (1996) 8731–8737.

113. V. D. Huan, K. Ohtani, R. Kasai, K. Yamasaki, and N. V. Tuu, Sweet pregnane glycosides from *Telosma procumbens*, *Chem. Pharm. Bull.*, 49 (2001) 453–460.
114. T. Warashina and T. Noro, Steroidal glycosides from the aerial part of *Asclepias incarnate*, *Phytochemistry*, 53 (2000) 485–498.
115. F. Abe, Y. Mori, and T. Yamauchi, Steroidal constituents from roots stem of *Asclepias fruticosa*, *Chem. Pharm. Bull.*, 42 (1994) 1777–1783.
116. T. Kanchanapoom, R. Kasai, K. Ohtani, M. Andriantsiferana, and K. Yamasaki, Pregnane and pregnane glycosides from the malagasy plant, *Cynanchum aphyllum*, *Chem. Pharm. Bull.*, 50 (2002) 1031–1034.
117. F. Abe and T. Yamauchi, Pregnane glycosides from the roots of *Asclepias tuberosa*, *Chem. Pharm. Bull.*, 48 (2000) 1017–1022.
118. R. Vleggaar, F. R. van Heerden, and L. A. P. Anderson, Toxic constituents of the *Asclepiadaceae*. Structure elucidation of Sarcovimiside A-C, pregnane glycosides of *Sarcostemma viminale*, *J. Chem. Soc. Perkin Trans. I* (1983) 483–487.
119. K. Kulowski, E. Wendt-Pienkowski, L. Han, K. Yang, L. C. Vining, and C. R. Hutchinson, Functional characterization of the jadI gene as a cyclase forming angucyclinones, *J. Am. Chem. Soc.*, 121 (1999) 1786–1794.
120. U. Rix, J. Zheng, L. L. Remsing Rix, L. Greenwell, K. Yang, and J. Rohr, The dynamic structure of jadomycin B and the amino acid incorporation step of its biosynthesis, *J. Am. Chem. Soc.*, 126 (2004) 4496–4497.
121. L. Wang, R. L. White, and L. C. Vining, Biosynthesis of the dideoxysugar component of jadomycin B: Genes in the jad cluster of *Streptomyces venezuelae* ISP5230 for L-digitoxose assembly and transfer to the angucycline aglycone, *Microbiology*, 148 (2002) 1091–1103.
122. Z. U. Ahmed and L. C. Vining, Evidence for a chromosomal location of the genes coding for chloramphenicol production in *Streptomyces venezuelae*, *J. Bacteriol.*, 154 (1983) 239–244.
123. T. Nakashima, M. Miura, and M. Hara, Tetrocarcin A inhibibits mitochondrial function of Bcl-2 and suppresses its anti-apoptotic activity, *Cancer Res.*, 60 (2000) 1229–1235 and references therein.
124. T. Tamaoki, M. Kasai, K. Shirahata, and F. Tomita, Tetrocarcins E1, E2, F and F-1, new antibiotics. Fermentation, isolation and characterization, *J. Antibiot.*, 35 (1982) 979–984.
125. Y. Igarashi, K. Takagi, Y. Kan, K. Fujii, K. Harada, T. Furumai, and T. Oki, Arisostatins A and B, new members of tetrocarcin class of antibiotics from *Micromonospora* sp. TP-A0316. II. Structure determination, *J. Antibiot.*, 53 (2000) 233–440.
126. S. Kitanata, K. Ogata, and M. Takido, Studies on the constituents of the leaves of *Cassia torosa* Cav I. The structure of two new *C*-glycosylflavones, *Chem. Pharm. Bull.*, 37 (1989) 2441–2444.
127. J. Rohr, C. Mendez, and J. A. Salas, The biosynthesis of aureolic acid group antibiotic, *Bioorg. Chem.*, 27 (1998) 41–54.
128. D. Rodríguez, L. M. Quirós, and J. A. Salas, MtmMII-mediated *C*-methylation during biosynthesis of the antitumor drug mithramycin is essential for biological activity and DNA-drug interaction, *J. Biol. Chem.*, 279 (2004) 8149–8158.
129. S. O'Connor, Aureolic acids: Similar antibiotics with different biosynthetic gene clusters, *Chem. Biol.*, 11 (2004) 8–10.
130. R. Katahira, Y. Uosaki, H. Ogawa, Y. Yamashita, H. Nakano, and M. Yoshida, UCH9, a new antitumor antibiotic produced by Streptomyces. II. Structure elucidation of UCH9 by mass and NMR spectroscopy, *J. Antibiot.*, 51 (1998) 267–274.
131. H. Jayasuriya, R. B. Lingham, P. Graham, D. Quamina, L. Herranz, O. Genilloud, M. Gagliardi, R. Danzeisen, J. E. Tomassini, D. L. Zink, Z. Guan, and S. B. Singh, Durhamycin A a potent inhibitor of HIV Tat transactivation, *J. Nat. Prod.*, 65 (2002) 1091–1095.

132. E. Fernandez, U. Weissbach, C. Sanchez Reillo, A. F. Brana, C. Mendez, J. Rohr, and J. A. Salas, Identification of two genes from *Streptomyces argillaceus* encoding glycosyltransferases involved in transfer of a disaccharide during biosynthesis of the antitumor drug mithramycin, *J. Bacteriol.*, 180 (1998) 4929–4937.
133. D. L. Banville, M. A. Keniry, and R. H. Shafer, NMR investigation of mithramycin A binding to d(ATGCAT)2: A comparative study with chromomycin A3, *Biochemistry*, 29 (1990) 9294–9304.
134. D. J. Silva, R. Goodnow Jr, and D. Kahne, The sugars in chromomycin A3 stabilize the Mg(2+)-dimer complex, *Biochemistry*, 32 (1993) 463–471.
135. M. Nur-e-Alam, C. Mendez, J. A. Salas, and J. Rohr, Elucidation of the glycosylation sequence of mithramycin biosynthesis: Isolation of 3A-deolivosylpremithramycin B and its conversion to premithramycin B by glycosyltransferase MtmGII, *Chembiochem.*, 6 (2005) 632–636.
136. L. L. Remsing, A. M. Gonzalez, M. Nur-e-Alam, M. J. Fernandez-Lozano, A. F. Brana, U. Rix, M. A. Oliveira, C. Mendez, J. A. Salas, J. Rohr. Mithramycin SK, a novel antitumor drug with improved therapeutic index, mithramycin SA, and demycarosyl-mithramycin SK: Three new products generated in the mithramycin producer *Streptomyces argillaceus* through combinatorial biosynthesis, *J. Am. Chem. Soc.*, 125 (2003) 5745–5753.
137. N. Menendez, M. Nur-e-Alam, A. F. Brana, J. Rohr, J. A. Salas, and C. Mendez, Biosynthesis of the antitumor chromomycin A3 in *Streptomyces griseus*: Analysis of the gene cluster and rational design of novel chromomycin analogs, *Chem. Biol.*, 11 (2004) 21–32.
138. W. Lu, C. Leimkuhler, M. Oberthür, D. Kahne, and K. Walsh, AknK is an L-2-deoxyfucosyltransferase in the biosynthesis of the anthracycline aclacinomycin A, *Biochemistry*, 43 (2004) 4548–4558 and references therein.
139. I. Fujii and Y. Ebizuka, Anthracycline biosynthesis in *Streptomyces galilaeus*, *Chem. Rev.*, 97 (1997) 2511–2524.
140. K. Raty, J. Kantola, A. Hautala, J. Hakala, K. Ylihonko, and P. Mantsala, Cloning and characterization of *Streptomyces galilaeus* aclacinomycins polyketide synthase (PKS) cluster, *Gene*, 293 (2002) 115–122.
141. F. Animati, F. Arcamone, M. Berettoni, A. Cipollone, M. Franciotti, and P. Lombardi, New anthracycline disaccharide. Synthesis of L-daunosaminyl-α(1→4)-2-deoxy-L-rhamnosyl and of L-daunosamyl-α(1→4)-2-deoxy-L-fucosyl daunorubicin analogues, *J. Chem. Soc. Perkin Trans. I* (1996) 1327–1329.
142. J. Gildersleeve, A. Smith, K. Sakurai, S. Raghavan, and D. Kahne, Scavenging byproducts in the sulfoxide glycosylation reaction: Application to the synthesis of ciclamycin 0, *J. Am. Chem. Soc.*, 121 (1999) 6176–6182.
143. G. Pratesi, M. E. Cesare, C. Caserini, P. Perego, L. D. Bo, D. Polizzi, R. Supino, M. Bigioni, S. Manzini, E. Iafrate, C. Salvatore, A. Casazza, F. Arcamone, and F. Zunino, Improved efficacy and enlarged spectrum of activity of a novel anthracycline disaccharide analogue of doxorubicin against human tumor xenografts, *Clin. Cancer Res.*, 4 (1998) 2833–2839.
144. S. Weber, C. Zolke, J. Rohr, and J. M. Beale, Investigations of the biosynthesis and structural revision of landomycin A, *J. Org. Chem.*, 59 (1994) 4211–4214.
145. W. R. Roush and R. J. Neitz, Studies on the synthesis of landomycin A. Synthesis of the originally assigned structure of the aglycone, landomycinone, and revision of structure, *J. Org. Chem.*, 69 (2004) 4906–4912.
146. B. Faust, D. Hoffmeister, G. Weitnauer, L. Westrich, S. Haag, P. Schneider, H. Decker, E. Kunzel, J. Rohr, and A. Bechthold, Two new tailoring enzymes, a glycosyltransferase and oxygenase, involved in biosynthesis of the angucycline antibiotic urdamycin A in *Streptomyces fradiae* Tü2717, *Microbiology*, 146 (2000) 147–154.

147. K. Rohn and J. Rohr, Angucyclines: Total syntheses, new structure and biosynthetic studies of an emerging new class of antibiotics, *Top. Curr. Chem.*, 188 (1997) 127–195.
148. Y.-B. Zhao, Y.-M. Shen, H.-P. He, Y.-M. Li, Q.-Z. Mu, and X.-J. Hao, Carbohydrates from *Cynanchum otophyllum*, *Carbohydr. Res.*, 339 (2004) 1967–1972.
149. M. Gilleron, A. Venisse, J. J. Fournie, M. Riviere, M. A. Dupont, N. Gas, and G. Puzo, Structural and immunological properties of the phenolic glycolipids from *Mycobacterium gastri* and *Mycobacterium kansasii*, *Eur. J. Biochem.*, 189 (1990) 167–173.
150. L. Rodriguez, D. Rodriguez, C. Olano, A. F. Brana, C. Mendez, and J. A. Salas, Functional analysis of OleY L-oleandrosyl 3-*O*-methyltransferase of the oleandomycin biosynthetic pathway in *Streptomyces antibioticus*, *J. Bacteriol.*, 183 (2001) 5358–5363.
151. J. W. Kim, H. Adachi, K. Shin-ya, Y. Hayakawa, and H. Seto, Apoptolidin, a new apoptosis inducer in transformed cells from *Nocardiopsis sp.*, *J. Antibiot.*, 50 (1997) 628–630.
152. Y. Hayakawa, J. W. Kim, H. Adachi, K. Shin-ya, K.-i. Fujita, and H. Seto, Structure of apoptolidin, a specific apoptosis inducer in transformed cells, *J. Am. Chem. Soc.*, 120 (1998) 3524–3525.
153. M. T. Crimmins and A. Long, Enantioselective synthesis of apoptolidin sugars, *Org. Lett.*, 7 (2005) 4157–4160.
154. H. Wehlan, M. Dauber, M. T. Mujica Fernaud, J. Schuppan, R. Mahrwald, B. Ziemer, M. E. Juarez Garciz, and U. Koert, Total synthesis of apoptolidin, *Angew. Chem. Int. Ed.*, 43 (2004) 4597–4601.
155. K. C. Nicolaou, Y. Li, K. Sugita, H. Monenschein, P. Guntupalli, H. J. Mitchell, K. C. Fylaktakidou, D. Vourloumis, P. Giannakakou, and A. O'Brate, Total synthesis of apoptolidin: Completion of the synthesis and analogue synthesis and evaluation, *J. Am. Chem. Soc.*, 125 (2003) 15443–15454.
156. I. Aguirrezabalaga, C. Olano, N. Allende, L. Rodriguez, A. F. Braña, C. Méndez, and J. A. Salas, Identification and expression of genes involved in biosynthesis of L-oleandrose and its intermediate L-olivose in the oleandomycin producer *Streptomyces antibioticus*, *Antimicrob. Agents Chemother.*, 44 (2000) 1266–1275.
157. C. Fischer, L. Rodriguez, E. P. Patallo, F. Lipata, A. F. Brana, C. Mendez, J. A. Salas, and J. Rohr, Digitoxosyltetracenomycin C and glucosyltetracenomycin C, two novel elloramycin analogues obtained by exploring the sugar donor substrate specificity of glycosyltransferase ElmGT, *J. Nat. Prod.*, 65 (2002) 1685–1689.
158. W. R. Roush and R. J. Brown, Total synthesis of carbohydrates. 3. Efficient enantioselective syntheses of 2,6-dideoxyhexoses, *J. Org. Chem.*, 48 (1983) 5093–5101.
159. W. R. Roush and R. J. Brown, Total synthesis of carbohydrates: Stereoselective syntheses of 2, 6-dideoxy-D-arabino-hexose and 2,6-dideoxy-D-ribo-hexose, *J. Org. Chem.*, 47 (1982) 1371–1373.
160. G. Fronza, C. Fuganti, P. Grasselli, G. Pedrocchi-Fantoni, and C. Zirotti, On the steric course of the addition of diallylzinc onto α,β-dialkoxy chiral carbonyl compounds: Stereospecific synthesis of 2,6-dideoxysugars of the L-series, *Tetrahedron Lett.*, 23 (1982) 4143–4146.
161. T. Ulven and P. H. J. Carlsen, Synthesis of all diastereomers of the 2-deoxypentoses and the 2,6-dideoxyhexoses from 2-phenyl-1,3-dioxan-5-one hydrate, *Eur. J. Org. Chem.* (2001) 3367–3374.
162. P. R. Andreana, J. S. McLellan, Y. Chen, and P. G. Wang, Synthesis of 2,6-dideoxysugars via ring-closing olefinic metathesis, *Org. Lett.*, 4 (2002) 3875–3878.
163. A. S. Pepito and D. C. Dittmer, Application of tellurium chemistry and sharpless asymmetric epoxidation to the synthesis of optically active boivinose, *J. Org. Chem.*, 59 (1994) 4311–4312.
164. D. Horton, T.-M. Cheung, and W. Weckerle, 2,6-Dideoxy-D-*ribo*-hexose (digitoxose), *Method. Carbohydr. Chem.*, 8 (1980) 195–199 and references therein.

165. D. Horton, T.-M. Cheung, and W. Weckerle, 2,6-Dideoxy-D-*lyxo*-hexose (2-deoxy-L-fucose), *Method. Carbohydr. Chem.*, 8 (1980) 201–205.
166. J. S. Brimacombe, R. Hann, M. S. Saeed, and L. C. N. Tucker, Convenient synthesis of L-digitoxose, L-cymarose, and L-ristosamine, *J. Chem. Soc. Perkin I*, 1 (1982) 2583–2587.
167. H. W. Pauls and B. Fraser-Reid, Stereocontrolled conversion of allylic alcohols into vicinal cis-diol, *J. Carbohydr. Chem.*, 4 (1985) 1–14.
168. C. M. Schafer and T. F. Molinski, Practical synthesis of 2,6-dideoxy-D-*lyxo*-hexose ("2-deoxyfucose") from D-galactose, *Carbohydr. Res.*, 310 (1998) 223–228.
169. P. A. Searle, R. K. Richter, and T. F. Molinski, Bengazoles C-G from the sponge *Jaspis* sp. Synthesis of the side chain and determination of absolute configuration, *J. Org. Chem.*, 61 (1996) 4073–4079.
170. A. F. Hadfield, L. Cunningham, and A. C. Sartorelli, The synthesis and cytotoxic activity of 1,3,4-tri-*O*-acetyl-2,6-dideoxy-L-*arabino*- and -L-*lyxo*-hexopyranose, *Carbohydr. Res.*, 72 (1979) 93–104.
171. K. Bock, I. Lundt, and C. Pedersen, The preparation of some bromodeoxy- and deoxy-hexoses from bromodeoxyaldonic acids, *Carbohydr. Res.*, 90 (1981) 7–16.
172. M. Bols, I. Lundt, and E. R. Ottosen, Preparation of 2-, 3-, and 4-deoxy derivatives of L-rhamnose, and derivatives of 2-azido-2-deoxy-L-rhamnose and 2,6-dideoxy-2-fluoro-L-glucose, for use in glycosylation reactions, *Carbohydr. Res.*, 222 (1991) 141–149.
173. A. P. Rauter, S. Lucas, T. Almeida, D. Sacoto, V. Ribeiro, J. Justino, A. Neves, F. V. Silva, M. C. Oliveira, M. J. Ferreira, M. S. Santos, and E. Barbosa, Synthesis, surface active and antimicrobial properties of new alkyl 2,6-dideoxy-L-*arabino*-hexopyranosides, *Carbohydr. Res.*, 340 (2005) 191–201.
174. S. Amann, G. Dräger, C. Rupprath, A. Kirschning, and L. Elling, (Chemo)enzymatic synthesis of dTDP-activated 2,6-dideoxysugars as building blocks of polyketide antibiotics, *Carbohydr. Res.*, 335 (2001) 23–32.
175. J. Thiem, in D. Horton, L.D. Hawkins, and G.J. McGarvey (Eds.), Approaches to deoxy-oligosaccharides of antibiotics and cytostatics by stereoselective glycosylations, Trends in Synthetic Carbohydrate Chemistry, *ACS Symp. Ser.*, American Chemical Society, Washington, 1989, pp. 131–149.
176. J. Thiem and S. Köpper, Syntheses of kijanimicin oligosaccharides, *Tetrahedron*, 46 (1990) 113–138.
177. S. Köpper and J. Thiem, One-pot-synthesis of α-linked deoxy sugar trisaccharides, *Carbohydr. Res.*, 260 (1994) 219–232.
178. K. Suzuki, G. A. Sulikowski, R. W. Friesen, and S. J. Danishefsky, Application of substituent-controlled oxidative coupling of glycals in a synthesis and structural corroboration of ciclamycin 0: New possibilities for the construction of hybrid anthracyclines, *J. Am. Chem. Soc.*, 112 (1990) 8895–8902.
179. S. Raghavan and D. Kahne, A one step synthesis of the ciclamycin trisaccharide, *J. Am. Chem. Soc.*, 115 (1993) 1580–1581.
180. J. Gildersleeve, A. Smith, K. Sakurai, S. Raghavan, and D. Kahne, Scavenging byproducts in the sulfoxide glycosylation reaction: Application to the synthesis of ciclamycin 0, *J. Am. Chem. Soc.*, 121 (1999) 6176–6182.
181. F. E. McDonald and M. Wu, Stereoselective synthesis of L-oliose trisaccharide via iterative alkynol cycloisomerization and acid-catalyzed glycosylation, *Org. Lett.*, 4 (2002) 3979–3981.
182. T. B. Durham and W. R. Roush, Stereoselective synthesis of functionalized precursors of the CDEF and CDE 2,6-dideoxy-tetra- and trisaccharide units of durhamycins A and B, *Org. Lett.*, 5 (2003) 1875–1878.

183. B. Lindberg, Components of bacterial polysaccharides, *Adv. Carbohydr. Chem.*, 48 (1990) 279–318.
184. L. Kenne and B. Lindberg, Bacterial polysaccharides, in G. O. Aspinall (Ed.), *The Polysaccharides*, Vol. 2, Academic Press, Inc., New York, 1983, pp. 287–363.
185. S. B. Svenson, M. Nurminen, and A. A. Lindberg, Artificial *Salmonella vaccines*: O-Antigenic oligosaccharide-protein conjugates induce protection against infection with *Salmonella typhimurium*, *Infect Immun.*, 25 (1979) 863–872.
186. S. B. Svenson and A. A. Lindberg, Artificial *Salmonella* vaccines: *Salmonella typhimurium* O-antigen-specific oligosaccharide-protein conjugates elicit protective antibodies in rabbits and mice, *Infect Immun.*, 32 (1981) 490–496.
187. H. J. Jorbeck, S. B. Svenson, and A. A Lindberg, Artificial *Salmonella* vaccines: *Salmonella typhimurium* O-antigen-specific oligosaccharide-protein conjugates elicit opsonizing antibodies that enhance phagocytosis, *Infect Immun.*, 32 (1981) 497–502.
188. D. C. Watson, J. B. Robbins, and S. C. Szu, Protection of mice against *Salmonella typhimurium* with an O-specific polysaccharide-protein conjugate vaccine, *Infect Immun.*, 60 (1992) 4679–4686.
189. K. A. Mattos, A. R. Todeschini, N. Heise, C. Jones, J. O. Previato, and L. Mendonça-Previato, Nitrogen-fixing bacterium *Burkholderia brasiliensis* produces a novel yersiniose A-containing O-polysaccharide, *Glycobiology*, 15 (2005) 313–321.
190. M. Gilleron, J. Vercauteren, and G. Puzo, Lipo-oligosaccharidic antigen from *Mycobacterium gastri*, complete structure of a novel C4-branched 3,6-dideoxy-alpha-*xylo*-hexopyranose, *Biochemistry*, 22 (1994) 1930–1937.
191. J. Alam, N. Beyer, and H. w. Liu, Biosynthesis of colitose: Expression, purification, and mechanistic characterization of GDP-4-keto-6-deoxy-D-mannose-3-dehydrase (ColD) and GDP-L-colitose synthase (ColC), *Biochemistry*, 43 (2004) 16450–16460.
192. X. He, G. Agnihotri, and H.-w. Liu, Novel enzymatic mechanisms in carbohydrate metabolism, *Chem. Rev.*, 100 (2000) 4615–4662.
193. P. F. Jezyk and D. Fairbairn, Ascarosides and ascaroside esters in *Ascaris lumbricoides* (Nematoda), *Comp. Biochem. Physiol.*, 23 (1967) 691–705.
194. N. Wisnewski, M. McNeil, R. B. Grieve, and D. L. Wassom, Characterization of novel fucosyl- and tyvelosyl-containing glycoconjugates from *Trinchinela spiralis* muscle stage larvae, *Mol. Biochem. Parasitol.*, 61 (1993) 25–35.
195. D. R. Bundle, E. Eichler, M. A. Gidney, M. Meldal, A. Ragauskas, B. W. Sigurskjold, B. Sinnott, D. C. Watson, M. Yaguchi, and N. M. Young, Molecular recognition of a *Salmonella* trisaccharide epitope by monoclonal antibody Se155-4, *Biochemistry*, 33 (1994) 5172–5182.
196. T. M. Weigel and H.-w. Liu, Synthesis of stereospecifically labeled carbohydrates II: Preparation of (3*S*) and (3*R*)-[3-$^2H_1$]abequose, *Tetrahedron Lett.*, 29 (1988) 4221–4224.
197. R. N. Russell, T. M. Weigel, O. Han, and H.-w. Liu, Synthesis of stereospecifically labelled 3, 6-dideoxyhexoses, *Carbohydr. Res.*, 201 (1990) 95–114.
198. H. N. Yu, Ch.-Ch. Ling, and D. R. Bundle, Synthesis of three *Salmonella* epitopes for biosensor studies of carbohydrates-antibody interactions, *Can. J. Chem.*, 80 (2002) 1131–1140.
199. H. N. Yu, P. Zhang, C.-C. Ling, and D. R. Bundle, A practical route to 3,6-dideoxyhexoses, *Tetrahedron Asymmetry*, 11 (2000) 465–479.
200. K. Zegelaar-Jaarsveld, S. C. van der Plas, G. A. van der Marel, and J. van Boom, Preparation of disaccharide haptens corresponding to *Salmonella* Serogroups B and D, *J. Carbohydr. Chem.*, 15 (1996) 665–689.
201. D. R. Bundle and S. Josephson, A facile synthesis of methyl 3,6-dideoxy-L-*xylo*-hexopyranoside (colitose), *Can. J. Chem.*, 56 (1978) 2686–2690.

202. O. Varela, A. Fernández Cirrelli, and R. M. de Lederkremer, β-Elimination in aldonolactones. Synthesis of 3,6-dideoxy-L-*arabino*-hexose (Ascarylose), *Carbohydr. Res.*, 70 (1979) 27–35.
203. O. Varela, A. Fernandez Cirelli, and R. M. de Lederkremer, A crystalline furanose derivative of ascarylose. Synthesis of 2,5-di-*O*-benzoyl-3,6-dideoxy-α-L-*arabino*-hexofuranose, *Carbohydr. Res.*, 83 (1980) 130–135.
204. J. C. Florent, C. Monneret, and Q. Khuong-Huu, Synthèse et réactivite du méthyl-2-*O*-benzoyl-3-bromo-3,6-didésoxy-α-L-altropyranoside et du méthyl-2-*O*-benzoyl-3-bromo-3,6-didésoxy-4-*O*-méthyl-α-L-altropyranoside, *Carbohydr. Res.*, 56 (1977) 301–314.
205. P. Jütten and H.-D. Scharf, A simple preparation of methyl 3,6-dideoxy-α-L-*arabino*-hexopyranoside via a photodeoxygenation, *J. Carbohydr. Chem.*, 9 (1990) 675–681.
206. Y. A. Knirel, S. N. Senchenkova, P. E. Jansson, A. Weintraub, M. Ansaruzzaman, and M. J. Albert, Structure of the O-specific polysaccharide of an *Aeromonas trota* strain cross-reactive with *Vibrio cholerae* O139 Bengal, *Eur. J. Biochem.*, 238 (1996) 160–165.
207. J. Muldoon, A. V. Perepelov, A. S. Shashkov, R. P. Gorshkova, E. L. Nazarenko, V. A. Zubkov, E. P. Ivanova, Y. A. Knirel, and A. V. Savage, Structure of a colitose-containing O-specific polysaccharide of the marine bacterium *Pseudomonas tetraodonis* IAM 14160^{T}, *Carbohydr. Res.*, 333 (2001) 41–46.
208. Y. A. Knirel, L. Paredes, P.-E. Jansson, A. Weintraub, G. Widmalm, and M. J. Albert, Structure of the capsular polysaccharide of Vibrio cholerae O139 synonym Bengal containing D-galactose 4,6-cyclophosphate, *Eur. J. Biochem.*, 232 (1995) 391–396.
209. S. Gunawardena, C. R. Fiore, J. A. Johnson, and C. A. Bush, The conformation of a rigid tetrasaccharide epitope in the capsular polysaccharide of *Vibrio cholerae* O139, *Biochemistry*, 38 (1999) 12062–12071.
210. J. A. Johnson, A. Joseph, and J. G. Morris, Capsular polysaccharide-protein conjugate vaccines against *Vibrio cholerae* O139 Bengal, *Bull. Inst. Pasteur*, 93 (1995) 285–290.
211. S. N. Senchenkova, A. S. Shashkov, Y. A. Knirel, E. Schwarzmüller, and H. Mayer, Structure of the O-specific polysaccharide of *Salmonella enterica* ssp. Arizonae O50 (Arizona 9a,9b), *Carbohydr. Res.*, 301 (1997) 61–67.
212. L. Kenne, B. Lindberg, E. Söderholm, D. R. Bundle, and D. W. Griffith, Structural studies of the O-antigens from *Salmonella greenside* and *Salmonella adelaide*, *Carbohydr. Res.*, 111 (1983) 289–296.
213. B. Lindberg, F. Lindh, J. L. A. Lindberg, and S. B. Svenson, Structural studies of the O-specific side-chain of the lipopolysaccharide from *Escherichia coli* O55, *Carbohydr. Res.*, 97 (1981) 105–112.
214. T. Iversen and D. R. Bundle, Synthesis of the colitose determinant of *Escherichia coli* O111 and 3,6-di-*O*-(α-D-galactopyranosyl)-α-D-glucopyranoside, *Can. J. Chem.*, 60 (1982) 299–303.
215. S. Oscarson, U. Tedebark, and D. Turek, Synthesis of colitose-containing oligosaccharide structures found in polysaccharides from *Vibrio cholerae* O139 synonym Bengal using thioglycoside donors, *Carbohydr. Res.*, 299 (1997) 159–164.
216. A. J. Reason, L. A. Ellis, J. A. Appleton, N. Wisnewski, R. B. Grieve, M. McNeil, D. L. Wassom, H. R. Morris, and A. Dell, Novel tyvelose-containing tri- and tetra-antennary *N*-glycans in the immunodominant antigens of the intracellular parasite *Trichinella spiralis*, *Glycobiology*, 4 (1994) 593–603.
217. L. A. Ellis, C. S. McVay, M. A. Probert, J. Zhang, D. R. Bundle, and J. A. Appleton, Terminal beta-linked tyvelose creates unique epitopes in *Trichinella spiralis* glycan antigens, *Glycobiology*, 7 (1997) 383–390.
218. M. Escalante, F. Romaris, M. Rodriguez, E. Rodriguez, J. Leiro, M. T. Garate, and F. M. Ubeira, Evaluation of *Trichinella spiralis* larva group 1 antigens for serodiagnosis of human trichinellosis, *J. Clin. Microbiol.*, 42 (2004) 4060–4066.

219. T. Iversen and D. R. Bundle, Antigenic determinants of *Salmonella* serogroups and D_1. Synthesis of trisaccharide glycosides for use as artificial antigens, *Carbohydr. Res.*, 103 (1982) 29–40.
220. G. I. Birnbaum and D. R. Bundle, Conformation of methyl 3,6-dideoxy-α-D-*arabino*-hexopyranoside, the inmunodominant sugar of *Salmonella* serogroup D_1: Crystal structure, ^{1}H nmr analysis, and semi-empirical calculations, *Can. J. Chem.*, 63 (1985) 739–744.
221. M. A. Probert, J. Zhang, and D. R. Bundle, Synthesis of α and β-linked tyvelose epitopes of the *Trichinella spiralis* glycan: 2-Acetamido-2-deoxy-3-*O*-(3,6-dideoxy-D-*arabino*-hexopyranosyl)-β-D-galactopyranosides, *Carbohydr. Res.*, 296 (1996) 149–170.
222. M. Nitz and D. R. Bundle, Efficient synthesis of 3,6-dideoxy-β-D-*arabino*-hexopyranosyl-terminated LacdiNac glycan chains of the *Trichinella spiralis* parasite, *J. Org. Chem.*, 65 (2000) 3064–3073.
223. B. Classon, P. J. Garegg, and B. Samuelson, Conversion of hydroxyl groups into bromo groups in carbohydrates with inversion of configuration, *Can. J. Chem.*, 59 (1981) 339–343.
224. R. P. Gorshkova, V. A. Zubkov, V. V: Isakov, and Y. S. Ovodov, Yersiniose, a new branched-chain sugar, *Carbohydr. Res.*, 126 (1984) 308–312.
225. R. P. Gorshkova, V. V. Isakov, V. A. Zubkov, and Yu. S. Ovodov, Structure of O-specific polysaccharide chain of the *Yersinia bercoviery* O:10 lipopolysaccharide, *Bioorg. Khim.*, 20 (1994) 1231–1235.
226. A. Sonesson and E. Jantzen, The branched-chain octose yersiniose A is a lipopolysaccharide constituent of *Legionella micdadei* and *Legionella maceachernii*, *J. Microbiol. Methods*, 15 (1992) 241–248.
227. V. A. Zubkov, R. P. Gorshkova, Y. S. Ovodov, A. F. Sviridov, and A. S. Shashkov, Synthesis of 3,6-dideoxy-4-*C*-(4^1-hydroxyethyl)hexopyranoses (yersinioses) from 1,6-anhydro-β-D-glucopyranose, *Carbohydr. Res.*, 225 (1992) 189–207.
228. H. Chen, Z. Guo, and H.-w. Liu, Biosynthesis of yersiniose: Attachment of the two-carbon branched-chain is catalyzed by a thiamine pyrophosphate-dependent flavoprotein, *J. Am. Chem. Soc.*, 120 (1998) 11796–11797.
229. Z. Zhao and H.-w. Liu, Synthesis of a deoxysugar dinucleotide containing an exo-difluoromethylene moiety as a mechanistic probe for studying enzymes involved in unusual sugar biosynthesis, *J. Org. Chem.*, 66 (2001) 6810–6815.
230. S. L. Ward, Z. Hu, A. Schirmer, R. Reid, W. P. Revill, C. D. Reeves, O. V. Petrakovsky, S. D. Dong, and L. Katz, Chalcomycin biosynthesis gene cluster from Streptomyces bikiniensis: Novel features of an unusual ketolide produced through expression of the *chm* polyketide synthase in *Streptomyces fradiae*, *Antimicrob. Agents Chemother.*, 48 (2004) 4703–4712 and references therein.
231. S. Omura, S. Namiki, M. Shibata, T. Muro, and H. Nakayoshi, Studies on the antibiotics from *Streptomyces spinichromogenes* var. *kujimyceticus*. II. Isolation and characterization of kujimycins A and B, *J. Antibiot.*, 22 (1969) 500–505.
232. E. I. Graziani, C. R. Overk, and G. T. Carter, Purification, structure determination, and antimicrobial activity of neutramycins B-G, *J. Nat. Prod.*, 66 (2003) 1149–1153.
233. S. Donadio, M. J. Staver, J. B. McAlpine, S. J. Swanson, and L. Katz, Biosynthesis of the erythromycin macrolactone and a rational approach for producing hybrid macrolides, *Gene*, 115 (1992) 97–103.
234. V. Pozsgay and A. Neszmélyi, Synthesis and carbon-13 n.m.r.-spectral study of methyl 2,6- and 3,6-dideoxy-α-L-*arabino*- and methyl 4,6-dideoxy-α-L-*lyxo*-hexopyranoside, *Carbohydr. Res.*, 85 (1980) 143–150.
235. P. G. Hultin and R. M. Buffie, Syntheses of methyl (4,6-dideoxy-α-L-*lyxo*-hexopyranosyl)-(1 → 3)- and (4-deoxy-4-fluoro-α-L-rhamnopyranosyl)-(1 → 3)-2-acetamido-2-deoxy-α-D-glucopyranosides, analogs of the mycobacterial arabinogalactan linkage disaccharide, *Carbohydr. Res.*, 322 (1999) 14–25.

236. T. Nishio, Y. Miyake, K. Kubota, M. Yamai, S. Miki, T. Ito, and T. Oku, Synthesis of the 4-, 6-deoxy, and 4,6-dideoxy derivatives of D-mannose, *Carbohydr. Res.*, 280 (1996) 357–363.
237. G. Siewert and O. Wastphal, Substitution sekundärer Tosylestergruppen durch Jod Synthese von 4-Desoxy- und 4,6-Didesoxy-D-*xylo*-hexose, *Liebigs Ann. Chem.*, 720 (1968) 161–170.
238. B. T. Lawton, W. A. Szarek, and J. K. N. Jones, A facile synthesis of 4,6-dideoxy-D-*xylo*-hexose, *Carbohydr. Res.*, 14 (1970) 255–258.
239. U. Küfner and R. R. Schmidt, Synthesis of 4,6-diedoxy-D- and -L-hexoses from racemic and *meso*-dipropenylglycol, *Carbohydr. Res.*, 161 (1987) 211–223.
240. H. Redlich and W. Roy, Synthese von β-Glycosiden und β-Glycosidisch Verknüpften Disacchariden der D-Chalcose. Anwendung der Entchlorierung mit Tributylzinnhydrid, *Carbohydr. Res.*, 68 (1979) 275–285.
241. S. Danishefsky and J. F. Kerwin Jr., A simple synthesis of dl-chalcose, *J. Org. Chem.*, 47 (1982) 1597–1598.
242. U. Küfner and R. R. Smidt, Synthesis of deoxyhexoses from divinylglycols. The synthesis of D- and L-chalcose, *Angew. Chem. Int. Ed.*, 25 (1986) 89.
243. K. Torsell and M. P. Tyagi, Synthesis of D,L-chalcose and D,L-chalcosamine, *Acta Chem. Scand. B*, 31 (1977) 7–10.
244. K. Kefurt, Z. Kefurtová, and J. Jarý, Synthesis of 4,6-dideoxy-D-*arabino*-hexose, 3,4, 6-trideoxy-D-*erythro*- and *threo*-hexoses, 2,4,6-trideoxy-D-*erythro*-hexose, and their derivatives, *Collect. Czechoslov. Chem. Commun.*, 40 (1975) 164–173.
245. W. Adam, C. R. Saha-Moller, and K. S. Schmid, Synthesis of 4,6-dideoxyfuranoses through the regioselective and diastereoselective oxyfunctionalization of a dimethylphenylsilyl-substituted chiral homoallylic alcohol, *J. Org. Chem.*, 66 (2001) 7365–7371.
246. K. Mitsudo, T. Kawaguchi, S. Miyahara, W. Matsuda, M. Kuroboshi, and H. Tanaka, Electro-oxidative glycosylation through C-S bond cleavage of 1-arylthio-2,3-dideoxyglycosides. Synthesis of 2′,3′-dideoxynucleosides, *Org. Lett.*, 7 (2005) 4649–4652.
247. M. H. Haukaas and G. A. O'Doherty, Enantioselective synthesis of 2-deoxy- and 2,3-dideoxyhexoses, *Org. Lett.*, 4 (2002) 1771–1774.
248. P.-T. Ho and S. Chung, Synthesis of 2,4-dideoxy-D-*erythro*-hexopyranose. An intermediate for synthesis of the lactone moiety of inhibitors of hydroxymethylglutaryl-coenzyme A reductase, *Carbohydr. Res.*, 125 (1984) 318–322.
249. S. Shibahara, S. Kondo, K. Maeda, H. Umezawa, and M. Ohno, Total syntheses of negamycin and the antipode, *J. Am. Chem. Soc.*, 94 (1972) 4353–4354.
250. K. Augustyns, A. Van Aerschot, and P. Herdwijn, Synthesis of 1-(2,4-dideoxy-β-D-*erythro*-hexopyranosyl)thymine and its incorporation into oligonucleotides, *Bioorg. Med. Chem. Lett.*, 2 (1992) 945–948.
251. W. M. Rockey, M. K. Dowd, P. J. Reilly, and A. D. French, Modeling of deoxy- and dideoxyaldohexopyranosyl ring puckering with MM3(92), *Carbohydr. Res.*, 335 (2001) 261–273.
252. C. L. Stevens, K. Nagarajan, and T. H. Haskell, The structure of amicetin, *J. Org. Chem.*, 27 (1962) 2991–3005.
253. F. Arcamone, W. Barbieri, G. Franceschi, S. Penco, and A. Vigevani, Axenomycin. I. The structure of chromophore and sugar moieties, *J. Am. Chem. Soc.*, 95 (1973) 2008–2009; F. Arcamone, G. Franceschi, B. Gioia, S. Penco, and A. Vigevani, Axenomycins. II. The structure of axenolide, *J. Am. Chem. Soc.*, 95 (1973) 2009–2011.
254. L. Xuan, S. Xu, H. Zhang, Y. Xu, and M. Chen, Dutomycin, a new anthracycline antibiotic from *Streptomyces*, *J. Antibiot.*, 45 (1992) 1974–1976.
255. I. Momose, W. Chen, H. Nakamura, H. Naganawa, H. Inuma, and T. Takeuchi, Polyketomycin, a new antibiotic from *Streptomyces* sp. MK277-AF1. II. Structure determination, *J. Antibiot.*, 51 (1998) 26–32.

256. C. L. Stevens, K. Nagarajan, and T. H. Haskell, The structure of amicetin, *J. Org. Chem.*, 27 (1962) 2991–3005.
257. D. S. Micalizzi, J. P. Dougherty, L. A. Noecker, G. R. Smith, and R. M. Giuliano, Synthesis of the polyketomycin disaccharide, *Tetrahedron Asymmetry*, 14 (2003) 3183–3188.
258. I. Momose, W. Chen, N. Kinoshita, H. Iinuma, M. Hamada, and T. Takeuchi, Polyketomycin, a new antibiotic from Streptomyces sp. MK277-AF1. I. Taxonomy, production, isolation, physico-chemical properties and biological activities, *J. Antibiot.*, 51 (1998) 21–25.
259. K. L. Rinehart Jr. and D. B. Borders, Streptolydigin. II. Ydiginic Acid, *J. Am. Chem. Soc.*, 85 (1963) 4037–4038.
260. H. Brockman and H. Greve, Zur Kenntnis der β-Rhodomycine, *Tetrahedron Lett.*, 16 (1975) 831–834.
261. K. Räty, A. Hautala, S. Torkkell, J. Kantola, P. Mäntsälä, J. Hakala, and K. Ylihonko, Characterization of mutations in aclacinomycin A-non-producing *Stretomyces galilaeus* strains with altered glycosylation patterns, *Microbiology*, 148 (2002) 3375–3384 and references therein.
262. C. L. Stevens, P. Blumbergs, and D. L. Wood, Stereochemical identification and synthesis of amicetose and the stereochemical identification of rhodinose and the sugar from *Streptolydigin*, *J. Am. Chem. Soc.*, 86 (1964) 3592.
263. E. L. Albano and D. Horton, Synthesis of methyl 2,3,6-trideoxy-α-D-*erythro*-hexopyranoside (methyl α-amicetoside), *J. Org. Chem.*, 34 (1969) 3519–3522.
264. G. S. Bethell and R. J. Ferrier, The relayed introduction of alkylthio-groups into carbohydrates derivatives: A novel synthesis of amicetose, *J. Chem. Soc. Perkin I* (1993) 1400–1405.
265. J. S. Brimacombe, L. W. Doner, and A. J. Rollins, Syntheses of methyl 2,3,6-trideoxy-α-L-*erythro*-hexopyranoside (methyl α-L-hexopyranoside) and methyl 2,3,4,6-tetradeoxy-4-(dimethylamino)-α-L-*threo*-hexopyranoside, *J. Chem. Soc. Perkin I* (1972) 2977–2979.
266. G. Berti, P. Caroti, G. Catelani, and L. Monti, Synthesis of D-amicetose and L-rhodinose from L-glutamic acid, *Carbohydr. Res.*, 124 (1983) 35–42.
267. S. Lajšić, D. Miljković, and G. Ċetkovi, An improved synthesis of D-amicetose, *Carbohydr. Res.*, 233 (1992) 261–264.
268. L. A. Noecker, J. A. Martino, P. J. Foley, D. M. Rush, R. M. Giuliano, and F. J. Villani Jr., Synthesis of amicetose by three enentioselective methods, *Tetrahedron Asymmetry*, 9 (1998) 203–212 and references therein.
269. A. H. Haines, The synthesis of trideoxy sugars. A preparation of rhodinose (2,3,6-trideoxy-α-L-*threo*-hexose) and methyl 2,3,6-trideoxy-α-L-*erythro*-hexopyranoside, *Carbohydr. Res.*, 21 (1972) 99–109.
270. E. J. Corey and R. A. E. Winter, A new, stereospecific olefin synthesis from 1,2-diols, *J. Am. Chem. Soc.*, 85 (1963) 2677–2678.
271. H. E. El Khadem and R. C. Cermak, Rhodinose derivatives suitable for the synthesis of anthracycline analogs, *Carbohydr. Res.*, 75 (1979) 335–339.
272. T. Ross Kelly and P. N. Kaul, A simple synthesis of rhodinose from (*S*)-ethyl lactate, *J. Org. Chem.*, 48 (1983) 2775–2777.
273. S. Servi, 2,2,5-Trimethyl-1,3-dioxolane-4-carboxaldehyde as a chiral synthon: Synthesis of the two enantiomers of methyl 2,3,6-trideoxy-α-L-*threo*-hex-2-enopyranoside, key intermediate in the synthesis of daunosamine and of (+)- and (-)-rhodinose, *J. Org. Chem.*, 50 (1985) 5865–5867.
274. A. Kirschning, U. Hary, and M. Ries, Synthesis of specifically labelled 2,3,6-trideoxyhexoses, *Tetrahedron*, 51 (1995) 2297–2304.
275. K. Kefurt, Z. Kefurtová, and J. Jarý, Synthesis of 4,6-dideoxy-D-*arabino*-hexose, 3,4,6-trideoxy-D-*erythro*- and *threo*-hexoses, 2,4,6-trideoxy-D-*erythro*-hexose, and their derivatives, *Collect. Czechoslov. Chem. Commun.*, 40 (1975) 164–173.

276. O. J. Varela, A. Fernández Cirelli, and R. M. de Lederkremer, β-Elimination in aldonolactones: Synthesis of 2-*O*-benzoyl-3,5,6-trideoxy-α-DL-*threo*-hexofuranose, *Carbohydr. Res.*, 100 (1982) 424–430.
277. M. Sznaidman, A. Fernández Cirelli, and R. M. de Lederkremer, Síntesis de 2-*O*-benzoil-3,4,6-tridesoxi-DL-*treo*-hexopiranosa, un nuevo ejemplo de síntesis de desoxiazúcares via reacciónes de eliminación-β en aldonolactonas, *Anales Asoc. Quím. Argentina*, 70 (1982) 341–348.

SUCROSE CHEMISTRY AND APPLICATIONS OF SUCROCHEMICALS

By Yves Queneau[a], Slawomir Jarosz[b], Bartosz Lewandowski[b] and Juliette Fitremann[c]

[a]Laboratoire de Chimie Organique; ICBMS; UMR 5246 CNRS; Université Lyon 1; INSA-Lyon; CPE Lyon; Institut National des Sciences Appliquées, Bâtiment Jules Verne, 20 Avenue Albert Einstein, 69621 Villeurbanne cedex, France
[b]Institute of Organic Chemistry, Polish Academy of Sciences, ul. Kasprzaka 44/52, 01-224 Warsaw, Poland
[c]Laboratoire des IMRCP (UMR 5623 CNRS-UPS), Bât 2R1, Université Paul Sabatier Toulouse 3, 118 Route de Narbonne, 31062 Toulouse cedex 4, France

ISBN: 978-0-12-373920-9
DOI: 10.1016/S0065-2318(07)61005-1

I. Introduction

1. Overview

Sucrose (β-D-fructofuranosyl α-D-glucopyranoside, see Scheme 1) is a natural disaccharide that is by far the most available of all low molecular weight carbohydrates. It is produced from sugar beet or sugar cane on the industrial scale (148 million tonnes in 2006), and its chemistry has attracted considerable interest. It could be used as an organic raw material, as it is cheap, pure, stable, and chemically reactive. Sucrochemistry, by analogy to petrochemistry, covers all processes using sucrose (and sometimes, by extension, using other available carbohydrates, notably glucose) as starting material and leads to materials or compounds of industrial interest, many for everyday applications and used in high tonnage. Being obtained from renewable agricultural resources, such simple carbohydrates thus constitute valuable starting compounds for replacing those produced from fossil resources, provided that economically viable processes can be developed. On the organic chemistry side, the need for synthetic efficiency presents a real challenge because of the structural complexity and functional richness of the sucrose molecule. Consequently organic chemists have been, already for many decades, interested in the chemistry of sucrose.

The purpose of this chapter is to provide an update on our understanding of the reactivity of sucrose and the selectivity of its transformations into functionalized derivatives that have become industrial realities because of their biodegradability or biocompatibility. It follows historical accounts on the topic: books on sucrose and its chemistry,[1–4] a series of books dealing with the use of carbohydrates (in general) as organic raw materials,[5–8] and also chapters and review articles.[9–24]

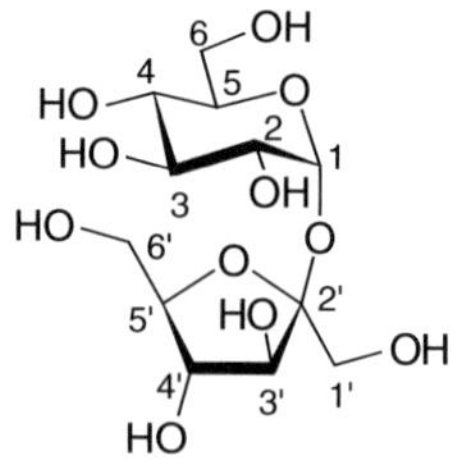

SCHEME 1. Sucrose, β-D-fructofuranosyl α-D-glucopyranoside, and atom numbering.

The use of sucrose as a chemical raw material was first motivated by the desire to increase the small proportion (less than 2%)[4] of the total production dedicated to applications of higher value, essentially for nonfood uses. The most efficient and large-scale of these was (and still remains) the production of ethanol by fermentation.[18,25] More recently, the global realization of an increasing shortage of fossil resources has provided further incentive for projects directed towards new and efficient processes for converting renewable starting materials, including sucrose and other abundant sugars and polysaccharides, into useful products.[5–8]

2. Reactivity of Sucrose and Its Control

Sucrochemistry can be considered from two viewpoints of equivalent importance. The first concerns the fundamental reactivity of sucrose and the control of selectivity in its chemical transformations. Unlike hydrocarbons, sucrose is a complex, polyfunctional molecule, highly oxygenated, and chemically sensitive. With eight reactive hydroxyl groups, two anomeric carbon atoms, and no chemical function being readily selected, most attempted transformations are prone to give complex mixtures. A main objective of sucrochemistry is therefore to understand the relative reactivity of the various functional groups of sucrose and to control their transformations. In the synthesis of functionalized sucrose derivatives, the ability to achieve selectivity allows structure–properties relationships to be established and, consequently, the design of sucrochemicals having tunable properties. The economic efficiency of chemical processes involving sucrose must deal with the complexity of reaction and afford target products of low-added value. This means essentially one-step transformations with simple purification procedures. All means to achieve selectivity, exploiting innate facets of sucrose reactivity along with the aid of catalytic (chemical or enzymatic) methods, contribute to the effectiveness by minimizing undesired (wasted) products and by facilitating purification procedures. Appropriate selection of the reaction medium needs to take into account both the solubility of the substrates and potential difficulties in removal of traces of solvent. Sucrose is reasonably soluble—to an extent compatible with preparative purposes—in only a limited number of solvents: for instance water, *N*,*N*-dimethylformamide (DMF), dimethyl sulfoxide (Me_2SO), and pyridine. In this respect, nontypical media such as water (even in the presence of water-sensitive reagents), supercritical carbon dioxide, and ionic liquids are of interest, as well as processes using no solvent. Possible alternatives to classical stirring or heating include such activation methods as ultrasonic irradiation, which increases phase mixing in heterogeneous systems (often observed because of the very high polarity of sucrose and sucrose solutions), as well as irradiation by microwaves, which provides rapid and uniform heating of the reaction mixture. Finally, biotransformations are attractive, as they are very selective and generally take place in aqueous media.

3. Targeted Synthesis from Sucrose

The second viewpoint is to define the targets. Two strategies may be employed, namely with or without retention of the carbohydrate skeleton. Among the derivatives that have been targeted from the early days of sucrochemistry, surfactants (constructed by attachment to sucrose of hydrophobic fatty alkyl chains through various linkages), and polymerizable compounds have probably been the most studied. Production of ethanol by fermentation of pure sucrose or molasses is also a major application among nonfood uses of sucrose. However, improved processes towards those identified goals are still to be established, notably in the context of the economic pressure occasioned by increasing oil prices. It is also important to consider new targets, exploiting the potential of sucroderivatives as high added-value, biocompatible, biodegradable, and nontoxic compounds in all fields of applications including food, cosmetic, or biomedical domains, or as sophisticated chiral synthons, when properly and selectively protected for organic syntheses.

II. Chemical Transformations

1. Structural and Theoretical Bases

The main goals of sucrochemistry evoked here are, for the most part, connected with selectivity. Whether it be for direct (ideally one step) transformations of unprotected sucrose, avoiding protection–deprotection strategies, or for the design of properly substituted derivatives to be used as synthons, the control of selectivity remains the major challenge. Sucrose is a nonreducing disaccharide in which the interglycosidic bond is quite acid-sensitive. This bond is hydrolyzed rapidly at pH 4, making difficult any acid-catalyzed transformations that would conserve the disaccharide backbone. Enzyme-catalyzed cleavage is able to convert sucrose efficiently into a mixture of glucose and fructose (the process known as inversion, from the change of sign of the optical rotation), or to other derivatives by transglycosylation. There is no equivalent chemical process acting at the anomeric carbons of sucrose because of the enhanced lability of fructose fragments. This makes sucrose a less competitive substrate as compared to glucose for chemical acid-catalyzed glycosylations (transglycosylations).

Unlike glucose, where the hemiacetal linkage can be targeted selectively, the hydroxyl groups of sucrose offer the main opportunity for its derivatization. However, the eight hydroxyl groups (3 primary (6, 1′, 6′) and 5 secondary (2, 3, 4, 3′, 4′)) are all available to react, and the possible combinations for substitution at all positions can produce as many as 255 different compounds. Fortunately, the relative rates of reaction of all OH groups depend on the type of transformation and on the reaction conditions, providing two types of selectivity, namely the degree of substitution and the regiochemistry. Both are key factors on which the properties of the products depend strongly (and consequently the possible applications), as illustrated in several examples in this chapter. Limiting the number of products is a major goal, as it simplifies also the purification processes. Here, following is an overview of the theoretical knowledge on the structure of sucrose and its physical properties, with emphasis on those factors that interfere with, or better, which might direct its reactivity as an unprotected polyol.[9,12–14,22,23,26]

The conformational structure of sucrose is essentially based on the intramolecular hydrogen-bond network that connects hydroxyl groups from the glucose and the fructose moieties. In the solid state, two main hydrogen bonds are found, the first between O-H-1′ (Fru) and O-2 (Glc), and a second, weaker, between O-H-6′ (Fru) and the pyranosyl oxygen O*p*-5 (Glc). In solution, this latter bond is rapidly cleaved, and only the O-2···H-O-1′ bond remains, with O-H-3′ competing with O-H-1′, thus leading to an equilibrium between two main conformations (Scheme 2) which have been fully rationalized by theoretical and spectroscopic studies.[27–32] Carbon atoms C-2, C-1′, and C-3′ are linked to either one or the other anomeric centers of sucrose. The electron-withdrawing effects and the hydrogen bonds connecting OH-2, OH-1″, and OH-3′ make these

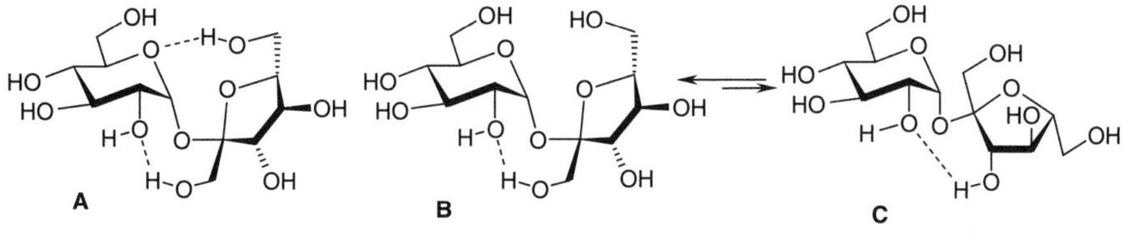

SCHEME 2. Conformational behavior of sucrose in the solid state (A) and in solution (B, C).[31]

latter more reactive. The conformational behavior of sucrose in water, involving a bridging water molecule between O-2 and OH-1′, was determined by NMR and molecular modeling.[33–35] Structural studies by deuterium NMR spectroscopy were also conducted on polydeuterated sucrose.[36]

Among all positions, OH-2 was shown to be the most acidic. Some scales of relative acidity of the hydroxyl groups of sucrose, converging on the highest acidity for OH-2, can be established either by semiempirical calculations[37] or be deduced from the distribution of the regioisomers after selective substitution, as illustrated by many examples in the following sections. This notable acidity of OH-2 makes sucrose significantly more acidic as compared to water or simple alcohols. The Merck Index[38] gives the p*K*a of sucrose as 12.62, but the literature data vary from 12.4 to 12.8, according to the method of measurement or calculation. A very precise value of 12.62 at 19 °C was proposed by Michaelis as early as 1913, using one of the first platinum electrodes.[39] Later in the 1930s,[40,41] two conductimetric studies established acidity constants for sucrose as $pKa_1 = 12.5$ and $pKa_2 = 13.5$ (at 25 °C), and $pKa_1 = 12.7$ and $pKa_2 = 13.7$ (at 23 °C). These values were confirmed by a polarimetric study, which gave pKa_1 as 12.5 and pKa_2 as 13.5.[42,43] Later,[44] a value of 12.43 was measured at 20 °C and potentiometric studies gave 12.75 or 12.80 at 25 °C (depending on the calculation method),[45] and $pKa_1 = 12.57$ and $pKa_2 = 13.45$ at 25 °C.[46] Finally, the value of 12.70 at 30 °C was determined by capillary electrophoresis.[47] The acidity of sucrose has also been discussed in various reviews dealing with the relationships between structure and properties (such as acidity and reactivity) of carbohydrates that point out the higher reactivity of an equatorial hydroxyl group vicinal to an axial oxygen atom, and pointing out the "α-glucoside" behavior for sucrose.[48–50]

Besides the special reactivity of the OH-2, OH-1′, and OH-3′ groups lies also the "classical" relative reactivity between the primary and secondary hydroxyl groups. Depending on the reaction conditions and the nature of the electrophilic species, it may be seen that these two types of possible reactivity can direct the reactivity of sucrose. Of course, the product distribution also depends on whether the transformations are kinetically or thermodynamically controlled. For those reactions under kinetic control, if there is enough difference in the rate of the first substitution at the most reactive hydroxyl group and the second one, then the regioselectivity also monitors the degree of substitution.

In accordance with all of these structural and electronic considerations, it is shown in the following sections that it would be wrong to simplify our understanding of the selectivity of sucrose transformations to merely the reactivity order of primary versus secondary hydroxyl groups.

Before going into details in the transformations of sucrose, it should be mentioned that, although sucrose is available in large quantity from plants, its complex structure has been for years a major challenge to synthetic chemists; although several syntheses of this disaccharide have been accomplished.[51–59]

2. Etherification

Because of the high stability of the ether function, etherification of unprotected sucrose leads to a kinetic distribution of products directly reflecting the relative reactivity of the hydroxyl groups. This reaction is therefore the best probe for reactivity studies at least for discussing the relative rates of the first substitution. The following substitutions are more difficult to compare, since supplemental factors (electronic and steric) arising from the first substitution interfere with the "natural" reactivity order of unprotected sucrose.

Reaction of unsubstituted sucrose with benzyl bromide in the presence of silver oxide or sodium hydride affords 2-*O*-benzylsucrose, obtained in 80% yield among other monosubstituted products together with small amounts of products etherified at positions 1′ and 3′ (Scheme 3).[60,61] Mixtures enriched in ethers at the same positions are also obtained in electrochemical etherification (Table I).[62]

However, when bulky electrophilic species are used, such as chlorotrimethylsilane or highly substituted silyl chlorides, the primary alcohols at positions 6 and 6′ react faster, the more hindered OH-1′ being less reactive because of its proximity to the quaternary carbon atom C-2.[63–65] The trisilyl ether of all three primary OH groups is formed when *tert*-butyldiphenylsilyl chloride (TBDPSCl) is used in excess. In the presence of 1.1 equiv. of TBDPSC, significant regioselectivity is observed in favor of 6-OH, since the 6′-monoether can be obtained

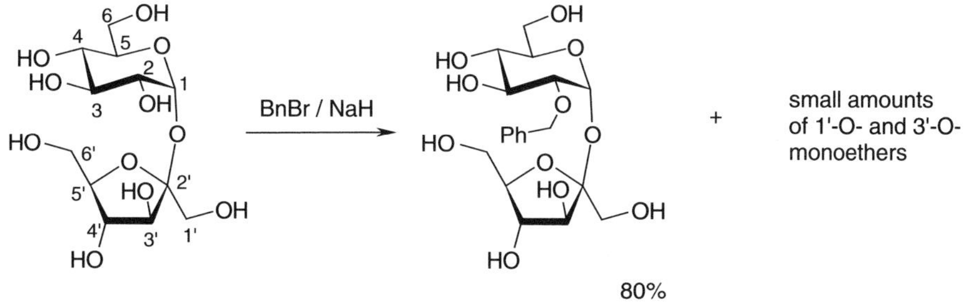

SCHEME 3. Preferential benzylation at OH-2.

TABLE I
Distribution of the Products of Alkylation of Free Sucrose

Substrate	Alkyl Halide	Product Distribution at C-positions (%)				Yield (%)
		2	6	1′	3′	
Sucrose	Benzyl bromide	48	–	39	13	48
Sucrose	Methyl iodide	54	–	27	19	63
Sucrose	Allyl bromide	39	–	61	–	28

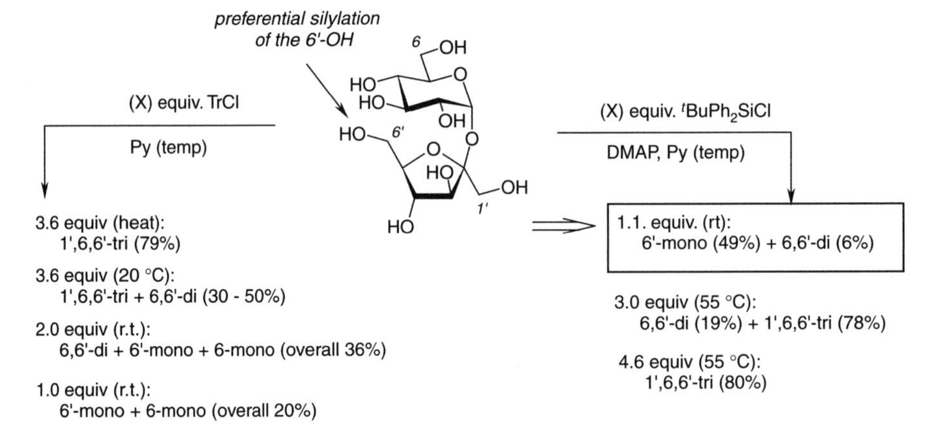

SCHEME 4. Partial tritylation and silylation of sucrose.

in 49% yield. This reaction can be used as a first step towards partially protected compounds that are then desilylated in order to have only primary OH groups available. For example, the silylated derivatives can be peracetylated; however, the removal of the silyl protecting groups by fluoride ion is usually accompanied by migration of an acetyl group from the O-4 to the O-6 position.[65,66]

Tritylation offers an alternative (and less expensive) to silylation. Tritylation of sucrose can be achieved by heating it with an excess of chlorotriphenylmethane (TrCl), leading to 1′,6,6′-tri-*O*-tritylsucrose as the only product in 79% yield.[67] The 6,6′-di-*O*-tritylated derivative is obtained at room temperature.[64,68] If this reaction is performed with 2.0 equiv. of TrCl, a mixture of both mono- (at C-6 and C-6′) and ditritylated (C-6,6′) derivatives is formed, but the overall yield decreases.[69] If 1.2 equiv. is used, a mixture of both monotritylated derivatives (at C-6 and C-6′) is formed in almost equal amounts, but in only 20% overall yield (Scheme 4).[63]

SCHEME 5. Liquid-crystal behavior of hydroxyalkyl ethers.

Amphiphilic hydroxyalkyl ethers are obtained by reaction with terminal epoxides of linear alkanes in Me_2SO in the presence of a base with major substitution occurring at O-2, O-1′. In water, the reaction is only slightly less selective, despite the perturbation due to solvation by water molecules, with more than 60% of the monosubstituted products being substituted at O-2 and O-1′.[70] This is consistent with the established structure of sucrose in water, which involves a bridging water molecule between O-2 and OH-1′.[34,35] In the addition of such epoxides to sucrose in aqueous medium, the best catalysts are tertiary amines or ammonium resins, which help to overcome the difficulties caused by the heterogeneity of the medium.[71–74] The monosubstituted hydroxydodecyl sucroses obtained from 1,2-epoxydodecane could be separated and the pure different regioisomers studied for their surfactant and liquid–crystal properties (Scheme 5) (see also Section III.1). Affecting the global geometry of the molecule by substitution of OH-2 or OH-3′ (known to be involved in the hydrogen-bonding network) led to significant variations in the thermotropic phase behavior as compared to other positions.[75]

The regiochemical outcome of this reaction has been compared for sucrose and trehalose. This pointed out the higher reactivity at O-2 in sucrose, with 76% substitution at O-2 of the glucose moiety, as compared with 24% for trehalose (Table II, entries 2 and 3).[74] The relative proportions of ethers at positions 3, 4, and 6 of the glucose moiety of sucrose or trehalose are identical, within experimental error (entries 4 and 5).

The reaction of sucrose with propylene oxide in aqueous basic medium affords 2-hydroxypropyl ethers.[76] Similar conditions gave sucrose glycerol–sucrose hybrids by reaction with glycidol.[77] Polymeric resins are obtained, starting from sucrose or partially esterified sucrose, when diepoxides are used.[78,79]

TABLE II

Regioisomeric Distribution (%) of Hydroxydodecyl Ethers for Sucrose and Trehalose[74]

Entry	Positions Compared	2	3	4	6	3′	1′	4′	6′
1	All positions in sucrose	41	1	5	7	14	25	4	4
2	Relative proportions for the 4 positions of the glucose moiety in sucrose	76	2	9	13	–	–	–	–
3	Relative proportions for the 4 positions in trehalose	24	9	30	37	–	–	–	–
4	Relative proportions for the positions 3, 4, 6 in the glucose moiety of sucrose	–	8	38	54	–	–	–	–
5	Relative proportions for the positions 3, 4, 6 in trehalose	–	12	40	48	–	–	–	–

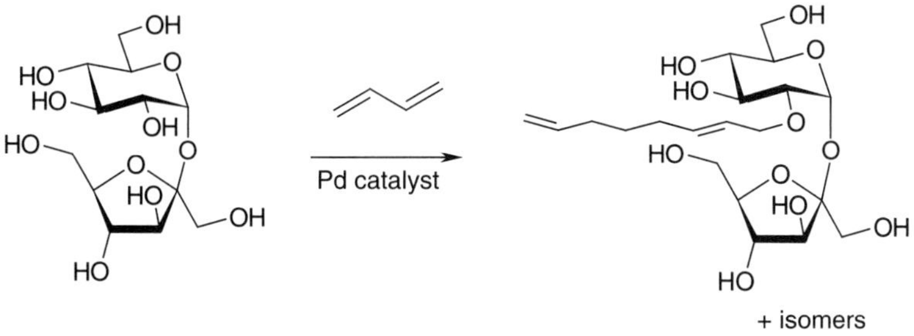

SCHEME 6. Telomerization of 1,3-butadiene in the presence of sucrose.

The carboxymethylation of sucrose can be achieved by reaction with sodium chloroacetate in water or water–2-propanol mixtures in basic medium. In this case, a similar regioselectivity is observed as for the reaction with epoxides or alkyl halides with major substitutions at secondary positions, notably at OH-2.[80] Cyanoethylated sucrose derivatives and the corresponding carboxylated compounds were also prepared.[81]

Alkyl ethers of sucrose have been prepared by reaction with long-chain alkyl halides to provide mixtures of regioisomers and products of different degree of substitution.[82,83] A similar reaction with chloromethyl ethers of fatty alcohols provides formaldehyde acetals.[84,85] Alkenyl ethers of various carbohydrates, and notably of sucrose, can also be obtained by palladium-catalyzed telomerization of butadiene (Scheme 6).[86–88] Despite a low-selectivity control, this simple and clean alternative to other reactions can be carried out in aqueous medium when sulfonated phosphines are used as water-soluble ligands.

3. Esterification

a. Carboxylic Esters of Sucrose.—Sucrose esters are of interest in such applications as fat substitutes,[89] bleaching boosters,[90] and emulsifiers in the food and cosmetic industries (compare Section III). Their properties depend strongly on their compositions in terms of degree of substitution and regiochemistry. Controlling the selectivity is therefore of interest for preparative purposes and studying structure–properties relationships is important for applications. Unlike ethers, however, esters are not stable under certain conditions, and it is not easy to relate the observed distribution in the products with the relative reactivity of the hydroxyl groups, because intramolecular transesterification (acyl group migration) readily occurs. When esterification or transesterification is performed under strongly basic conditions, mixtures of esters at the primary positions are the principal products. Peracetylation, often used as an isolation and identification method in carbohydrate chemistry, likewise applies with sucrose upon treatment with acetic anhydride in pyridine or in ionic liquids of low viscosity.[91]

(i) Esterification at Primary Positions. For the same reasons as previously noted in the etherification section, the primary groups are usually protected first when a bulky acylating reagent is used. Esterification of free sucrose with 2 equiv. of pivaloyl chloride affords the 6,6′-dipivalate (Scheme 7). When an excess of pivaloyl chloride is used, the remaining groups are protected in the following order: 1′-OH > 3′-OH > 2-OH > 3-OH > 4-OH. Reaction of sucrose with benzoyl chloride occurs preferentially at the 6-OH position.[92]

It is possible to achieve selective esterification at OH-6 in good yield even with acetyl chloride by employing low temperature in the presence of a controlled amount of acylating agent.[13,93] Some tin derivatives (such as dibutyltin oxide) also show significant selectivity towards position 6 (Scheme 7).[94,95] A polymer-bound stannyl reagent was also applied for the selective C-6 acetylation of sucrose.[96]

When esterification is achieved under the Mitsunobu conditions (diethyl azodicarboxylate, Ph_3P), only esters at positions 6 and 6′ are produced, and isolation of the 6-monoester, which is formed faster is possible. Thus diesters can be efficiently prepared.[97–99] In this type of reaction, when the carboxylic acid is

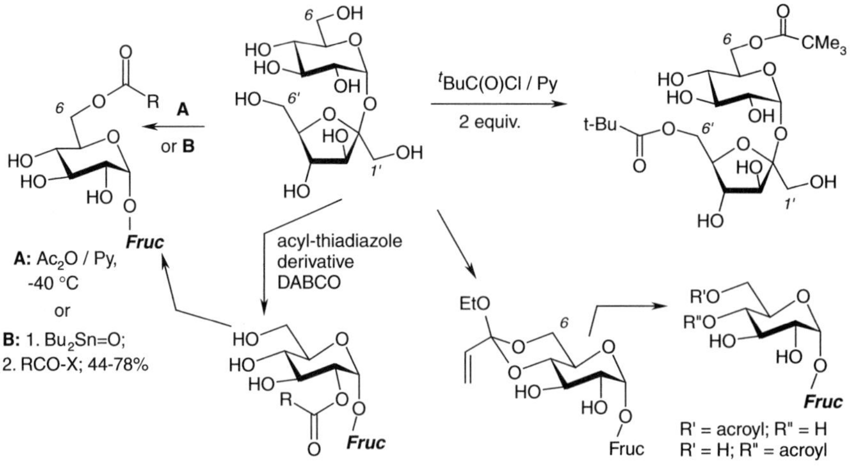

SCHEME 7. Partial esterification of sucrose.

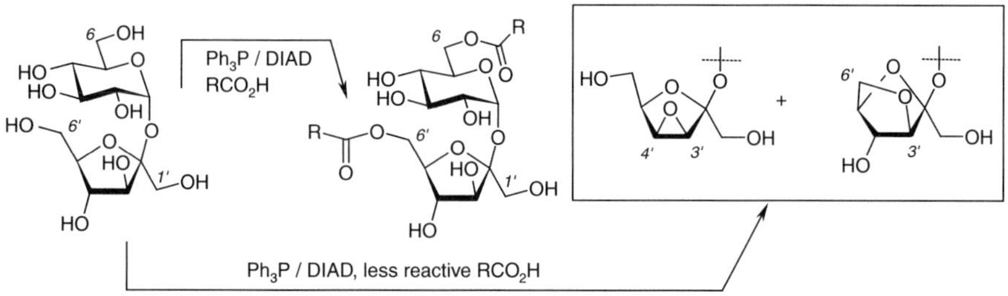

SCHEME 8. Mitsunobu esterification of sucrose.

not reactive enough, intramolecular etherification can compete with the desired intermolecular esterification leading to 6-*O*-acyl-3′,4′- or 3′,6′-anhydrosucrose.[100] The Mitsunobu reaction was also used to prepare sucrose phosphates and phosphonates (Scheme 8).[101,102]

(ii) Esterification at the Secondary Positions. Esterification of sucrose can take place at secondary alcohol functions, sometimes with good selectivity at OH-2 when an acylating agent such as *N*-acylthiazolidinethione is used, either giving the 2-monoester (using catalytic NaH as the base) or controlled migration (using 1,8-diazabicyclo[5.4.0]undec-7-ene, DBU) towards the primary OH-6.[99,103,104] It was shown that, in sucrose having a fatty acyl chain at position 2, the alkyl chain adopts a perpendicular arrangement with respect to the disaccharidic skeleton.[105]

Selective esterification at OH-2 was also observed in the formation of tosyl derivatives.[106] Esterification of metal chelates based on the known affinity of carbohydrates for metal cation species[107] also afforded variations in the regioisomeric distribution, with the 3- or 3′-esters as major compounds[108] or in the degree of substitution.[109]

(iii) Enzymatic Esterifications. A major alternative to the classical basic catalysis is the use of enzymes for esterification, in particular with proteases and lipases.[110–112] To make these enzymes, which normally hydrolyze amide or ester linkages, work in the reverse direction of esterification, the reactions have to be performed in organic media, with only the small amount of water necessary to preserve their active conformation. In such reactions, the difficulty is to find those conditions of solvent and temperature compatible with both the solubility of the substrates and the stability and the activity of the enzyme.[113,114] In the case of sucrose (Scheme 9), most proteases lead selectively to monoesters at position 1′.[111,115,116] These reactions are often performed in DMF, but examples in Me_2SO, which is much less toxic, have also been reported, despite the ability

SCHEME 9. Enzyme-catalyzed esterifications of sucrose.

of Me_2SO to denature proteins.[117–120] Variations in selectivity can sometimes be observed, as in the case of the reaction of sucrose with vinyl dodecanoate catalyzed by protease AL-89, which afforded the 6-*O*-substituted derivative, while substilin A catalyzed formation of a monoester at the 1′-position.[121] Some examples of a surprising selectivity at OH-2 have been described.[121,122]

Once bearing some substituents, the decrease of polarity of the sucrose derivatives makes them soluble in less-polar solvents, such as acetone or *tert*-butanol, in which some lipases are able to catalyze esterifications. Unlike proteases, which necessitate most often the use of an activated acyl donor (such as vinyl or trifluoroethyl esters), lipases are active with simple esters and even the parent carboxylic acids in the presence of a water scavenger. The selectivity of the lipase-catalyzed second esterification is specific for OH-6′ allowing the synthesis of mixed 1′,6′-diesters.[123,124] For some lipases, a chain-length dependence on the regiochemistry was observed.[125] Selectively substituted monoesters were thus prepared and studied for their solution and thermotropic behavior.[126,127] Combinations of enzyme-mediated and purely chemical esterifications led to a series of specifically substituted sucrose fatty acid diesters with variations in the chain length, the level of saturation, and the position on the sugar backbone. This allowed the impact of structural variations on thermotropic properties to be demonstrated (compare Section III.1).[128]

(iv) Esterification in Aqueous Medium. Sucrose can also be esterified in aqueous medium. Thanks to the peculiar reactivity of OH-2, which is significantly more acidic than water, even such water-sensitive acylating agents, as acyl chlorides can be used. Of course, the reactivity gap is not large enough to totally prevent hydrolysis of both the acyl donor and the ester products. Different methods allow these undesired competitive hydrolyses to be minimized. One possibility is to start from very concentrated solutions, thus limiting the amount

of reactive water molecules, as shown by many studies on the structure of sugar syrups. However in such a case, the medium becomes very lipophobic increasing the tendency towards polysubstitution when fatty acid acylating agents are used. Decreasing the strength of the cohesive energy density of water by adding a co-solvent such as tetrahydrofuran (THF) or 2-propanol, and adding an acylation catalyst such as 4-dimethylaminopyridine (DMAP), which helps incorporation of the acyl chain within the aqueous phase, are also ways to limit polyesterification and low substitution. Esters and mixed carbonates were thus obtained by reaction with acid chlorides and alkyl chloroformates.[129,130] During these reactions, the progress of the regioisomeric distribution could be followed by HPLC, and revealed that significant initial esterification takes place at secondary hydroxyl groups, with rapid and almost total subsequent migration towards primary positions.[131] Similar trends were observed in the case of the preparation of carbamates by reaction with isocyanates although with much lower migration rates.[132]

Further esterification of selectively substituted mono- or di-esters with galloyl residues led to a series of polyesters which were prepared and studied for their antioxidant properties.[133–135] Several similar derivatives containing one gallate unit (at the 6′-O, 4′-, 6-, 1′-, and 2-positions) were isolated from natural sources.[136]

(v) Partially Esterified Sucrose by Deprotection of Sucrose Derivatives. Another method for the preparation of partially esterified sucroses involves selective deacetylation of sucrose octaacetate. Hexa- and hepta-*O*-acetyl sucroses were thus prepared under either basic catalysis with Al_2O_3–K_2CO_3[137,138] primary amines[139] or by using enzymes.[140–143] Selective hydrolysis of the 6-ester of 6,6′-diesters by enzymatic hydrolysis has been shown to provide the 6′-monoester.[144] A multistep route, based on selective desilylations of trisilylated sucrose derivatives, was shown to provide heptaesters with the 1′-position unprotected.[145]

Finally, it may be noted that 4,6-orthoesters obtained for example by reaction of sucrose with ethyl orthoacrylate can be opened to form either the 6- or 4-esters,[146] providing convenient starting materials for the preparation of biodegradable polymers (see Scheme 7).

b. Sucrose Esters Other Than Carboxylic Esters.—Sulfuric esters of sucrose have attracted much attention because of their physiological properties, a notable example being the aluminum salt of the octasulfate (Sucralfate), which is used as an antiulcer drug.[147] The octasulfate[148,149] is prepared by reaction of free sucrose with the SO_3–pyridine complex, in either pyridine or DMF (Scheme 10).

Partial sulfation of sucrose can be also achieved. Thus, reaction of sucrose with $SOCl_2$ afforded a mixture of diastereoisomeric cyclic sulfites, which are readily oxidized to sulfates. Treatment of these latter compounds with fatty acids and potassium carbonate provided the ester with the sulfate group placed on O-4. The sulfate function can be also introduced at other positions as demonstrated by reaction of 6-*O*-acylsucrose and 1′-*O*-acylsucrose with SO_3–pyridine (Scheme 11).[150]

Sulfonylation of free sucrose with 3 equiv. of *p*-toluenesulfonyl chloride in pyridine affords the 1′,6,6′-tri-tritosylated sucrose in moderate yield,[151,152] accompanied by the 6,6′-di-substituted as well as penta- and tetra-substituted derivatives. Using the more bulky mesitylenesulfonyl chloride, the 1′,6,6′-tri-sulfonylated derivative was obtained in 55% yield.[152] Direct regioselective 2-*p*-toluenesulfonylation of sucrose with *N*-(*p*-toluenesulfonyl)imidazole has been reported.[106]

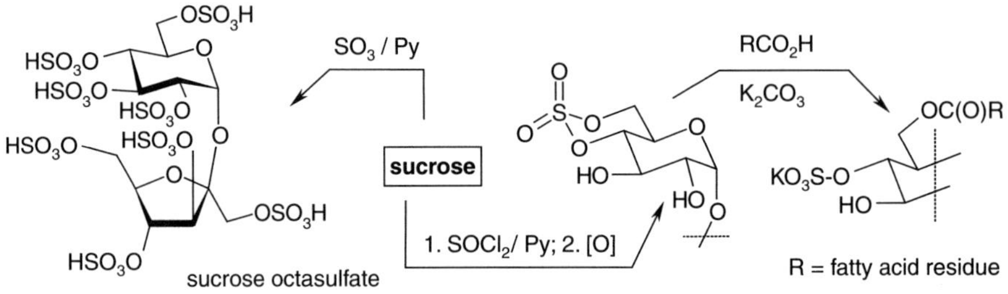

SCHEME 10. Sucrose sulfates.

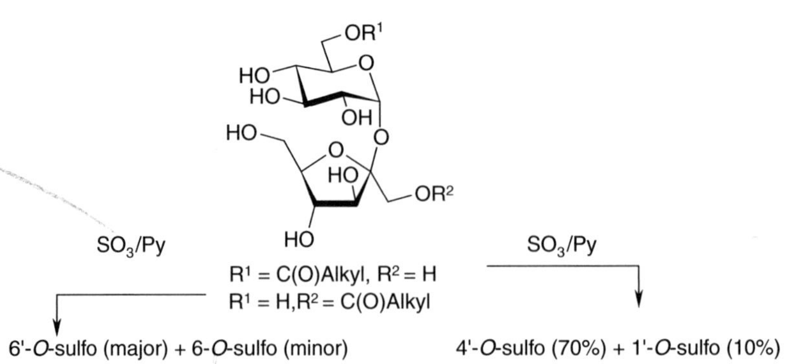

SCHEME 11. Sulfation of sucrose partial esters.

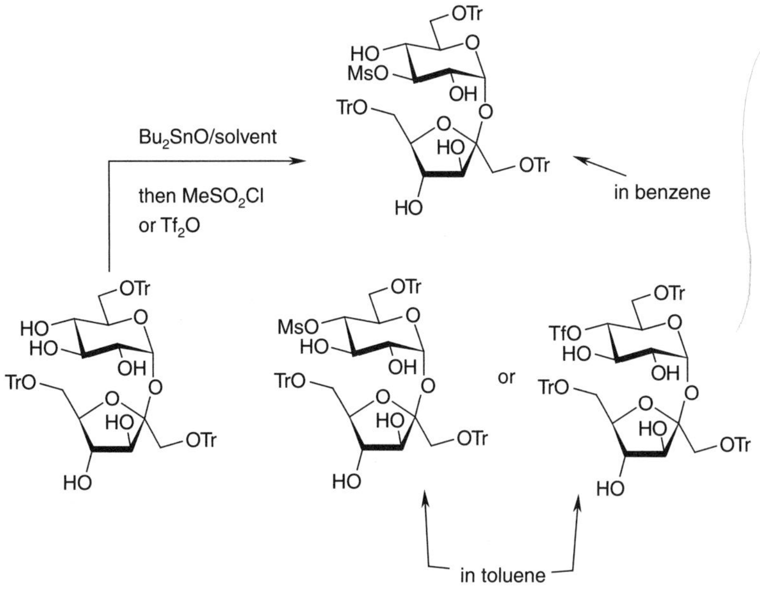

SCHEME 12. Sulfonation at secondary positions.

Selective sulfonylation of 1′,6,6′-tri-*O*-tritylsucrose (activated by Bu_2SnO) with methanesulfonyl chloride in benzene afforded the 3-mesylated derivative, whereas the same process performed in toluene surprisingly provided the 4-mesylate in 50% yield. If triflic anhydride is used as the sulfonating reagent, the 4-triflate was obtained in 40% yield (Scheme 12).[153]

4. Acetalation

The hydroxyl groups of carbohydrates react with carbonyl groups to give, under acid catalysis, cyclic acetals or ketals.[154,155] In the case of sucrose, the sensitivity of the glycosidic bond to acids limits the use of this reaction to a small number of very reactive carbonyl substrates. Acetone, or its acetalation equivalents (2-methoxypropene and dimethoxymethane) provide good yields of isopropylidenated products. Carboxylated derivatives at the 4,6-positions were thus obtained.[156] The reaction (which develops under thermodynamic control) is very selective towards the 4,6-diol group, which is the only diol able to produce a *trans*-decalin type of cyclic acetal (two 6-membered rings having one bond in common). Interestingly, when the reaction time is prolonged, a second isopropylidene ring is constructed engaging OH-2 and OH-1′ (those OH groups favorably oriented and reactive). In this case, both monosaccharide moieties of sucrose are doubly connected through an 8-membered ring.[157,158] This diacetal can undergo monodeprotection of the 4,6-diol leading—in moderate yields—to the 1′,2-monoacetal.[159] These sucrose isopropylidene acetals (mono and di) have been used in many multistep sequences, involving esterifications, and etherifications in particular.

In reaction with benzaldehyde or its dimethyl acetal, or benzylidene dibromide, sucrose undergoes conversion into its 4,6-acetal in moderate yield (28–35%).[160,161] The transacetalation reaction of dialkyl acetals prepared from unsaturated or aromatic aldehydes can take place under very mild acidic conditions. For example, sucrose-derived monomers or surfactants (variously substituted 4,6-*O*-methylidenesucroses) have been prepared[146,162] and 4,6-monoacetals incorporating masked aromas or fragrances derived from β-citral or α- or β-ionone were prepared directly from sucrose in high yields.[163,164] These latter molecules, whose unsaturated carbonyl system is very susceptible to oxidation, are thus temporarily protected and are readily released under very mild conditions (Scheme 13).

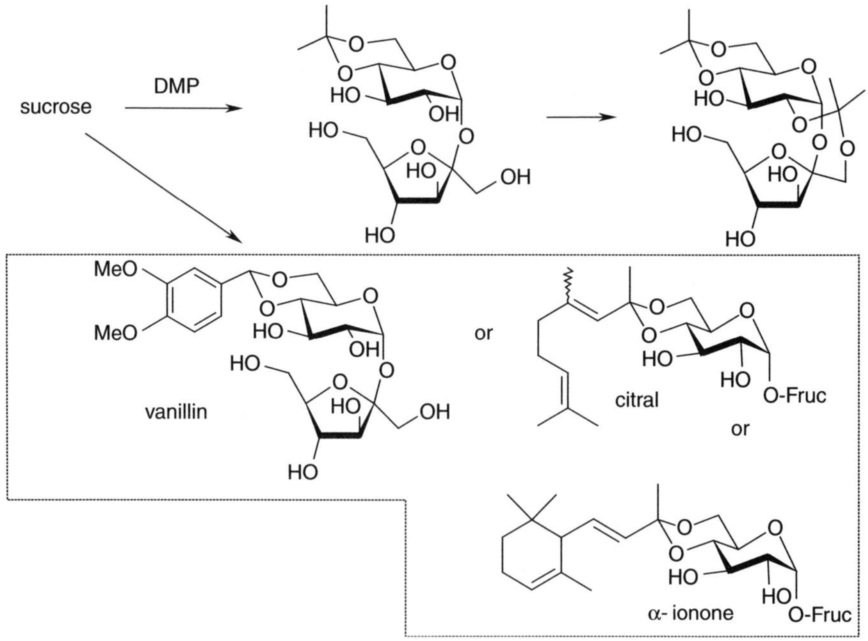

SCHEME 13. Examples of sucrose acetals (DMP = 2,2-dimethoxypropane).

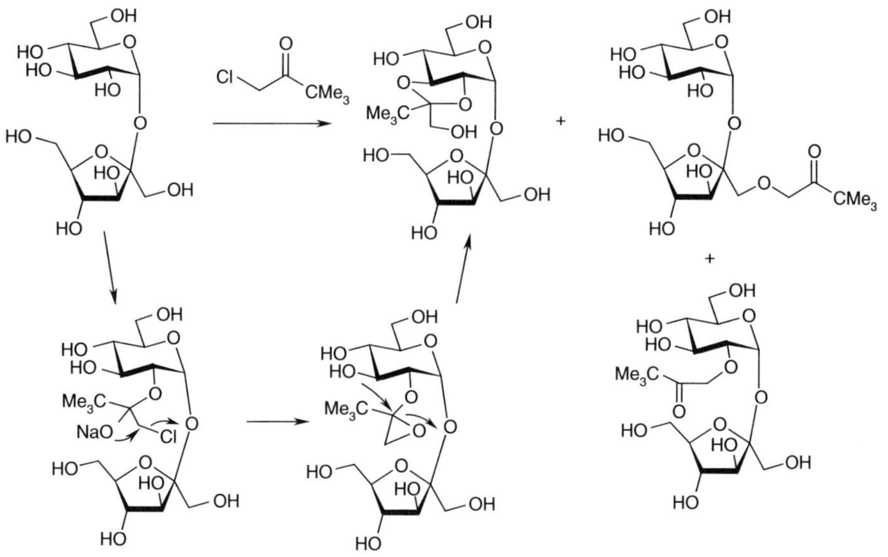

SCHEME 14. Acetalation at OH-2 and OH-3 by reaction with *t*-butyl chloromethyl ketone.

It is also possible to obtain acetals under basic conditions by the reaction of sucrose with *gem*-dihalo compounds,[155] or with an α-chloromethyl ketone. For example, the reaction of sucrose with *tert*-butyl chloromethyl ketone provides, as the major product, an acetal that involves positions 2 and 3. The peculiar reactivity of OH-2 towards electrophilic species, in this case the carbonyl group of the chloromethyl ketone, is again the starting point leading to an intermediate alcoholate that is immediately transformed in a 3-membered-ring cyclic acetal. This latter is opened by OH-3 to generate the new acetal (Scheme 14), together with the simple ethers at positions 2 and 1′ arising from the classical Williamson reaction by displacement of the chlorine atom.[165,166]

5. Oxidation

a. Carboxylated Derivatives.—In addition to glycol cleavage of sucrose by periodate or lead tetraacetate, affecting either one or both of the glucosyl or fructosyl moieties,[167–177] various methods permit the conversion of alcohol groups into ketones or carboxylates with conservation of the sucrose skeleton (Scheme 15). The Pt/O_2 oxidation system provides derivatives with carboxyl groups at the primary positions.[178–182] “6-Monocarboxysucrose” undergo hydrolysis by invertase, producing D-fructose and D-glucuronic acid, whereas 6′-monocarboxysucrose and 6,6′-dicarboxysucrose are competitive inhibitors of invertase.[15,181,182] A difficulty in this reaction is control of the number of oxidized alcohol functions and the achievement of selectivity among the mono-, di-, and tricarboxy derivatives. Catalyst poisoning is also problematic. This difficultly could be solved by continuous extraction of the monocarboxysucroses from the medium.[178,184,185]

Electrocatalytic oxidation also led to derivatives having a carboxy group at the initial primary positions although complex mixtures were obtained including more-degraded compounds.[186] Use of the NaOCl–TEMPO (tetramethylpiperidinyloxy) system, which was applied to unprotected carbohydrates and polysaccharides by Van Bekkum *et al.*[187,188] was studied with sucrose with a focus on the effect of ultrasound on the formation of "tricarboxysucrose."[189–198] With excess hypochlorite, overoxidation to a pentacarboxy side-product was observed, probably arising from C-3′ to C-4′ cleavage. Explanations of the sonochemical effect were first supported by a study performed on methyl α-D-glucopyranoside, which possesses only one primary hydroxyl group. Sodium (methyl α-D-glucopyranosid)uronate was obtained, and showed that rate acceleration was related to the ultrasound characteristics (power, frequency, probe diameter). It was also shown that the reaction could be performed in the absence of sodium bromide (at a much lower rate) thus decreasing the environmental impact of the method. Furthermore, the possible competition between oxidation of primary and secondary OH groups because of the presence of hypobromite is improved in favor of the primary ones. In this case, the actual role of bromide ion in the mechanism (already a matter of question in the NaOCl–NaBr system in the absence of TEMPO), would not be to oxidize hypochlorite ions to

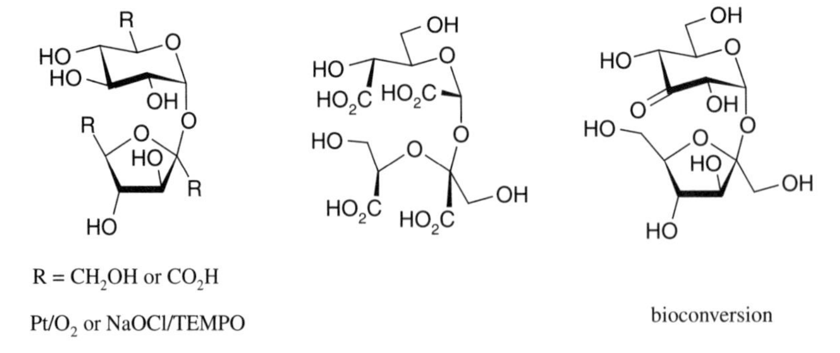

SCHEME 15. Main products of sucrose oxidation.

hypobromite ions, themselves able to oxidize TEMPO to its nitrosonium species.[195,199–201] Under ultrasound, other oxidant species, such as chlorine molecules, cations, or radicals can be produced, and would directly oxidize TEMPO, whereas without ultrasound no recycling of the catalyst would occur. This is supported by the increased initial rate observed when a sonication period is applied prior to the addition of the carbohydrate substrate, due to an accumulation of nitrosonium ions in the medium. Other variations of the TEMPO method, such as the presence of silver catalysts were described (Scheme 15).[202]

b. Oxo-sucroses.—The oxidation of sucrose to "3-oxo-sucrose" by various methods involving the D-glucoside 3-dehydrogenase of *Agrobacterium tumefaciens* has been widely investigated. Novel new derivatives of sucrose were prepared by reduction (to the sucrose 3-epimer) or by reaction with nitrogen-containing nucleophilic species.[203–212] Alternatively, chemical oxidation of sucrose with bromine gives the 2-oxo derivative as the major product, together with the 4- and 3′-oxo sucroses.[213] Finally, there is the dehydrogenation–redeuteration process using deuterium and Raney nickel, which produces deuterio analogues of carbohydrates.[214–217] When applied to sucrose, different levels of deuterium incorporation were observed depending on the conditions.[36,218,219] In this reaction, retention of configuration at the deuteration sites is generally observed, although some epimerization is difficult to prevent.[220]

All types of oxidized sucrose species constitute useful synthons that have been extensively used in multistep sequences, as described in Section II.8.

6. Isomerizations and Bioconversions

Many chemical products are obtained by fermentation, notably for the utilization of the amount of sucrose remaining in molasses (Scheme 16).[221] Thus sucrose and molasses are preferred carbon sources for the preparation of

$$\text{Sucrose} \xrightarrow[H_2O]{\textit{Saccharomyces cerevisiae}} 4\ C_2H_5OH + 4\ CO_2$$

SCHEME 16. Conversion of sucrose into ethanol by *Saccharomyces cerevisiae*.

polyhydroxyalkanoates.[222] Other carbohydrates, especially D-glucose, may be fermented to produce "biofuel," as well as starch. Cellulose and hemicelluloses have future potential. Ethanol production by fermentation has become a major industrial target due to the increasing shortage of fossil resources.

Specific microorganisms, yeasts, and bacteria can also convert sucrose into other alcohols, as well as organic acids, amino acids, and vitamins (Scheme 15). All these biological processes have been improved with the help of modern biotechnology, making them more chemically and economically efficient and to direct them towards new and useful chemical products. The "biocracking" of carbohydrates has been shown to be a way for the production of "biohydrogen."[223]

Other bioconversions include isomerizations and synthesis of oligo- and polysaccharides. With sucrose, such reactions (with *Protoaminobacter rubrum CB5 574.77*) lead to new disaccharides such as isomaltulose and trehalulose. These disaccharides are constructed from one glucose and one fructose residue, but with glycosidic connections different from that in sucrose. Isomaltulose, also referred to as palatinose is connected α-(1→6), whereas trehalulose has an α–(1→1) one and leucrose an α–(1→5) link. All these disaccharides are reducing sugars, existing in various pyranose or furanose forms,[224] and have therefore a supplemental reactivity at the hemiacetal position. After development of the isomaltulose process on the industrial scale, mainly for its use as noncariogenic and reduced calorie sweetener, by itself or as its hydrogenated derivative isomalt (Palatinit®), there has been intense research on the use of isomaltulose, in a way comparable to that was previously conducted with sucrose towards classical targets.[225–228] This has included the formation and uses of new bicyclic lactones.[198,229–231] Engineered microorganisms can now direct the reaction towards either isomaltulose or trehalulose with improved selectivity and efficiency. Some uses of leucrose as a starting material have been described.[232] Heating isomaltulose in the presence of an acidic resin in Me_2SO provides 5-(α-D-glucosylmethyl)furfural (Schemes 17 and 18).[233]

sucrose
protoaminobacter rubrum
CB5 574.77
HO HO HO O HO O 6′ O OH HO 2′ OH 1′ HO
120 °C/Me_2SO
H^+ resin
HO HO HO O HO O O O

SCHEME 17. Conversion of sucrose into isomaltulose then to GMF.

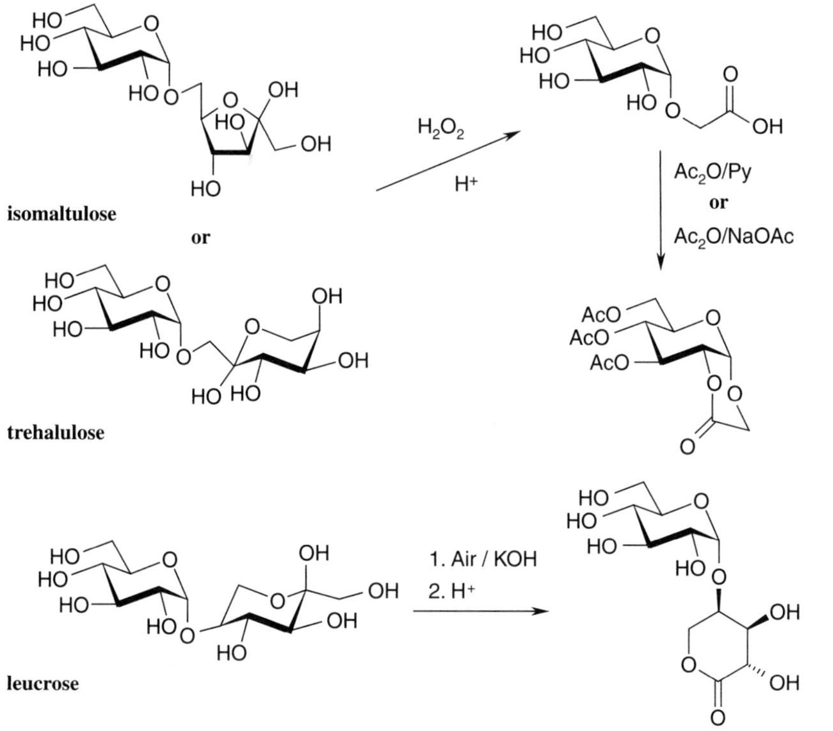

SCHEME 18. Oxidation of isomaltulose, trehalulose, and leucrose.

Disaccharides analogous to isomaltulose, but containing other sugars in place of α-D-glucose, can be obtained by similar bioconversions from different disaccharides.[234] Sucrose analogues (galactose–fructose, xylose–fructose, fucose–fructose) have also been prepared by isomerization, catalyzed by an engineered fructosyltransferase.[235,236]

Oligo- and polysaccharides can be obtained by the use of transglycosidases or glycosyltransferases that recognize sucrose, either as the receiving sugar or as the provider. Many reports by Monsan and co-workers have described how powerful these processes are and interesting applications may be envisaged for specifically engineered polysaccharides.[237–241] Sucrose acts also as a glucose donor in polymerization reactions and for the synthesis of glucosides, as in the case of flavonoid glucosides.[242] Polymers constructed only with glucose units, having controlled and regular glycosidic linkages, or with only fructose (inulin like) can thus be produced, the remaining monosaccharide moiety being used by the microorganism as an energy source.[243–245] Some fructooligosaccharides having a DP of 5–10 have been shown to be bifidogenic and are used as food additives.[227,246]

Oxidative bioconversions include notably the oxidation of sucrose to "3-keto-sucrose" (Scheme 18),[25,203,204,207] a synthon that has been used as starting point for synthesis of many sucrose derivatives (see Sections II.5 and II.8).

Finally, there is the important biotechnological hydrolysis of sucrose to a 1:1 mixture of glucose and fructose by the use of invertase (so-called because the dextrorotatory sucrose gives a levorotatory final mixture). Enzyme-mediated esterifications of sucrose have been discussed in the "esterification" section.

7. Miscellaneous

a. Heterosubstitutions.—One of the most useful processes for conversion of alcohols into halides (mostly chlorides) is the Appel reaction.[247] Treatment of free sucrose with Ph_3P and CCl_4 in pyridine affords 6,6′-dichloro-6,6′-dideoxysucrose in 91% yield.[248,249] Surprisingly, a similar procedure performed on a penta-*O*-benzylsucrose was much less selective.[250] Other dihalosucroses (bromides, iodides) are also accessible by reaction with modified Appel or related reagents.[251,252]

Treatment of free sucrose with phthalimide under Mitsunobu conditions affords modified derivatives in which the primary 6-OH and 6′-OH groups are replaced by a phthalimido moiety, with concomitant epoxide formation at C-3′ and C-4′ (Scheme 19).[253] Chloro-anhydro derivatives were formed similarly under Mitsunobu conditions in the presence of zinc chloride.[254]

6,6′-Dibromo-6,6′-dideoxysucrose was used as a starting material for the preparation of various sulfides of sucrose.[255–257] The 6,6′-dithiol can be oxidized

X = Cl (for Ph_3P / CCl_4 / py; yield 91%)
or X = Br, I

SCHEME 19. 6,6′-Dihalosucrose and 6,6′-dideoxy-6,6′-diphthalimido-3′,4′-anhydrosucrose.

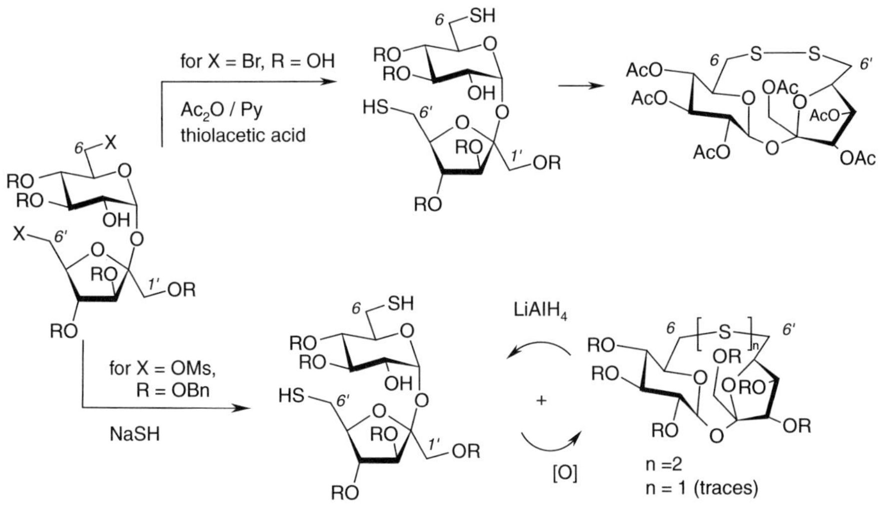

SCHEME 20. Sucrose sulfides.

by air into disulfide. The same process can be conducted on protected hexa-*O*-benzylsucrose. The S_N2 substitution of the 6,6′-dimesylates with NaSH provided the corresponding dithiol, contaminated with the disulfide. Traces of a cyclic monosulfide were also detected (Scheme 20).[258] Mono-6-deoxy-6-fluoro- and 6,6′-dideoxy-6,6′-difluorosucroses were prepared by similar multistep strategies, as well as 4,6-dideoxy-6-iodosucrose.[259–261] Some of these compounds were found to exhibit inhibitory activity towards glucansucrases.

b. Carbamates, Carbonates, and Thiocarbonates.—The reaction of carbohydrates with diisocyanates is well documented, as it affords a route to polyurethane materials. However, the controlled and selective synthesis of *N*-alkylcarbamoyl carbohydrate derivatives has been much less studied.[262–269]

The preparation of carbamoyl derivatives of sucrose under conditions similar to those used for esterifications has been reported.[103,270] Similar trends in the regiochemical outcome was observed ether in organic solvents or in aqueous medium. The increased stability of the carbamate linkage as compared to carbonates or esters limits the rate of group migration from one position of the sugar to another, and therefore makes the reaction more selective.[122] These compounds might find application as detergents, because of their increased stability under basic conditions.

The reaction of sucrose with chloroformates lead to cyclic carbonates or alkyloxycarbonyl derivatives.[271–276] These reactions exhibit behavior similar to

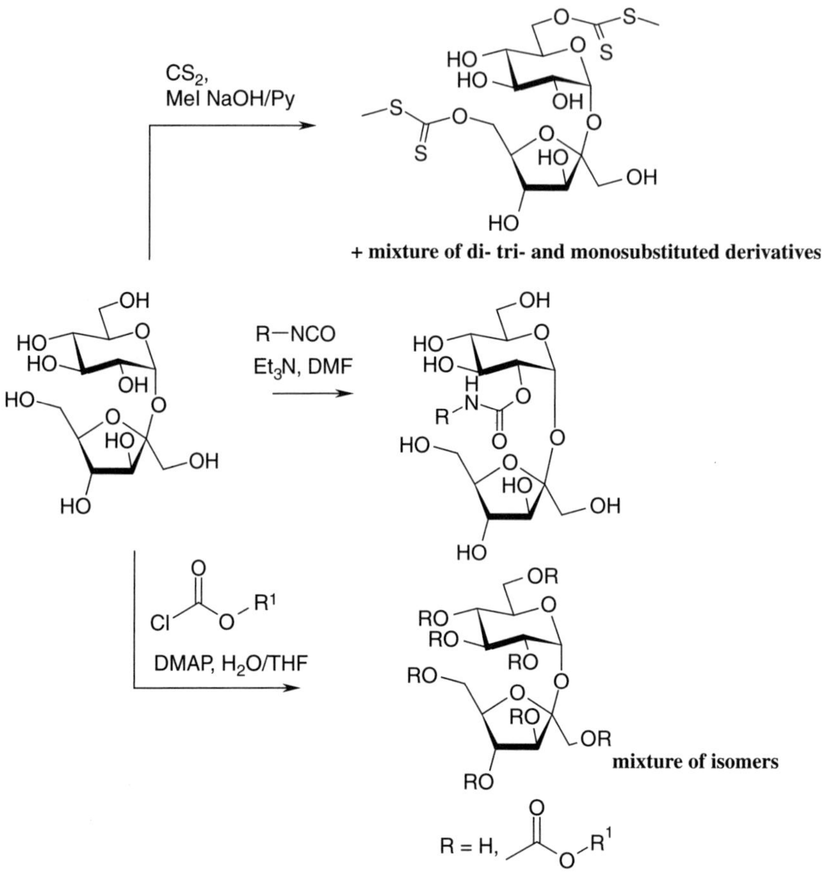

SCHEME 21. Sucrose thiocarbonates, carbamates, and carbonates.

that observed for other acylations and can be performed in organic or aqueous media. For example, the addition of allyl chloroformate (0.25 equiv.) to an aqueous solution of sucrose led to good yields of monosubstituted allyloxycarbonylsucrose derivatives.[130,131] The reverse addition mode led mostly to more-substituted derivatives.[272]

Dithiocarbonates of sucrose were obtained by reaction with carbon disulfide and alkyl halides under basic conditions.[277] Subsequent treatment by amines led to the corresponding mixed thiocarbamates (Scheme 21).

c. Isotopically Labeled Sucroses.—A ^{13}C isotopic variation of sucrose has been prepared by biosynthesis under an atmosphere of $^{13}CO_2$.[278,279] C-Deuterio analogues have been described, notably 2- and 3-monodeuteriosucroses obtained by reduction of "2- and 3-oxo-sucroses."[205,280] Poly-*C*-deuterated sucrose analogues have been prepared by Raney nickel-catalyzed deuteration of sucrose in refluxing deuterium oxide.[36,219,281] An oxidation–reduction sequence achieved on a pentabenzylated sucrose led to the hexadeuterio analogue having all primary positions labeled.[282] A sucrose monotritiated at position 6 was obtained by reduction of the "6-aldehydosucrose" derived from the hepta-*O*-benzoyl derivative having only OH-6 available.[283]

d. Anhydro Derivatives.—As mentioned in the preceding sections, anhydro derivatives of sucrose are often observed during Mitsunobu-type reactions (see Scheme 7).[98–100,253,254,284–286]

Other examples include, bridged interglycosidic anhydro derivatives that were identified upon treatment of sucrose with HF[287–291] or by basic treatment of intermediate tosyl derivatives.[137,293–296]

8. Targeted Multistep Synthesis from Sucrose

Various multistep sequences are described in this section. Generally, great care is taken to preserve the readily cleaved glycosidic bond in sucrose.[297] However, there are some useful examples that include a hydrolysis step. Synthesis of several imino sugars—glycosidase inhibitors—from sucrose was reported by Stütz. 6,6′-Dichlorosucrose (see Section II.7) was converted into the known diazide,[249] and subsequently hydrolyzed, providing a mixture of monosaccharide azides; both were used as starting materials for the preparation of 1-deoxymannonojirimycin.[298] Alternatively, the sucrose diazide was benzylated and then hydrolyzed, affording a mixture of perbenzylated monosaccharide azides. The fructose-derived intermediate was then converted in several steps into the corresponding imino sugars (Scheme 22).[299]

Hydrolysis of sucrose in acidic 2-chloroethanol provided the corresponding glycoside, basic cyclization of which furnished a spiro derivative.[15,300]

Sucrose was also used as starting material for the preparation of morpholine derivatives. The first step involved a controlled oxidation of either ring of sucrose (Scheme 23).[173]

Disaccharides analogous to sucrose, but having different rings might have useful biological properties. "Galactosucrose" was obtained by a multistep sequence from sucrose in 27% overall yield.[235] An enzymatic route to this derivative, involving the fructosyltransferase (FTF) from *Bacillus subtilis* NCIMB

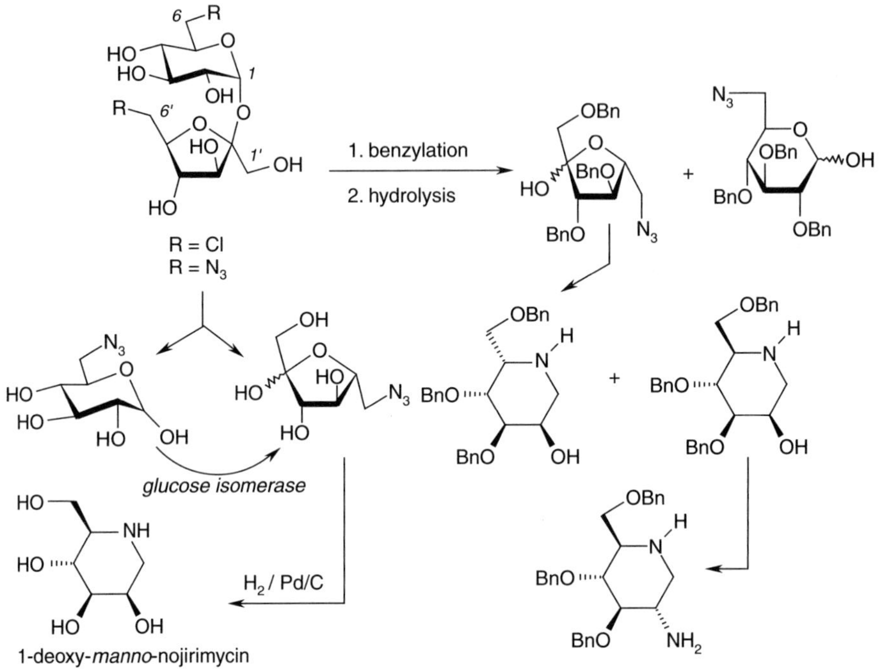

SCHEME 22. Synthesis of imino sugars from sucrose.

11871, replacing the glucose moiety by a galactose one, proved to be much more efficient (Scheme 24).

Condensation of partially protected sucroses with the phosphonic acid under Mitsunobu conditions was the key step in the preparation of sucrose phosphonates. A monoclonal antibody having good binding properties showed a regioselective esterase activity towards the 6-heptylphosphonate as compared with 6′-octanoylsucrose (Scheme 25).[102]

Treatment of silyl ethers of sucrose with the Vilsmeier reagent gave various sucrose formates. When only one silyl protecting group was present, the reaction afforded the corresponding monoesters in high yields. However, when di- or trisilylated sucroses were used as substrates, a mixture of mono- and di-formates were obtained (Scheme 26).[301]

An approach to sucrose derivatives galloylated at the terminal positions was proposed by Barros *et al.* Selective desilylation at the 1′-position, performed on 1′,6,6′-tri-*O*-silylated penta-*O*-acetylsucrose was a key step in the preparation of 1′-*O*-galloylated analogues (Scheme 27).[135]

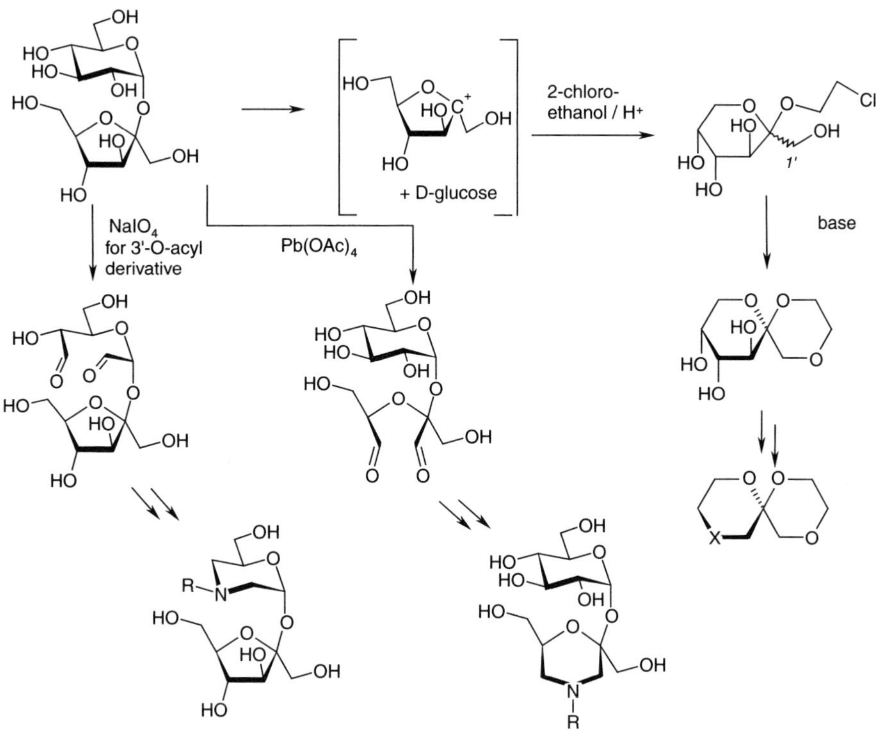

SCHEME 23. Preparation of morpholine and spiro derivatives from sucrose.

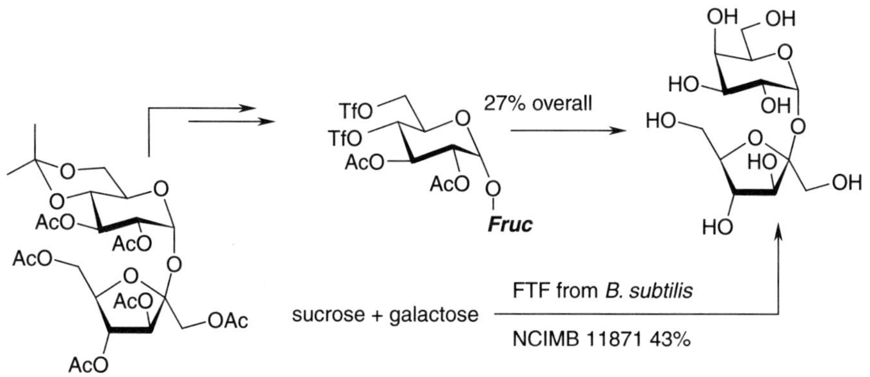

SCHEME 24. Synthesis of "galactosucrose" by chemical and enzymatic methods.

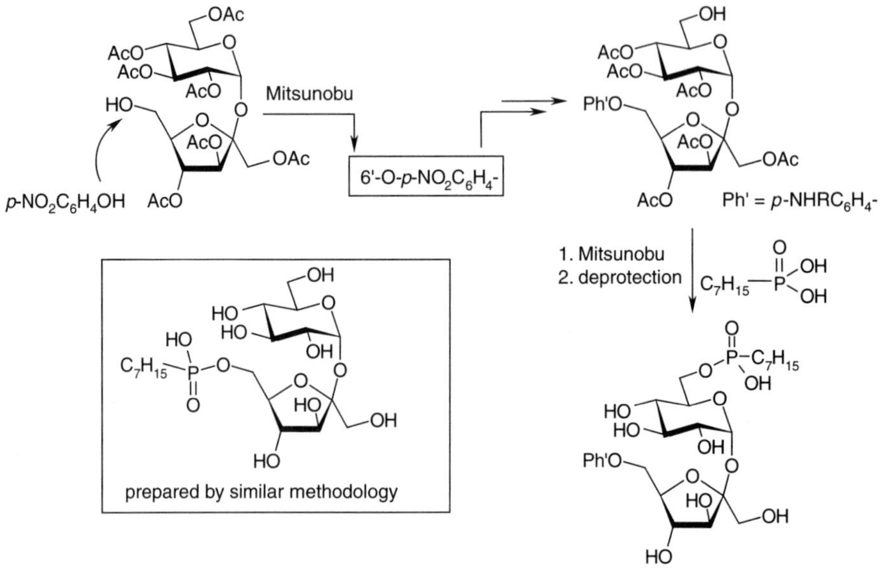

SCHEME 25. Preparation of sucrose phosphonates.

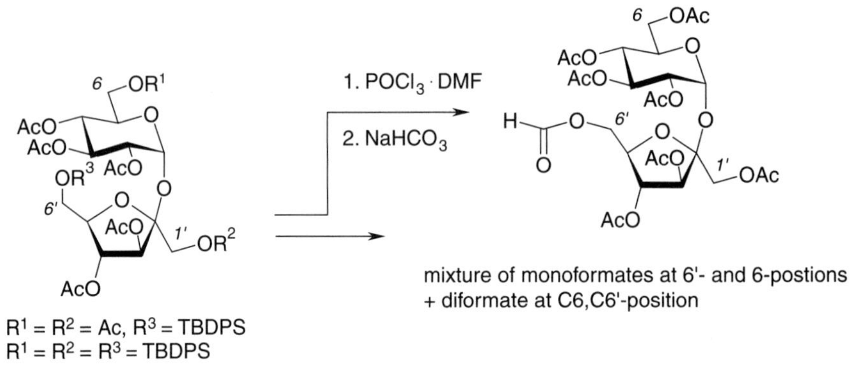

SCHEME 26. Synthesis of sucrose formates.

Treatment of tri-*O*-silyl-penta-*O*-benzoylated sucrose with HF removed the silyl blocks from the C-6 and C-6′ positions, leading to the 1′-*O*-monosilylated derivative in 96% yield. This compound can be esterified to afford heptaesters (Scheme 28).[145]

Regioselective silylation of sucrose afforded the 6′-protected derivative, which was fully benzylated and then deprotected at the 6′-position, and further

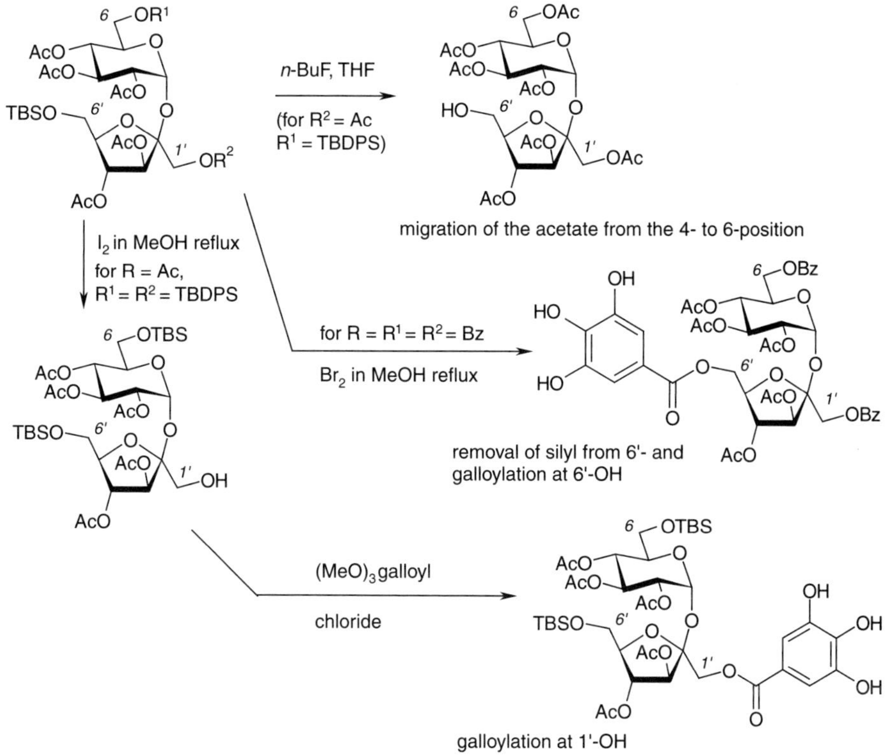

SCHEME 27. Synthesis of galloylated sucrose derivatives.

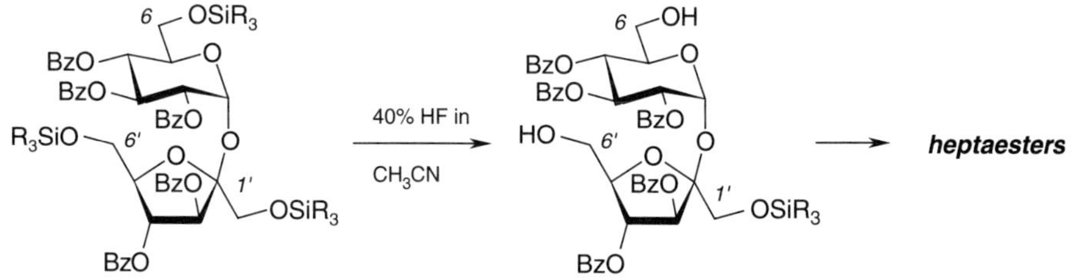

SCHEME 28. Selective desilylation of trisilylated sucrose.

converted into unsaturated esters.[66] These latter were transformed into copolymers with styrene or methyl methacrylate (Scheme 29).[302]

Functionalization of sucrose molecule can be also performed regioselectively at any of the terminal positions: C-1′ or C-6, or C-6′ with the secondary

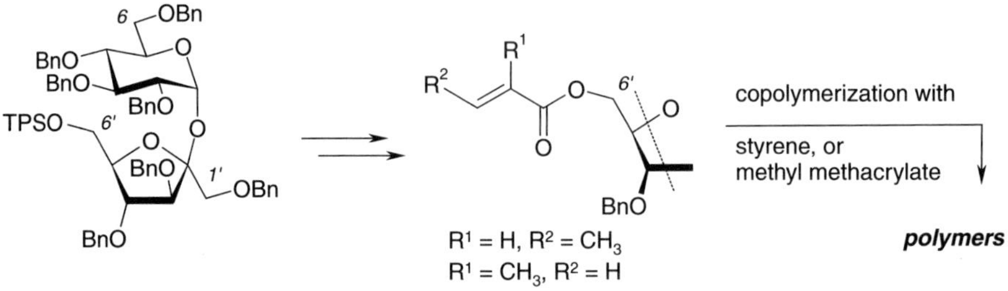

SCHEME 29. Synthesis of unsaturated 6′-esters of sucrose as monomers.

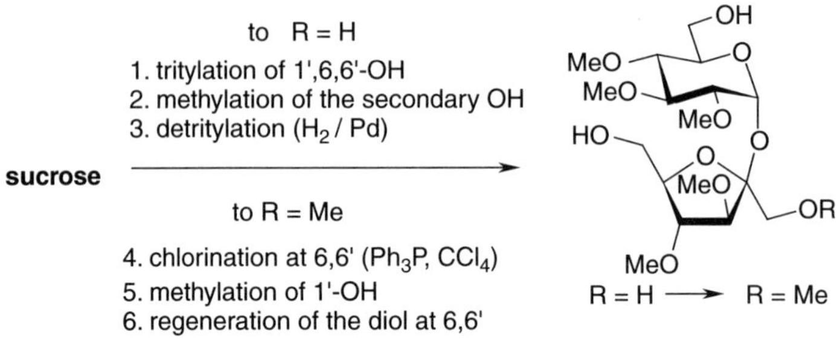

SCHEME 30. Penta-*O*-methyl and hexa-*O*-methyl sucroses.

positions remaining intact. Such a goal can be realized by placing temporary, but stable, protecting groups (Tr or SiR_3; see Section II.2) on the terminal hydroxyl groups followed by protection of the secondary ones and final deprotection of the primary functions.

Penta-*O*-benzoyl-tri-*O*-trityl- or penta-*O*-acetyl-sucroses were prepared by esterification of the 1′,6,6′-tri-*O*-trityl derivative, followed by removal of the trityl groups under acidic conditions.[303] However, migration of the acetyl group from the O-4 to O-6 was observed. Penta-*O*-methylsucrose can be prepared from the same tri-*O*-tritylated precursor by methylation of all secondary hydroxyl groups followed by reductive removal of the temporary trityl blocks.[304] This compound was further transformed into the hexa-*O*-methyl derivative by: conversion of the 6- and 6′-OH groups into the dihalide (Ph_3P/CCl_4), etherification of the remaining 1′-OH, and regeneration of a diol function by displacement of chlorine by cesium acetate, followed by hydrolysis (Scheme 30). Treatment of this diol with methacryloyl chloride afforded an unsaturated diester.[305]

Benzylation of 1,6,6′-tri-*O*-tritylsucrose afforded the fully protected derivative from which the trityl protecting groups were removed under controlled acidic conditions to provide 2,3,3′,4,4′-penta-*O*-benzylsucrose.[306,307] The primary hydroxyl groups were further differentiated, furnishing any of the three monoalcohols with OH-1′, OH-6, or OH-6′ free.[308] The compound having OH-1′ free was prepared by esterification at the most reactive 6,6′-OH groups with *p*-nitrobenzoic acid under Mitsunobu conditions. Protection of the remaining OH-1′ as its methoxymethyl (MOM) ether[308] or a benzyloxymethyl (BOM) ether[309] followed by removal of the ester blocks afforded the diol(s) having 6,6′-OH free. Differentiation of these two hydroxyl groups was readily accomplished by use of silylating reagents. For example, treatment of the diol with 1 equiv. of *tert*-butylchlorodiphenylsilane afforded the O-6′-monoprotected derivative in 80% yield. No formation of the O-6 monoprotected regioisomer was detected. The latter was prepared in 68% yield by selective deprotection at O-6′ of the disilylated derivative, which was obtained by reaction of the diol with an excess of Bu^tPh_2SiCl in 95% yield (Scheme 31).[309]

Such compounds served as convenient precursors of sucrose-derived uronic acids and amines (with the functional group being placed at the desired position), which were obtained by standard methods (Scheme 32).[310] Each hydroxymethylene group was readily oxidizable to an aldehyde[308,309] or convertible into phosphonates that were applied in the preparation of "higher sucroses."[311] The benzyl groups blocking the secondary position of the sucrose backbone were readily removed, leading to new modified sucroses. Alternatively, the heterocyclic ring can be constructed at the terminal position of sucrose.[310]

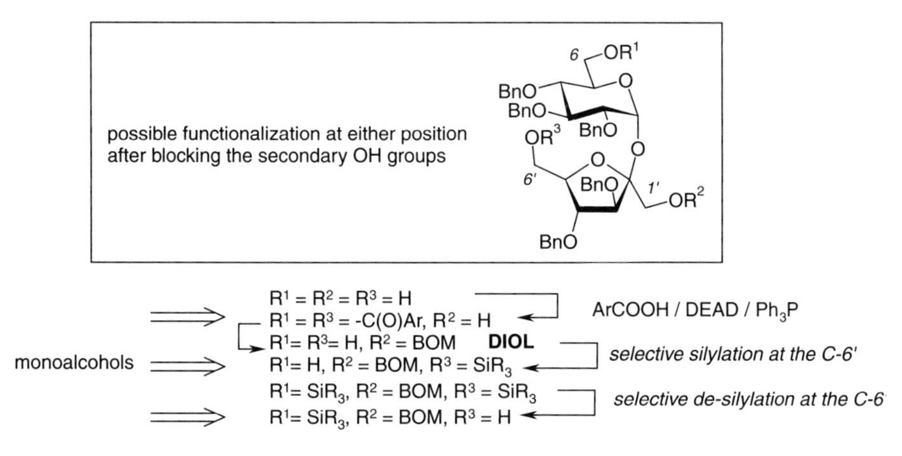

SCHEME 31. Differentiation of the primary hydroxyl groups in 2,3,4,3′,4′-penta-*O*-benzylsucrose.

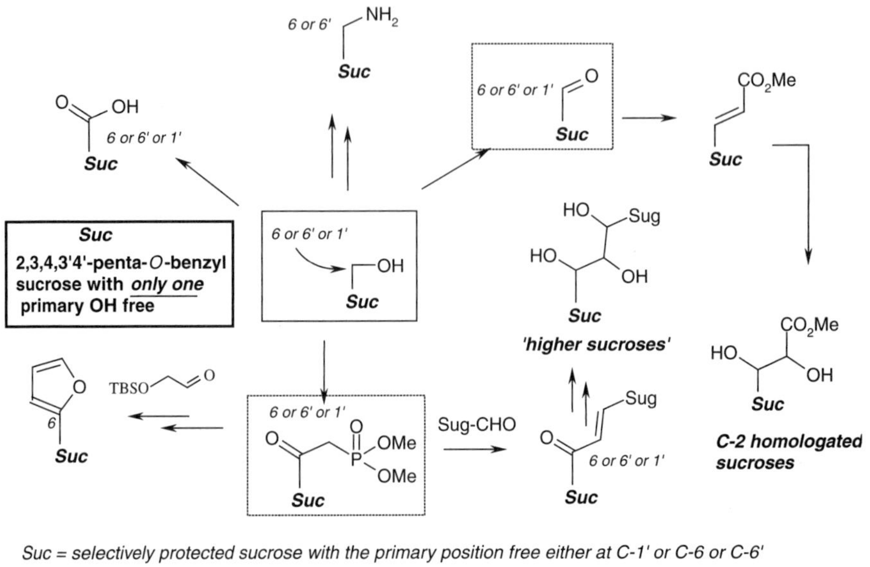

Suc = selectively protected sucrose with the primary position free either at C-1′ or C-6 or C-6′

SCHEME 32. Application of selectively protected sucrose in the synthesis of analogues.

New derivatives incorporating sucrose inside the backbone of a macrocycle have been prepared. The C-6 and C-6′ "ends" in free sucrose are close to each other (see Scheme 2 in Section II.1). They obviously are also close in partially protected sucrose (1′-*O*-MOM or 1′-*O*-BOM derivatives of penta-*O*-benzylsucrose) since they can be connected together by (at least) a C_4 carbon bridge by a Williamson-type coupling of the sucrose diol with 1,4-diiodobutane (Scheme 33).[310]

This opened interesting possibilities for the synthesis of macrocyclic receptors based on sucrose. The diols reacted with polyglycol ditosylates under basic conditions to afford crown ether analogues in yields up to 50%.[312,313] 1′,2,3,3′,4,4′-Hexa-*O*-benzylsucrose, prepared from 6,6′-dichlorosucrose, was even more convenient for the synthesis of macrocycles.[258] Receptors having different cavities and heteroatoms other than oxygen in the macrocyclic ring were prepared.[313] It was possible to remove the benzyl (and also BOM) groups by simple hydrogenation to release the free macrocyclic analogue (Scheme 34).[312] These compounds were tested as receptors for the alkali metal cations and also ammonium ion, but they exhibited very weak complexing properties (K_a up to 260 M in acetone-d_6).[313]

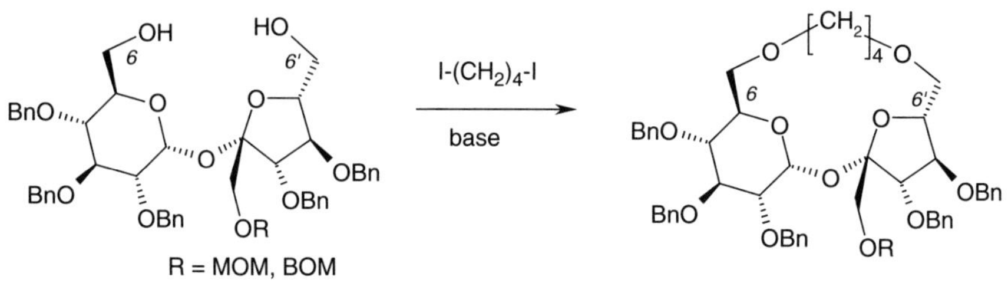

SCHEME 33. Coupling at the 6,6′-positions of partially protected sucrose by Williamson etherification.

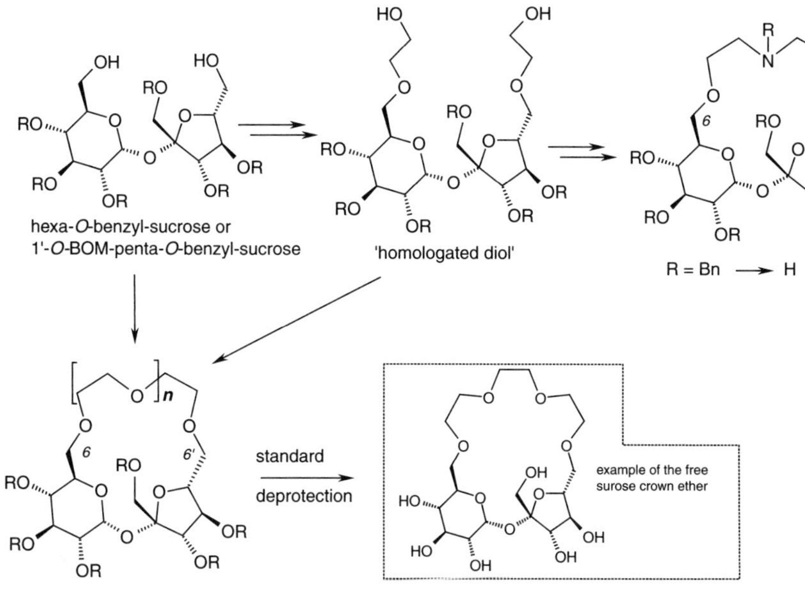

SCHEME 34. Synthesis of sucrose-based macrocyclic receptors.

Ring-closing metathesis (RCM) of the 6,6′-diallyl ether afforded macrocyclic alkenes,[314] which can be reduced (H_2/Pd) to saturated compounds with simultaneous removal of all benzyl protecting groups (Scheme 35).[68]

Macrocycles of higher symmetry might possess better complexing properties than "simple crowns." A model study on the synthesis of C_2-symmetrical derivatives containing the sucrose molecule was reported. The synthesis was initiated from hexa-*O*-benzylsucrose, the crucial step being the selective silylation of the 6′-OH group. Further standard transformation provided two intermediate

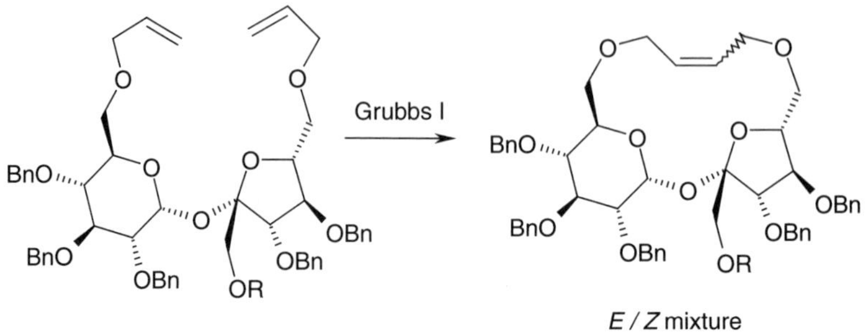

SCHEME 35. Coupling of the 6,6′-position of sucrose by ring-closing metathesis (RCM).

synthons. The first one, a 6′-monoalcohol having a mesylated hydroxyethyl ether ready for cyclization under the basic conditions surprisingly provided the monomeric crown-ether derivative instead of the expected C_2-symmetrical derivative.[315] Indeed, previous observation showed that the bridge should have at least four atoms to connect the C-6,6′ positions (Scheme 36).[310]

The second intermediate, a 6′-*O*-acroyl 6-*O*-allyl-dialkene was subjected to RCM (with Grubbs' II catalyst) and afforded small amounts of the C_2-symmetrical derivative, accompanied by *E*- and *Z*-cyclic monomers.[316]

Another combination of silylation at the primary positions, benzylation at all five secondary hydroxyl groups and desilylation provides the triol, which could be oxidized to the tricarboxylic ester and then reduced by lithium aluminium deuteride. This sequence was designed for the synthesis of 1′,1′,6,6,6′,6′-hexadeuteriosucrose (+6 mass units), an analogue of sucrose having a clear signature in mass spectrometry, as well as its possible degradation products (+2 for glucose and +4 for fructose).[282] The oxidation step of the triol could be achieved either by the NaOCl/TEMPO method followed by esterification with methyl iodide or, better, by direct oxidation of the primary alcohols to the *t*-butyl esters by the pyridinium dichromate/*t*-butanol/Ac_2O method.[317,318] This reaction is based on the *in situ* oxidation of the intermediate hemiacetal formed by *t*-butanol and the aldehyde first produced. Reduction and debenzylation led to the desired hexadeuterio analogue of sucrose (Scheme 37).

The primary hydroxyl groups in 2,3,3′,4,4′-penta-*O*-benzylsucrose can also be readily oxidized to a trialdehyde, which can be trapped with a phosphorane ($Ph_3P = CHCO_2Me$) to afford the corresponding trialkene (Scheme 38).[307]

Oxidation of OH-2, 3, or 4 of sucrose has also been used as a first step towards many other derivatives. When 2-*O*-benzyl sucrose is transformed to its heptabenzoate, selective debenzylation can provide the isolated 2-OH group. This

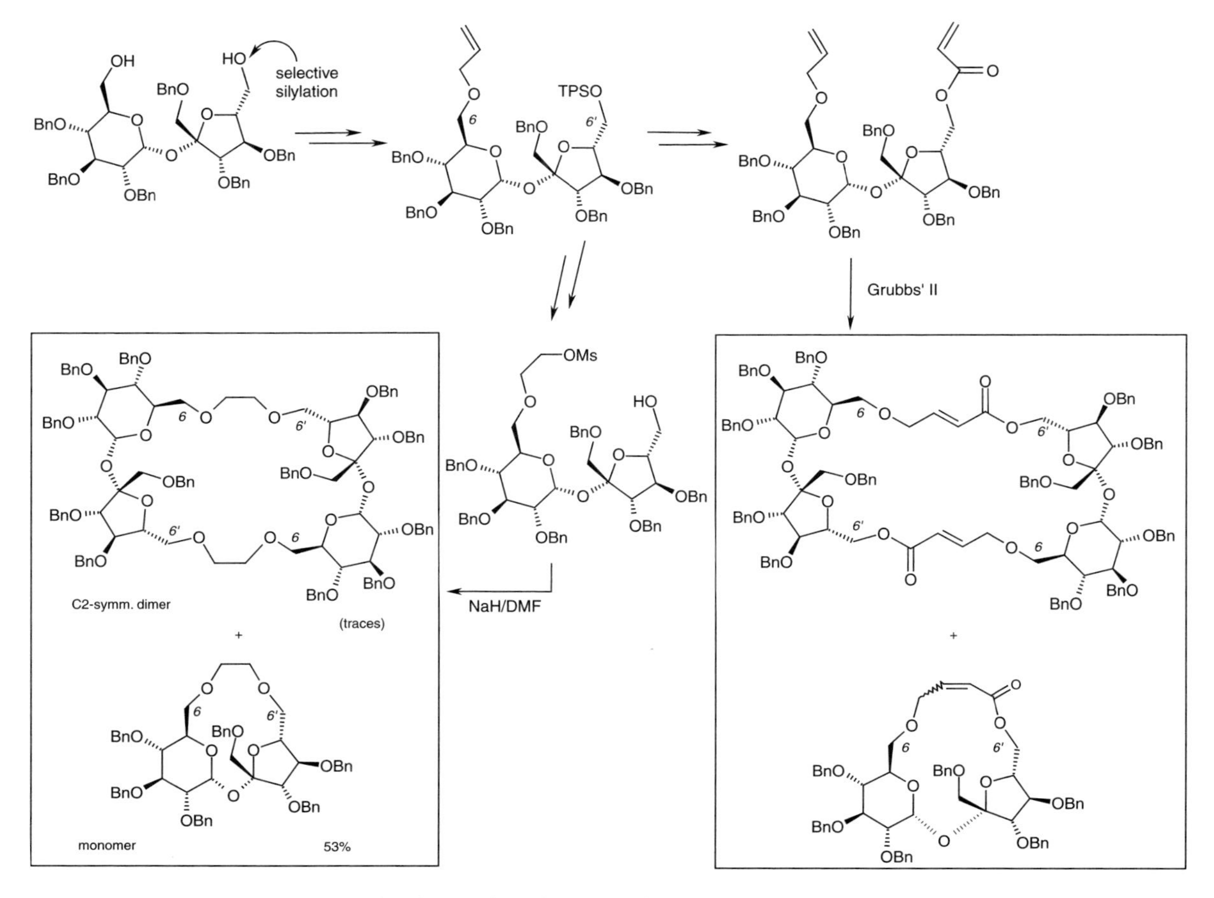

SCHEME 36. Approach to C_2-symmetrical receptors containing sucrose.

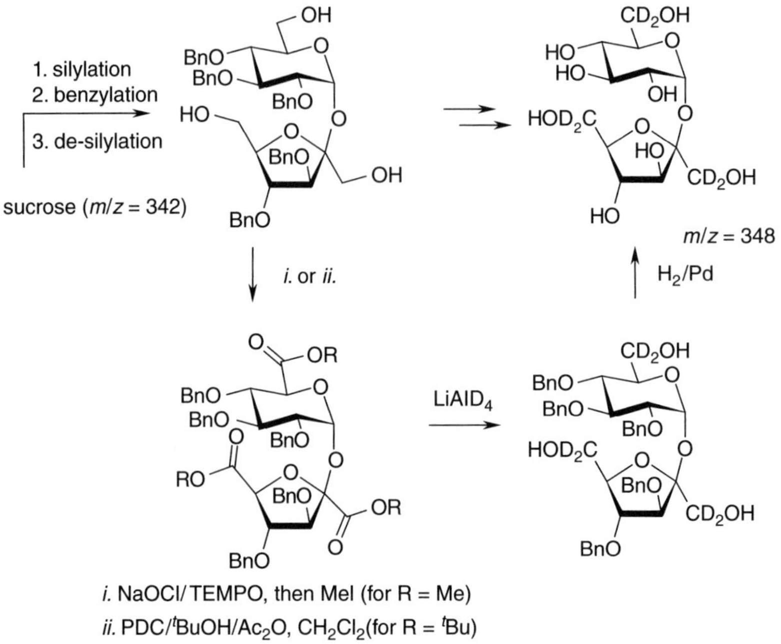

SCHEME 37. Preparation of sucrose hexadeuterated at the 1′,6,6′-positions.

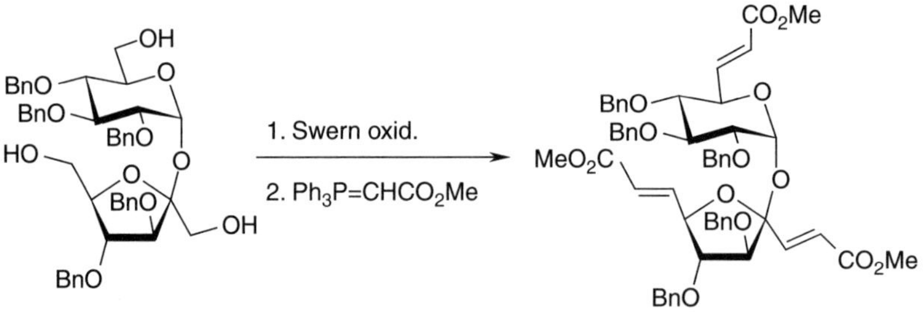

SCHEME 38. Synthesis of sucrose homologated at all three primary positions.

latter can be oxidized (with concomitant elimination) to a very elaborated enone,[61] or reduced to 2-deoxysucrose.[319] Triflation at O-2 followed by S_N2 displacement led to the C-2 epimer or its aminodeoxy analogue, as well as the 2,3-epoxide and its ring-opening products (Scheme 39).[320]

Incomplete pivaloylation leaving OH-4 as the only free hydroxyl group[321,322] followed by oxidation at C-4 led to a series of sucrose analogues modified at

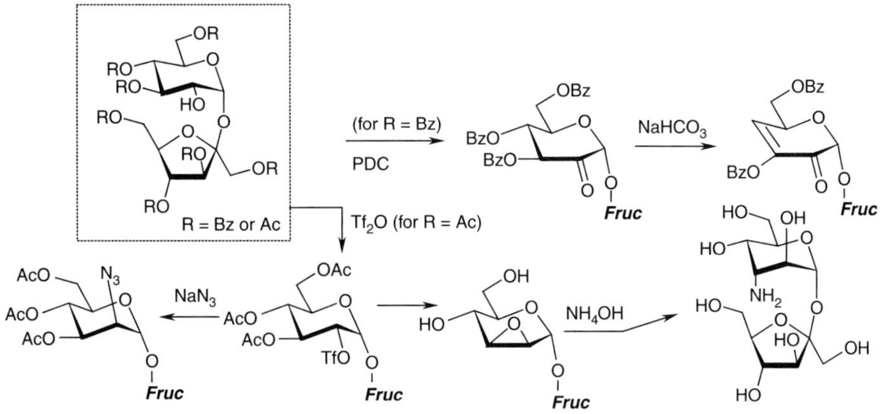

SCHEME 39. Synthesis of sucrose analogues modified at the secondary positions.

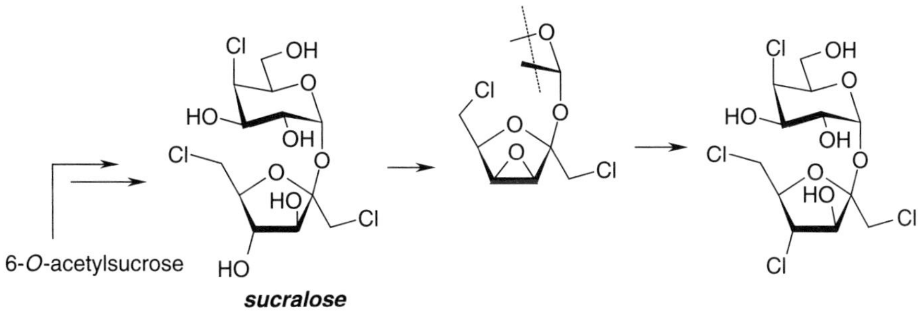

SCHEME 40. Synthesis of polyhalogenated analogues of sucrose.

position 4, including a protected 4-episucrose, and such 4-deoxysucrose analogues as 4-fluoro, 4-azido, 4-amino, and 4-thio derivatives.[323] Many sucrose derivatives have been prepared from "3-oxosucrose" which is obtained selectively by biocatalytic oxidation (see Section II.5 on oxidation).[205–208,210–212,324] Amino acid–sucrose hybrids were prepared by this route.[325]

Finally, a valuable example of multistep modifications of sucrose is the synthesis of Sucralose (1′,4,6′-trideoxy-1′,4,6′-trichloro-*galacto*-sucrose)—a compound 650 times more sweeter than sucrose,[15,326,327] which was obtained by treatment of 6-*O*-acetylsucrose[93] with sulfuryl chloride in pyridine.[251,252] Further reaction of this derivative with triphenylphosphine and diethyl azodicarboxylate afforded an epoxide from which a tetrachloro-derivative was obtained.[15] Other chlorodeoxysucroses (both: 1′- and 6′-mono-, and also 1′,6′-di-) and "*galacto*-sucroses" (4-mono, 1′,4-di, 1′,4,6′-tri-, or 1′,4,4′,6′-tetra-) are also available (Scheme 40).[328] Starting from 3′,4′,6′-tri-*O*-benzoyl-1′,2,4,6-di-*O*-isopropylidenesucrose, the free 3-OH group was transformed by various means leading to 3-oxo, 3-*allo*, 3-*allo*-triflate, 3-deoxy-3-fluoro, and 3-deoxy analogues.[329,330]

The Mitsunobu conditions, applied without any carboxylic acid, were shown to provide anhydro (3′,4′-epoxide)[284–286] and dianhydro sucrose derivatives.[331] Some of these compounds were further transformed by reduction (leading to dehydrosucroses) or ring-opening leading to sucrose epimers and dehydrohalo- or amino sucroses (see also Scheme 7).[332]

9. Hydrolysis, Alcoholysis, Thermolysis, and Degradation Reactions

The fragility of the glycosidic bond of sucrose is a typical aspect of its chemistry. It is sensitive to heat under a wide range of pH conditions. For example, it can be cleaved under rather mild acidic conditions, such as 0.1 HCl in MeOH in 30 min, providing glucose and methyl fructosides.[297] In this reaction, the glycosidic oxygen atom is retained by glucose, and the fructosyl oxonium cation reacts with the available nucleophilic species, an alcohol in the case of such an alcoholysis.[333,334] It is noteworthy that the formation of fructosides and glucose or glucosides from sucrose by reaction with alcohols under acidic conditions (transglycosylation reaction) is not straightforward because of the sensitivity to acid of fructose and fructosides themselves. Indeed, when less-reactive alcohols are used in this process, the fructosides formed in first instance are degraded,[335] leading to complex mixtures containing colored products, and to anhydromonosaccharides and polymers arising from the self-glycosylation (acid reversion) of the monosaccharide residues. The acid-catalyzed thermal polymerization of sucrose can however lead to relatively well-defined fructoglucan polymers (Scheme 41).[336]

The most efficient method for the clean hydrolysis of sucrose is by the use of invertase, leading to an equimolar mixture of glucose and fructose (invert sugar). The presence of salts increases the rate of thermal degradation of sucrose.[337] The reaction is also possible in the presence of such heterogeneous acidic catalysts as zeolites.[338] The hydrolysis of the glycosidic bond is the first step of a number of subsequent reactions that can occur on the glucose and fructose residues, such as dehydrations, combinations with amino acids (Maillard reaction), and many other chemicals or fermentation processes.[339]

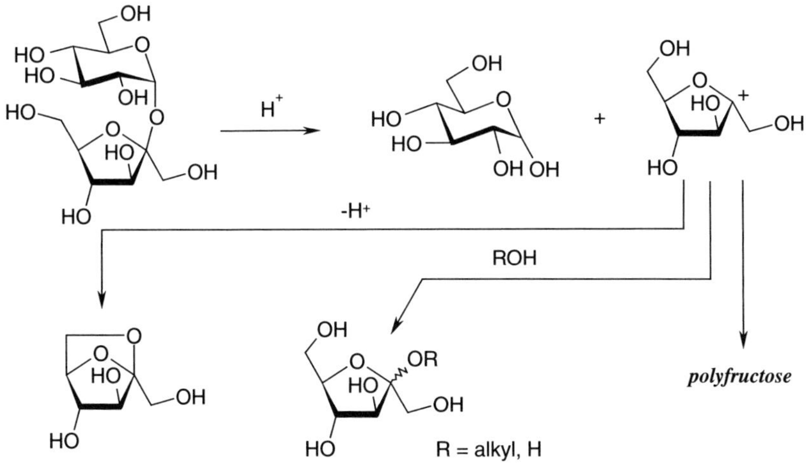

SCHEME 41. Degradation of sucrose under acidic conditions.

Among the important processes taking place when sucrose is heated is caramelization. Caramel is actually a very complex mixture. Some constituents are volatile, such as furan derivatives, pyrones, aldehydes, and low molecular weight carboxylic acids, and these account for ~10% of the mass.[340] The main constituents of the nonvolatile component have also been identified; these include notably variously glucosylated difructose dianhydrides (Scheme 42).[289,290,292] Depending on the process (acid- or base-catalyzed, leading to "aromatic" or "color" caramel) and on the particular sugar substrate (sucrose, fructose), different proportions and types of these molecules can be found.[291] Some of the caramel components are similar to those observed when fructose and sucrose react in hydrogen fluoride, notably bisglycosylated difructose dianhydrides.[287,341]

Some of these processes can be controlled and used for the preparation of chemicals of industrial interest, such as 5-(hydroxymethyl-2-furaldehyde, HMF). HMF is the product of a triple dehydration of fructose, which is itself one of the first degradation products of sucrose. The preparation and the uses of HMF have been extensively studied and reviewed by Lichtenthaler.[342] It can be obtained under various degradative conditions, including acid catalysis and catalysis by lanthanide ions.[343] Processes for the production of HMF on the multi-ton scale have been developed as well as many uses, notably polycondensations.[344,345] The polymerization of HMF has been shown to produce complex

SCHEME 42. Some of the compounds found during the caramelization of sucrose.

materials having useful optical or conductivity properties,[346] and the use of a protic ionic liquid as solvent and catalyst has been described for the formation of HMF.[347]

Stepwise degradation of sugars under catalytic conditions leads to shorter-chain sugars and such by-products as CO, CO_2, and H_2O depending on the conditions (see also next section).[348] Sucrose or other carbohydrates have been used for the preparation of new solid acidic catalysts by partial carbonization followed by sulfonation. This leads to sheets of amorphous carbon bearing hydroxyl, carboxyl, and sulfonic groups. These have been used for the production of "biodiesel" fuel.[349]

10. Processes, Solvents, and Methods of Activation

For sucrose to be useful as a raw material for chemical technology, synthetic methods must be satisfactory in terms of economic and of environmental aspects. Bioconversions are cleaner methodologies, but are not necessarily economically viable, and there is a need to find media and methodologies that propose alternatives to expensive and/or toxic solvents, as well as procedures that require heating.

Most transformations of sucrose are achieved in solution, in the few common solvents in which it is adequately soluble namely DMF, Me_2SO, pyridine, and water. Water may be considered as an incompatible medium when water-sensitive reagents are concerned. However, it has been shown in many instances—reaction with acid chlorides, chloroformates, epoxides, and isocyanates—that the significantly higher acidity of sucrose as compared to water allows the use of water or water–cosolvent mixtures with only limited competitive hydrolysis of the eletrophilic reagents. Nevertheless, when these reagents are not only water-sensitive but also hydrophobic, then difficulties arise because of heterogeneity of the medium. As regards DMF, regulatory aspects are very strict, notably for products having applications in the food or cosmetic industries. Also the rather high boiling point of DMF makes its removal by evaporation difficult. Dimethyl sulfoxide is much less toxic and remains approved for the synthesis of sucrose derivatives for food applications. However, elimination of traces of this solvent in the final products is costly. Certainly the best solvent is "no solvent," but as yet there has been no success in this domain for the industrial-scale chemical conversion of sucrose. What has been studied the most are those reactions leading to fatty acid esters of sucrose, which are already on the market as food and cosmetic emulsifiers. Here, the presence of solid–solid interphases (sucrose, basic catalyst) and of undiluted reagents makes a very complex heterogeneous medium, highly viscous, with reactivity outcome very different from that observed in solution. Higher temperature is necessary to compensate the lack of diffusion, leading to increased degradation, coloring, and consequently more difficult purifications.[350,351] Some studies of the use of reactive extrusion for the same sucrose ester synthesis was reported.[352] Also, a microemulsion process could be used.[353]

Supercritical CO_2 is not considered a satisfactory solvent for unprotected carbohydrates, but some peracetylated sugar were stated to act as emulsifiers in CO_2–water or CO_2–organic solvent mixtures.[354] Ionic liquids have also been shown to solvate carbohydrates at a level allowing preparative processes,[355,356] for example for the acetylation sucrose.[91] This list of unusual solvents should include hydrogen fluoride, which has been used in the synthesis of sugar dianhydrides.[288]

Concerning the methods of enhancing reactivity, comprehensive accounts on the use of ultrasound and microwave heating in carbohydrate chemistry have been published.[197,357–359] Their application to sucrose chemistry remains rather limited (see Section II.5) to the use of sonocatalysis in the TEMPO-mediated oxidation of sucrose. Microwave heating is generally useful for very energy-demanding reactions, whereas sucrose chemistry is often handicapped by competitive degradation because of its thermal instability. Pyrolysis, thermal cracking, and hydrocracking of plant biomass (50–70% carbohydrate content) procedures are of current interest for the preparation of "biogas" or other chemicals.[360] Because of the higher functional complexity of these materials as compared to that of hydrocarbons, new catalysts that are more stable and more selective need to be designed. The vapor cracking of sugars leads to hydrogen and such by-products as CO, CO_2, methane, and other alkanes.[361,362]

It may be noted that sucrose solutions enhanced exhibit hydrophobic effects, leading to increased rate accelerations of such reactions as the Diels–Alder and related processes.[363]

11. Conclusion

The nucleophilic reactivity of sucrose follows particular rules, and various hydroxyl groups of the molecule can be targeted according to the type of electrophilic reagents or the reaction conditions, sometimes with very good control of selectivity. Besides the classical functionalization of the 4,6-diol by acetalation, three other main processes are typical of sucrose reactivity: the electronically favored reactivity of OH-2, the sterically favored reactivity of OH-6 and OH-6′, and the peculiar reactivity of OH-1′ or OH-6′ in enzyme-mediated esterifications using proteases or lipase, respectively. This chemoselectivity allows the level of substitution to be limited to mono- or disubstituted derivatives. Combinations of various standard reactions such as acetalation, etherification, and esterification can give access to specifically protected sucrose derivatives that can be integrated in synthetic schemes towards more complex targets.

III. Applications of Sucrochemicals

The most useful property of sucrose is probably its high polarity, which is shared with many sucrose derivatives,[18,364] with practical implications as surfactants, polymers, or complexing agents. Other sucrose derivatives have found applications as bleaching boosters, and sucrose can be used as a complex diol in condensations with diisocyanates, leading to polyurethane foams. Degradation or fermentation products of sucrose have applications that are not described here. Apart from these latter and from the direct use of sucrose as a cheap polyol, there remain few applications that have reached industrial development. Aside from bioconversions leading to ethanol and to oligo- and polysaccharides, three chemical success stories are noteworthy: sucrose fatty acid esters of low degree of substitution, referred to as "sucroesters," which serve as emulsifiers in the food and cosmetic industries; the persubstituted fatty acid esters of sucrose, behaving as a noncaloric fat (Olestra®) and approved in some food applications; and trichlorogalactosucrose (Sucralose®), a powerful sweetener. These three examples, which constitute the *showcase* of industrial sucrochemistry, have reached success thanks to the determination of a few pioneers in the use of carbohydrates as organic raw materials (see references in Section I.1). It may be hoped that the increased efficiency of the processes described in the preceding sections will facilitate other applications with motivation from the shortage of fossil resources and their increasing cost.

Because of the very extensive patent literature on the topic, this chapter cannot be exhaustive, but it gives an overview of the most important applications of functional derivatives made from sucrose.

1. Surfactants

There has been long interest in using sucrose as the hydrophilic part of a nonionic surfactant by taking advantage of its high polarity.[365] Esters of sucrose and fatty acids, often named "sucroesters," are prepared on the industrial scale for applications as food and cosmetic additives, essentially as emulsifiers. They are prepared by base-catalyzed transesterification from fatty acid methyl esters, or directly from triglycerides (leading then to mixtures of glycerol, mono- and diglycerides, and sucrose esters called "sucroglycerides").[366–372] Although the synthesis of sucrose esters is generally performed in a solvent such as Me_2SO, various other processes in microemulsion or without any solvent have been reported.[351–353] Some selective approaches towards defined regioisomers are discussed in Section II.3.

Sucroesters are prepared and used as mixtures of regioisomers and of mono-, di-, and triesters, and their physicochemical properties depend mainly on the average degree of substitution, as well as on the presence of impurities arising from the fabrication process (essentially remaining fatty acid and soaps), thus making their referencing by the calculated hydrophilic–lipophilic balance (HLB) rather uncertain. An alternative using Winsor systems was shown to provide a more precise method for comparisons of sucroester properties.[373]

By choosing the appropriate level of substitution (mixtures enriched in mono-, di-, or triesters) and the length of the alkyl chain of the fatty acid, a wide range of surfactant properties can be reached, while maintaining the simplicity of the ester connection. Notably, very polar emulsifiers can be obtained (low level of substitution) without the supplemental ethoxylation step as needed in other monosaccharide-based compounds. The key property is their absolute absence of toxicity. Reviewing all of the academic and patent literature on the applications in food and cosmetics of sucrose esters as emulsifiers (Scheme 43) would warrant a full article, and thus is not detailed here. Among reviews on this area are those by Nelen and Cooper,[374] and Desai and Lowicki.[375] Taking into account their good surfactant and ecotoxicological properties,[376] these compounds promise a good market future since, in addition, both the polar and the hydrophobic parts of the surfactant are obtained from renewable resources. Interesting gelling properties of a sucrose–aminoacid–fatty acid hybrid have also been described.[377]

Most studies of the properties of fatty acid sucrose esters relate to their applications as emulsifiers, and are often conducted on mixtures of regioisomers and various levels of substitution. It is clear that the chain length and the number of chains have the largest influence on the surfactant properties.[150,373,378–389]

Critical micellar concentrations (CMCs) of sucrose fatty acid esters have been determined by various methods, for mixtures and for pure derivatives (Table III).[95,126,390–395] Studies on the lyotropic properties of sucrose

RCO_2Me, base, Me_2SO

+ regioisomers + diesters + triesters

SCHEME 43. Sucrose fatty acid esters as emulsifiers.

TABLE III

Critical Micellar Concentration of Sucrose Monoesters as a Function of the Chain Length and the Position on the Sugar

Chain Length	Position	CMC (mmol/L)	Reference
8	M	6.5	390
	6	33	126
10	M	0.57	390
	M or 6	4.18	126
12	M	0.29	390
	M	0.28	126
		0.34	391
	6	0.35	392
	6′	0.21	390
	6	0.52	95
	6	0.4	393
	6	0.25	394
	6	0.46	395
	1′	0.15	393
14	M	0.022	126
	6′	0.021	390
	6	0.08	95
	6	0.13	393
	1′	0.091	393
16	M	0.004	390
	6′	0.0041	390
	6	0.017	85
	6	0.028	394

M, Mixture of regioisomers.

octadecanoate demonstrate the occurrence of fully developed liquid–crystalline lamellar phases.[396]

The influence on solution behavior has been clearly demonstrated of the position of the chain on the sucrose backbone.[392,393] Indeed, variations in the attachment position of the chain lead to changes in the properties, as well as the effect of unstauration. This has been shown by measurements of diffusion coefficients,[126] which are related to the size and shape of micelles or aggregates. Micelles of hexadecanoyl sucrose exhibited significant shape changes for the 1′-ester as compared to the 6- or 6′-esters. These data were obtained by a Pulse Gradient Field Spin-Echo (PGFSE) NMR method, in which the precise concentration of sucrose and sucrose ester solutions was determined by the ERETIC (electronic reference to access in vivo concentrations) NMR method.[397] Enhanced hydrophilicity of sucrose esters may be obtained by sulfation,[150] by oxidation at primary positions to carboxyl analogues,[398] or by enzymatic glycosylation.[399]

Larger differences were observed in the thermotropic properties of these materials according to the attachment position of the chain on the sugar backbone. A series of sucrose esters of fatty acids, varying in their chain length, degree of substitution, level of saturation, and position (prepared by chemical and/or enzymatic methods) were studied by polarized-light microscopy, differential scanning calorimetry, and small-angle X-ray diffraction.[127,389] These studies pointed out that, among other parameters, the position of the chain on the sucrose backbone influences significantly the stability of the liquid–crystalline phases. For example, 1′,6′-diesters exhibit lower mesophase stability as compared to 6,6′-diesters. Also, an unusual phase transition within the smectic A* phase of 6,6′-sucrose bis(octadecanoate) was observed. The aggregation behavior of sucrose diesters was studied because of their structural similarity to such biological materials as cord factor, the dimycolic ester of trehalose, which is associated with virulent strains of tubercle bacilli.[97,128,400] For the series of monoesters, a model based on the intramolecular hydrogen bonds exhibits ion channels within the columnar phases. A general review on carbohydrate liquid crystals has been published by Jeffrey.[401]

Among esters of sucrose, some derivatives of fluorinated fatty acids have been prepared for evaluation as surfactants specifically designed for fluorocarbons containing emulsions used for biomedical applications, notably as blood substitutes.[98,402] Some sucrose aspartate surfactants have been described.[403]

Mixed carbonates and alkyl carbamates having long alkyl chains were prepared in aqueous alcoholic medium by reaction with chloroformates and isocyanates, respectively, demonstrating that water-sensitive reagents can be used in aqueous media on account of the significantly higher reactivity of sucrose as compared to water and simple alcohols.[130,132] The higher stability towards basic conditions of the carbamates as compared to esters or carbonates is such that these materials can be used in basic pH formulations.

Ethers of sucrose having a low degree of substitution have also been prepared for their potential surfactant properties.[82–85] Unlike esters, ethers can be used under a wider pH range, therefore extending the applications to domains other than emulsification. Sucrose ethers can be prepared either by reaction with alkyl halides or long-chain epoxides. In this latter instance, the reaction is a simple addition process and does not require stoichiometric quantities of base. The reaction of sucrose with long-chain epoxides was shown to proceed in Me_2SO or in water, in the presence of homogeneous or heterogeneous basic catalysts,

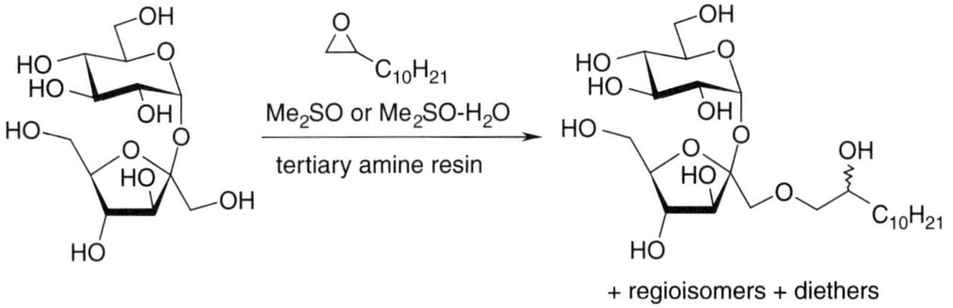

SCHEME 44. Amphiphilic sucrose ethers from epoxides.

notably tertiary amines (Scheme 44).[73,404] This study, which was also extended to isomaltitol and trehalose, afforded monoethers that were compared for their physicochemical properties, in particular their foaming properties.[74] Careful separation of the different regioisomers of hydroxyalkyl ethers of sucrose obtained by reaction with 1,2-epoxydodecane permitted a study of the influence of chain position on thermotropic properties. The importance of the hydrogen-bonding network in establishing the conformation of the disaccharide component was emphasized by the different mesophase organization observed for the derivatives substituted at OH-2 and OH-3′ having a columnar arrangement as compared to all other isomers, which exhibited smectic A* phases.[75]

Long-chain acetals, like sucrose ethers, are also compatible with basic pH formulations. They were prepared by a transacetalation reaction and were also shown to exhibit surfactant properties.[162]

2. Polymers

The introduction of sugars into polymeric molecules can bestow new properties, such as increased polarity, chirality, and biodegradability or biocompatibility. Sucrose-containing polymers can be of various types, either having the carbohydrate backbone within the polymer chain or having pendant sucrose residues serving as functional polar groups.

The polymerization of sucrose derivatives (esters, ethers, acetals) bearing a carbon–carbon double bond has been studied (Scheme 45). Polymers can be obtained by polymerization or copolymerization.[146,404–414] The monomers are prepared either by multistep synthesis, leading to defined compounds and subsequently rather well-controlled polymerization processes,[302,415,416] or by direct functionalization of unprotected sucrose, leading to mixtures of isomers and

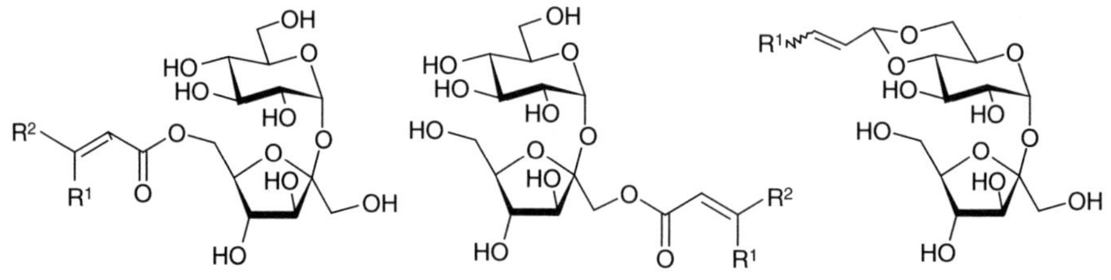

$R^1, R^2 = H, alkyl$

SCHEME 45. Unsaturated sucrose derivatives used as monomers for polymerization.

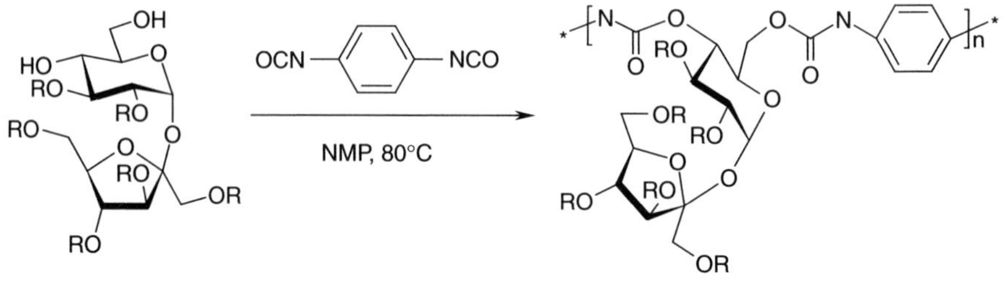

OCN—C6H4—NCO, NMP, 80°C

SCHEME 46. Polyurethanes from sucrose.

therefore to more complex polymers. Notably, monosubstituted monomers lead to linear polymers, whereas polysubstituted derivatives lead to cross-linked materials. Hydrogels can thus be prepared.[417] Acrylic monomers in particular have been prepared by various methods.[119,120] These polymers have properties related to their high polarity such as water absorption, and can serve as rheological additives, thickening agents, chromatography gels, and materials for biology.[418,419]

Sucrose reacts with diisocyanates leading to polyurethanes, which are used as thermal insulating foams, notably in cars. Partially protected sucrose esters can be used for the synthesis of better-defined polymers (Scheme 46).[265] A first step of hydroxypropylation is sometimes necessary to obtain sufficient miscibility with the diisocyanate derivative, as well as for tuning the physicochemical properties of the polyurethane foams.[78,305,420]

Some polysaccharides, for instance glucans, fructoglucans, or fructans, can be produced by either thermal or biocatalyzed polymerization.[237–245,336] Such polysaccharides are complementary to all other natural or engineered polysaccharides for which a vast number of applications are known. Polyhydroxyalkanoates can also be obtained.[222,421]

Polyethers prepared by reaction with diepoxides have been described.[79] Such reactions have been achieved from sucrose or partially esterified sucrose for the preparation of polymeric resins used as coating agents.[78] Polysucrose (Ficoll400®), a water-soluble copolymer prepared from sucrose and epichlorhydrin, has been developed on the industrial scale. It is used as a biocompatible additive for cell separation and diagnostics, and is also described as having valuable properties as a food ingredient.[18] Sucrose dicarboxylates can react with diamines to provide polyamides.[183]

3. Food Additives and Pharmaceutical Compounds

In the field of food additives, several sucrose derivatives have found applications.[18] The use of sucrose itself as a food ingredient on account of its sweet taste and its properties as a bulking agent is not detailed here. Some fructooligosaccharides having a DP of 5–10 have been shown to be bifidogenic, and are used as food additives (Scheme 47).[227,246] Also, most of the modified

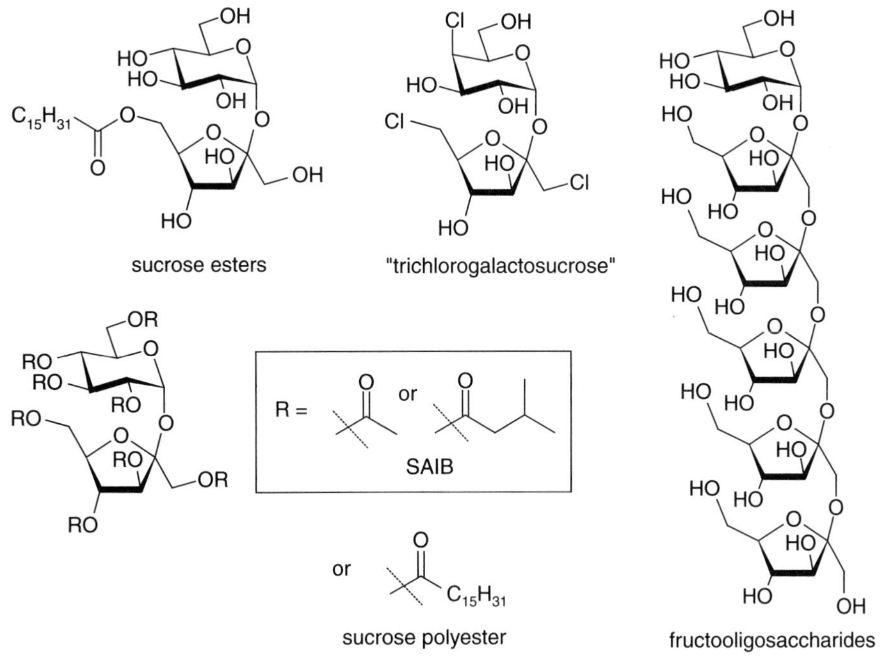

sucrose esters

"trichlorogalactosucrose"

SAIB

sucrose polyester

fructooligosaccharides

Scheme 47. Sucrose-derived food additives.

polysaccharides obtained by chemical or biocatalyzed transformations of sucrose (see the bioconversion and polymer sections) have applications as food additives or as materials connected with their physicochemical properties. Isomaltulose and its derived alditol, isomaltitol, referred to as Palatinit®, behave as good substitutes for sucrose, while providing lower caloric intake (half of sucrose: 2 Cal/g instead of 4), and being also less cariogenic (see the "bioconversion" section). Thermal and acid- or base-catalyzed degradations of sucrose leads to various types of caramel, with many uses in the food industry either as an aroma (often obtained by acid-catalyzed processes) or as a coloring agent (base-catalyzed processes), depending on the proportions of the main components (oligosaccharidie anhydrides and 2-(hydroxymethyl) furfural derivatives) (see Section II.9).

The use of sucrose fatty acid esters of low degree of substitution as emulsifiers has been detailed in the surfactant section. They are referenced as E 473 in the European system of food additives, whereas E 472 refers to sucroglycerides.

Highly substituted sucrose fatty acid esters (penta- to octaesters), also referred to as sucrose polyesters (SPE), behave as fats and can substitute for triglycerides in cooking and frying applications. Not being recognized as substrates by digestive lipases, they are not metabolized and can therefore serve as noncaloric fat substitutes. They are produced under the trade name Olestra®, and are approved in North America for cooking "diet" snacks.[89] A beneficial effects of SPE on the intestinal absorption of LDL cholesterol has also been reported.[422,423] SPE have also been used as oral contrasting agents for medical magnetic resonance imaging.[424] In common with other fat substitutes, SPE have been shown to extract fat-soluble vitamins. A deuterated analogue of sucrose octa(hexadecanoate) has been prepared for studies of molecular motion.[219] A nonsolvent process starting from sucrose octaacetate has been described, involving acyl exchange with hexadecanoyl chains.[425]

Mixed acetyl-isobutanoyl sucrose esters (SAIB) are used as phase densifiers in beverages.[426,427] Sucralose (trichlorogalactosucrose) is produced at the industrial scale and marketed in the UK and other countries as a sweetener, having a 650 times the sweetening power of sucrose.[428] Being more stable to heat as compared to other synthetic sweeteners, it can be used in cooking. It has also been shown that the sweet taste of sucralose is based on interactions with both subunits of the sweet-taste receptor.[429]

Galloyl esters having antioxidant properties have been prepared, but have not reached significant application.[133–135]

Among pharmaceutical application, sucrose octasulfate (as its aluminum salt) exhibits anti-ulcer properties.[147–149,430] Modified sucrose octasulfate lacking one or two sulfate groups and having ether substituents have been shown to interact with fibroblast growth factors.[431] "Polysucrose" (Ficoll400®) is also used in biotechnological preparations. Some sucrose-based hydrogels made from sucrose acrylic monomers can be used for controlled drug delivery.[412] In the same field, sucrose acetate isobutyrate (SAIB) is used as a biodegradable gel for the controlled release of oral or injected drugs.[432]

4. Additives for Materials and Chemical Intermediates

Sucrose, like other simple carbohydrates, is used as additive in mineral suspensions, concrete, and ceramics. The hygroscopic character of the carbohydrates and their ability to interact with colloidal inorganic particles and ions modifies the hydration phenomena within the suspensions, and consequently on the rheology of the medium and the kinetics (most often slowing down) of the setting process.[433–435]

The SAIB and also sucrose polybenzoate (degree of substitution, DS 6–8) are used as additives in the formulation of cosmetic varnishes, and as plastifying and filmogenic agents, for surface coating.[436,437] Sucrose octaacetate is also used as a repulsive agent in various plastics, adhesives, and other products because of its very bitter taste. Some sucrose acetates[90,438] or other esterified carboxylated sucrose derivatives[184,398] are used as bleaching boosters to replace EDTA. Polyesters of shorter-chain acids have also been studied for their properties as bleaching boosters.[90]

Carbonized sugar derivatives are used as solid acid catalysts for the production of biodiesel fuel,[349] and carbonized sucrose treated with ethylene and then pyrolyzed provides materials used as hard-carbon anodes for lithium-ion batteries.[439]

Among the products arising from sucrose degradation or thermolysis, HMF is a useful chemical that can be integrated in various synthetic schemes.[228,342,346,440,441]

5. Complexation Properties

Sucrose-derived crown ethers (see the multistep section) have been prepared and their ion complexation studied.[22,314] In addition to the complexation of metal ions by carbohydrates,[107] oxidized carbohydrates and polysaccharides find use as detergent builders or co-builders because of their ion chelating and

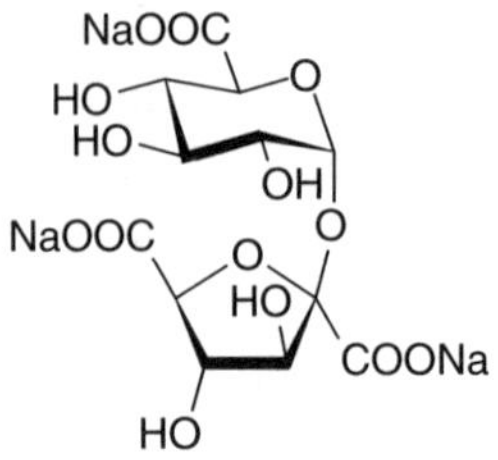

SCHEME 48. Sucrose tricarboxylate (sodium salt).

sequestering properties, substituting for the banned tripolyphosphates. Oxidized derivatives of sucrose, either mono-, di- or tricarboxysucrose obtained by oxidation of the primary alcohol groups or the acyclic compounds arising from glycol cleavage are polyhydroxycarboxylates that have potential sequestering properties.[442,443] Although there has been little mention in the literature, it is likely that similar properties could be observed for carboxysucroses and sucrose carboxymethyl ethers. Complexing properties (as well as bleaching booster and surfactant properties) of acetylated carboxysucroses and carboxymethyl ethers of sucrose have also been patented (Scheme 48).[184,398,444]

IV. CONCLUSION

Sucrochemistry is already more than 50 years old, and has become a field of carbohydrate chemistry on its own. Indeed, considerable progress has been achieved in the monitoring of the chemical reactivity of sucrose, with the efforts of many research teams who have built on the steps of a few pioneers. Many sucrose derivatives can now be prepared, and sophisticated synthons as well as simple substituted compounds have been reported. However, only a few examples have yet reached the level of the industrial development, and these are mainly in the field of food and cosmetic additives and surfactants. Various polymers, additives for materials, and some chemical intermediates have also been produced. Bioconversions are certainly a major avenue for using sucrose as a starting material, and ethanol production will increase as a consequence of high oil prices. Current awareness of the shortage of fossil resources emphasizes the potential for chemical transformations of sucrose in providing new uses of this abundant natural resource.

Acknowledgments

The authors dedicate this chapter to Professor Gérard Descotes for his accomplishments in the field of sucrochemistry, and they thank him for fruitful discussions during the preparation of this chapter.

References

1. V. Kollonitsch, Sucrose chemicals: A critical review of a quarter-century of research by the sugar research foundation, Int. Sugar Res. Found., Bethesda, MD, 1970.
2. J.L. Hickson (Ed.), *Sucrochemistry*, ACS Symposium Series, Vol. 41, ACS, Washington, DC, 1977.
3. M. A. Clarke and M. A. Godshall, *Chemistry and Processing of Sugar Beet and Sugar Cane*, Elsevier, Amsterdam, 1988.
4. M. Mathlouthi and P. Reiser (Eds.), *Sucrose: Properties and Applications*, Blackie, Glasgow, 1995.
5. F. W. Lichtenthaler (Ed.), *Carbohydrates as Organic Raw Materials*, Vol. 1, VCH, Weinheim, 1991.
6. G. Descotes (Ed.), *Carbohydrates as Organic Raw Materials*, Vol. 2, VCH, Weinheim, 1993.
7. H. van Bekkum, H. Röper H, and A. G. J. Voragen (Eds.), *Carbohydrates as Organic Raw Materials*, Vol. 3, VCH, Weinheim, 1996.
8. W. Praznik and H. Huber (Eds.), *Carbohydrates as Organic Raw Materials*, Vol. 4, WUV-Univ, Vienna, 1998.
9. R. Khan, The chemistry of sucrose, *Adv. Carbohydr. Chem. Biochem.*, 33 (1976) 235–294.
10. L. Hough, Selective substitution of hydroxyl groups in sucrose, in J.L. Hickson (Ed.), *Sucrochemistry*, ACS Symposium Series, Vol. 41, ACS, Washington, DC, 1977, pp. 4–21.
11. R. Khan, Some fundamental aspects of the chemistry of sucrose, in J.L. Hickson (Ed.), *Sucrochemistry*, ACS Symposium Series, Vol. 41, ACS, Washington, DC, 1977, pp. 40–61.
12. M. R. Jenner, Sucrose, in C. K. Lee (Ed.), *Developments in food carbohydrates-2*, Applied Science Publishers, London, 1980, pp. 91–143.
13. R. Khan, Chemistry and new uses of sucrose: How important? *Pure Appl. Chem.*, 56 (1984) 833–844.
14. C. E. James, L. Hough, and R. Khan, Sucrose and its derivatives, *Prog. Chem. Org. Nat. Prod.*, 55 (1989) 117–184.
15. L. Hough, Applications of the chemistry of sucrose, in F. W. Lichtenthaler (Ed.), *Carbohydrates as Organic Raw Materials*, Vol. 1, VCH, Weinheim, 1991, pp. 32–55.
16. R. Khan, Sucrose: Its potential as a raw material for food ingredients and for chemicals, in M. Mathlouthi and P. Reiser (Eds.), *Sucrose: Properties and Applications*, Blackie, Glasgow, 1995, pp. 264–278.
17. F. W. Lichtenthaler and P. Pokinskyj, Exploitation of sucrose towards products with industrial profiles, in W. Praznik and H. Huber (Eds.), *Carbohydrates as Organic Raw Materials*, Vol. 4, WUV-Univ, Vienna, 1998, pp. 9–19.
18. M. A. Godshall, Future directions for the sugar industry, *Int. Sugar J.*, 103 (2001) 378–384.
19. S. Jarosz, The chemistry of sucrose, *Pol. J. Chem.*, 70 (1996) 972–987.
20. G. Descotes, Y. Queneau, A. Bouchu, J. Gagnaire, P. Salanski, S. Belniak, A. Wernicke, S. Thévenet, and N. Giry-Panaud, Preparation of sucrose esters, ethers and acetals, *Pol. J. Chem.*, 73 (1999) 1069–1077.

21. G. Descotes, J. Gagnaire, and Y. Queneau, Recent developments in sucrochemistry, *Recent Res. Devel. Synth. Org. Chem.*, 3 (2000) 27–50.
22. S. Jarosz and M. Mach, Regio- and stereoselective transformations of sucrose at the terminal positions, *Eur. J. Org. Chem.*, 5 (2002) 769–780.
23. Y. Queneau, J. Fitremann, and S. Trombotto, The chemistry of sucrose: The selectivity issue, *C. R. Chimie*, 7 (2004) 177–188.
24. J.F. Robyt, Modifications, in *Essentials of Carbohydrate Chemistry*, Springer Verlag, New York, Ch. 4, 1998, pp. 76–141.
25. H. Röper, Renewable raw materials in Europe-Industrial utilisation of starch and sugar, *Starch/Stärke*, 54 (2002) 89–99.
26. F. W. Lichtenthaler, P. Pokinskyj, and S. Immel, Sucrose as a renewable organic raw material: New, selective entry reactions via computer simulation of its solution conformations and its hydroxyl groups reactivities, *Zuckerind.*, 121 (1996) 174–190.
27. K. Bock and R. U. Lemieux, The conformational properties of sucrose in aqueous solution: Intramolecular hydrogen-bonding, *Carbohydr. Res.*, 100 (1982) 63–74.
28. D. C. McCain and J. L. Markley, The solution, conformation of sucrose: Concentration and temperature dependence, *Carbohydr. Res.*, 152 (1986) 73–80.
29. J. C. Christofides and D. B. Davies, Co-operative and competitive hydrogen bonding in sucrose determined by simple ^{1}H NMR spectroscopy, *J. Chem. Soc. Chem. Commun.* (1985) 1533–1534.
30. S. Pérez, The structure of sucrose in the crystal and in solution, in M. Mathlouthi and P. Reiser (Eds.), *Sucrose: Properties and Applications*, Blackie, Glasgow, 1995, pp. 11–32.
31. F. W. Lichtenthaler, S. Immel, and U. Kreis, Old roots—new branches: Evolution of the structural representation of sucrose, in F. W. Lichtenthaler (Ed.), *Carbohydrates as Organic Raw Materials*, Vol. 1, VCH, Weinheim, 1991, pp. 1–32.
32. B. Bernet and A. Vasella, Intra- and intermolecular H-bonds of alcohols in DMSO. ^{1}H-NMR analysis of inter-residue H-bonds in selected oligosaccharides: Cellobiose, lactose, N,N″-diacetylchitobiose, maltose, sucrose, agarose, and hyaluronates, *Helv. Chim. Acta*, 83 (2000) 2055–2071.
33. C. Hervé du Penhoat, A. Imberty, N. Roques, V. Michon, J. Mentech, G. Descotes, and S. Pérez, Conformational behaviour of sucrose and its deoxy analogue in water as determined by NMR and molecular modelling, *J. Am. Chem. Soc.*, 113 (1991) 3720–3727.
34. S. B. Engelsen, C. Hervé du Penhoat, and S. Pérez, Molecular relaxation of sucrose in aqueous solution: How a nanosecond molecular dynamics helps to reconcile NMR data, *J. Phys. Chem.*, 99 (1995) 13334–13351.
35. S. Immel and F. W. Lichtenthaler, The molecular conformation of sucrose in water: A molecular dynamics approach, *Liebigs. Ann. Chem.* (1995) 1925–1937.
36. P. M. Tyrell and J. H. Prestegard, Stuctural studies of carbohydrates by deuterium NMR: Sucrose, *J. Am. Chem. Soc.*, 108 (1986) 3990–3995.
37. S. Houdier and S. Perez, Assessing sucrose hydroxyl acidities through semiempirical calculations, *J. Carbohydr. Chem.*, 14 (1995) 1117–1132.
38. *The Merck Index*, 12th edn., in S. Budavari (Ed.), Merck Editions, pp. 1517–1518.
39. L. Michaelis and P. Rona, Die Dissoziationskonstanten einiger sehr schwacher Säuren, insbesondere des Kohlenhydrate, gemessen auf electrometrischem Wege, *Biochem. Z.*, 49 (1913) 232–248.
40. P. Hirsch and R. Schlags, The determination of the alkali-combining quantity of the most important sugars, *Z. Phys. Chem., Abt. A*, 141 (1929) 387–412.
41. A. E. Stearn, The polybasicity of several common sugars, *J. Phys. Chem.*, 35 (1931) 2226–2236.
42. K. Smolenski and W. Kozlowski, Optical rotation of alkali solutions of sucrose, *Roczniki Chem.*, 16 (1936) 270–279; *Chem. Abstr.*, 30 (1936) 65307.

43. K. Smolenski and S. Porejko, The pH of aqueous solutions of calcium hydroxide and that of solutions of sucrose, *Roczniki Chem.*, 16 (1936) 281–287; *Chem. Abstr.*, 30 (1936) 65308.
44. E.S. Lygin, Solubility of sucrose in water in relation to pH, *Sakharnaya Promyshlennost*, 42 (1968) 16–17; *Chem. Abstr.*, 69 (1968) 54747.
45. E. M. Wooley, J. Tomkins, and L. G. Hepler, Ionization constants for very weak organic acids in aqueous solution and apparent ionization constants for water in aqueous organic mixtures, *J. Solution Chem.*, 1 (1972) 341–351.
46. F. Coccioli and M. Vicedomini, Dissociation of sucrose in alkaline solution, *Ann. Chim. (Rome)*, 64 (1974) 369–375.
47. X. M. Fang, F. Y. Gong, J. N. Ye, and Y. Z. Fang, Determination of ionization constants of saccharides by capillary zone electrophoresis with amperometric detection, *Chromatographia*, 46 (3/4) (1997) 137–140.
48. A. H. Haines, Relative reactivities of hydroxyl groups in carbohydrates, *Adv. Carbohydr. Chem. Biochem.*, 33 (1976) 11–109.
49. B. Capon and W. G. Overend, Constitution and physicochemical properties of carbohydrates, *Adv. Carbohydr. Chem.*, 15 (1960) 32.
50. J.A. Rendleman Jr., Ionization of carbohydrates in the presence of metal hydroxides and oxides, in *Carbohydrates in Solution*, Advances in Chemistry Series, Vol. 117, ACS, Washington, DC, 1973, pp. 51–69.
51. R. U. Lemieux and G. Huber, A chemical synthesis of sucrose, *J. Am. Chem. Soc.*, 75 (1953) 4118.
52. R. U. Lemieux and G. Huber, A chemical synthesis of sucrose, *J. Am. Chem. Soc.*, 78 (1956) 4117–4119.
53. A. Klemer and B. Dietzel, Synthese der octa-O-methyl sucrose, *Tetrahedron Lett.*, 11 (1970) 275–278.
54. R. K. Ness and H. G. Fletcher, Synthesis of sucrose and of α-D-glucopyranosyl α-D-fructofuranoside through the use of 1,3,4.6-tetra-*O*-benzyl-D-fructofuranose, *Carbohydr. Res.*, 17 (1971) 465–470.
55. D. E. Iley and B. Fraser-Reid, New synthesis of sucrose which demonstrates a novel approach to the synthesis of alpha-linked disaccharides, *J. Am. Chem. Soc.*, 97 (1975) 2563–2565.
56. B. Fraser-Reid and D. E. Iley, A stereoselective synthesis of sucrose. Part II. Theoretical and chemical considerations, *Can. J. Chem.*, 57 (1979) 645–652.
57. B. Fraser-Reid and D. E. Iley, A stereoselective synthesis of sucrose. Part III. Spectroscopic analyses of key intermediates, *Can. J. Chem.*, 57 (1979) 653–661.
58. A. G. M. Barrett, L. M. Melcher, and C. B. Bezuidenhoudt, Redox glycosylation: The use of Nozoki-Takai methylenation in a highly stereoselective synthesis of sucrose, *Carbohydr. Res.*, 232 (1992) 259–272.
59. S. Oscarson and F. W. Sehgemleble, A novel β-directing fructofuranosyl donor concept. Stereospecific synthesis of sucrose, *J. Am. Chem. Soc.*, 122 (2000) 8869–8872.
60. E. Reinefeld and K. D. Heincke, Selektive *O*-Alkylierung von Saccharose, *Chem. Ber.*, 104 (1971) 265–269.
61. F. W. Lichtenthaler, S. Immel, and P. Pokinskyj, Selective 2-*O*-benzylation of sucrose: A facile entry to its 2-deoxy- and 2-keto-derivatives and to sucrosamine, *Liebigs Ann. Chem.* (1995) 1938–1947.
62. C. H. Hamann, S. Fischer, H. Polligkeit, and P. Wolf, The alkylation of mono- and disaccharides via an initialising electrochemical step, *J. Carbohydr. Chem.*, 12 (1993) 173–190.
63. T. Otake, Studies on Tritylated Sucrose. I. Mono-*O*-tritylsucroses, *Bull. Chem. Soc. Jpn.*, 43 (1970) 3199–3205.
64. L. Hough, K. S. Mufti, and R. Khan, Sucrochemistry. II. 6,6′-Di-*O*-tritylsucrose, *Carbohydr. Res.*, 21 (1972) 144–147.

65. F. Franke and R. D. Guthrie, 6,6′-Di-*O*-t-butyldimethylsilylsucrose: Studies on the rearrangements accompanying deblocking of such silyl ethers, *Aust. J. Chem.*, 31 (1978) 1285–1290.
66. H. Karl, C. K. Lee, and R. Khan, Synthesis and reactions of tert-butyl diphenyl silyl ethers of sucrose, *Carbohydr. Res.*, 101 (1982) 31–38.
67. K. Josephson, Uber Triphenylmethyl-aether einiger Di- und Trisaccharide. Ein Beitrag zur Kenntnis der Konstitution der Maltose, Saccharose und Raffinose, *Liebigs Ann. Chem.*, 472 (1929) 230–240.
68. S. Jarosz, M. Mach, and A. Listkowski, Crown ether analogues from sucrose, *J. Carbohydr. Chem.*, 20 (2001) 485–493.
69. T. Otake, Studies on tritylated sucrose. II. Di-*O*-tritylsucroses, *Bull. Chem. Soc. Jpn.*, 45 (1972) 2895–2898.
70. J. Gagnaire, G. Toraman, G. Descotes, A. Bouchu, and Y. Queneau, Synthesis in water of amphiphilic sucrose hydroxyalkyl ethers, *Tetrahedron Lett.*, 40 (1999) 2757–2760.
71. E. Gérard, H. Götz, S. Pellegrini, Y. Castanet, and A. Mortreux, Epoxide–tertiary amine combinations as efficient catalysts for methanol carbonylation into methyl formate in the presence of carbon dioxide, *Appl. Catal., A*, 170 (1998) 297–306.
72. J. Gagnaire, A. Cornet, A. Bouchu, G. Descotes, and Y. Queneau, Study of the competition between homogeneous and interfacial reactions during the synthesis of surfactant sucrose hydroxyalkyl ethers in water, *Colloids Surf. A*, 172 (2000) 125–138.
73. R. Pierre, I. Adam, J. Fitremann, F. Jerome, A. Bouchu, G. Courtois, J. Barrault, and Y. Queneau, Catalytic etherification of sucrose with 1,2-epoxydodecane: Investigation of heterogeneous and homogeneous catalysts, *C. R. Chimie*, 7 (2004) 151–160.
74. N. Villandier, I. Adam, F. Jérôme, J. Barrault, R. Pierre, A. Bouchu, J. Fitremann, and Y. Queneau, Selective synthesis of amphiphilic hydroxyalkylethers of disaccharides over solid basic catalysts. Influence of the superficial hydrophilic–lipophilic balance of the catalyst, *J. Mol. Catal. A: Chem.*, 259 (2006) 67–77.
75. Y. Queneau, J. Gagnaire, J. J. West, G. Mackenzie, and J. W. Goodby, The effect of molecular shape on the liquid crystal properties of the mono-*O*-(2-hydroxydodecyl)sucroses, *J. Mater. Chem.*, 11 (2001) 2839–2844.
76. M. P. Yadav, J. N. BeMiller, and Y. Wu, O-(hydroxypropyl)sucrose, *J. Carbohydr. Chem.*, 14 (1994) 991–1001.
77. M. Danel, A. Bouchu, and Y. Queneau, unpublished results.
78. R.N. Faulkner, Surface coating sucrose resin development, in J.L. Hickson (Ed.), *Sucrochemistry*, ACS Symposium Series, Vol. 41, ACS, Washington, DC, 1997, pp. 176–197.
79. G. C. Spila Riera, H. F. Azurmndi, M. E. Ramia, H. E. Bertorello, and C. A. Martin, Synthesis of cross-linked polymers by reaction of sucrose and diepoxide monomers: Characterization and nuclear magnetic resonance study, *Polymer*, 39 (1998) 3515–3521.
80. A. Wernicke and Y. Queneau, unpublished results.
81. H. Bazin, A. Bouchu, and G. Descotes, Hydrolysis of cyanoethylated carbohydrates: Synthesis of new carboxylic derivatives of sucrose, D-glucose and D-fructose, *J. Carbohydr. Chem.*, 14 (1995) 1187–1207.
82. V. R. Gaertner, Sucrose ether- and ester-linked surfactants, *J. Am. Oil Chem. Soc.*, 38 (1961) 410–418.
83. M. A. El-Nokaly and M. A. El-Taraboulsy, Preparation of *O*-dodecyl sucrose surfactant, *J. Disp. Sci. Technol.*, 1 (1980) 373–392.
84. G. R. Ames, H. M. Blackmore, and T. A. King, Alkyloxymethyl ethers of sucrose and glucose as surfactants, *J. Appl. Chem.*, 14 (1964) 245–249.
85. G. R. Ames, H. M. Blackmore, and T. A. King., Reaction of sugars with halomethyl ethers, *J. Appl. Chem.*, 14 (1964) 503–506.

86. K. Hill, B. Gruber, and K. J. Weese, Palladium catalyzed telomerization of butadiene with sucrose: A highly efficient approach to novel sucrose ethers, *Tetrahedron Lett.*, 35 (1994) 4541–4542.
87. I. Pennequin, J. Meyer, I. Suisse, and A. Mortreux, A further application of TPPTS in catalysis: Efficient sucrose-butadiene telomerization using palladium catalysts in water, *J. Mol. Catal. A: Chem.*, 120 (1997) 139–142.
88. V. Desvergnes-Breuil, C. Pinel, and P. Gallezot, Green approach to substituted carbohydrates: Telomerisation of butadiene with sucrose, *Green Chem.*, 3 (2001) 175–177.
89. C. C. Akoh, Lipid-based synthetic fat substitutes, *Food Sci. Technol. (NY, USA)*, 117 (2002) 695–727.
90. I. Janicot, A. Bouchu, G. Descotes, and E. Wong, Correlation structure-activity in the bleaching properties of peracetylated carbohydrates, *Tenside Surfactants Deterg.*, 33 (1996) 290–296.
91. S. A. Forsyth, D. R. MacFarlane, R. J. Thomson, and M. von Itzstein, Rapid, clean and mild *O*-acetylation of alcohols and carbohydrates in a ionic liquid, *Chem. Commun.* (2002) 714–715.
92. D. M. Clode, D. McHale, J. B. Sheridan, G. G. Birch, and E. B. Rathbone, Partial benzoylation of sucrose, *Carbohydr. Res.*, 139 (1985) 141–146.
93. R. Khan and K. S. Mufti, 4,1′,6′-Trichloro-4,1′,6′-trideoxygalactosucrose, UK Patent 2,079,749; *Chem. Abstr.*, 96 (1982) 163112j.
94. T. Ogawa and M. Matsui, A new approach to regioselective acylation of polyhydroxy compounds, *Carbohydr. Res.*, 56 (1977) C1–C6.
95. I. R. Vlahov, P. I. Vlahova, and R. J. Lindhardt, Regioselective synthesis of sucrose monoesters as surfactants, *Carbohydr. Res.*, 16 (1997) 1–10.
96. W. M. Macindoe, A. Williams, and R. Khan, Tin(IV)-functionalised polymer supports; non-toxic and practical reagents for regioselective acetylation of sucrose, *Carbohydr. Res.*, 283 (1996) 17–25.
97. S. Bottle and I. A. Jenkins, Improved synthesis of 'cord factor' analogues, *J. Chem. Soc., Chem. Commun.* (1984) 385.
98. S. Abouhilale, J. Greiner, and J. G. Riess, One-step preparation of 6-perfluoroalkylalkanoates of trehalose and sucrose for biomedical uses, *Carbohydr. Res.*, 212 (1991) 55–64.
99. K. Baczco, C. Nugier-Chauvin, J. Banoub, P. Thilbault, and D. Plusquellec, A new synthesis of 6-*O*-acylsucroses and of mixed 6,6′-di-*O*-acylsucroses, *Carbohydr. Res.*, 269 (1995) 79–88.
100. V. Molinier, J. Fitremann, A. Bouchu, and Y. Queneau, Sucrose esterification under Mitsunobu conditions: Evidence for the formation of 6-*O*-acyl-3′,6′-anhydrosucrose besides mono and diesters of fatty acids, *Tetrahedron: Asymmetry*, 15 (2004) 1753–1762.
101. J. G. Buchanan, D. A. Cummerson, and D. M. Turner, The synthesis of sucrose 6′-phosphate, *Carbohydr. Res.*, 21 (1972) 283–292.
102. M. C. Scherrmann, A. Boutboul, B. Estramareix, A. S. Hoffmann, and A. Lubineau, Binding properties and esterase activity of monoclonal antibodies elicited against sucrose 6-heptylphosphonate, *Carbohydr. Res.*, 334 (2001) 295–307.
103. C. Chauvin, K. Baczko, and D. Plusquellec, New highly regioselective reactions of unprotected sucrose. Synthesis of 2-*O*-acylsucroses and 2-*O*-(*N*-alkylcarbamoyl)sucroses, *J. Org. Chem.*, 58 (1993) 2291–2295.
104. V. Molinier, K. Wiesniewski, A. Bouchu, J. Fitremann, and Y. Queneau, Transesterification of sucrose in organic medium: Study of acyl group migrations, *J. Carbohydr. Chem.*, 22 (2003) 657–669.
105. C. Hervé du Penhoat, S. B. Engelsen, D. Plusquellec, and S. Pérez, A structural study of 2-*O*-lauroylsucrose with molecular modeling and NMR methods, *Carbohydr. Res.*, 305 (1998) 131–145.

106. K. Teranishi, Direct regioselective 2-*O*-(p-toluenesulfonylation) of sucrose, *Carbohydr. Res.*, 337 (2002) 613–619.
107. S. J. Angyal, Complexes of metal cations with carbohydrates in solution, *Adv. Carbohydr. Chem. Biochem.*, 47 (1989) 1–43.
108. J. L. Navia, R. A. Roberts, and R. E. Wingard Jr., Study on the selectivity of benzoylation of metal chelates of sucrose, *Carbohydr. Res.*, 14 (1995) 465–480.
109. E. Avela, S. Aspelund, B. Holmbom, B. Melander, H. Jalonen, and C. Peltonen, Selective substitution of sucrose hydroxyl groups via chelates, in J.L. Hickson (Ed.), *Sucrochemistry*, ACS Symposium Series, Vol. 41, ACS, Washington, DC, 1977, pp. 62–76.
110. A. M. Klibanov, Enzymes that work in organic solvents, *Chem. Tech.* (1986) 354–359.
111. S. Riva, J. Chopineau, A. P. G. Kieboom, and A. M. Klibanov, Protease-catalyzed regioselective esterification of sugars and related compounds in anhydrous dimethylformamide, *J. Am. Chem. Soc.*, 110 (1988) 584–589.
112. J. S. Dordick, Designing enzymes for use in organic solvents, *Biotechnol. Prog.*, 8 (1992) 259–267.
113. S. Riva, in C. C. Akoh and B. G. Swanson (Eds.), *Carbohydrate Polyesters as Fat Substitutes*, Marcel Dekker, New York, 1994, pp. 37–64.
114. M. Woudenberg-van Oosterom, F. van Rantwijk, and R. A. Sheldon, Regioselective acylation of disaccharides in tert-butyl alcohol catalyzed by Candida antarctica lipase, *Biotechnol. Bioeng.*, 49 (1996) 328–333.
115. G. Carrea, S. Riva, F. Secundo, and B. Danieli, Enzymatic synthesis of various 1′-*O*-sucrose and 1-*O*-fructose esters, *J. Chem. Soc., Perkin Trans.*, 1 (1989) 1057–1061.
116. F. J. Plou, M. A. Cruces, M. Bernabe, M. Martin-Lomas, J. L. Parra, and A. Ballesteros, Enzymic synthesis of partially acylated sucroses, *Ann. N.Y. Acad. Sci.*, 750 (1995) 332–337.
117. F. J. Plou, M. A. Cruces, M. Ferrer, G. Fuentes, E. Pastor, M. Bernabé, M. Christensen, F. Comelles, J. L. Parra, and A. Ballesteros, Enzymatic acylation of di- and tri-saccharides with fatty acids: Choosing the appropriate enzyme, support and solvent, *J. Biotechnol.*, 96 (2002) 55–66.
118. N. R. Pedersen, P. J. Halling, L. H. Pedersen, R. Wimmer, R. Matthiesen, and O. R. Veltman, Efficient transesterification of sucrose catalysed by the metalloprotease thermolysin in dimethylsulfoxide, *FEBS Lett.*, 519 (2002) 181–184.
119. P. Potier, A. Bouchu, G. Descotes, and Y. Queneau, Proteinase N-catalysed transesterifications in DMSO-water and DMF-water: Preparation of sucrose methacrylate, *Tetrahedron Lett.*, 41 (2000) 3597–3600.
120. P. Potier, A. Bouchu, J. Gagnaire, and Y. Queneau, Proteinase N-catalysed regioselective esterification of sucrose and other mono- and disaccharides, *Tetrahedron: Asymmetry*, 12 (2001) 2409–2419.
121. N. R. Pedersen, R. Wimmer, R. Matthiesen, L. H. Pedersen, and A. Gessesse, Synthesis of sucrose laurate using a new alkaline protease, *Tetrahedron Asymmetry*, 14 (2003) 667–673.
122. L. Ferreira, M. A. Ramos, M. H. Gil, and J. S. Dordick, Exquisite regioselectivity and increased transesterification activity of an immobilized *Bacillus subtilis* protease, *Biotechnol. Prog.* (2002) 986–993.
123. P. Potier, A. Bouchu, G. Descotes, and Y. Queneau, Lipase-catalysed selective synthesis of sucrose mixed diesters, *Synthesis* (2001) 458–462.
124. M. Ferrer, M. A. Cruces, M. Bernabé, A. Ballesteros, and F. J. Plou., Lipase-catalyzed regioselective acylation of sucrose in two-solvent mixtures, *Biotechnol. Bioeng.*, 65 (1999) 10–16.
125. N. R. Pedersen, R. Wimmer, J. Emmersen, P. Degn, and L. H. Pedersen, Effect of fatty acid chain length on initial reaction rates and regioselectivity of lipase-catalysed esterification of disaccharides, *Carbohydr. Res.*, 337 (2002) 1179–1184.

126. V. Molinier, B. Fenet, J. Fitremann, A. Bouchu, and Y. Queneau, PGFSE-NMR study of the self diffusion of sucrose fatty acid monoesters in water, *J. Colloid Interface Sci.*, 286 (2005) 360–368.
127. V. Molinier, P. J. J. Kouwer, J. Fitremann, A. Bouchu, G. Mackenzie, Y. Queneau, and J. W. Goodby, Self-organizing properties of monosubstituted sucrose fatty acid esters: The effects of chain length and unsaturation, *Chem. Eur. J.*, 12 (2006) 3547–3557.
128. V. Molinier, P. J. J. Kouwer, J. Fitremann, A. Bouchu, G. Mackenzie, Y. Queneau, and J. W. Goodby, Shape dependence in the formation of condensed phases exhibited by disubstituted sucrose esters, *Chem. Eur. J.*, 13 (2007) 1763–1775.
129. S. Thévenet, G. Descotes, A. Bouchu, and Y. Queneau, Hydrophobic effect driven esterification of sucrose in aqueous medium, *J. Carbohydr. Chem.*, 16 (1997) 691–696.
130. A. Wernicke, S. Belniak, S. Thévenet, G. Descotes, A. Bouchu, and Y. Queneau, Synthesis of sucrose carbonates in aqueous medium, *J. Chem. Soc., Perkin Trans.*, 1 (1998) 1179–1181.
131. S. Thévenet, A. Wernicke, S. Belniak, G. Descotes, A. Bouchu, and Y. Queneau, Esterification of unprotected sucrose with acid chlorides in aqueous medium: Kinetic reactivity versus acyl- or alkyloxycarbonyl-group migrations, *Carbohydr. Res.*, 318 (1999) 52–66.
132. D. Christian, J. Fitremann, A. Bouchu, and Y. Queneau, Preparation of amphiphilic sucrose carbamates by reaction with alkyl isocyanates in water–alcohol mixtures, *Tetrahedron Lett.*, 45 (2004) 583–586.
133. P. Potier, V. Maccario, M. B. Giudicelli, Y. Queneau, and O. Dangles, Gallic esters of sucrose as a new class of antioxidants, *Tetrahedron Lett.*, 40 (1999) 3387–3390.
134. C. Dufour, E. Da Silva, P. Potier, Y. Queneau, and O. Dangles, Gallic esters of sucrose as efficient radical scavengers in lipid peroxidation, *J. Agric. Food Chem.*, 50 (2002) 3425–3430.
135. M. T. Barros, C. D. Maycock, F. Sinerez, and C. Thomassigny, Fast galloylation of a sugar moiety: Preparation of three monogalloylsucroses as references for antioxidant activity. A method for the selective deprotection of tert-butyldiphenylsilyl ethers, *Tetrahedron*, 56 (2000) 6511–6516.
136. Y. Kashiwada, G. I. Nonaka, and I. Nishioka, Galloylsucroses from rhubarbs, *Phytochemistry*, 27 (1988) 1469–1472.
137. K. Capek, T. Vydra, M. Ranny, and P. Sedmera, Structure of hexa-*O*-acetylsucroses formed by deacteylation of sucrose octa-*O*-acetate, *Collect. Czech. Chem. Commun.*, 50 (1985) 2191–2200.
138. K. Capek, M. Vodrazkova-Medonosova, J. Moravlova, and P. Sedmera, Partially acetylated sucrose. Structure of hepta-*O*-acetylsucroses formed by deactetylation of octa-*O*-acetylsucrose, *Collect. Czech. Chem. Commun.*, 51 (1986) 1476–1485.
139. A. H. Haines, P. Konowicz, and H. F. Jones, Selective deacetylation of sucrose octa-acetate with primary amines to give 2,3,4,6,1′,6′-hexa-*O*-acetylsucrose, *Carbohydr. Res.*, 205 (1990) 406–409.
140. M. Kloosterman, J. G. J. Weijnen, N. K. de Vries, J. Mentech, I. Caron, G. Descotes, H. E. Schoemaker, and E. M. Meijer, Octa-*O*-acetylsucrose: Regioselective deacetylation by lipolytic enzymes, *J. Carbohydr. Chem.*, 8 (1989) 693–704.
141. G. T. Ong, K. Y. Chang, S. H. Wu, and K. T. Wang, Selective deacetylation on the glucosyl moiety of octa-*O*-acetylsucrose by enzymic hydrolysis: Formation of 2,1′,3′,4′,6′-penta-*O*-acetylsucrose, *Carbohydr. Res.*, 241 (1993) 327–333.
142. K. Y. Chang, S. H. Wu, and K. T. Wang, Preparation of hepta-*O*-acetyl sucroses and hexa-*O*-acetyl sucroses by enzymatic hydrolysis, *J. Carbohydr. Chem.*, 10 (1991) 251–261.
143. K. Y. Chang, S. H. Wu, and K. T. Wang, Regioselective enzymic deacetylation of octa-*O*-acetyl-sucrose: Preparation of hepta-*O*-acetylsucroses, *Carbohydr. Res.*, 222 (1991) 121–129.
144. C. Chauvin and D. Plusquellec, A new chemoenzymatic synthesis of 6′-*O*-acylsucroses, *Tetrahedron Lett.*, 32 (1991) 3495–3498.

145. M. T. Barros, C. D. Maycock, and C. Thomassigny, Preparation of sucrose heptaesters unsubstituted at the C-1 hydroxyl group of the fructose moiety via selective *O*-desilylation, *Carbohydr. Res.*, 328 (2000) 419–423.
146. E. Fanton, C. Fayet, J. Gelas, D. Jhurry, A. Deffieux, and M. Fontanille, Ethylenic acetals of sucrose and their copolymerization with vinyl monomers, *Carbohydr. Res.*, 226 (1992) 337–343.
147. M. Namekata, T. Tanaka, N. Sakamoto, and K. Moro, Oligosaccharide sulfates and monosaccharide sulfates for medical purposes. Antiulcerogenic properties of sucrose sulphate-aluminium complex, *Yakugaku Zasshi*, 87 (1967) 889–893; *Chem. Abstr.*, 68 (1968) 20831d.
148. K. Ochi, Y. Watanabe, K. Okui, and M. Shindi, Crystalline salts of sucrose octasulfate, *Chem. Pharm. Bull.*, 28 (1980) 638–641.
149. W. Szeja, Convenient synthesis and application of sucrose sulfates, in F. W. Lichtenthaler (Ed.), *Carbohydrates as Organic Raw Materials*, Vol. 1, VCH, Weinheim, 1991, pp. 117–125.
150. H. G. Bazin, T. Polat, and R. J. Linhardt, Synthesis of sucrose-based surfactants through regioselective sulfonation of acylsucrose and the nucleophilic opening of a sucrose cyclic sulfate, *Carbohydr. Res.*, 309 (1998) 189–205.
151. R. U. Lemieux and J. P. Barrette, A chromatographic analysis of the product from the tritosylation of sucrose: Crystalline 6,6′-di-*O*-tosylsucrose, *Can. J. Chem.*, 38 (1960) 656–662.
152. J. M. Ballard, L. Hough, S. P. Phadnis, and A. C. Richardson, Selective tetratosylation of sucrose: Isolation of the 2,6,1′,6′-tetrasulphonate, *Carbohydr. Res.*, 83 (1980) 138–141.
153. A. S. M. Sofian, C. K. Lee, and A. Linden, Regioselective sulfonylation of 6,1′,6′-tri-*O*-tritylsucrose through dibutyl-stannylation: Synthesis of 4′-*O*-sulfonyl derivatives of sucrose, *Carbohydr. Res.*, 337 (2002) 2377–2381.
154. J. Gelas, The reactivity of cyclic acetals of aldoses and aldosides, *Adv. Carbohydr. Chem. Biochem.*, 39 (1981) 71–156.
155. P. Calinaud and J. Gelas, Synthesis of isopropylidene, benzylidene, and related acetals, in S. Hanessian (Ed.), *Preparative Carbohydrate Chemistry*, Marcel Dekker, New York, 1997, pp. 3–32.
156. S. Carbonel, C. Fayet, and J. Gelas, Introduction of a carboxyl group through an acetal as a new route to carboxylic acid derivatives of sugars, *Carbohydr. Res.*, 319 (1999) 63–73.
157. R. Khan and K. S. Mufti, Sucrochemistry. XVII. Synthesis and reactions of 1′,2:4,6-di-*O*-isopropylidenesucrose, *Carbohydr. Res.*, 43 (1975) 247–253.
158. E. Fanton, J. Gelas, D. Horton, H. Karl, R. Khan, C. K. Lee, and G. Patel, Kinetic acetonation of sucrose: Preparative access to a chirally substituted 1,3,6-trioxacyclooctane system, *J. Org. Chem.*, 46 (1981) 4057–4060.
159. R. Khan and H. Lindseth, Selective deacetalation of 1′,2,4,6-di-*O*-isopropylidenesucrose tetraacetate, *Carbohydr. Res.*, 71 (1979) 327–330.
160. R. Khan, Sucrochemistry. Part XIII. Synthesis of 4,6-benzylidenesucrose, *Carbohydr. Res.*, 32 (1974) 375–379.
161. R. Khan, K. S. Mufti, and M. R. Jenner, Synthesis and reactions of 4,6-acetals of sucrose, *Carbohydr. Res.*, 65 (1978) 109–113.
162. E. Fanton, C. Fayet, and J. Gelas, Long-chain acetals derived from sucrose as a new class of surfactants, *Carbohydr. Res.*, 298 (1997) 85–92.
163. P. Salanski, G. Descotes, A. Bouchu, and Y. Queneau, Monoacetalization of unprotected sucrose with citral and ionones, *J. Carbohydr. Chem.*, 17 (1998) 129–142.
164. S. Porwanski, P. Salanski, G. Descotes, A. Bouchu, and Y. Queneau, Selective synthesis of 4,6-*O*-alkenylidene and -benzylidene acetals from unprotected sucrose by lanthanide(III) resin-catalyzed transacetalization, *Synthesis* (2000) 525–528.
165. N. Giry-Panaud, G. Descotes, A. Bouchu, and Y. Queneau, Consequences of the preeminent reactivity of 2-OH in sucrose: Cyclic acetalation at 2-OH and 3-OH under basic conditions, *Eur. J. Org. Chem.* (1999) 3393–3398.

166. S. Porwanski, P. Salanski, N. Panaud, G. Descotes, A. Bouchu, and Y. Queneau, Regioselectivity in acid- or base-catalysed acetalation of sucrose: Selection of [OH-2; OH-3] or [OH-4; OH-6] diols, *Top. Catal.*, 13 (2000) 335–338.
167. P. Fleury and J. Courtois, Oxidation of sucrose by periodic acid, *C. R. Acad. Sci.*, 214 (1942) 366–368.
168. P. Fleury and J. Courtois, Oxidation of sucrose by periodic acid, *Bull. Soc. Chim. Fr.*, 10 (1943) 245–251.
169. P. Fleury and J. Courtois, Action of periodic acid on sucrose. II. Hydrolysis at 100° of the acid obtained by oxidation by means of periodic acid followed by bromine, *Bull. Soc. Chim. Fr.*, 12 (1945) 548–553.
170. R. C. Hockett and M. Zief, Lead tetraacetate oxidations in the sugar group. XI. The oxidation of sucrose and preparation of glycerol and glycol, *J. Am. Chem. Soc.*, 72 (1950) 2130–2132.
171. A. K. Mitra and A. S. Perlin, The reaction of sucrose with glycol-cleaving oxidants, *Can. J. Chem.*, 37 (1959) 2047–2052.
172. K. J. Hale, L. Hough, and A. C. Richardson, The cyclisation of the di- and tetra-aldehydes derived from sucrose with nitroalkanes, *Tetrahedron Lett.*, 28 (1987) 891–894.
173. K. J. Hale, L. Hough, and A. C. Richardson, Morpholino-glucosides: New potential sweeteners derived from sucrose, *Chem. Ind. (London)* (1988) 268–269; *Chem. Abstr.*, 109 (1988) 36785t.
174. D. de Wit, F. van Rantwijk, and A. P. G. Kieboom, Sucrose derived aldehydes: The course of the periodate oxidation as studied by HPLC, *Recl. Trav. Chim. Pays-Bas*, 108 (1989) 335–338.
175. D. de Wit, F. van Rantwijk, and A. P. G. Kieboom, The effect of pH and temperature on the periodate oxidation of sucrose, *Recl. Trav. Chim. Pays-Bas*, 109 (1990) 518–522.
176. D. de Wit, R. van den Berg, L. Anders, A. M. Johansson, F. van Rantwijk, L. Maat, and A. P. G. Kieboom, The periodate oxidation of sucrose in aqueous N,N-dimethylformamide, *Carbohydr. Res.*, 226 (1992) 253–260.
177. A. Badel, G. Descotes, and J. Mentech, Oxydation periodique de dérivés du saccharose, *Carbohydr. Res.*, 205 (1990) 323–331.
178. M. Kunz and C. Recker, A new continuous oxidation process for carbohydrates, *Carbohydr. Eur.*, 13 (1995) 11–15.
179. W. Fritsche-Lang and E.I. Leupold, Preparation of sucrose tricarboxylic acid, Patent DE 3555720 (1985); *Chem. Abstr.*, 107 (1987) 59408v.
180. E.I. Leupold, K.H. Schoenwealder, W. Fritsche-Lang, M. Schlingmann, A.H. Linkies, W. Gohla, and F.G. Dany, Preparation of sucrose oxidation products containing tricarboxy derivative for use in detergents, Patent DE 3900677 (1989); *Chem. Abstr.*, 113 (1990) 214316h.
181. L. A. Edye, G. V. Meehan, and G. N. Richards, Platinum catalysed oxidation of sucrose, *J. Carbohydr. Chem.*, 10 (1991) 11–23.
182. L. A. Edye, G. V. Meehan, and G. N. Richards, Influence of temperature and pH on the platinum catalyzed oxidation of sucrose, *J. Carbohydr. Chem.*, 13 (1994) 273–283.
183. M. Kunz, A. Schwarz, and J. Kowalczyk, Process and apparatus for continous manufacture of di- and higher-oxidized carboxylic acids from carbohydrates or their derivatives or primary alcohols, Patent DE 19542287 (1997); *Chem. Abstr.*, 127 (1997) 52504.
184. S. Ehrhardt, M. Kunz, and M. Munir, Surface-active acylated sucrose monocarboxylic acids, German Patent DE 19542303 (1997); *Chem. Abstr.*, 127 (1997) 52518.
185. M. Kunz, H. Puke, C. Recker, L. Scheiwe, and J. Kowalczyk, Process and apparatus for preparation of monooxidized products from carbohydrates, carbohydrate derivatives and primary alcohols, German Patent DE 4307388 (1994); *Chem. Abstr.*, 122 (1994) 56411q.
186. P. Parpot, K. B. Kokoh, B. Beden, and C. Lamy, Electrocatalytic oxidation of saccharose in alkaline medium, *Electrochim. Acta*, 38 (1993) 1679–1683.

187. A. E. J. De Nooy, A. C. Besemer, and H. van Bekkum, Highly selective TEMPO mediated oxidation of primary alcohol groups in polysaccharides, *Recl. Trav. Chim. Pays-Bas*, 113 (1994) 165–166.
188. A. E. J. De Nooy, A. C. Besemer, and H. van Bekkum, Highly selective nitroxyl radical-mediated oxidation of primary alcohol groups in water-soluble glucans, *Carbohydr. Res.*, 269 (1995) 89–98.
189. A. E. J. De Nooy, A. C. Besemer, and H. van Bekkum, On the use of stable organic nitroxyl radicals for the oxidation of primary and secondary alcohols, *Synthesis* (1996) 1153–1174.
190. K. Schnatbaum and H. J. Schäfer, Electroorganic synthesis 66: Selective anodic oxidation of carbohydrates mediated by TEMPO, *Synthesis* (1999) 864–872.
191. E. M. Belgsir and H. J. Schäfer, Selective oxidation of carbohydrates on Nafion®–TEMPO-modified graphite felt electrodes, *Electrochem. Commun.*, 3 (2001) 32–35.
192. M. Schämann and H. J. Schäfer, TEMPO-mediated anodic oxidation of methyl glycosides and 1-methyl and 1-azido disaccharides, *Eur. J. Org. Chem.* (2003) 351–358.
193. D. L. Verraest, J. A. Peters, and H. van Bekkum, Oxidation and carboxymethylation of sucrose and inulin, *Zuckerind.*, 120 (1995) 799–803.
194. S. Lemoine, C. Thomazeau, D. Joannard, S. Trombotto, G. Descotes, A. Bouchu, and Y. Queneau, Sucrose tricarboxylate by sonocatalysed TEMPO-mediated oxidation, *Carbohydr. Res.*, 326 (2000) 176–184.
195. S. Brochette-Lemoine, D. Joannard, G. Descotes, A. Bouchu, and Y. Queneau, Sonocatalysis of the TEMPO-mediated oxidation of glucosides, *J. Mol. Catal. A: Chem.*, 150 (1999) 31–36.
196. S. Brochette-Lemoine, S. Trombotto, D. Joannard, G. Descotes, A. Bouchu, and Y. Queneau, Ultrasound in carbohydrate chemistry: Sonophysical glucose oligomerisation and sonocatalysed sucrose oxidation, *Ultrason. Sonochem.*, 7 (2000) 157–161.
197. G. Descotes and Y. Queneau, Microwaves and ultrasound in carbohydrate chemistry, in W. Praznik and H. Huber (Eds.), *Carbohydrates as Organic Raw Materials*, Vol. 4, WUV-Univ, Vienna, 1998, pp. 39–63.
198. S. Trombotto, E. Viollet-Courtens, L. Cottier, and Y. Queneau, Oxidation of two major disaccharides: Sucrose and isomaltulose, *Top. Catal.*, 27 (2004) 31–37.
199. G. Fleche, Oxidation of sugars with hypohalides in preparation of carboxylates used in detergents formulation, Patent EP 0798 310 A1 (1996); *Chem. Abstr.*, 127 (1997) 307619p.
200. P. L. Bragd, A. C. Besemer, and H. van Bekkum, Bromide-free TEMPO-mediated oxidation of primary alcohol groups in starch and methyl α-D-glucopyranoside, *Carbohydr. Res.*, 328 (2000) 355–363.
201. A. C. Besemer and H. van Bekkum, Dicarboxy starch by sodium hypochlorite/bromide oxidation and its calcium binding properties, *Starch/Stärke*, 46 (1994) 95–101.
202. H. Kochkar, L. Lassalle, M. Morawietz, and W. F. Hölderich, Regioselective oxidation of hydroxyl groups of sugar and its derivatives using silver catalysts mediated by TEMPO and peroxodisulfate in water, *J. Catal.*, 194 (2000) 343–351.
203. M. J. Bernaerts, J. Furnelle, and J. De Ley, The preparation of some new disaccharides and D-allose from 3-ketoglycosides, *Biochim. Biophys. Acta*, 69 (1963) 322–330.
204. J. Van Beeumen and J. De Ley, Hexopyranoside: Cytochrome c oxidoreductase from *Agrobacterium tumefaciens*, *Eur. J. Biochem.*, 6 (1968) 331–343.
205. L. Hough and E. O'Brien, α-D-Allopyranosyl β-D-fructofuranoside (allo-sucrose) and its derivatives, *Carbohydr. Res.*, 84 (1980) 95–102.
206. E. Stoppok, K. Matalla, and K. Buchholz, Microbial modification of sugars as building blocks for chemicals, *Appl. Microbiol. Biotechnol.* (1992) 604–610.
207. M. Pietsch, M. Walter, and K. Buchholz, Regioselective synthesis of new sucrose derivatives via 3-ketosucrose, *Carbohydr. Res.*, 254 (1994) 183–194.

208. K. Buchholz, Synthesis of saccharides by enzymatic modifications of sugar, *Zuckerind.*, 120 (1995) 692–699.
209. C. Simiand, E. Samain, O. R. Martin, and H. Driguez, Sucrose analogues modified at position 3: Chemoenzymatic synthesis and inhibition studies of dextransucrases, *Carbohydr. Res.*, 267 (1995) 1–15.
210. E. Stoppok, J. Walter, and K. Buchholz, The effect of pH and oxygen concentration on the formation of 3-ketodisaccharides by *Agrobacterium tumefaciens*, *Appl. Microbiol. Biotechnol.*, 43 (1995) 706–712.
211. J. Walter and H.-J. Jödening, Kinetic model of disaccharide oxidation by *Agrobacterium tumefaciens*, *Biotechnol. Bioeng.*, 48 (1995) 12–16.
212. K. Buchholz, R. Buczys, H.-P. Lieker, and V. Timme, Process for producing allitol from sucrose and its use as a sweetener, Patent WO 9742339 (1997); *Chem. Abstr.*, 128 (1997) 2975.
213. R. Andersson, O. Larm, E. Scholander, and O. Theander, Bromine oxidation of 1,2-*O*-isopropylidene-α-D-glucofuranose and sucrose, *Carbohydr Res.*, 78 (1980) 257–265.
214. H. J. Koch and R. S. Stuart, A novel method for specific labelling of carbohydrates with deuterium by catalytic exchange, *Carbohydr. Res.*, 59 (1977) C1–C6.
215. F. Balza, N. Cyr, G. K. Hamer, A. S. Perlin, H. J. Koch, and R. S. Stuart, Applications of catalytic, hydrogen-deuterium exchange in ^{13}C-NMR spectroscopy, *Carbohydr. Res.*, 59 (1977) C7–C11.
216. H. J Koch. and R. S. Stuart, The catalytic C-deuteration of some carbohydrate derivatives, *Carbohydr. Res.*, 67 (1978) 341–348.
217. E. A. Cioffi and J. H. Prestegard, Deuterium labelling of a glycosphingolipid using an ultrasonicated nickel catalyst, *Tetrahedron Lett.*, 27 (1986) 415–418.
218. E. A. Cioffi, R. H. Bell, and B. Le, Microwave-assisted C–H bond activation using a commercial microwave oven for rapid deuterium exchange labeling (C–H → C–D) in carbohydrates, *Tetrahedron: Asymmetry*, 16 (2005) 471–475.
219. G. B. Buchanan, G. McManus, and H. C. Jarell, Solid state molecular motion in sucrose octapalmitate as studied by deuterium NMR spectroscopy, *Chem. Phys. Lipids*, 104 (2000) 23–34.
220. F. Balza and A. S. Perlin, Some stereochemical characteristics of C-^{1}H-C-^{2}H exchange-reactions with Raney nickel catalyst in deuterium oxide, *Carbohydr. Res.*, 107 (1982) 270–278.
221. D. Wilke, Raw materials for fermentation, in H. van Bekkum, H. Röper, and A. G. J. Voragen (Eds.), *Carbohydrates as Organic Raw Materials*, Vol. 3, VCH, Weinheim, 1996, pp. 115–128.
222. F. Liu, W. Li, D. Ridgway, T. Gu, and Z. Shen, Production of polyhydroxybutyrate on molasses by recombinant *Escherichia coli*, *Biotechnol. Lett.*, 20 (1998) 245–248.
223. P. C. Hallenbeck and J. R. Benemann, Biological hydrogen production: Fundamentals and limiting processes, *Int. J. Hydrogen Energy*, 27 (2002) 1185–1193.
224. F. W. Lichtenthaler and S. Rönninger, α-D-Glucopyranosyl-D-fructoses: Distribution of furanoid and pyranoid tautomers in water, dimethyl sulphoxide, and pyridine. Part 4. Studies on ketoses, *J. Chem. Soc., Perkin Trans*, 2 (1990) 1489–1497.
225. R. Weidenhagen and S. Lorenz, Ein neues bakterielles Umwandlungsprodukt der Saccharoses, *Angew. Chem.*, 69 (1957) 641.
226. R. Weidenhagen and S. Lorenz, Process for the production of palatinose, German Patent DE 1,049,800 (1957); *Chem. Abstr.*, 55 (1961) 2030b.
227. H. Schiweck, M. Munir, K. M. Rapp, B. Schneider, and M. Vogel, New developments in the use of sucrose as an industrial bulk chemical, in F. W. Lichtenthaler (Ed.), *Carbohydrates as Organic Raw Materials*, Vol. 1, VCH, Weinheim, 1991, pp. 57–94.
228. F. W. Lichtenthaler and S. Peters, Carbohydrates as green raw materials for chemical industry, *C. R. Chimie*, 7 (2004) 65–90.

229. S. Trombotto, A. Bouchu, G. Descotes, and Y. Queneau, Hydrogen peroxide oxidation of palatinose and trehalulose: Direct preparation of carboxymethyl α-D-glucopyranoside, *Tetrahedron Lett.*, 41 (2000) 8273–8277.
230. S. Trombotto, M. Danel, J. Fitremann, A. Bouchu, and Y. Queneau, Straighforward route for anchoring a glucosyl moiety onto nucleophilic species: Reaction of amines and alcohols with carboxymethyl 3,4,6-tri-*O*-acetyl-α-D-glucopyranoside 2-*O*-lactone, *J. Org. Chem.*, 68 (2003) 6672–6678.
231. A. Le Chevalier, R. Pierre, S. Chambert, A. Doutheau, and Y. Queneau, Preparation of amide-linked pseudo-disaccharides by the carboxymethylglycoside lactone (CMGL) strategy, *Tetrahedron Lett.*, 47 (2006) 2431–2434.
232. D. Schwengers, Leucrose, a ketodisaccharide of industrial design, in F. W. Lichtenthaler (Ed.), *Carbohydrates as Organic Raw Materials*, Vol. 1, VCH, Weinheim, 1991, pp. 183–195.
233. F. W. Lichtenthaler, D. Martin, T. Weber, and H. Schiweck, 5-(α-D-Glucosylmethyl)furfural: Preparation from isomaltulose and exploration of its ensuing chemistry, *Liebigs Ann. Chem.* (1993) 967–974.
234. D. Martin and F. W. Lichtenthaler, Versatile building blocks from disaccharides: Glycosylated 5-hydroxymethylfurfurals, *Tetrahedron: Asymmetry*, 17 (2006) 756–762.
235. J. Seibel, R. Moraru, and S. Gotze, Biocatalytic and chemical investigations in the synthesis of sucrose analogues, *Tetrahedron*, 61 (2005) 7081–7086.
236. J. Seibel, R. Moraru, S. Götze, K. Buchholz, S. Na'amnieh, A. Pawlowski, and H.-J. Hecht, Synthesis of sucrose analogues and the mechanism of action of *Bacillus subtilis* fructosyltransferase (levansucrase), *Carbohydr. Res.*, 341 (2006) 2335–2349.
237. M. Remaud-Simeon, R. M. Willemot, P. Sarçabal, G. Potocki de Montalk, and P. Monsan, Glucansucrases: Molecular engineering and oligosaccharide synthesis, *J. Mol. Catal. B: Enzym.*, 10 (2000) 117–128.
238. M. Remaud-Simeon, C. Albenne, G. Joucia, E. Fabre, S. Bozonnet, S. Pizzut, P. Escalier, G. Potocki-Veronese, and P. Monsan, Glucan sucrases: Structural basis, mechanistic aspects, and new perspectives for engineering, *ACS Symp. Ser.*, 849 (2003) 90–103.
239. F. J. Plou, M. T. Martin, A. Gomez de Segura, M. Alcalde, and A. Ballesteros, Glucosyltransferases acting on starch or sucrose for the synthesis of oligosaccharides, *Can. J. Chem.*, 80 (2002) 743–752.
240. E. Fabre, G. Joucla, C. Moulis, S. Edmond, G. Richard, G. Potocki-Veronese, P. Monsan, and M. Remaud-Simeon, Glucansucrases of GH family 70: What are the determinants of their specificities? *Biocatal. Biotransform.*, 24 (2006) 137–145.
241. B. A. van der Veen, L. K. Skov, G. Potocki-Veronese, M. Gajhede, P. Monsan, and M. Remaud-Simeon, Increased amylosucrase activity and specificity, and identification of regions important for activity, specificity and stability through molecular evolution, *FEBS J.*, 273 (2006) 673–681.
242. A. Bertrand, S. Morel, F. Lefoulon, Y. Rolland, P. Monsan, and M. Remaud-Simeon, Leuconostoc mesenteroides glucansucrase synthesis of flavonoid glucosides by acceptor reactions in aqueous-organic solvents, *Carbohydr. Res.*, 341 (2006) 855–863.
243. J. F. Robyt, Properties and uses of dextrans and related glucans, *Polym. Mater. Sci. Eng.*, 72 (1995) 139–140.
244. M. A. Clarke, A. V. Bailey, E. J. Roberts, and W. S. Tsang, Polyfructose: A new microbial polysaccharide, in F. W. Lichtenthaler (Ed.), *Carbohydrates as Organic Raw Materials*, Vol. 1, VCH, Weinheim, 1991, pp. 169–181.
245. L. De Leenheer, Production and use of inulin: Industrial reality with a promising future, in H. van Bekkum, H. Röper, and A. G. J. Voragen (Eds.), *Carbohydrates as Organic Raw Materials*, Vol. 3, VCH, Weinheim, 1996, pp. 68–92.

246. B.C. Tungland, Fructooligosaccharides and other fructans: Structures and occurrence, production, regulatory aspects, food applications and nutrional health significance, in *Oligosaccharides in Food and Agriculture*, ACS Symposium Series, Vol. 849, ACS, Washington, DC, 2003, pp. 135–152.
247. R. Appel, Tertiary phosphane/tetrachloromethane a versatile reagent for chlorination, dehydration and P-N linkage, *Angew. Chem. Int. Ed. Engl.*, 14 (1975) 801–811.
248. R. L. Whistler and A. K. M. Anisuzzaman, Preferential halogenation of the primary hydroxyl group, *Meth. Carbohydr. Chem.*, 8 (1980) 227–231.
249. L. Hough and K. S. Mufti, Sucrochemistry. Part VI. Further reactions of 6,6′-di-*O*-tosylsucrose and a comparison of reactivity of the 6 and 6′ positions, *Carbohydr. Res.*, 25 (1972) 497–503.
250. Z. Ciunik and S. Jarosz, Synthesis and crystal structure of 6,6′-dichloro-6,6′-dideoxy-2,3,4,3′,4′-penta-*O*-benzyl-sucrose, *Pol. J. Chem.*, 71 (1997) 207–212.
251. A. K. M. Anissuzzaman and R. L. Whistler, Selective replacement of primary hydroxyl groups in carbohydrates: Preparation of some carbohydrate derivatives containing halomethyl groups, *Carbohydr. Res.*, 61 (1978) 511–518.
252. P. J. Garegg and B. Samulesson, Novel reagent system for converting a hydroxy group into an iodo group in carbohydrates with inversion of configuration, *J. Chem. Soc., Chem. Commun.* (1979) 978–980.
253. L. Amariutei, G. Descotes, C. Kugel, J. P. Maitre, and J. Mentech, Sucrochimie IV. Synthèse régiosélective de dérivés amines du saccharose via la reaction de Mitsunobu, *J. Carbohydr. Chem.*, 7 (1998) 21–31.
254. G. Descotes, J. Mentech, and S. Veesler, Identification of anhydrosucrose derivatives formed by Mitsunobu chlorination of sucrose, *Carbohydr. Res.*, 190 (1989) 309–312.
255. L. Hough, L. V. Sinchareonkul, A. C. Richardson, F. Akhtar, and M. G. B. Drew, Bridged derivatives of sucrose: The synthesis of 6,6′-dithiosucrose, 6,6′-epidithiosucrose, and 6,6′-epithiosucrose, *Carbohydr. Res.*, 174 (1988) 145–160.
256. W. J. Lees and G. M. Whitesides, Interpretation of the reduction potential of 6,6′-dithiosucrose cyclic disulfide, 6,6′-dithiosucrose, and sucrose in aqueos solution, *J. Am. Chem. Soc.*, 115 (1993) 1860–1869.
257. D. Bhar and S. Chandrasekaran, A simple synthesis of sugar disulfides using tetrathiomolybdate as a sulfur-transfer reagent, *Carbohydr. Res.*, 301 (1997) 221–224.
258. S. Jarosz and A. Listkowski, Synthesis of macrocyclic derivatives containing sucrose unit, *J. Carbohydr. Chem.*, 22 (2003) 753–763.
259. S. H. Eklund and J. F. Robyt, Synthesis of 6-deoxy-6-fluorosucrose, and its inhibition of Leuconostoc and Streptococcus D-glucansucrases, *Carbohydr. Res.*, 177 (1988) 253–258.
260. J. N. Zikopoulos, S. H. Eklund, and J. F. Robyt, Synthesis of 6,6′-dideoxy-6,6′-difluorosucrose, *Carbohydr. Res.*, 104 (1982) 245–251.
261. A. Tanriseven and J. F. Robyt, Synthesis of 4,6-dideoxysucrose, and inhibition studies of Leuconostoc and Streptococcus D-glucansucrases with deoxy and chloro derivatives of sucrose modified at carbon atoms 3, 4, and 6, *Carbohydr. Res.*, 186 (1989) 87–94.
262. F. Bachmann, M. Ruppenstein, and J. Thiem, Synthesis of aminosaccharide-derived polymers with urea, urethane, and amide linkages, *J. Polym. Sci. Part A: Polym. Chem.*, 39 (2001) 2332–2341.
263. F. Bachmann, J. Reimer, M. Ruppenstein, and J. Thiem, Synthesis of novel polyurethanes and polyureas by polyaddition reactions of dianhydrohexitol configurated diisocyanates, *Macromol. Chem. Phys.*, 202 (2001) 3410–3419.
264. K. Kurita, N. Hirakawa, and Y. Iwakura, Synthetic polymers containing sugar residues, 7. Synthesis and properties of polyurethanes from D-cellobiose and diisocyanates, *Makromol. Chem.*, 181 (1980) 1861–1870.

265. D. Jhurry and A. Deffieux, Sucrose-based polymers: Polyurethanes with sucrose in the main chain, *Eur. Polym. J.*, 33 (1997) 1577–1582.
266. D. Plusquellec, F. Roulleau, and E. Brown, Chimie des sucres sans groupements protecteurs: Synthese de carbamates, d'urées et de thiourées en position 1 du lactose, *Tetrahedron Lett.*, 25 (1984) 1901–1904.
267. D. Plusquellec and M. Lefeuvre, Sugar chemistry without protecting groups: A regioselective addition of the primary hydroxyl of monosaccharides to alkylisocyanates, *Tetrahedron Lett.*, 28 (1987) 4165–4168.
268. I. Hirao, K. Itoh, N. Sakairi, Y. Araki, and Y. Ishido, Bis(tributyltin) oxide—phenyl isocyanate system for regioselective (phenyl-carbamoyl)ation of purine and pyrimidine ribonucleosides, *Carbohydr. Res.*, 109 (1982) 181–205.
269. E. Ulsperger, Isocyanate und Epoxyverbindungen als hydrophobe Rohstoffe fuer die Herstellung von Tensinden, *Tenside Surfactants Deterg.*, 3 (1966) 1–6.
270. R. Kholstrung, M. Kunz, A.R. Haji Begli, and F.W. Lichtenthaler, Poster at the 20th International Carbohydrate Symposium, August 2000, Hamburg, Germany.
271. L. Hough and J. E. Priddle, Carbonate derivatives of methyl α–D-mannopyranoside and of D-mannose, *J. Chem. Soc.* (1961) 3178–3181.
272. L. Hough, J. E. Priddle, and R. S. Theobald, Carbohydrate carbonates. Part II. Their preparation by ester-exchange methods, *J. Chem. Soc.* (1962) 1934–1938.
273. W. M. Doane, B. S. Shasha, E. I. Stout, C. R. Russel, and C. E. Rist, A facile route to trans cyclic carbonates of sugars, *Carbohydr. Res.*, 4 (1967) 445–451.
274. C. J. Gray, K. Al-Dulaimi, and S. A. Barker, Carbonate derivatives of methyl D-glucopyranosides, *Carbohydr. Res.*, 47 (1976) 321–325.
275. H. Komura, T. Yoshino, and Y. Ishido, Preparation of cyclic carbonates of sugar derivatives with some carbonylating agents, *Carbohydr. Res.*, 40 (1975) 391–395.
276. R. S. Theobald, Carbionic esters of sucrose. Part III. The direct preparation of polysucrose carbonates from sucrose, *J. Chem. Soc.* (1961) 5370–5376.
277. G. Descotes, G. Muller, and J. Mentech, Synthèse et étude structurale de dithiocarbonates et thiocarbamates du saccharose, *Carbohydr. Res.*, 134 (1984) 313–320.
278. V. H. Kollman, J. L. Hanners, J. Y. Hutson, T. W. Whaley, D. G. Ott, and C. T. Gregg, Large scale photosynthetic production of carbon-13 labeled sugars: The tobacco leaf system, *Biochem. Biophys. Res. Comm.*, 50 (1973) 826–831.
279. V. H. Kollman, J. L. Hanners, R. E. London, E. G. Adame, and T. E. Walker, Photosynthetic preparation and characterization of ^{13}C-labeled carbohydrates in agmenellum quadruplicatum, *Carbohydr. Res.*, 73 (1979) 193–202.
280. L. Hough and E. O'Brien, β-D-Fructofuranosyl α-D-arabino-hexopyranosid-2-ulose (2-ketosucrose), *Carbohydr. Res.*, 92 (1981) 314–317.
281. E. A. Cioffi, R. H. Bell, and B. Le, Microwave-assisted C-H bond activation using a commercial microwave oven for rapid deuterium exchange labelling (C-H → C-D) in carbohydrates, *Tetrahedron: Asymmetry*, 16 (2005) 471–475.
282. M.-H. Gouy, M. Danel, M. Gayral, A. Bouchu, and Y. Queneau, Synthesis of 6,6,1′,1′,6′,6′-hexadeutero sucrose, *Carbohydr. Res.*, 342 (2007) 2303–2308.
283. J.F. Robyt, Modifications, in *Essentials of Carbohydrate Chemistry*, Springer Verlag, New York, Ch. 4, 1998, p. 132.
284. R. D. Guthrie, I. D. Jenkins, and R. Yamasaki, Epoxidation with triphenylphosphine and diethyl azodicarboxylate: Epoxide derivatives of sucrose, *Carbohydr. Res.*, 85 (1980) C5–C6.
285. R. D. Guthrie, I. D. Jenkins, S. Thang, and R. Yamasaki, Derivatives of sucrose 3′,4′-epoxide, *Carbohydr. Res.*, 121 (1983) 109–117.

286. R. D. Guthrie, I. D. Jenkins, S. Thang, and R. Yamasaki, Formation of 3′,6′-anhydrosucrose by Mitsunobu dehydration of sucrose, *Carbohydr. Res.*, 176 (1988) 306–308.
287. J. Defaye, A. Gadelle, and C. Pedersen, The behaviour of D-fructose and inulin towards anhydrous hydrogen fluoride, *Carbohydr. Res.*, 136 (1985) 53–65.
288. J. Defaye and C. Pedersen, Hydrogen fluoride, solvent and reagent for carbohydrate conversion technology, in F. W. Lichtenthaler (Ed.), *Carbohydrates as Organic Raw Materials*, Vol. 1, VCH, Weinheim, 1991, pp. 247–265.
289. J. Defaye and J. M. Garcia-Fernandez, Protonic and thermal activation of sucrose and the oligosaccharide composition of caramel, *Carbohydr. Res.*, 256 (1994) C1–C4.
290. J. Defaye and J. M. Garcia-Fernandez, The oligosaccharide components of caramel, *Zuckerind.*, 120 (1995) 700–704.
291. V. Ratsimba, J. M. Garcia Fernadez, J. Defaye, H. Nigay, and A. Voilley, Qualitative and quantitative evaluation of mono- and disaccharides in D-fructose and sucrose caramels by gas-liquid chromatography-mass spectrometry; di-D-fructose dianhydrides as tracers of caramel authenticity, *J. Chromatogr. A*, 844 (1999) 283–293.
292. J. M. Garcia Fernandez, A. Gadelle, and J. Defaye, Difructose dianhydrides from sucrose and fructooligosaccharides and their use as building blocks for the preparation of amphiphiles, liquid crystals and polymers, *Carbohydr. Res.*, 265 (1994) 249–269.
293. K. Capek, E. Cadova, and P. Sedmera, α-D-glucopyranosyl 1,3-anhydro-D-xylo-hexulofuranoside: An intermediate in the alkaline hydrolysis of α-D-glucopyranosyl 3,4-anhydro-β-D-lyxo-hexulofuraoside, *Carbohydr. Res.*, 238 (1993) 321–325.
294. K. Capek, E. Cadova, and P. Sedmera, 2,3′-anhydrosucrose, *Carbohydr. Res.*, 205 (1990) 161–171.
295. K. Capek and T. Vydra, Oxirane-oxetane-1,4-dioxane anhydro-ring migration in sucrose derivatives, *Carbohydr. Res.*, 168 (1987) C1–C4.
296. K. Capek, T. Vydra, and P. Sedmera, Structure of penta-*O*-acetylsucroses formed by deacetylation of octa-*O*-acetylsucrose. Reaction of 2,3,4,6,6′-penta-*O*-acetylsucrose, *Collect. Czech. Chem. Commun.*, 53 (1998) 1317–1330.
297. C. B. Purves and C. S. Hudson, The analysis of gamma-methylfructoside mixtures by means of invertase. IV. Behaviour of sucrose in methyl alcohol containing hydrogen chloride, *J. Am. Chem. Soc.*, 56 (1934) 1973–1977.
298. A. De Raadt and A. E. Stutz, A simple convergent synthesis of the mannoside inhibitor 1-deoxymannonojirimycin from sucrose, *Tetrahedron Lett.*, 33 (1992) 189–192.
299. G. Gradnig, G. Legler, and A. E. Stutz, A novel approach to the 1-deoxynojirimycin system: Synthesis from sucrose of 2-acetamido-1,2-dideoxynojirimycin, as well as some 2-N-modified derivatives, *Carbohydr. Res.*, 287 (1996) 49–57.
300. J. Y. C. Chan, L. Hough, and A. C. Richardson, The synthesis of (*R*)- and (*S*)-spirobi-1,4-dioxane and related spirobicycles from D-fructose, *J. Chem. Soc., Perkin Trans*, 1 (1985) 1457–1462.
301. M. M. Andrade and M. T. Barros, Facile conversion of O-silyl protected sugars into their corresponding formates using $POCl_3$/DMF complex, *Tetrahedron*, 60 (2004) 9235–9243.
302. M. T. Barros, K. T. Petrova, and A. M. Ramos, Regioselective copolymerization of acryl sucrose monomers, *J. Org. Chem.*, 69 (2004) 7772–7775.
303. G. G. McKeown, R. S. E. Serenius, and L. D. Hayward, Selective substitution in sucrose. I. The synthesis of 1′,4,6′-tri-O-methyl sucrose, *Can. J. Chem.*, 35 (1957) 30–38.
304. G. W. O'Donnell and G. N. Richards, Synthesis of some partially methylated sucrose derivatives, *Aust. J. Chem.*, 25 (1972) 407–412.
305. N. D. Sachinvala, W. P. Niemczura, and M. H. Litt, Monomers derived from sucrose, *Carbohydr. Res.*, 218 (1991) 237–245.
306. S. Jarosz and I. Kościołowska, Preparation of 2,3,4,3′,4′-penta-*O*-benzylsaccharose, Polish patent PL 177187 B1 (1999); *Chem. Abstr.*, 134 (2001) 281070v.

307. S. Jarosz, Selective reactions of the free hydroxyl groups of 2,3,4,3′,4′-penta-*O*-benzyl-sucrose, *J. Carbohydr. Chem.*, 15 (1996) 73–79.
308. S. Jarosz and M. Mach, Higher sucrose analogues: Homologation of a glucose unit by two carbon atoms, *J. Carbohydr. Chem.*, 16 (1997) 1111–1122.
309. S. Jarosz, M. Mach, and J. Frelek, Synthesis and structural analysis of the higher analogues of sucrose, *J. Carbohydr. Chem.*, 19 (2000) 693–715.
310. S. Jarosz and M. Mach, Synthesis of sucrose derivatives modified at the terminal carbon atoms, *Pol. J. Chem.*, 73 (1999) 981–988.
311. M. Mach and S. Jarosz, Reaction of sugar phosphonates with sucrose aldehydes. Synthesis of higher analogues of sucrose, *J. Carbohydr. Chem.*, 20 (2001) 411–424.
312. M. Mach, S. Jarosz, and A. Listkowski, Crown ether analogues from sucrose, *J. Carbohydr. Chem.*, 20 (2001) 485–493.
313. S. Jarosz, A. Listkowski, B. Lewandowski, Z. Ciunik, and A. Brzuszkiewicz, Macrocyclic receptors containing sucrose skeleton, *Tetrahedron*, 61 (2005) 8485–8492.
314. S. Jarosz, A. Listkowski, and M. Mach, Coupling of the C6 and C6′ positions of sucrose by metathesis reaction, *Polish J. Chem.*, 75 (2001) 683–687.
315. S. Jarosz and A. Listkowski, Sugar derived crown ethers and their analogues: Synthesis and properties, *Curr. Org. Chem.*, 10 (2006) 643–662.
316. S. Jarosz and A. Listkowski, Towards C-2 symmetrical macrocyles with incorporated sucrose unit, *Can. J. Chem.*, 84 (2006) 492–496.
317. E. J. Corey and B. Samuelsson, One-step conversion of primary alcohols in the carbohydrate series to the corresponding carboxylic tert-butyl esters, *J. Org. Chem.*, 49 (1984) 4735.
318. D. Magaud, C. Grandjean, A. Doutheau, D. Anker, V. Shevchik, N. Cotte-Pattat, and J. Robert-Baudouy, Synthesis of the two monomethyl esters of the disaccharide 4-*O*-α-D-galacturonosyl-D-galacturonic acid and of precursors for the preparation of higher oligomers methyl uronated in definite sequences, *Carbohydr. Res.*, 314 (1998) 189–199.
319. S. Delrot, N. Roques, G. Descotes, and J. Mentech, Recognition of some deoxy derivatives of sucrose by the sucrose transporter of the plant plasma membrane, *Plant Physiol. Biochem.*, 29 (1991) 25–29.
320. F. W. Lichtenthaler and S. Mondel, *Manno-* and *altro*-sucrose, and some amino-analogues, *Carbohydr. Res.*, 303 (1997) 293–302.
321. M. S. Chowdary, L. Hough, and A. C. Richardson, Selective esterification of sucrose using pivaloyl chloride, *J. Chem. Soc., Chem. Commun.* (1978) 664–665.
322. M. S. Chowdary, L. Hough, and A. C. Richardson, Sucrochemistry. Part 33. The selective pivaloylation of sucrose, *J. Chem. Soc., Perkin Trans.*, 1 (1984) 419–427.
323. C. Simiand and H. Driguez, Synthesis of sucrose analogues modified at position 4, *J. Carbohydr. Chem.*, 14 (1995) 977–983.
324. C. Simiand, E. Samain, O. R. Martin, and H. Driguez, Sucrose analogues modified at position 3: Chemoenzymatic synthesis and inhibition studies of dextransucrases, *Carbohydr. Res.*, 267 (1995) 1–15.
325. J. Anders, R. Buczys, E. Lampe, M. Walter, E. Yaacoub, and K. Buchholz, New regioselective derivatives of sucrose with aminoacid and acrylic groups, *Carbohydr. Res.*, 341 (2006) 322–331.
326. B. Helferich, G. Sprock, and E. Besler, A D-glucose 5,6-dichlorohydrin, *Ber. Dtsch. Chem. Ges.*, 58 (1925) 886–891.
327. L. Hough, S.P. Phadnis, R. Khan, and M.R. Jenner, Chlorinated sucrose sweeteners, UK Patent 1,543,167 (1977); *Chem. Abstr.*, 87 (1977) 202019v.
328. W. Szarek, Deoxyhalogenosugars, *Adv. Carbohydr. Chem. Biochem.*, 28 (1973) 225.
329. T. P. Binder and J. F. Robyt, Synthesis of 6-thiosucrose, and an improved route to 6-deoxysucrose, *Carbohydr. Res.*, 132 (1984) 173–177.

330. T. P. Binder and J. F. Robyt, Synthesis of 3-deoxy- and 3-deoxy-3-fluorosucrose, and α-D-allopyranosyl β-D-fructofuranoside, *Carbohydr. Res.*, 147 (1986) 149–154.
331. J. G. Buchanan and D. A. Cummerson, 1′,4′:3′,6′-dianhydrosucrose, *Carbohydr. Res.*, 21 (1972) 293–296.
332. R. Khan, M. R. Jenner, H. Lindseth, K. S. Mufti, and G. Patel, Ring-opening reactions of sucrose epoxides: Synthesis of 4′-derivatives of sucrose, *Carbohydr. Res.*, 162 (1987) 199–207.
333. G. N. Richards and F. Shafizadeh, Mechanism of thermal degradation of sucrose. A preliminary study, *Aust. J. Chem.*, 31 (1978) 1825–1832.
334. W. Moody and G. N. Richards, Thermolysis of sucrose in the presence of alcohols. A novel method for the synthesis of D-fructofuranosides, *Carbohydr. Res.*, 97 (1981) 247–255.
335. S. Brochette-Lemoine, A. Bouchu, and Y. Queneau, unpublished results.
336. M. Manley-Harris and G. N. Richards, A novel fructoglycan from the thermal polymerisation of sucrose, *Carbohydr. Res.*, 240 (1993) 183–196.
337. G. Eggleston, J. R. Vercellotti, L. A. Edye, and M. A. Clarke, Effects of salts on the initial thermal degradation of concentrated aqueous solutions of sucrose, *J. Carbohydr. Chem.*, 15 (1996) 81–94.
338. C. Moreau, R. Durand, F. Aliès, M. Cotillon, T. Frutz, and M. A. Théoleyre, Hydrolysis of sucrose in the presence of H-form zeolites, *Ind. Crops Prod.*, 11 (2000) 237–242.
339. M. A. Clarke, Technological value of sucrose in food products, in M. Mathlouthi and P. Reiser (Eds.), *Sucrose: Properties and Applications*, Blackie, Glasgow, 1995, pp. 223–247.
340. L. Cottier, G. Descotes, C. Neyret, and H. Nigay, Pyrolysis of carbohydrates, analysis of industrial caramel vapors, *Ind. Aliment. Agric.*, 106 (1989) 567–570; *Chem Abstr.*, 113 (1990) 117350.
341. A. Bouchu, J. Chedin, J. Defaye, D. Lafont, and E. Wong, Preparation of branched oligo- and polyglycosides, especially from sucrose, Patent WO 8707275 (1987); *Chem. Abstr.*, 109 (1988) 95053a.
342. F. W. Lichtenthaler, Unsaturated *O*- and *N*-heterocycles from carbohydrate feedstocks, *Acc. Chem. Res.*, 35 (2002) 728–737.
343. K. Seri, Y. Inoue, and H. Ishida, Highly efficient catalytic activity of lanthanide(III) ions for conversion of saccharides to 5-hydroxymethyl-2-furfural in organic solvents, *Chem. Lett.* (2000) 22–23.
344. M. Kunz, From sucrose to semisynthetical polymers, in G. Descotes (Ed.), *Carbohydrates as Organic Raw Materials*, Vol. 2, VCH, Weinheim, 1993, pp. 135–161.
345. B. F. M. Kuster, 5-Hydroxyfurfural (HMF). A review focusing on its manufacture, *Starch/Stärke*, 42 (1990) 314–321.
346. U. Schöberl, J. Salbeck, and J. Daub, Precursor compounds for organic metals and nonlinear optical devices from carbohydrate-derived hydroxymethylfurfuraldehyde: Furanoid electron donors and push/pull substituted compounds, *Adv. Mater.*, 4 (1992) 41–44.
347. C. Moreau, A. Finiels, and L. Vanoye, Dehydration of fructose and sucrose into 5-hydroxymethylfurfural in the presence of 1-H-3-methyl imidazolium chloride acting both as solvent and catalyst, *J. Mol. Catal. A: Chem.*, 253 (2006) 165–169.
348. M. A. Andrews, S. A. Klaeren, and G. L. Gould, The discovery and design of transition-metal catalyst for the homogeneous hydrogenation of carbohydrate C-C bonds, in G. Descotes (Ed.), *Carbohydrates as Organic Raw Materials*, Vol. 2, VCH, Weinheim, 1993, pp. 3–25.
349. M. Toda, A. Takagaki, M. Okamura, J. N. Kondo, S. Hayashi, K. Domen, and M. Hara, Green chemistry: Biodiesel made with sugar catalyst, *Nature*, 438 (2005) 178.
350. K.J. Parker, K. James, and J. Hurford, Sucrose esters surfactants—a solventless process and the products thereof, in J.L. Hickson (Ed.), *Sucrochemistry*, ACS Symposium Series, Vol. 41, ACS, Washington, DC, Ch. 7, 1977, pp. 97–114.

351. A. L. Le Coënt, M. Tayacout-Fayolle, F. Couenne, S. Briançon, J. Lieto, J. Fitremann-Gagnaire, Y. Queneau, and A. Bouchu, Kinetic parameter estimation and modelling of sucrose esters synthesis without solvent, *Chem. Eng. Sci.*, 58 (2003) 367–376.
352. H.J.W. Nieuwenhuis and G.M. Vianen, Esters of a non-reducing sugar and one or more fatty acids, Patent EP 190779 (1986).
353. L.I. Osipow and W. Rosenblatt, Transesterification in the presence of a transparent emulsion, US Patent 3 644 333; DE 1916886 (1972).
354. P. Raveendran and S. L. Wallen, Sugar acetates as novel, renewable CO_2-philes, *J. Am. Chem. Soc.*, 125 (2002) 7274–7275.
355. Q. Liu, M. H. A. Janssen, F. van Rantwijk, and R. A. Sheldon, Room-temperature ionic liquids that dissolve carbohydrates in high concentrations, *Green Chem.*, 7 (2005) 39–42.
356. S.K. Spear, A.E.Visser, and R.D. Rogers, Ionic liquids: Green solvents for carbohydrate studies, *Proc. Sugar Process. Res. Conf.*, Sugar Processing Research Institute, New Orleans, 2002, pp. 336–340.
357. N. Kardos and J. L. Luche, Sonochemistry of carbohydrate compounds, *Carbohydr. Res.*, 332 (2001) 115–131.
358. A. Loupy, A. Petit, J. Hamelin, F. Texier-Boullet, P. Jacquault, and D. Mathe, New solvent-free organic synthesis using focused microwaves, *Synthesis*, 9 (1998) 1213–1234.
359. A. Corsaro, U. Chiacchio, V. Pistara, and G. Romeo, Microwave-assisted chemistry of carbohydrates, *Curr. Org. Chem.*, 8 (2004) 511–538.
360. A. Oasmaa, E. Kuoppala, and Y. Solantausta, Fast pyrolysis of forestry residue. 2. Physicochemical composition of product liquid, *Energy Fuels*, 17 (2003) 433–443.
361. M. Marquevich, S. Czernik, E. Chornet, and D. Montané, Hydrogen from biomass: Steam reforming of model coumponds of fast-pyrolysis, *Energy Fuels*, 13 (1999) 1160–1166.
362. R. D. Cortright, R. R. Davda, and J. A. Dumesic, Hydrogen from catalytic reforming of biomass-derived hydrocarbons in liquid water, *Nature*, 418 (2002) 964–967.
363. A. Lubineau, J. Augé, H. Bienaymé, Y. Queneau, and M. C. Scherrmann, Aqueous sugars solutions as solvent in organic synthesis: New reactivity and selectivity, in G. Descotes (Ed.), *Carbohydrates as Organic Raw Materials*, Vol. 2, VCH, Weinheim, 1993, pp. 99–112.
364. H. Schiweck, K. Rapp, and M. Vogel, Utilization of sucrose as an industrial bulk chemical-state of the art and future implications, *Chem. Ind.* (1988) 228–234.
365. K. Hill and O. Rhode, Sugar-based surfactants for consumer products and technical applications, *Fett-Lipid.*, 101 (1999) S25–S33.
366. T. Kosaka and T. Yamada, New plant and application of sucrose esters, in J.L. Hickson (Ed.), *Sucrochemistry*, ACS Symposium Series, Vol. 41, ACS, Washington, DC, 1977, pp. 85–96.
367. L. Bobichon, A sugar ester process and its application in calf feeding and human food additives, in J.L. Hickson (Ed.), *Sucrochemistry*, ACS Symposium Series, Vol. 41, ACS, Washington, DC, 1977, pp. 115–120.
368. T. Polat and R. J. Linhardt, Syntheses and applications of sucrose-based esters, *J. Surfactants Deterg.*, 4 (2001) 415–421.
369. S. Piccicuto, C. Blecker, J. C. Brohée, A. Mbampara, G. Lognay, C. Deroanne, M. Paquot, and M. Marlier, Les esters de sucres: Voies de synthèse et potentialités d'utilisation, *Biotechnol. Agron. Soc. Environ.*, 5 (2001) 209–219; *Chem. Abstr.*, 137 (2002) 21675.
370. L. I. Osipow, F. D. Snell, W. C. York, and A. Finchler, Methods for the preparation of fatty acid esters of sucrose, *Ind. Eng. Chem.*, 48 (1956) 1459–1462.
371. H.B. Hass, F.D. Snell, W.C. York, and L.I. Osipow, Process for producing sugar esters, US Patent 2 893 990 (1959).
372. K. Wada, K. Onuma, T. Ushikubo, and T. Ito, Process for preparing sucrose fatty acid esters. Patent EP 275939 (1988); *Chem. Abstr.*110 (1989) 173962v.

373. A. S. Muller, J. Gagnaire, Y. Queneau, M. Karaoglanian, J. P. Maitre, and A. Bouchu, Winsor behaviour of sucrose fatty acid esters: Choice of the cosurfactant and effect of the surfactant composition, *Colloids Surf.*, A 203 (2002) 55–66.
374. B. A. P. Nelen and J. M. Cooper, Sucrose esters, in R. J. Whitehurst (Ed.), *Emulsifiers in Food Technology*, Blackwell Publishing Ltd., Oxford, 2004, pp. 131–161.
375. N. B. Desai and N. Lowicki, New sucrose esters and their applications in cosmetics, *Cosmet. Toiletries*, 100 (1985) 55–59.
376. I. J. A. Baker, B. Matthews, H. Suares, I. Krodkiewska, D. N. Furlong, and C. J. Drummond, Sugar fatty acid ester surfactants: Structure and ultimate aerobic biodegradability, *J. Surfactants Deterg.*, 3 (2000) 1–11.
377. J. Fitremann, A. Bouchu, and Y. Queneau, Synthesis and gelling properties of *N*-palmitoyl-L-phenylalanine sucrose esters, *Langmuir*, 19 (2003) 9981–9983.
378. A. Guerrero, P. Partal, and C. Gallegos, Linear viscoelastic properties of sucrose ester-stabilized oil-in-water emulsions, *J. Rheol.*, 42 (1998) 1375–1388.
379. N. Garti, V. Clement, M. Leser, A. Aserin, and M. Fanun, Sucrose ester microemulsions, *J. Mol. Liq.*, 80 (1999) 253–296.
380. M. A. Bolzinger-Thevenin, J. L. Grossiord, and M. C. Poelman, Characterization of a sucrose ester microemulsion by freeze fracture electron micrograph and small angle neutron scattering experiments, *Langmuir*, 15 (1999) 2307–2315.
381. M. H. Kabir, M. Ishitobi, and H. Kunieda, Emulsion stability in sucrose monoalkanoate system with addition of cosurfactants, *Colloid Polym. Sci.*, 280 (2002) 841–847.
382. Y. Li, S. Zhang, J. Jang, and Q. Wang, Relationship of solubility parameters to interfacial properties of sucrose esters, *Colloids Surf. A*, 248 (2004) 127–133.
383. Y. Li, S. Zhang, Q. Wang, and J. Yang, Study on surface activity and critical aggregation concentration of sucrose esters containing different isomers of mono-, di- and polyesters, *Tenside Surfactants Deterg.*, 41 (2004) 26–30.
384. V. M. Sadtler, M. Guely, P. Marchal, and L. Choplin, Shear-induced phase transitions in sucrose ester surfactant, *J. Colloid Interface Sci.*, 270 (2004) 270–275.
385. T. Rades and C. C. Müller-Goymann, Electron and light microscopical investigation of defect structures in mesophases of pharmaceutical substances, *Colloid Polym. Sci.*, 275 (1997) 1169–1178.
386. G. Garofalakis and B. S. Murray, Surface pressure isotherms, dilatational rheology, and brewster angle microscopy of insoluble monolayers of sugar monoesters, *Langmuir*, 18 (2002) 4765–4774.
387. T. Kawaguchi, T. Hamanaka, Y. Kito, and H. Machida, Structural studies of a homologous series of alkyl sucrose ester micelle by X-ray scattering, *J. Phys. Chem.*, 95 (1991) 3837–3846.
388. T. Kawaguchi, T. Hamanaka, and T. Mitsui, X-ray structural studies of some nonionic detergent micelles, *J. Colloid Interface Sci.*, 96 (1983) 437–453.
389. V. Molinier, P. J. J. Kouwer, Y. Queneau, J. Fitremann, G. Mackenzie, and J. W. Goodby, A bilayer to monolayer phase transition in liquid crystal glycolipids, *J. Chem. Soc., Chem. Commun.* (2003) 2860–2861.
390. G. Garofalakis, B. S. Murray, and D. B. Sarney, Surface activity and critical aggregation concentration of pure sugar esters with different sugar headgroups, *J. Colloid Interface Sci.*, 229 (2000) 391–398.
391. T. M. Herrington and S. S. Sahi, Temperature dependence of the micellar aggregation number of aqueous solutions of sucrose monolaurate and sucrose monooleate, *Colloids Surf.*, 17 (1986) 103–113.
392. F. A. Husband, D. B. Sarney, M. J. Barnard, and P. J. Wilde, Comparison of foaming and interfacial properties of pure sucrose monolaurates, dilaurate and commercial preparations, *Food Hydrocolloids*, 12 (1998) 237–244.

393. T. Polat, H. G. Bazin, and R. J. Linhardt, Enzyme catalyzed regioselective synthesis of sucrose fatty acid surfactants, *J. Carbohydr. Chem.*, 16 (1997) 1319–1325.
394. M. Ferrer, F. Comelles, F. J. Plou, M. A. Cruces, G. Fuentes, J. L. Parra, and A. Ballesteros, Comparative surface activities of di- and trisaccharide fatty acid esters, *Langmuir*, 18 (2002) 667–673.
395. I. Södeberg, C. J. Drummond, D. N. Furlong, S. Godkin, and B. Matthews, Non-ionic sugar-based surfactants: Self assembly and air/water interfacial activity, *Colloids Surf. A*, 102 (1995) 91–97.
396. C. Calahorro, J. Munoz, M. Berjano, A. Guerrero, and C. Gallegos, Flow behaviour of sucrose stearate/water systems, *J. Am. Oil Chem. Soc.*, 69 (1992) 660–666.
397. V. Molinier, B. Fenet, J. Fitremann, A. Bouchu, and Y. Queneau, Concentration measurements of sucrose and sugar surfactants solutions by using the ^{1}H NMR ERETIC method, *Carbohydr. Res.*, 341 (2006) 1890–1895.
398. S. Ehrhardt, A.H. Begeli, and M. Kunz, Acylated carboxyalkyl saccharide preparation for use in detergents. Eur. Pat. Appl. EP 814088 (1997); *Chem. Abstr.*, 128 (1997) 103627.
399. K. Tachikawa, J. Sasaki, H. Asano, and S. Nakata, Glycosyl sucrose fatty esters enzymic manufacture, JP Patent 2002105193; *Chem. Abstr.*, 136 (2002) 278508.
400. I. D. Jenkins and M. B. Goren, Improved synthesis of cord factor analogues via the Mitsunobu reaction, *Chem. Phys. Lipids*, 41 (1986) 225–235.
401. G. A. Jeffrey, Carbohydrate liquid crystals, *Acc. Chem. Res.*, 19 (1986) 168–173.
402. J. G. Riess and J. Greiner, Carbohydrate and related polyol-derived fluorosurfactants: An update, *Carbohydr. Res.*, 327 (2000) 147–168.
403. G. Kretzschmar, Large scale synthesis and properties of the novel sucrose aspartate surfactants, *J. Carbohydr. Chem.*, 19 (2000) 879–889.
404. J. Fitremann-Gagnaire, G. Toraman, G. Descotes, A. Bouchu, and Y. Queneau, Synthesis in water of amphiphilic sucrose hydroxyalkylethers, *Tetrahedron Lett.*, 40 (1999) 2757–2760.
405. H. Gruber, Hydrophile Polymergele mit reaktiven Gruppen, I, Herstellung und Polymerisation von Glucose- und Saccharosemethacrylate, *Monatsh. Chem.*, 112 (1981) 273–285.
406. H. Gruber, Hydrophile Polymergele mit reaktiven Gruppen II. Saccharosemethacrylaten, *Monatsh. F. Chem.*, 112 (1981) 445–457.
407. H. Gruber and S. Knaus, Synthetic polymers based on carbohydrates: Preparation, properties and applications, *Macromol. Symp.*, 152 (2000) 95–105.
408. H. Gruber and G. Greber, Reactive sucrose derivatives, in F. W. Lichtenthaler (Ed.), *Carbohydrates as Organic Raw Materials*, Vol. 1, VCH, Weinheim, 1991, pp. 95–116.
409. J. Klein and D. Herzog, Poly(vinylsaccharide)s 2, *Makromol. Chem.*, 188 (1987) 1217–1232.
410. G. Wulff, J. Schmid, and T. Venhoff, The synthesis of polymerizable vinyl sugars, *Macromol. Chem. Phys.*, 197 (1996) 259–274.
411. X. Chen, A. Johnson, J. S. Dordick, and D. G. Rethwisch, Chemoenzymatic synthesis of linear poly(sucrose acrylate): Optimization of enzyme activity and polymerization conditions, *Macromol. Chem. Phys.*, 195 (1994) 3567–3578.
412. J. Chen and K. Park, Synthesis of fast-swelling, superporous sucrose hydrogels, *Carbohydr. Polym.*, 41 (2000) 259–268.
413. D. Jhurry, A. Deffieux, M. Fontanille, I. Betremieux, J. Mentech, and G. Descotes, Sucrose-based polymers. 1. Linear polymers with sucrose side-chains, *Makromol. Chem.*, 193 (1992) 2997–3007.
414. L. Ferreira, M. M. Vidal, C. F. G. C. Geraldes, and M. H. Gil, Preparation and characterisation of gels based on sucrose modified with glycidyl methacrylate, *Carbohydr. Polym.*, 41 (2000) 15–24.
415. D. R. Patil, J. S. Dordick, and D. G. Rethwisch, Chemoenzymatic synthesis of novel sucrose-containing polymers, *Macromolecules*, 24 (1991) 3462–3463.

416. D. R. Patil, D. G. Rethwisch, and J. S. Dordick, Enzymic synthesis of a sucrose-containing linear polyester in nearly anhydrous organic media, *Biotechnol. Bioeng.*, 37 (1991) 639–646.
417. N. S. Patil, J. S. Dordick, and D. G. Rethwisch, Macroporous poly(sucrose acrylate) hydrogel for controlled release of macromolecules, *Biomaterials*, 17 (1996) 2343–2350.
418. P. Galgali, A. J. Varma, U. S. Puntambekar, and D. V. Gokhale, Towards biodegradable polyolefins: Strategy of anchoring minute quantities of monosaccharides and disaccharides onto functionalized polystyrene and their effects on facilitating polymer biodegradation, *Chem. Commun.* (2002) 2884–2885.
419. X. C. Liu and J. S. Dordick, Sugar acrylate-based polymers as chiral molecularly imprintable hydrogels, *J. Polym. Sci., Part A: Polym. Chem.*, 37 (1999) 1665–1671.
420. K.C. Frich and J.E. Kresta, An overview of sugars in urethanes, in J.L. Hickson (Ed.), *Sucrochemistry*, ACS Symposium Series, Vol. 41, ACS, Washington, DC, 1977, pp. 238–256.
421. S. Y. Lee, Y. Lee, and F. Wang, Chiral compounds from bacterial polyesters: Sugars to plastics to fine chemicals, *Biotechnol. Bioeng.*, 65 (1999) 363–368.
422. J. R. Crouse and S. M. Grundy, Effects of sucrose polyester on cholesterol metabolism in man, *Metab., Clin. Exp.*, 28 (1979) 994–1000.
423. R. J. Jandacek, M. M. Ramirez, and J. R. Crouse, Effects of partial replacement of dietary fat by olestra on dietary cholesterol absorption in man, *Metab., Clin. Exp.*, 39 (1990) 848–852.
424. R. Ballinger, R. L. Malin, and A. G. Webb, Sucrose polyester: A new oral contrast agent for MRI, *Magn. Reson. Med.*, 19 (1991) 199–202.
425. C. C. Akoh and B. G. Swanson, Optimized synthesis of sucrose polyesters: Comparison of physical properties of sucrose polyesters, raffinose polyesters and salad oils, *J. Food. Sci.*, 55 (1990) 236–243.
426. R. C. Reynolds and C. I. Chappel, Sucrose acetate isobutyrate (SAIB): Historical aspects of its use in beverages and a review of toxicity studies, *Food Chem. Toxicol.*, 36 (1998) 81–93.
427. C. Henry, Sucrose acetate isobutyrate special grade for beverage applications, *International Food Ingred.* (1995) 47–49.
428. M. R. Jenner, The uses and commercial development of sucralose, in T.H. Grenby (Ed.), *Advances in Sweeteners*, 1996, pp. 253–262.
429. Y. Nie, S. Vigues, J. R. Hobbs, G. L. Conn, and S. D. Munger, Distinct contributions of T1R2 and T1R3 taste receptor subunits to the detection of sweet stimuli, *Curr. Biol.*, 15 (2005) 1948–1952.
430. D. B. Volkin, A. M. Verticelli, K. E. Marfia, C. J. Burke, H. Mach, and C. R. Middaugh, Sucralfate and soluble sucrose octasulfate bind and stabilize acidic fibroblast growth factor, *Biochim. Biophys. Acta*, 1203 (1993) 18–26.
431. T. Polat, M. Mohammadi, and R. J. Linhardt, Synthesis of sulfosucrose derivatives for evaluation as regulators of fibroblast growth factot activity, *Tetrahedron Lett.*, 43 (2002) 8047–8049.
432. J. Lee, G. I. Jallo, M. B. Penno, K. L. Gabrielson, G. D. Young, R. M. Johnson, E. M. Gillis, C. Rampersaud, B. S. Carson, and M. Guarnieri, Intracranial drug-delivery scaffolds: Biocompatibility evaluation of sucrose acetate isobutyrate gels, *Toxicol. Appl. Pharmacol.*, 215 (2006) 64–70.
433. N. L. Thomas and J. D. Birchall, The retarding action of sugars on cement hydration, *Cem. Concr. Res.*, 13 (1983) 830–842.
434. M. C. Garci Juenger and H. M. Jennings, New insights into the effects of sugars on the hydration and microstructure of cement pastes, *Cem. Concr. Res.*, 32 (2002) 393–399.
435. R. N. Das, Nanocrystalline ceramics from sucrose process, *Mater. Lett.*, 47 (2001) 344–350.
436. E.P. Lira and R.F. Anderson, Sucrose benzoate-the unique modifier, in J.L. Hickson (Ed.), *Sucrochemistry*, ACS Symposium Series, Vol. 41, ACS, Washington, DC, 1977, pp. 223–233.

437. C.H. Coney, SIAB in coatings, in J.L. Hickson (Ed.), *Sucrochemistry*, ACS Symposium Series, Vol. 41, ACS, Washington, DC, 1977, pp. 213–222.
438. J. Mentech, R. Beck, and F. Burzio, Sucrose derivatives as bleaching boosters for the detergent industry, in G. Descotes (Ed.), *Carbohydrates as Organic Raw Materials*, Vol. 2, VCH, Weinheim, 1993, pp. 185–201.
439. E. Buiel and J. R. Dahn, Reduction of the irreversible capacity in hard-carbon anode materials prepared from sucrose for Li-ion batteries, *J. Electrochem. Soc.*, 145 (1998) 1977–1981.
440. A. Gandini and M. N. Belgacem, Furans in polymer chemistry, *Prog. Polym. Sci.*, 22 (1997) 1203–1379.
441. J. Streith, A. Boiron, A. Frankowski, D. Le Nouen, H. Rudyk, and T. Tschamber, On the way to glycoprocessing inhibitors: A genaral one-pot synthesis of imidazolosugars, *Synthesis* (1995) 944–946.
442. A. C. Besemer and H. van Bekkum, Calcium sequestering agents based on carbohydrates, in H. van Bekkum, H. Röper, and A. G. J. Voragen (Eds.), *Carbohydrates as Organic Raw Materials*, Vol. 3, VCH, Weinheim, 1996, pp. 273–293.
443. H. Bazin, G. Descotes, A. Bouchu, and M. Petit-Ramel, Comparison of calcium complexation of some carboxylic acids derived from D-glucose and D-fructose, *Can. J. Chem.*, 73 (1995) 1338–1347.
444. M. Kunz, J. Kowalczyk, and S. Ehrhardt, Use of acylated oxidized carbohydrates as bleach activators and sequestrants in detergent formulations, Eur. Pat. Appl. EP 731161 (1996); *Chem. Abstr.* 125 (1996) 279255.

ANTI-INFLUENZA DRUG DISCOVERY: ARE WE READY FOR THE NEXT PANDEMIC?

BY TASNEEM ISLAM AND MARK VON ITZSTEIN

Institute for Glycomics, Griffith University (Gold Coast Campus), PMB50, Gold Coast Mail Centre, Queensland 9726, Australia

I. INTRODUCTION

1. Influenza Virus and the Disease

Epidemics and pandemics of influenza have been documented for centuries, with the first clear record dated back in 1580.[1] It is an acute respiratory disease which results in significant morbidity and mortality throughout the world. The pandemic of 1918–1919 was responsible for the death of 20 million people.[2] In

ISBN: 978-0-12-373920-9
DOI: 10.1016/S0065-2318(07)61006-3

recent years, the avian influenza virus of the H5N1 subtype has raised concerns of the threat of a pandemic and has spurred efforts to develop plans for its control. Although vaccination is the primary strategy for the prevention of influenza, it is not very effective because of the continual antigenic variation of the virus. Options for the therapeutic treatment of influenza are Amantadine and Rimantadine, which exert their antiviral effect by inhibiting the M2 protein ion-channel function and act indirectly on hemagglutinin.[3,4] Since influenza B do not possess an M2 protein, they are effective only against influenza A.[3–7] Furthermore, the clinical use of these agents is limited because of significant side effects and rapid emergence of resistance strains.[4–7]

In 1999 the US Food and Drug Administration (FDA) approved two new influenza drugs, Relenza, **1** (zanamivir-ZMV by Glaxo Wellcome/Biota) and Tamiflu, **2** (oseltamivir-OMV by Hoffman-La Roche/Gilead).[8,9] Different from M2 inhibitor drugs, Zanamivir and Oseltamivir target the virus surface glycoprotein sialidase and were found to be effective against all strains of influenza.

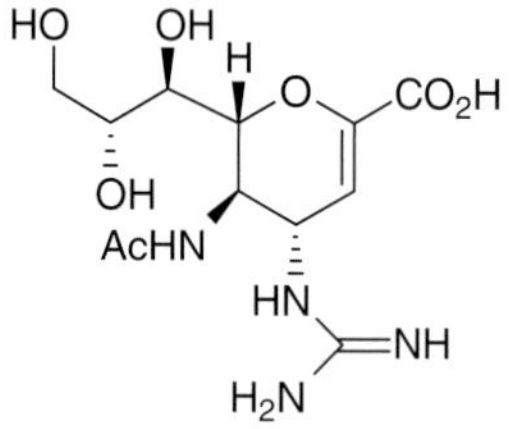

HO OH H O CO_2H OH AcHN HN =NH H_2N

1

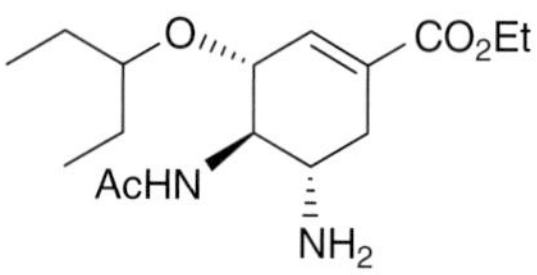

O CO_2Et AcHN NH_2

2

Because of the essential role of sialidase in influenza virus replication and the highly conserved enzyme active-site in influenza viruses A and B, most interest has been focused on the development of selective inhibitors of this enzyme.[10] This chapter discusses the tremendous progress that has been made in the discovery of this new class of anti-influenza agents, in addition to some background information to understand the fundamental role of influenza.

a. Influenza Virus.—Influenza viruses are usually transmitted via air droplets, and subsequently contaminate the mucosa of the respiratory tract. They are able to penetrate the mucin layer of the outer surface of the respiratory tract, entering respiratory epithelial cells, as well as other cell types. Replication is very quick: after only 6 h the first influenza viruses are shed from infected cells. It causes a broad range of illnesses, from symptomless infection through various respiratory syndromes, disorders affecting the lung, heart, brain, liver, kidneys, and muscles to secondary bacterial pneumonia.

Influenza viruses are enveloped single-stranded RNA viruses with a pleomorphic appearance, and an average diameter of 120 nm. They are members of the orthomyxovirus family and can be classified into three serologically distinct types A, B, and C. In humans only influenza A and B viruses are of epidemiological interest.[11,12] Based on the antigenic properties of its surface glycoproteins hemagglutinin (H or HA) and sialidase (neuraminidase, N or NA), influenza A viruses are further divided into sixteen H (H1–H16) and nine N (N1–N9) subtypes.[2,13] Numerous combinations of hemagglutinin and sialidase subtypes are found in avian species; in human, the three pandemics of the twentieth century were caused by viruses containing H1N1 in 1918, H2N2 in 1957, and H3N2 in 1968. The avian influenza that currently threatens a human-to-human transmission and causes a pandemic is H5N1.

R = NHAc

Neu5Ac **3**

Both hemagglutinin and sialidase are carbohydrate-recognizing proteins and in humans recognize the well-known sialic acid, *N*-acetylneuraminic acid (Neu5Ac, **3**), associated as the terminal carbohydrate unit of upper respiratory tract and lung-associated glycoconjugates.[12,14]

HA is a homotrimeric integral membrane glycoprotein anchored to the virus' lipid membrane. It is shaped like a cylinder, and is approximately 135 Å long (Fig. 1a).[15] The three identical monomers that constitute HA are constructed into a central α-helix coil; three spherical heads contain the sialic acid binding-sites. Two significant roles have been associated with this glycoprotein. The first function is to provide an initial point of contact for the virus to adhere to the target host cell-surface glycoconjugates via α-ketosidically-linked terminal sialic acid residues.[16–18] The second key role of this glycoprotein is its involvement in the internalization process of the virus through fusion of the viral envelope with the host cell.[18,19] The membrane fusion is thought to be mediated by a conformational change in the HA, triggered at a low pH (between 5 and 6) of the endosome.[20,21] The HA molecule is the major surface antigen of the virus against which neutralizing antibodies are produced and as a consequence undergoes antigenic variation leading to recurrent epidemics and respiratory diseases.[15,22]

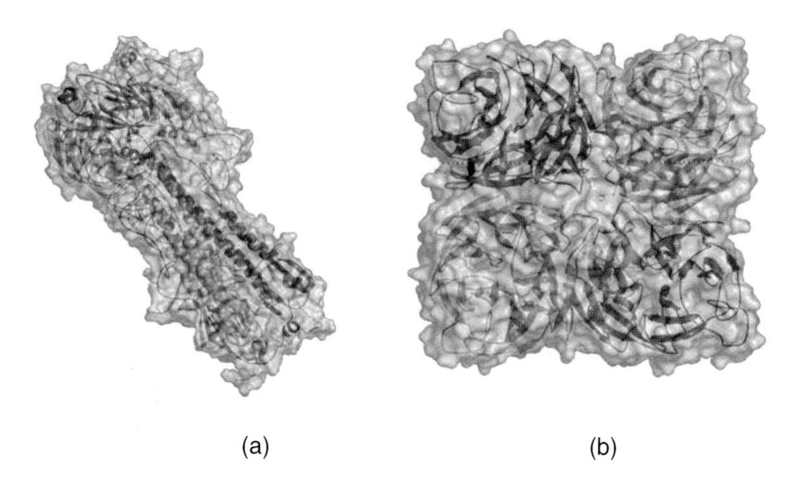

FIG. 1. (a) A view of the influenza virus hemagglutinin trimer showing *N*-acetylneuraminic acid (**3**, in CPK form) bound. (b) The tetrameric unit of influenza A virus sialidase. The figures were generated using the PyMOL Molecular Graphics System (Delano, W.L. (2002) at http://www.pymol.org).

Influenza virus sialidase (EC 3.2.1.18) is a tetramer consisting of four identical disulfide-linked subunits (Fig. 1b).[23] The tetrameric head is anchored to the viral surface by a thin long stalk, 100 Å in length. Each monomeric unit has a proteinfold that consists of a symmetric arrangement of six four-stranded antiparallel β-sheets arranged like the blades of a propeller, with approximate sixfold symmetry passing through the center of the subunit.[24,25]

The biochemistry of influenza virus sialidase has been studied extensively.[26,27] The enzyme catalyzes the cleavage of the sialic acid (Neu5Ac, **3**) residues from glycoproteins, glycolipids, and oligosaccharides via the oxocarbonium intermediate (see later).[28,29] This catalytic activity is essential for influenza virus replication and infectivity, since it liberates the viral progeny from the surface of the infected cells (Fig. 2).[30,31] In addition, removal of these terminal sugars prevents the HA of one virion binding to the sialic acids on an adjacent virion, thus reducing the propensity of the virus particles to self aggregate. Finally, the action of NA on heavily sialylated viral binding glycoproteins (mucins) in the respiratory tract may reduce the viscosity of mucus in the host environment, facilitating movement of the virus in this medium and expediting the virus to the target epithelial cell.[14,32]

The essential roles, although controversial at times, played by both of these surface glycoproteins in the infectious life cycle of the virus, coupled with a

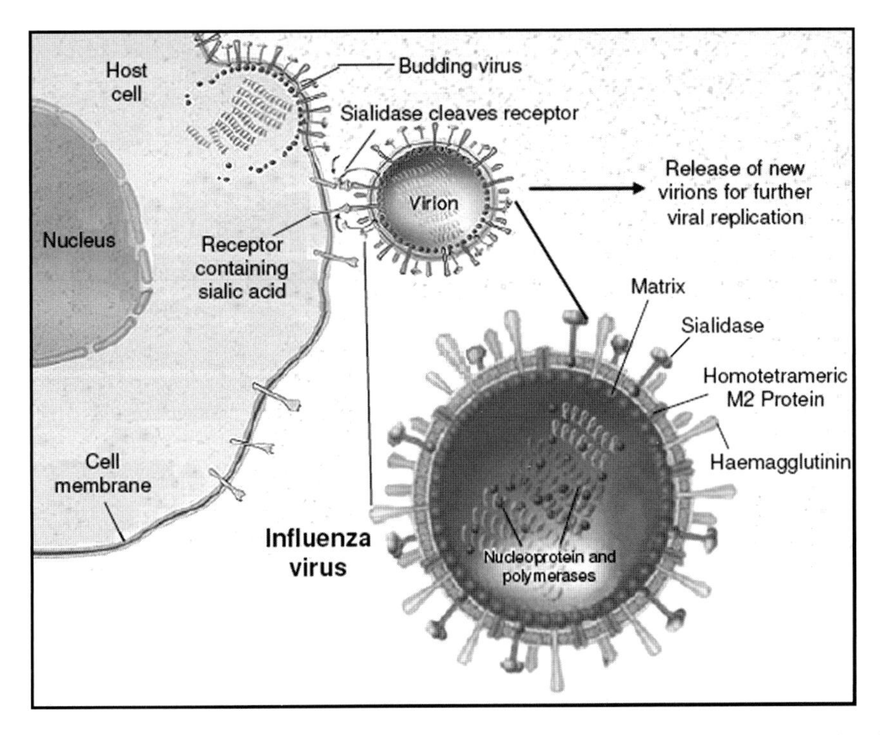

FIG. 2. Schematic representation of the influenza virus and the action of sialidase in the continued replication cycle (modified from Moscona[33]).

significant body of available structural information, has provided very many exciting opportunities for rational structure-based drug discovery of anti-influenza agents over the years.[19] To date the most successful structure-based anti-influenza drug discovery has arisen from targeting the sialidase function.

II. The Influenza Virus Sialidase

1. Influenza Virus Sialidase: Mechanism of Catalysis

As mentioned, sialidase is believed to play two critical roles in the life cycle of the virus. At the molecular level, sialidase cleave terminal α-ketosidically linked sialic acid residues. The mechanism by which sialidases exert their hydrolytic activity remains an active field of research.[34–37] In the absence of structural information, it was originally proposed that in the first step, the normally ${}^{2}C_{5}$

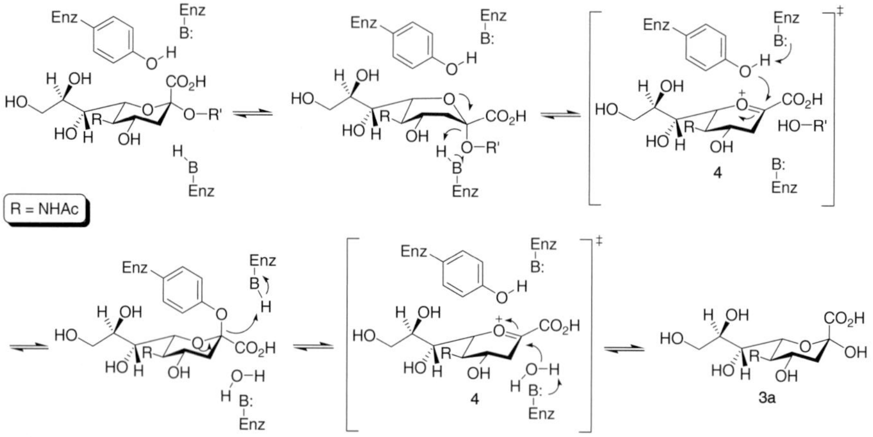

SCHEME 1. The reaction catalyzed by influenza virus sialidase.

configured α-sialoside binds to the enzyme and is distorted by the highly charged environment of the active site from the chair conformation that is preferred in solution into an α-boat conformation (Scheme 1).[37] The X-ray structures of influenza virus sialidase in complex with α-Neu5Ac (**3a**) later confirmed both distortion of the substrate upon binding and the formation of a salt bridge between the carboxylate and triarginyl cluster.[38,39] The resulting conformational strain induces the cleavage of the glycosidic bond, the departure of the attached sugar, and the formation of an oxocarbonium ion intermediate (**4**), which has been identified by kinetic isotope-effect measurements and molecular-modeling studies. This cationic intermediate is believed to be stabilized by a negatively charged environment within that region of the sialidase catalytic site.[40] A water molecule then reacts in a stereoselective manner with intermediate **4** to afford α-Neu5Ac (**3a**) which mutarotates to the thermodynamically more favored β anomer.[40,41] It is now known that the formation of a glycosyl–enzyme covalent intermediate is a common feature of most retaining glycosidases.[42–44]

2. Influenza Virus Sialidase: Substrate Binding and Active Site

The X-ray crystal structure of virus sialidase was first solved in early 1980s.[24,25] While the active site could be identified, resolution was insufficient to determine the orientation of the bound ligand (Neu5Ac) within the catalytic center. Subsequent efforts in protein crystallography allowed further refinement of the structure of the apo-enzyme and yielded high-resolution structures of NA from both influenza A and B virus sialidases (NA(A), NA(B)) complexed with *N*-acetylneuraminic acid **3** or the naturally occurring inhibitor 2-deoxy-2,3-dehydro-*N*-acetylneuraminic acid (Neu5Ac2en, **5**).[38,39,45–48]

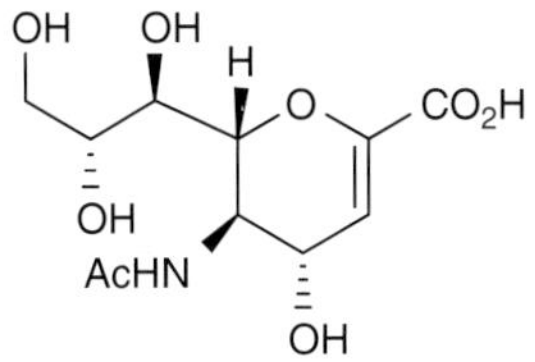

5

The sialidase active site is highly polar, consisting of ten Arg, Asp, and Glu residues and four hydrophobic residues. To facilitate the description of the binding modes of all known substrates and competitive inhibitors, the active site of influenza virus NA is divided into five regions or subsites (S1 through S5).[49] Active site and subsite topology is discussed here in the context of the inhibitor Neu5Ac2en (**5**), which is known to bind in an analogous manner to Neu5Ac (numbering used throughout reflects that reported for influenza virus N2 sialidase). The positively charged subsite S1 consists of Arg118, Arg292, and Arg371, termed the triarginyl cluster or arginyl triad, and interacts with the carboxylate residue (C-1 of Neu5Ac and Neu5Ac2en) via charge–charge interaction and hydrogen bonding. The two glutamate residues, Glu227 and Glu119 make up the negatively charged binding pocket S2, with Glu119 interacting with the hydroxyl group at C-4. Additional peripheral functionalities in S2 include Trp178, Asp151, and a water molecule. The *N*-acetyl moiety attached at C-5 acetamido function makes a hydrogen bond to Arg152 at subsite S3 through both its amide nitrogen and carbonyl oxygen atoms, and the methyl group is situated against a hydrophobic pocket consisting of Ile 222 and Trp178. Subsite S4 is not occupied by any portion of the inhibitor Neu5Ac2en and glycosides of Neu5Ac, and is primarily a hydrophobic region derived from the side-chain of Ile 222, Ala246, and the hydrophobic face of Arg224. Finally, the subsite S5 is comprised of Glu276 (trans configuration), which hydrogen bonds with C-8 and C-9 hydroxyl groups, and Arg224 makes hydrophobic contacts with the glycerol side-chain of Neu5Ac or Neu5Ac2en. Glu276 can exist in an alternative gauche conformation with its carboxylate ion-paired with Arg224. When Glu276 adopts this orientation, it combines with Ala246 to make S5 considerably more hydrophobic in character. A comparison of the crystal structures of influenza A and B

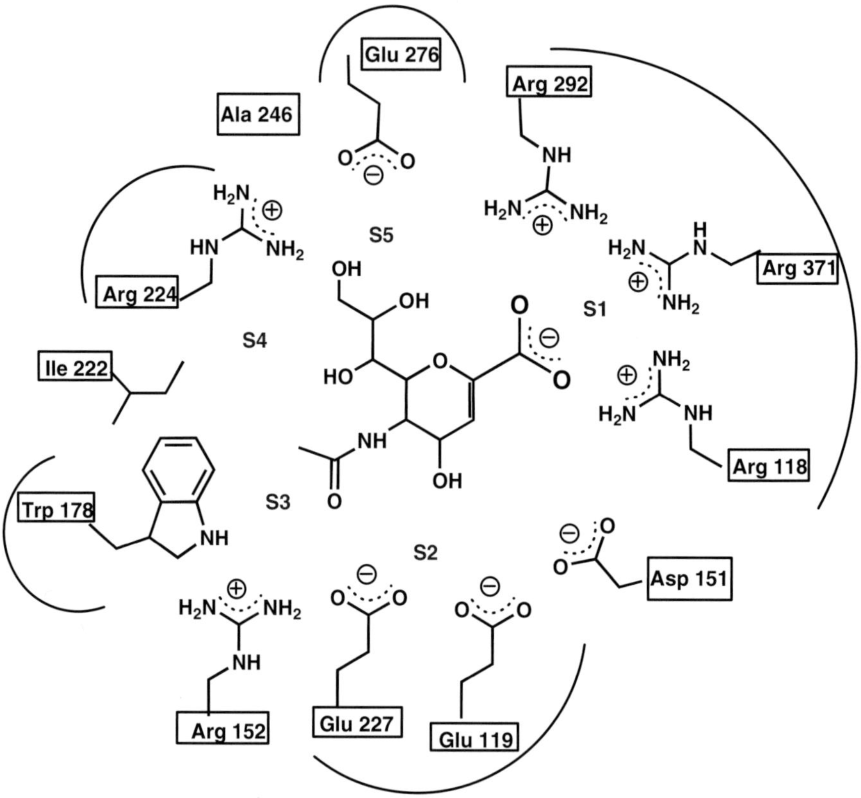

Fig. 3. A schematic representation of some key interactions of Neu5Ac2en **5** with conserved influenza virus A sialidase active site residues.

virus sialidases suggest that the active sites in each case orient the substrate in a similar fashion (Fig. 3).[38,39,47]

Key active site residues occupy similar positions, with only minor variation in the position of Glu276 in S5 and some bound water molecules evident among the different subtypes.

3. Influenza Virus Sialidase as a Drug-Discovery Target

Influenza virus sialidase poses a highly attractive target for antiviral drug design due to its prominent position at the surface of the virion and its profound role in viral pathogenesis. Random screening programs identified a number of

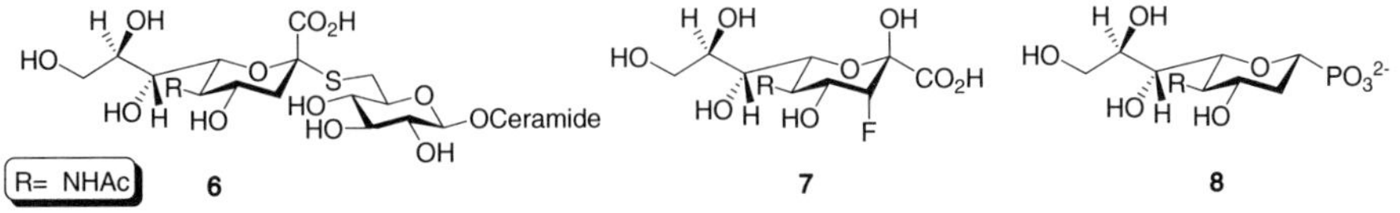

FIG. 4. A selection of substrate mimetics evaluated as inhibitors of influenza virus sialidase.

low affinity inhibitors which were not selective for the influenza virus sialidase.[50–52] Efforts to develop substrate mimetics as influenza virus sialidase inhibitors have met with limited success. Of these many examples, some of those which exhibit reasonable *in vitro* inhibition include thioglycoside ($K_i = 2.8 \times 10^{-6}$ M) (**6**), 3-deoxy-3-fluoro ($K_i = 8 \times 10^{-6}$ M) (**7**), and phosphonate ($IC_{50} \sim 1 \times 10^{-4}$ M) (**8**) analogues of Neu5Ac, are shown in Fig. 4.[53–55]

In the late 1960s, investigators following the classical mechanistic route, namely designing a possible transition-state analogue which resulted in a promising lead compound, Neu5Ac2en (**5**, $K_i \sim 10^{-6}$ M), with an $\sim 10^3$-fold greater binding affinity than the product, Neu5Ac ($K_i = 5 \times 10^{-3}$ M).[56,57] Improvements in potency and specificity were sought by synthesizing a number of Neu5Ac2en analogues, but the most potent derivative was identified as the 2,3-dehydro-2-deoxy-*N*-trifluoroacetylneuraminic acid ($K_i = 8 \times 10^{-7}$ M).[58] Although both these molecules showed that in cell culture it retarded virus shedding, they failed as effective antiviral agents in animal models, reportedly due to the rapid renal excretion of the compounds.[31,59–61] Moreover, these compounds were not suitable as therapeutic agents for influenza virus infection since they were essentially nonselective, inhibiting bacterial and mammalian, as well as viral sialidases.[57,58]

4. Design and Synthesis of Zanamivir

In 1983, Colman and colleagues solved the X-ray crystal structure of influenza virus sialidase, providing the structural information required for the design of potential inhibitors.[24,25,38,39,48] Utilizing the data from the crystallographic studies and in particular, those featuring the bound ligands α-Neu5Ac (**3a**) and Neu5Ac2en (**5**), allowed von Itzstein and colleagues to set about a rational structure-based approach to drug design.[62] Using computational chemistry, in particular the software program GRID,[63] the enzyme active site was probed to assess the introduction of new functionality on the Neu5Ac2en scaffold. A series of functional-group probes (ammonium, carboxylate, hydroxyl, methyl, and phosphate) were used in an effort to determine the energetically favored interactions between the various probes and the residues within the binding pocket.[64] It emerged from these studies that an amino group, 4-amino-4-deoxy-Neu5Ac2en (**9**), was potentially better suited than the C-4 hydroxyl group, due to the formation of a salt bridge with conserved Glu119. Further inspection of the C-4 hydroxyl binding domain revealed a conserved pocket of sufficient size to accommodate a larger basic functional group in place of the amine. Thus, it was predicted that a bifunctional group such as a guanidino substituent to give 4-deoxy-4-guanidino-Neu5Ac2en (**1**) would form strong interactions with two carboxylate residues (Glu119 and Glu227) via both of its terminal nitrogen atoms.

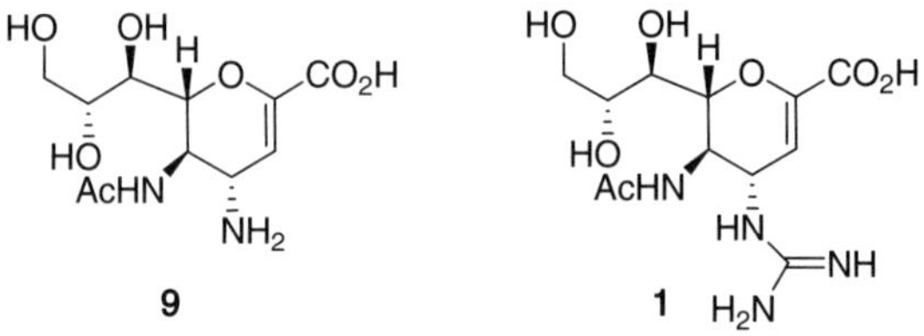

HO OH H O CO_2H HO AcHN NH_2 **9**

HO OH H O CO_2H HO AcHN HN =NH **1** H_2N

The two target molecules **9** and **1** were readily prepared from the oxazoline **10** as depicted in Scheme 2.[65,66] Briefly, reaction of the peracetylated Neu5Ac2en with $BF_3 \cdot Et_2O$[67] resulted in the oxazoline **10** which, when treated with azide, afforded the fully protected 4-azido-4-deoxy-Neu5Ac2en **11**. Reduction of the azide **11**, and subsequent deprotection gave the first target compound, 4-amino-4-deoxy-Neu5Ac2en (**9**). Treatment of compound **9** with the guanidinylating reagent aminoiminomethanesulfonic acid in base resulted in the second target compound **1**. Modifications and improvements in the synthesis of **1** have now been reported.[68]

The amine and guanidino derivatives, **9** and **1** respectively, were evaluated against NA from influenza virus A (N1 and N2) and B. Both **9** and **1** were

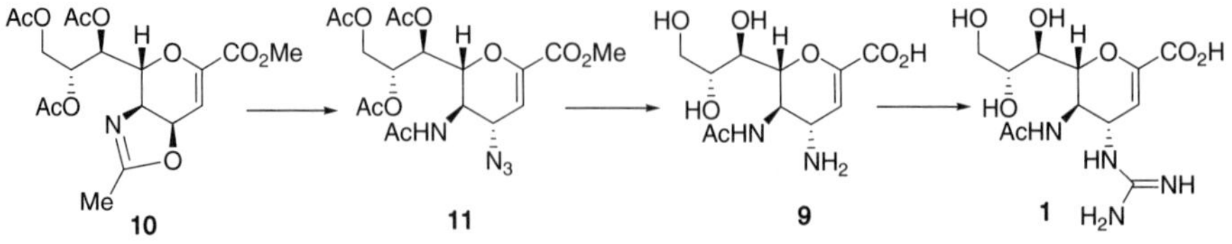

AcO AcO H O CO_2Me AcO N O Me **10** → AcO AcO H O CO_2Me AcO AcHN N_3 **11** → HO OH H O CO_2H HO AcHN NH_2 **9** → HO OH H O CO_2H HO AcHN HN =NH **1** H_2N

SCHEME 2. Synthesis of the FDA approved drug Relenza (**1**).

determined to be highly potent competitive inhibitors of the enzyme *in vitro* (**9** $K_i = 4 \times 10^{-8}$ M, **1** $K_i = 3 \times 10^{-11}$ M) and *in vivo* and prevented viral replication for all influenza A and B virus strains.[57,62,69] A crystal structure of the NA-4-deoxy-4-guanidino-Neu5Ac2en complex confirmed that 4-deoxy-4-guanidino-Neu5Ac2en has the same binding mode as Neu5Ac2en and generally validated the structure-based drug discovery (Fig. 5). A deviation from the initial-modeled structure was the orientation of the 4-guanidino substituent, which did not form hydrogen bonds to Glu119 as designed. One of the primary NH groups of the guanidino substituent is hydrogen bonded to a backbone of carbonyl oxygen

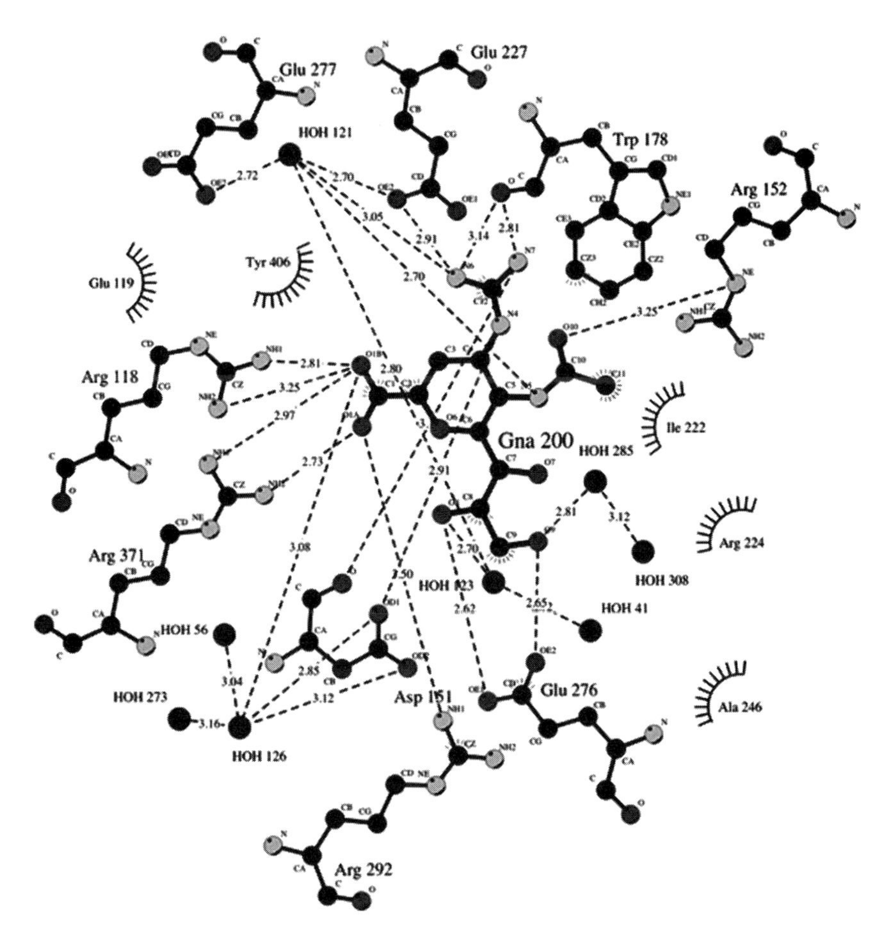

FIG. 5. The important interactions of Zanamivir (**1**) with the active site of influenza virus A sialidase.[70] This figure was generated from crystal structure data (PDB–1NNC) using LIGPLOT.[63]

(Trp178), a water molecule, and a carboxylate (Glu227), while the other primary nitrogen interacts with the backbone carbonyl oxygens of both Trp178 and Asp151. The secondary nitrogen in **1** occupies the place of the amine in **9**, and has been shown to interact with Glu119 via a salt bridge. In addition, a slow-binding kinetic was observed, which was accounted as a result of the expulsion of a water molecule from subsite S2 by the larger guanidino substituent, giving rise to an increase in the binding interactions. Importantly, it was found that the relatively bulky and basic guanidinyl moiety at C-4 allows selective inhibition of influenza virus sialidase over those of mammalian or bacterial origin.

The more potent compound, 4-deoxy-4-guanidino-Neu5Ac2en (**1**) was adopted as a lead drug candidate by Glaxo-Welcome under the generic name zanamivir (sometimes referred in literature as GG167) and in 1999 was commercialized as the first sialidase-targeting anti-influenza drug (Relenza^{TM}). Owing to its highly polar nature and consequent rapid excretion, Relenza^{TM} is delivered directly to the primary site of infection via inhalation. Since the development of Zanamivir (**1**) significant work has been undertaken in the design and the development of influenza virus sialidase inhibitors and this chapter provides an overview of more recent advances in this field.

III. Influenza Virus Sialidase Inhibitors

The discovery of Zanamivir as a potent and selective inhibitor of influenza virus sialidase prompted several researchers to investigate the synthesis and structure–activity relationship studies of Neu5Ac2en-based compounds as potential sialidase inhibitors. Exploration of these SAR studies were undertaken to optimize inhibitory activity and to improve the physicochemical properties of the sialic acid-based influenza virus sialidase inhibitor. A few *in vitro* assays are commonly employed to measure the effectiveness of influenza virus sialidase inhibitors. The first involves a fluorometric assay that measures release of a synthetic fluorophore following its cleavage from Neu5Ac by sialidase. Dye-uptake assay, such as the Neutral Red uptake assay, measures the uptake of a vital stain, Neutral Red in cell culture. The process requires intact membranes and active metabolism in the cell, and is expressed as percent protective rate against virus infection. The plaque-reduction assay is used to measure sialidase inhibition indirectly in cell culture, and provides some measure of the inhibitor's effect on the viability of the influenza virus. *In vitro* and *in vivo* systems for analysis of inhibitors of influenza virus enzymes have been reviewed.[71]

1. Inhibitors Based on the Neu5Ac2en Scaffold

a. C-1 Modifications.—Originally Meindl and Tuppy reported that the methyl ester derivative of Neu5Ac2en (Neu5Ac2en1Me, **12**) failed to inhibit influenza virus sialidase.[56] In contrast to these results, Flashner and coworkers demonstrated that **12** was indeed a competitive inhibitor of sialoside hydrolysis catalyzed by both N1 and N2 variants of influenza A virus NA.[72] However, Neu5Ac2en1Me was a considerably less potent inhibitor (~500-fold) than Neu5Ac2en, emphasizing the importance of the electrostatic interactions involving the arginyl triad and C1 carboxylate.

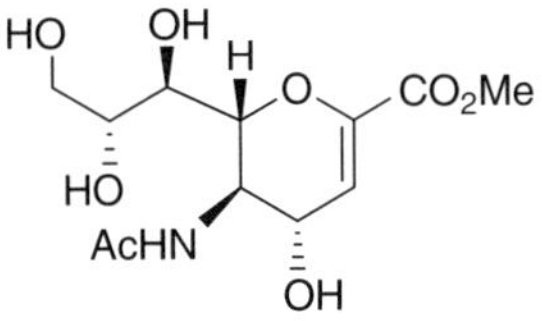

12

b. C-4 Modifications.—The success of the C-4 modification in Zanamivir instigated a thorough investigation into the structure–activity relationships of different substituents at that position. Complete removal of the substituent at C-4 (**13**) resulted in only a twofold drop in inhibition relative to Neu5Ac2en, adding support to computational work which indicated no major interaction between the C-4 hydroxyl group and the protein (Fig. 6).[64,73] The significance of the configuration at C-4 was also investigated. Azido, amino, and hydroxy-substituted C-4 epimers of Neu5Ac2en (**14**–**16**) resulted in a reduction in efficacy by 10–50-fold than the corresponding equatorially substituted derivatives.[57,64,65,72] Molecular-modeling studies with 4-amino-4-deoxy-*epi*-Neu5Ac-2en **15** indicated that the amine substituent could potentially engage in some favorable interaction with the protein at an alternative subsite, but that these contacts are not as significant as those created by interactions with glutamate residues in S2 by the equatorial substituents.

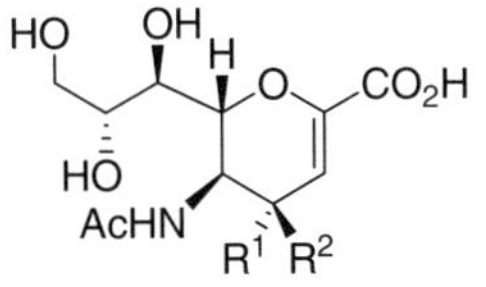

	R^1	R^2	K_i relative to **5**
13	H	H	2
14	H	N_3	7
15	H	NH_2	0.1
16	H	OH	55
17	N_3	H	0.5

FIG. 6. Structures and inhibitory activity of C-4 modified Neu5Ac derivatives against influenza A virus sialidase. K_i values are normalized against that of Neu5Ac2en ($K_i = 1.0$).

The C-4 hydroxy group of Neu5Ac2en was oxidized to the ketone **18**, resulting in a drop in potency by a factor of over 100.[72] Cyanomethyl ether **19**, a key intermediate towards the synthesis of the inhibitors **20** and **21**, which intended to mimic Zanamivir at C-4, was evaluated against NA.[74,75] These modifications were not particularly successful, as **19–21** showed only 7–30% inhibition of catalysis at 1.0 mM compared to Zanamivir, which inhibits NA activity by over 90%.

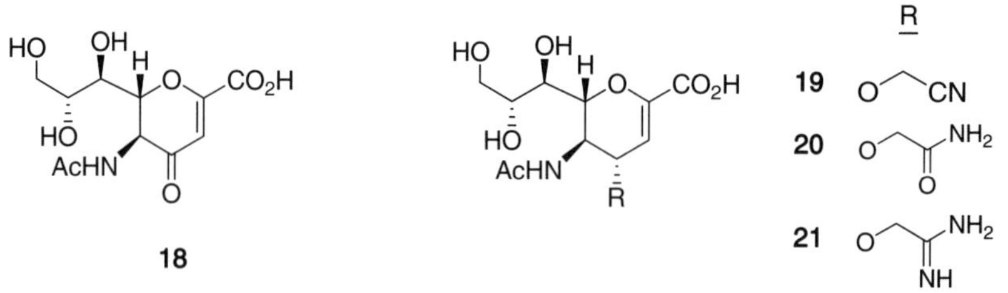

18

19–**21**

The majority of other modifications examined at the C-4 position of Neu5Ac2en involved *N*-alkylation or *N*-acylation of 4-amino-4-deoxy-Neu5Ac-2en or modification of the guanidino substituent of Zanamivir. *N*-Methylamino (**22**), *N*-allyl (**23**), and *N*,*N*-dimethylamino (**24**) derivatives of amine **9** were effective inhibitors of NA from influenza A and B viruses ($K_i = 10^{-5}$–10^{-7} M) though they showed diminished activity against a number of bacterial sialidases.[57,65]

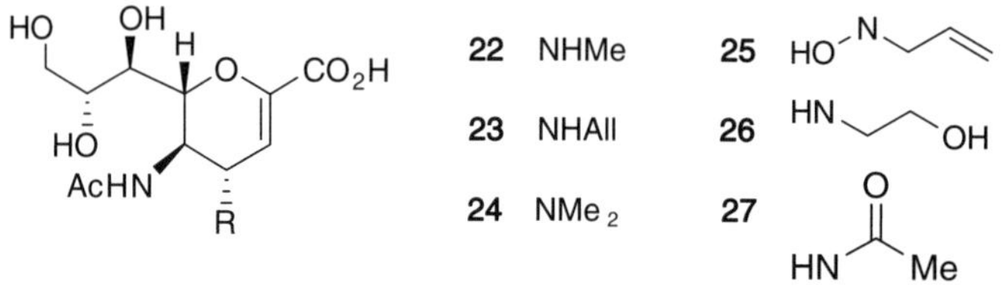

22 NHMe
23 NHAll
24 NMe_2
25
26
27

N-Allyl-*N*-hydroxy (**25**) and *N*-(1-hydroxy)ethyl (**26**) variants demonstrated inhibition values against influenza A virus sialidase at micromolar levels which was again lower relative to **9** ($K_i = 4 \times 10^{-8}$) and similar to that of Neu5Ac2en. The *N*-acetylated 4-amino derivative (**27**) showed the largest drop in potency among the derivatives studied, most likely as a result of the relatively low basicity of the amide nitrogen atom.[57]

Modifications of the guanidino function of Zanamivir (**1**) at the primary nitrogen resulted in a dramatic reduction in the inhibition of NA and of viral replication *in vivo*. N^3-Nitro, ethoxycarbonyl, methyl, hydroxy, and amino derivatives **28–32** inhibited the catalytic action of NA between 300 and 3000 times less than the parent guanidino derivative **1**.[68] Interestingly, the relative activities of these compounds were different in plaque-reduction and NA-inhibition assays, perhaps reflecting the importance of lipophilicity.

	R
28	NO_2
29	CO_2Et
30	Me
31	OH
32	NH_2

A series of fourteen 4-triazole-modified Zanamivir analogues were synthesized from the 4-azido-4-deoxy-Neu5Ac2en derivative using 'click' chemistry, and their inhibitory potencies against influenza virus sialidase were determined. These modifications were not that successful as **33–35** demonstrated only 6–40% protective rate at a concentration of 50 μM against the virus infection compared to Zanamivir which shows 86% at the same concentration. The best result was obtained for compound **36**, which showed a protective rate of >61%.[76]

	R
33	Cyclopropyl
34	$CH_2N(C_2H_4)_2O$
35	C_6H_4-4-OC_3H_7
36	$CH(OH)C_2H_5$

These results were rationalized in terms of their interaction between the substituent and the polar region of subsite **2**. In the case of bulky substituents such as **35**, it was suggested, based on molecular-modeling studies, that the R group is too large to be accommodated in the subsite S2, resulting in the six-membered ring flip over, and hence taking a reverse orientation of Zanamivir, as seen in Fig. 7. Being devoid of hydrogen bonding, this series of compounds were generally less potent than Zanamivir.[76]

Despite rather intensive efforts to design superior inhibitors to Zanamivir or 4-amino-4-deoxy-Neu5Ac2en by replacement of the group at C-4, no improvements have been realized. This reflects the importance and complexity of both electrostatic interactions and the extensive multi-atom intermolecular hydrogen-bond networks that assure the high affinities observed for **9** and **1**.

Fig. 7. Schematic representation of **35** in the active site of influenza virus sialidase.

c. C-5 Modifications.—X-Ray crystallographic structures of all sialidases solved to date, regardless of enzyme origin, have a well-defined binding pocket that can accommodate the C-5 acetamido group of Neu5Ac.[77] As described previously, interactions between the C-5 acetamido function and strictly conserved residues in the influenza virus NA active site are mediated via hydrogen bonding of the carbonyl oxygen to Arg152, by hydrogen bonding of the NH moiety to a bound water molecule, as well as by van der Waals contacts between the methyl group and the hydrophobic patch formed by Trp178 and Ile222 in subsite S3. The spatial constraints of S3 dictate that only small *N*-acyl substituents are accommodated, as observed in the earlier study.[58]

Interest in modification of the C-5 substituent of Neu5Ac2en arose in the 1960s when a group of eighteen *N*-acyl-modified Neu5Ac2en derivatives were screened against influenza A and B virus sialidases.[58] Among these compounds, trifluoroacetamide **37** and difluoroacetamide **38** were identified as the first inhibitors of

NA with comparable activity to Neu5Ac2en ($IC_{50} = 1 \times 10^{-5}$ M (**5**), 5×10^{-6} M (**37**), 1×10^{-5} M (**38**)). Similar replacement of the acetamido group of 4-deoxy-4-guanidino-Neu5Ac2en with a trifluoroacetamide, **39**, resulted in slightly weaker (approximately fivefold) inhibition of both influenza virus A and B sialidases, and also of viral replication *in vitro*. Removal of the C-5 acetamido group from either 4-amino-4-deoxy or 4-deoxy-4-guanidino-Neu5Ac2en (**40** and **41**, respectively) resulted in a marked decrease in the affinity of the enzyme in each case, with a 25,000-fold reduced affinity for **41** relative to Zanamivir.[78]

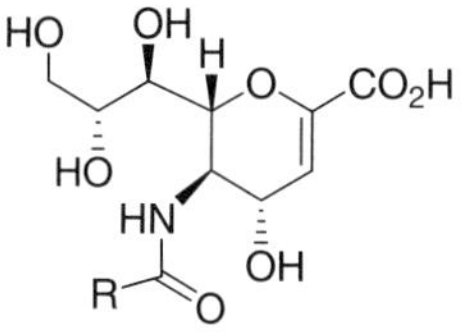

37 R = CF_3
38 R = CHF_2

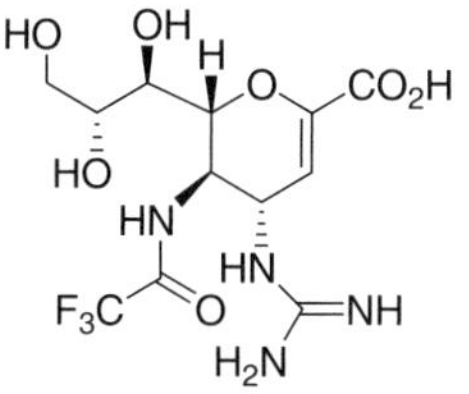

39

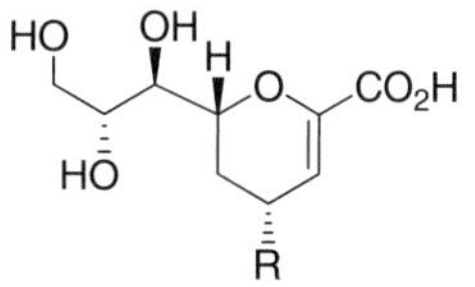

40 R = NH_2
41 R = $NHC(NH)NH_2$

In a study from *Vibrio cholerae*, a 1000-fold drop in inhibitory potency was observed for **42** relative to Neu5Ac2en, indicating that the acyl group contributes heavily to binding.[79] Interestingly, similar modification in the 4-amino-4-deoxy-Neu5Ac2en to give compound **43** and applied to influenza virus sialidase, inhibition was reduced only 10-fold relative to its *N*-acyl counterpart **9**.[65]

It was anticipated that successful replacement of the C-5 acetamide with alternative *N*-acyl groups would be governed by the stringent steric requirements imposed by subsite S3. These expectations led to the evaluation of a series of such inhibitors (**44–48**) related to Zanamivir and its 4-amino-4-deoxy precursor.[80]

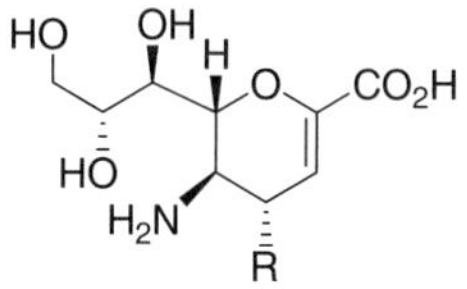

42 R = OH
43 R = NH_2

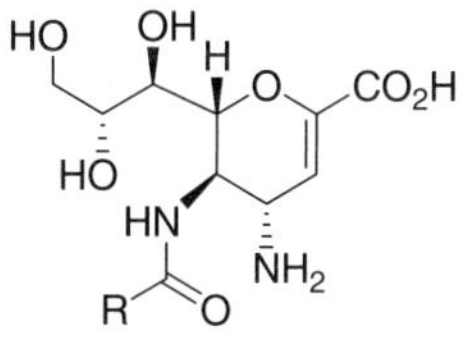

44 R = H
45 R = Et
46 R = cyclopropyl

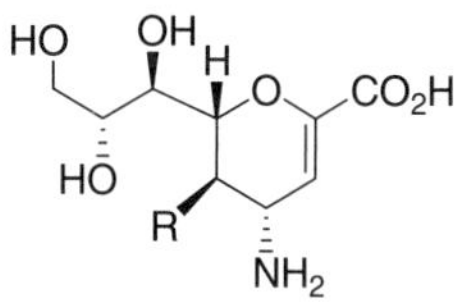

47 R = MeC(O)N(Me)
48 R = 2-oxopyrrolidin-1-yl

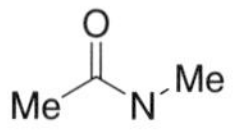

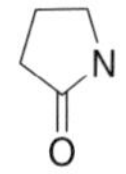

Replacement of the acetamido function on the 4-amino-4-deoxy-Neu5Ac2en scaffold with formamido (**44**), propanamido (**45**), or cyclopropylamido (**46**) groups resulted in a reduction in enzyme inhibition by factors of approximately 10, 100, and over 1000, respectively.

The replacement of the amido hydrogen atom with an alkyl group represented a potential means to displace a bound water molecule from the active site and achieve associated entropy-related enhancement of binding energy. However, *N*-methyl-*N*-acetyl (**47**) and cyclic lactam (**48**) derivatives proved highly ineffective against NA (IC_{50}~500 μM).[80] The installation of an ethanesulfonamide (**49**) or methanesulfonamide (**50** or **51**) in place of the acetamido group resulted in no improvement in activity.[80]

49 R^1 = Et, R^2 = NH_2
50 R^1 = Me, R^2 = NH_2
51 R^1 = Me, R^2 = NHC(NH)NH_2

d. C-6 Modifications.—Molecular-modeling studies demonstrated that the oxygen atom within the dihydropyran ring in Neu5Ac2en contribute marginally to ligand binding, and this position was thus targeted for modification.[41] The thio-isosteres (**52** and **53**) of 4-amino-4-deoxy and 4-deoxy-4-guanidino-Neu5Ac2en were prepared via a sequence analogous to that shown in Scheme 2 for the preparation of Zanamivir. The sulfur isostere **52** showed a reduction in activity of approximately 10-fold relative to **9**, while the 6-*S*-4-guanidino derivative **53** was found to have inhibitory activity comparable to that of the parent compound. This suggests that any impact of this heteroatom exchange on overall conformation is small. Carbocyclic analogues of Neu5Ac2en, formally substituted cyclohexenes, are discussed separately.

52 R = NH_2
53 R = NHC(NH)NH_2

e. C-7 Modifications.—The hydroxyl group at C-7 of Neu5Ac2en is the only polar group on this scaffold that has no direct interaction with the active site, and is exposed to bulk solvent. This observation allowed the development of a range of C-7-substituted derivatives of Zanamivir, in an effort to improve the physicochemical properties by mainly introducing lipophilic groups or attaching ligands to a multivalent scaffold.

The effect of the replacement of the C-7 hydroxy group of 4-deoxy-4-guanidino-Neu5Ac2en with hydrogen (**54**), fluorine (**55**), azide (**56**), or amine (**57**) has been evaluated. These derivatives were prepared through chemo-enzymatic synthesis from the corresponding C-4-modified *N*-acetylmannosamines and subsequent manipulation of the C-7-modified sialic acid to form the 4-deoxy-4-guanidino-Neu5Ac2en analogues. For the most part, these derivatives maintained both their ability to inhibit the enzyme and to prevent influenza virus replication *in vitro*.[81] In particular, the fluoro and azido derivatives **55** and **56** demonstrated superior characteristics to Zanamivir. Acylation of the amine **57** at C-7 proved to be effective inhibitors when shorter alkyl groups (<8 carbons) were employed. The sixteen carbon amide **58** and the *gem*-difluoro analogue of Zanamivir **60** saw the greatest reduction of potency both in enzyme inhibition and plaque-reduction assays (Fig. 8).

Modification of the substituent at C-7 in 4-deoxy-4-guanidino-Neu5Ac2en by O-alkylation has also been reported.[81,82] Ethers containing a linear alkyl chain ranging from one to fourteen carbons in length were evaluated against influenza

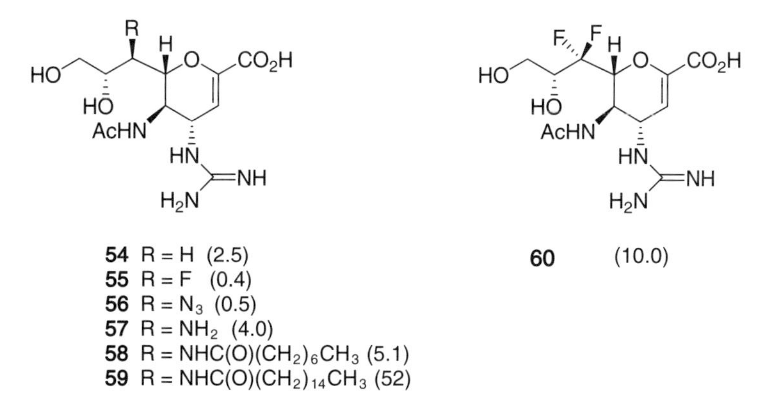

Fig. 8. Zanamivir derivatives modified by replacement of the C-7 hydroxyl. Values in parentheses reflect the relative ability of the compounds to inhibit influenza A virus sialidase. The reference value for Zanamivir **1** is 1.0.

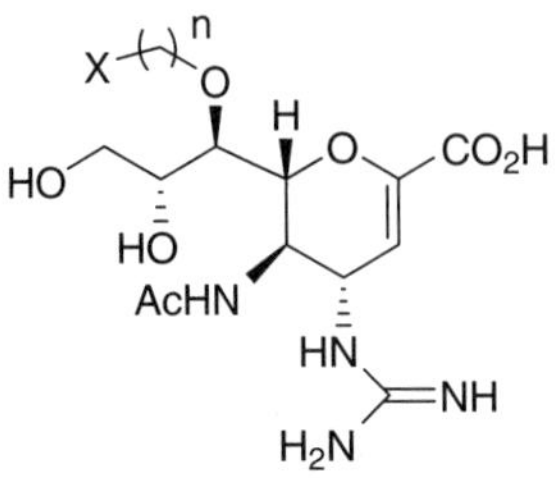

61 n = 2, X = OH (0.21)
62 n = 2, X = NH_2 (0.63)
63 n = 2, X = N_3 (0.13)
64 n = 2, X = NHAc (0.20)
65 n = 2, X = H (0.14)
66 n = 3, X = H (0.35)

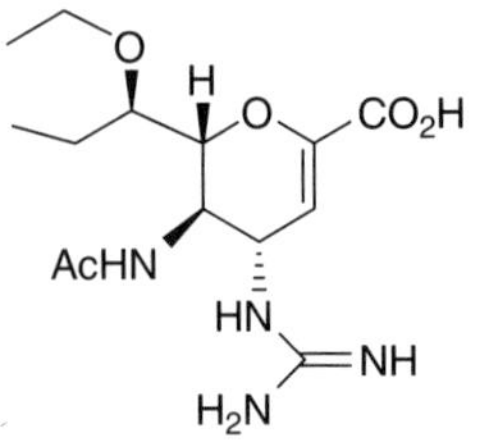

67 (1.9)

FIG. 9. Zanamivir derivatives modified through alkylation of the C-7 hydroxyl group. Values in parentheses reflect IC_{50} values against influenza A virus in a plaque-reduction assay, and are relative to a reference value of 1.0 for Zanamivir (**1**).

A virus in both sialidase inhibition and plaque-reduction assays. Alkyl ethers up to twelve carbon atoms in length exhibited similar inhibitory activity to Zanamivir against influenza A virus sialidase, however, showed a pronounced improvement in plaque-reduction assay compared to the parent triol **1**. Alkylation of the C-7 hydroxyl with two-carbon substituents bearing terminal hydroxyl, amino, azido, and acetamido groups yielded inhibitors **61**–**64** and did not significantly affect the binding and had similar potency to that of ethyl or propyl ethers **65** or **66** (Fig. 9).

Removal of the C-8 and C-9 hydroxy groups in order to produce a hydrophobic side-chain, gave such derivatives as **67**, which were evaluated against influenza A and B virus sialidases and used in a viral-replication assay.[82] While these molecules displayed similar efficacy to the parent triol against influenza A virus sialidase, they were relatively ineffective against influenza B virus.

A series of 7-*O*-carbamoyl derivatives of 4-amino-4-deoxy and 4-deoxy-4-guanidino-Neu5Ac2en, with predominantly hydrophobic *N*-alkyl groups, were prepared and screened for their ability to inhibit sialidase activity and viral replication (influenza virus A and B).[83] Acylation of the C-7 hydroxyl group with a carbamate linkage was chosen over an ester linkage, due to the known tendency of acyl migration along the glycerol side-chain of sialic acids. Potent inhibitors **68** and **69** were identified that rivaled the activity of Zanamivir **1** in enzyme assays, but proved slightly less effective in plaque-reduction assays (plaque IC_{50} influenza virus B: **68** 0.023 μg/mL, **69** 0.009 μg/mL, **70** 0.52 μg/mL, **1** 0.002 μg/mL). In general, monosubstitution of the carbamate nitrogen proved superior to disubstitution, and bulky and/or lipophilic *N*-substituents were inferior to those bearing polar or ionizable substituents. These results were rationalized in terms of interactions between the substituents and polar amino acid residues on the surface of the protein or with solvent. In keeping with trends observed for other Neu5Ac2en derivatives, 7-*O*-carbamoyl C-4 guanidino derivatives were significantly more potent inhibitors than the corresponding 7-*O*-carbamoyl C-4 amine derivatives. The ability to derivatize the C-7 hydroxyl group of Neu5Ac2en and related compounds without affecting binding to the NA active site has allowed for the development of a carbamate-based Zanamivir–biotin conjugate for potential application as an influenza diagnostic.[84]

68 $R^1 = H, R^2 = (CH_2)_6NH_2$
69 $R^1 = H, R^2 = (CH_2)_6CH_3$
70 $R^1 = R^2 = (CH_2)_2CH_3$

The numerous demonstrations that modification at C-7 of Neu5Ac2en-based inhibitors has minimal impact on binding suggested a starting point for the design of multimeric NA inhibitors. Both the carbamate and the ether linkages from the C-7 oxygen have been used to prepare multimeric conjugates of 4-deoxy-4-guanidino-Neu5Ac2en (**1**). These divalent to polyvalent assemblies varied with respect to the type and length of linker, symmetry, valency, and the presence or absence of a carrier molecule. Polymeric Zanamivir conjugates constructed using either a poly-L-glutamic acid or a modified polysaccharide backbone demonstrated decreased *in vitro* NA inhibition relative to the monomer.[35,85–87] However, these same conjugates displayed significantly greater antiviral activity than the monomer against influenza virus in cell-culture assays. An example of the ether-linked polymeric inhibitors (**71**) evaluated in a mouse model of influenza virus infection (intranasal administration) also showed better *in vivo* efficacy than Zanamivir (seven out of seven survivors at 10 days after infection, compared to one out of eight).

71

Multimeric conjugates of Zanamivir with defined valency have also demonstrated enhanced antiviral activity (100-fold increase), although their ability to inhibit NA in an enzyme assay was reduced by 10–20-fold relative to Zanamivir.[87] Among these, dimers formed by attachment of hydrocarbon linkers of varying sizes to the C-7 hydroxyl group have been particularly well studied.[88,89] While Zanamivir maintained greater activity than either of the dimeric conjugates **72** or **73** in isolated enzyme inhibition assays, the effect of the dimers on viral replication approached (**72**) or greatly exceeded (**73**) that of Zanamivir on a valency-corrected basis. Activity depended on the number of atoms bridging the two monomers, and from these data the means by which such conjugates attained higher potency against influenza virus could be inferred. Linkers comprised of between 14 and 18 atoms are able to affect intervirion and intersubunit cross-linking, resulting in viral aggregation and reduced infectivity. The introduction of more rigid linkers (**74**) resulted in a slight relative decrease in efficacy.

72 $(CH_2)_6$

73 $(CH_2)_{12}$

74 H_2C–C_6H_4–C_6H_4–CH_2

f. C-8 and C-9 Modifications.—Crystal structures of influenza virus sialidase in complex with Neu5Ac2en (**5**) identified interactions between Glu276 and hydroxyl groups at both C-8 and C-9, and hydrophobic contacts between the carbon backbone of the glycerol side-chain and Arg224.[38,39,48,62,90] The importance of the stereochemistry at C-8 and C-9 were determined by replacing the glycerol side-chain with a number of conformationally restricted cyclic ether moieties such as tetrahydropyran-2-yl, tetrahydrofuran-2-yl, and oxepan-2-yl groups. Not surprisingly, ethers (**75–77**), in which the positional and stereochemical arrangement of oxygen atoms mimicked that of the original side-chain triol, were effective inhibitors of NA activity and viral growth. The diol moiety was suggested through crystallography to engage in a bidentate hydrogen bond with Glu276, precluding rearrangement of the enzyme active-site. Although the activities of these diols against influenza B were not reported, they are expected to inhibit NA from either strain of the virus with comparable efficiency. The bicyclic inhibitor **76** was twofold more effective against influenza A virus than Zanamivir in a plaque-reduction assay and demonstrated to have good oral bioavailability in murine models of influenza infection. Diol **78** displayed enzyme inhibitory capacity that was reduced by 20 times relative to its positional isomer **76**. Inversion of configuration at C-7 (**79**) further reduced the potency of these compounds against influenza virus NA.

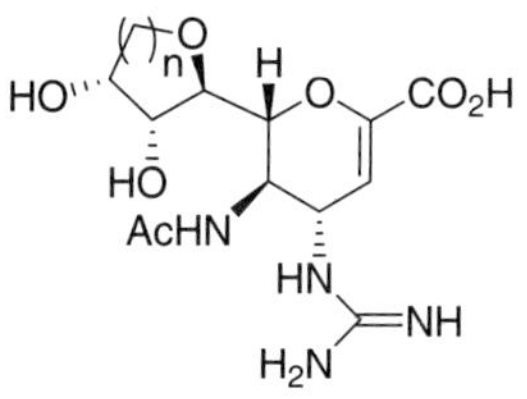

75 n = 1
76 n = 2
77 n = 3

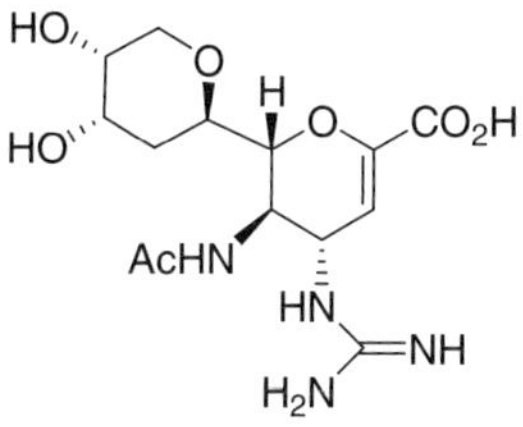

78

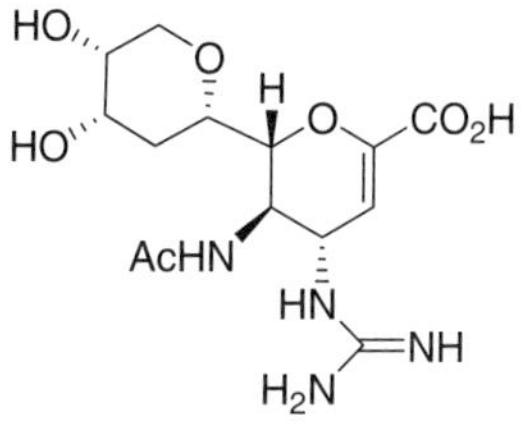

79

The importance of the C-9 hydroxyl group of Neu5Ac2en was also examined by replacing it with derivatives of nitrogen, with the expectation of significant electrostatic interaction between the carboxylate of Glu276 and a terminal amine/ammonium group. Both the amine **80** ($K_i = 400\,\mu M$) and diamine **81** ($K_i = 15\,\mu M$) demonstrated a reduction in inhibition on the order of 100–400-fold relative to their 9-hydroxy precursors **5** and **9** ($K_i = 4\,\mu M$ and $0.04\,\mu M$, respectively).[91] Computational evaluation of the observed binding-energies suggested that the major contributing factor to the weaker inhibition of the 9-amino derivatives **80** and **81** was the greater energy required to desolvate the ligand, relative to the 9-hydroxy analogues. A 9-amino-9-deoxy derivative (**82**) of trifluoroacetamide **38** was also prepared and displayed high-micromolar inhibition of influenza A virus sialidase activity ($K_i = 525\,\mu M$), a 100-fold change over that reported for the parent tetrol **38**.[58,92] Converting the C-9 amine to the acetamide **83** and trifluoroacetimidate **84** showed improved inhibition by 50–100-fold in comparison to the free amine.[92]

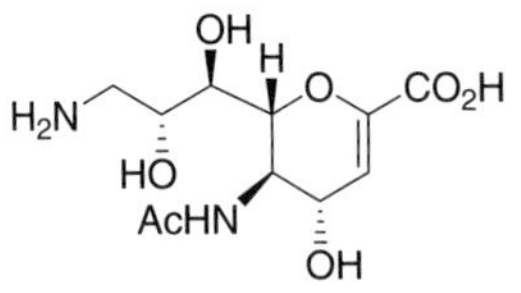

80 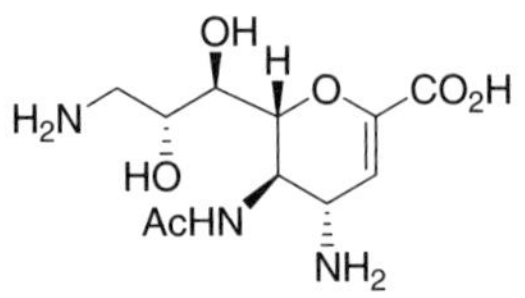 **81**

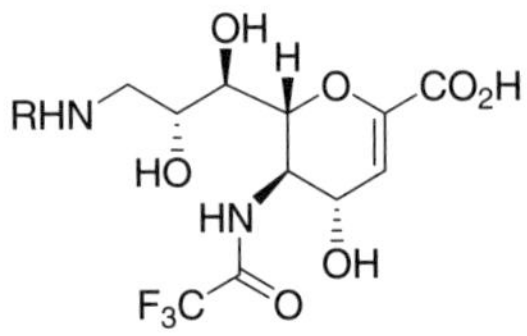

82 R = H
83 R = C(O)CH_3
84 R = C(O)CF_3

2. Mimetics of Neu5Ac2en and Zanamivir

a. Glycerol Side-Chain Modification.—A mutant NA strain in which the Glu276 carboxylate residue of subsite S5 was replaced (Glu276Gln) was shown to be devoid of hydrolytic activity, suggesting that its interaction with the diol motif is likely to significantly influence substrate binding.[93] Furthermore, the displacement of two water molecules from the NA active site by the C-7–C-9 arm upon binding highlights the potential contribution of this region to the entropic component of binding free energy.

Bamford and coworkers sought to determine the overall contribution to binding made by the C-7–C-9 triol section of 4-amino-4-deoxy and 4-deoxy-4-guanidino-Neu5Ac2en through iterative removal of hydroxymethyl units.[94] Six, seven, and eight carbon homologues (**85**–**90**) of 4-amino-4-deoxy-Neu5Ac2en (**9**) and its 4-deoxy-4-guanidino derivative (**1**) were prepared and assayed as inhibitors of influenza virus sialidase and of replication of influenza virus in cell culture (Fig. 10). Inhibition of enzyme activity was found, in an isolated NA(A) enzyme assay, to decrease in a stepwise manner with removal of each hydroxymethyl group. Seven- and eight-carbon Zanamivir homologues **90** and **89** displayed similar potencies to Neu5Ac2en (**5**, $IC_{50} = 9\,\mu M$) and 4-amino-4-deoxy-Neu5Ac2en (**9**, $IC_{50} = 0.32\,\mu M$), respectively, while complete removal of the side-chain resulted in considerable abrogation of inhibition.

85 R = NH_2 (13 μM)
88 R = NHC(NH)NH_2 (0.55 μM)
86 R = NH_2 (270 μM)
89 R = NHC(NH)NH_2 (9 μM)
87 R = NH_2 (>1000 μM)
90 R = NHC(NH)NH_2 (130 μM)

FIG. 10. Analogues of 4-amino-4-deoxy-Neu5Ac2en (**9**) and 4-deoxy-4-guanidino-Neu5Ac2en (**1**) modified by removal of C-9 through C-7. IC_{50} values for **85**–**90** against isolated influenza A virus sialidase are given in parentheses.

91
92 R= H
93 R= C_6H_5
94 X = NH_2
95 X = NHC(NH)NH_2
R^1, R^2 = H, alkyl, aryl

SCHEME 3. Synthesis of side-chain modified compounds **94** and **95**.

A number of C-6 carboxamide analogues were prepared in which the stereochemically complex glycerol side-chain of Zanamivir or 4-amino-4-deoxy-Neu5Ac2en was replaced entirely by secondary or tertiary amides with the general structure of **94** or **95**.[95,96] These series of compounds were prepared from the C-6 carboxylate **92** (prepared by periodate oxidation of the glycerol side-chain of **91**) by amide couling via the activated pentafluorophenyl ester **93** (Scheme 3).

Initial efforts focused on alkyl amides with polar termini (**96** and **97**) that were expected to maintain hydrogen-bonding interactions with Glu276 (Fig. 11). These amides demonstrated only weak inhibition of NA. The introduction of simple alkyl or aralkyl groups (**98**, **99**, or **100**) resulted in a drop in inhibitory activity of approximately 100-fold against influenza A virus, both in enzyme assay and in plaque-reduction assay.[96] However, the tertiary amides were considerably more potent inhibitors than their secondary counterparts (**99** versus **98**). More surprising was the observation that while some tertiary amides had IC_{50} values below 10 nM against influenza A sialidase, none of the compounds displayed IC_{50} values below 500 nM for the influenza B sialidase. Other lead

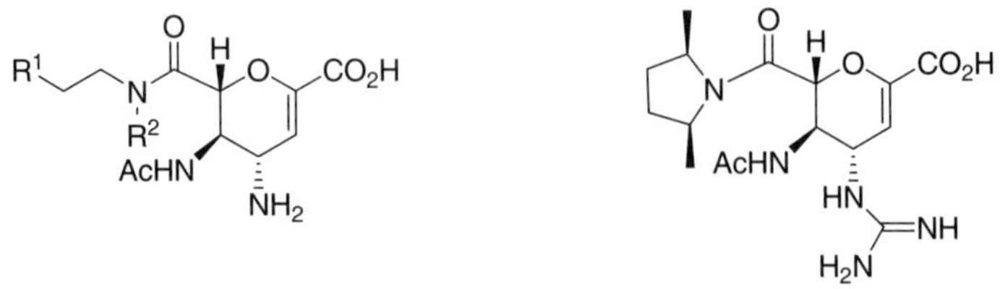

96 $R^1 = NH_2$, $R^2 = H$ 390 μM)
97 $R^1 = OH$, $R^2 = H$ 420 μM)
98 $R^1 = CH_3$, $R^2 = H$ 19 μM)
99 $R^1 = CH_3$, $R^2 = CH_3$ 0.18 μM)
100 $R^1 = Ph$ $R^2 = CH_3CH_3$ 0 005 μM)

101 0.005 μM)

FIG. 11. C-6 Carboxamide derivatives of 4-amino-4-deoxy-Neu5Ac2en (**9**) assayed as inhibitors of influenza virus sialidase. IC_{50} values against sialidase from influenza A are given.

compounds derived from an 80-member library prepared by automated synthesis included the 3,5-dialkyl pyrrolidinamides. Further investigation of this scaffold identified a strict requirement for the *cis*-orientation of alkyl groups, and 3,5-dimethylpyrrolidine derivative **101** offered the best NA inhibition.

4-Deoxy-4-guanidino variants of the C-6 amides generally displayed higher activity than their corresponding 4-amino-4-deoxy counterparts, but the degree of improvement was considerably lower than that observed between the analogous amine and guanidino derivatives of Neu5Ac2en. In keeping with these findings, modifications at C-4 were found to be tolerated well in the carboxamide series.[97] Most interestingly, there was virtually no difference between the *N*-(2-phenylethyl)-*N*-(1-propyl)carboxamides with a hydroxyl (**102**), an amine (**103**), a guanidino (**104**), or no substituent (**105**) at C-4 in inhibitory potency against influenza virus A sialidase. The axial alcohol **106** (17 nM) was only threefold less potent in an NA(A) inhibition assay than the epimeric equatorial alcohol **102** (7 nM) (Fig. 12). The equivalent pair of epimeric alcohols in the Neu5Ac2en series, **17** and **5**, were reported to vary in their ability to inhibit NA(A) by 20–50 times.[72]

Replacement of the C-5 acetamido substituent with trifluoroacetamide or propanamide was well tolerated. However, larger acyl groups hastened a significant drop in inhibition which again highlighted the tight steric requirements generally imposed by the residues that comprise subsite S3 surrounding this area.[97]

The differential inhibitory effect of the carboxamides on the sialidases from the influenza virus A and B was determined by extensive structure–activity

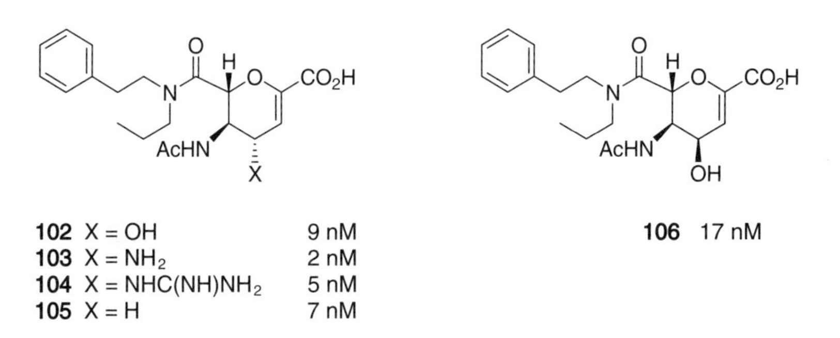

FIG. 12. C-6 Phenethylpropyl carboxamide derivatives of C-4-modified Neu5Ac2en (**5**).

studies, crystallography and molecular dynamics.[98,99] It was demonstrated that lipophilic amides induced a conformational reorganization of the binding pocket of both influenza A and B NA. Reorientation of Glu276 resulted in its engagement in a salt bridge with the guanidino side-chain of Arg224 and established a small lipophilic pocket, as well as an extended lipophilic cleft formed between Ile222 and Ala246 at the enzyme surface (Fig. 13). The relative sizes of these two regions readily explained the optimal substitution pattern identified by library screening, where amides with one large and one small (ethyl or propyl) substituent were found to be the best inhibitors of NA activity and viral replication.

The success of such C-6 amides in this series against influenza A NA was in part due to a match between the spatial arrangement of these two binding pockets and the favored orientation of the amide rotamer that featured a *trans*-relationship between the larger substituent and the pyran ring. In influenza B the formation of a salt bridge with Arg224 was still observed, however, reorientation of the Glu276 residue in influenza B NA induced considerable distortion of the protein backbone structure, in contrast to the minimal reorganization seen with influenza A, which is energetically less favorable and reflected in the lower binding affinities of the carboxamides.

A comparison of diethyl amide **107**, 3-pentyl ketone **108**, 3-pentyl ether **109**, and alcohol **110** was conducted in order to examine the effects of the side-chain conformation on both potency and selectivity in the carboxamide series of inhibitors (Fig. 14).[100] The ketone, ether, and alcohol resembled the amide in that all showed selective inhibitory activity towards influenza A NA. An sp^2-hybridized center was found to contribute to the potency of **107** and **108**.

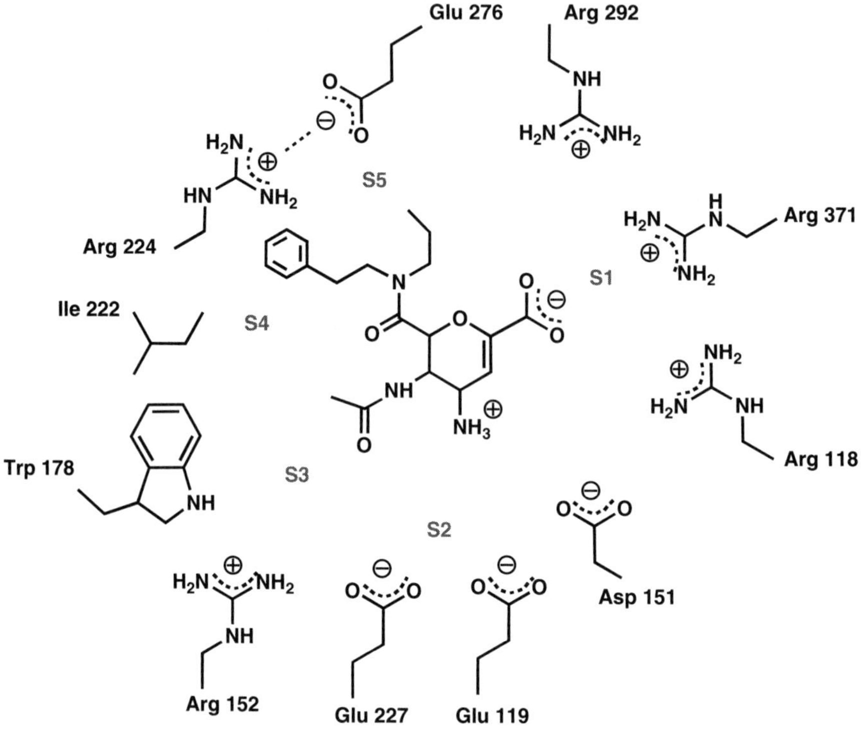

FIG. 13. A schematic representation of some key interactions of carboxamide **103** with conserved influenza virus A sialidase active site residues.

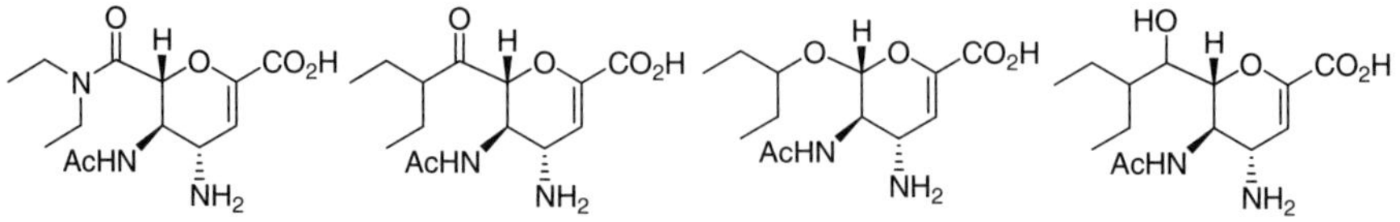

	107	**108**	**109**	**110**
IC_{50} NA(A) (nM)	3	2	770	39
IC_{50} NA(B) (nM)	360	470	42000	7400

FIG. 14. A comparison of influenza virus sialidase inhibition by C-6 diethylcarboxamide **107** with related C-6-keto, -ether, and -hydroxy derivatives **108–110**.

Presumably the ketone experiences less rotational restriction than the amide, but any enhancement in flexibility had little apparent effect on efficacy in this case. The relatively poor inhibitory capacity of the ether **109** was explained in terms of the energy penalty created by differences between the bound and solution conformation of the dihydropyran ring.

b. Heterocyclic Side-Chains.—Replacement of the glycerol side-chain with a number of substituted triazole and oxadiazole groups was selected as a potential means to realize affinity gains via enhanced hydrophobic contacts in the area encompassing subsites S4 and S5.[101] C-6 oxidiazoles (**111**) were prepared via the pentafluorophenyl ester of the Boc-protected 4-amino-6-carboxylate **93** by treatment with the appropriate amidoxime, and C-6 triazoles **112** were prepared via the hydrazide derivative of **93** by treatment with *S*-methyl thioamide and cyclization to the triazole. Both **111** and **112** again exhibited selectivity for influenza A virus sialidase, however, **112** proved to be a superior inhibitors of NA compared to **111**. On the basis of molecular-modeling studies and these enzyme-inhibition data the authors inferred that the heterocyclic side-chains induced the same active site redistribution generated by the C-6 amide series of inhibitors. It is also suggested that the differential inhibition demonstrated by the oxadiazole and triazole derivatives is analogous to the different activities of secondary and tertiary amides at C-6.

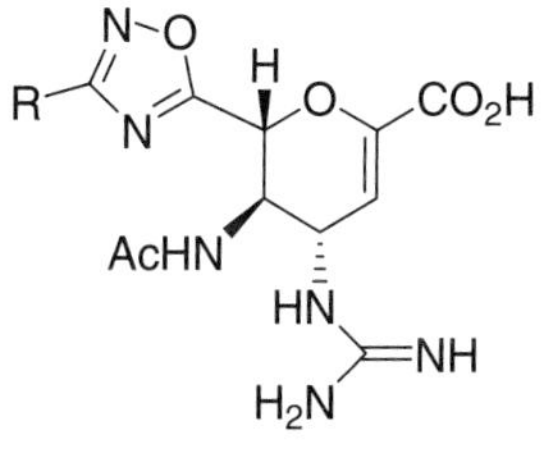

111
R = Me, Bu, Ph, Bn

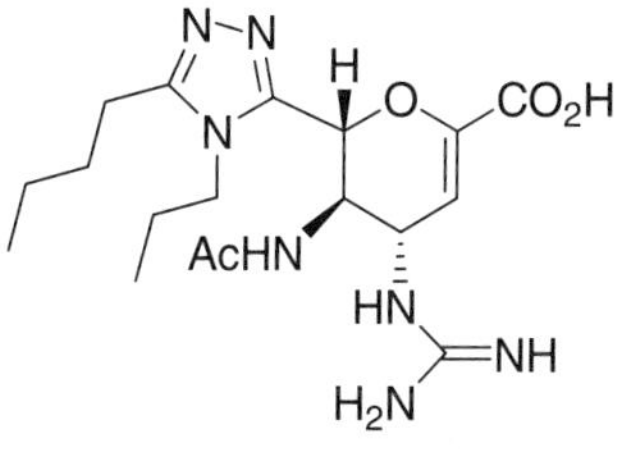

112

3. GlcNAc Glycosides as Mimetics of Neu5Ac2en

2-Acetamido 2,4-dideoxy-β-D-*threo*-hex-4-enopyranosiduronic acids (**113**) exhibit structural similarities to Neu5Ac2en (**5**), a feature that has been exploited in the design of a number of Neu5Ac2en mimetics.[100,102–104] The use of the relatively inexpensive 2-acetamido-2-deoxy-D-glucuronosiduronic scaffold in NA inhibitor design allows for complete replacement of the C-7–C-9 side-chain of sialic acid-based inhibitors with either functionalized or unfunctionalized side-chains.

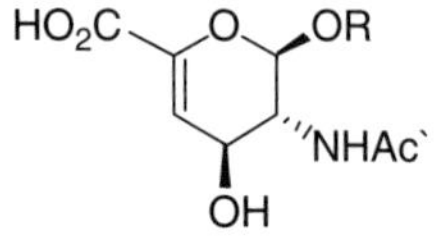

113

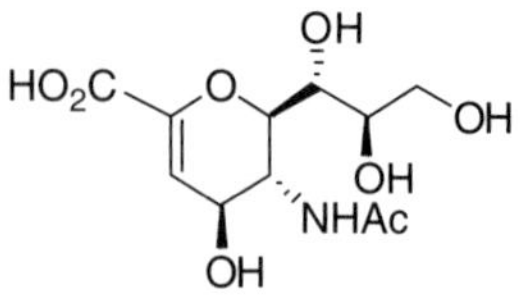

Neu5Ac2en, 5

A number of simple hydrophobic groups have been attached to the unsaturated 2-acetamido-2-deoxy-D-glucuronate template, and the glycosides (**114**–**117**) have been evaluated for their ability to inhibit influenza A NA from multiple sources (Fig. 15).[103,105] For the most part, candidates were found to inhibit the enzyme at a comparable level to Neu5Ac2en. Preliminary crystallographic findings suggest that binding of **114** is accompanied by a rearrangement of active-site residues similar to that observed for C-6 amides, triazoles, oxadiazoles, and GS4071 (see Sections 2 and 5b).[106] Docking studies suggest that the benzyl glycoside **117** does not interact significantly with the hydrophobic surface of the protein, and is instead oriented towards bulk solvent, accounting for its reduced potency.[103]

N-Acetyl-6-sulfo-D-glucosamine has also been advanced as a mimetic of Neu5Ac wherein the C-1 carboxylate is replaced by a sulfate ester at C-6 of D-glucosamine.[107,108] Molecular-modeling studies have suggested this apparent substrate mimetic approach may benefit from adoption by 6-sulfo-D-GlcNAc of a solution conformation that closely resembles enzyme-bound Neu5Ac.[108] A series of alkyl and aryl β-glycosides of 6-sulfo-D-GlcNAc were screened against influenza virus NA and were shown to inhibit the enzyme at micromolar concentrations. The supremacy of glycosides with hydrophobic aglycons in this series suggests that the inhibitor is bound in the expected orientation and that the aglycon may interact with hydrophobic regions of subsites S4 and S5.

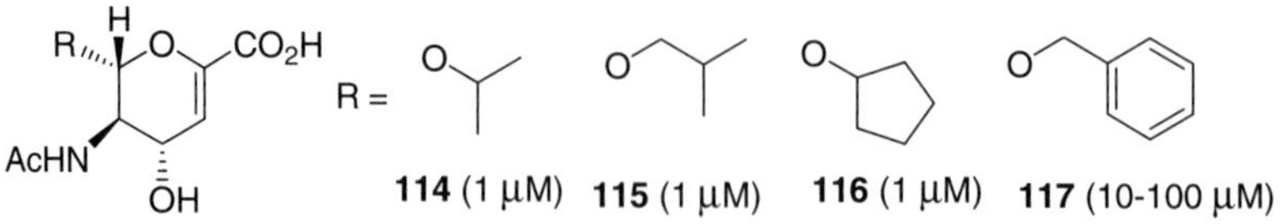

114 (1 μM) **115** (1 μM) **116** (1 μM) **117** (10-100 μM)

FIG. 15. 2-Acetamido-2,4-dideoxy-β-D-*threo*-hex-4-enopyranosiduronic acids **114**–**117** were evaluated for their ability to inhibit influenza A virus sialidase. Numbers in parentheses are approximate K_i values.

4. Influenza Virus Sialidase Inhibitors Based on an Aromatic Scaffold

The emerging knowledge about the promising effects of sialidase inhibitors in treatment of influenza infection initiated a variety of attempts to mimic *N*-acetylneuraminic acid by derivatives of benzoic acid.[109–116] The approach of using the benzene ring as a template was based on the observation that in the Neu5Ac2en (**5**)–NA crystal structure, all substituents attached to the dihydropyran ring are positioned in an equatorial position. Therefore, replacement of the dihydropyran ring with the benzene ring was predicted to generate minimal disturbance for the overall binding mode. In addition, the aromatic ring offered other benefits such as enhanced stability, reduced molecular complexity, and being synthetically much more accessible.

Early investigations into the Zanamivir analogue with the benzene template afforded compound **118**, which did not exhibit appreciable interaction with sialidase.[117] Interestingly, the benzoic acid analogue **119** (BANA113, BCX-140), without the glycerol side-chain, showed NA activity comparable to Neu5Ac2en and proved to be the lead compound. X-Ray structure determination of **119** bound to influenza virus sialidase revealed that the guanidino group in **119** was oriented in the glycerol binding-pocket at S5, engaged in charge–charge interactions with Glu276.[110–112] Efforts to occupy the S2 subsite by addition of a guanidino group at C-5 of the benzoic acid template (**120**) resulted in even worse inhibitory activity ($IC_{50} = 70\,\mu M$),[112] and attributed to inefficient interaction of the newly introduced guanidino group with Glu119 and Glu227. Addition of a hydrophobic residue at C-5 (**121**, R = $OCH(C_2H_5)_2$) led to activity ($IC_{50} = 3\,\mu M$) that did not exceed that of **119** ($IC_{50} = 2.5\,\mu M$).[114,116] Computational methods were used in the design of a second generation of benzoic acid-based NA inhibitors, including di-, tri- and tetrasubstituted derivatives.[111,114–116]

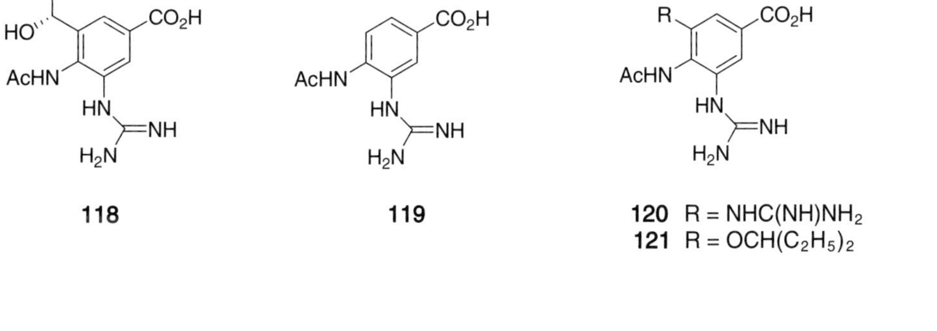

118 **119**

120 R = $NHC(NH)NH_2$
121 R = $OCH(C_2H_5)_2$

Investigations into tetrasubstituted benzene and pyridine derivatives led to a series of compounds such as **122** (4-*N*-acetyl-3-guanidino-5-hydroxyethylbenzoic acid).[118] Although it displayed poor *in vitro* activity ($IC_{50} = 0.4$ mM), analysis of the **122**–NA co-crystal structure revealed that the hydroxyethyl group (R^1) occupied subsite S5, the preferred site of the glycerol side-chain of Neu5Ac2en. Replacement of the hydroxymethyl substituent with an oxime afforded **123**, and resulted in a 10-fold increase in activity. The pyridine-based analogue of **119** was also prepared, and inhibition of NA(A) by **124** was found to be in the low μM range, but did not exceed the potency of the original carbocycle ($IC_{50} = 6$ μM for **124**, compared to $IC_{50} = 2.5$ μM for **119**). Both **119** and **124** were found to bind to NA in a similar orientation. The tetrasubstituted pyridine derivative **125** ($IC_{50} = 4$ μM) also showed no significant increase in activity compared to the benzene series. Overall the interactions between the pyridine nitrogen and the active site were not relevant to activity.[118]

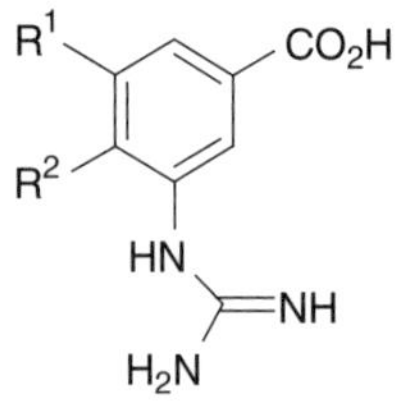

122 $R^1 = CH_2CH_2OH$, $R^2 = NHCOCH_3$
123 $R^1 = CH{=}NOH$, $R^2 = NHSO_2CH_3$

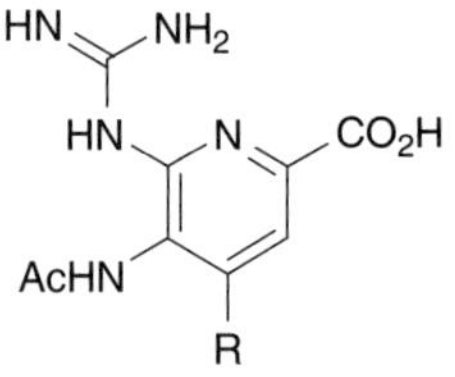

124 R = H
125 $R = NH_2$

Replacement of the guanidino group of **119** with various 2-aminoimidazole residues gave structures of type **126**.[113] The activity of the reported compounds (R = aromatic) was only moderate, though comparable to **119**, suggesting that neither the imidazole ring nor its substituents achieved additional binding contacts with NA. However, X-ray analysis of **126** (R = Ph) with influenza B virus sialidase revealed that in addition to the expected contacts made between the 2-amino-imidazole nitrogen atoms and those residues that bind the guanidino group in **119**, the phenyl group occupied a hydrophobic cleft formed by Ile220 and Ala244.[113]

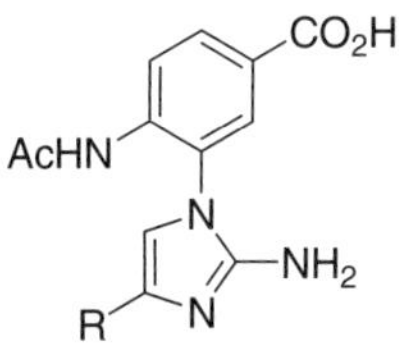

126

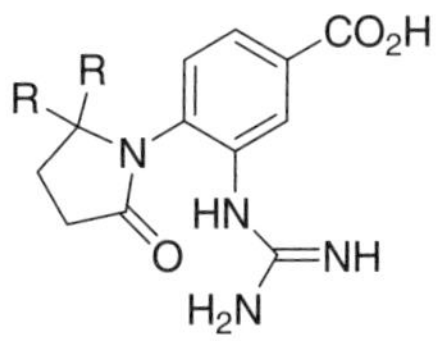

127 R = H
128 R = CH_2OH

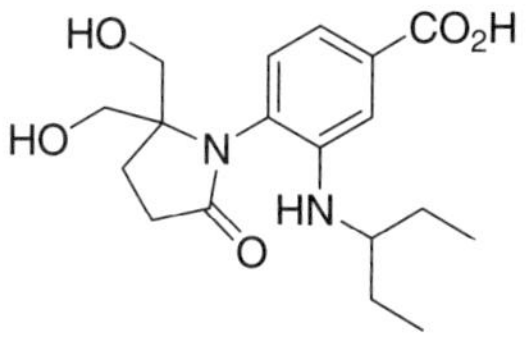

129

Replacement of the C-4 acetamido moiety of **119** with, for example, $-NHSO_2CH_3$, $CONHCH_3$, SO_2NH_2, or $-CH_2SOCH_3$ did not enhance activity, although these groups were each found to bind to the same pocket as the acetamido group, albeit in a different manner.[111] Another group of benzoic acid-based sialidase inhibitors in which the acetamide of **119** was replaced with substituted 2-pyrrolidinones was disclosed. Compound **127** was 10–15-fold less active than its acetamido counterpart **119**, consistent with the loss of a hydrogen-bonding interaction from the acetamide NH to an ordered water molecule. Introduction of a hydroxymethyl substituent onto C-5′ of the pyrrolidinone to give compound **128** resulted in lower μM activity against both NA(A) and NA(B) ($IC_{50} = 5$ and $8\,\mu M$, respectively).[119] X-Ray crystallography indicated that the methylene groups of the 2-pyrrolidinone moiety interacted with the hydrophobic portion of the S3 pocket, the hydroxymethyl group displaced a buried water molecule buried in S2, and the guanidino group remained bound in subsite S5. From these findings it was anticipated that exchange of the polar guanidino substituent with a more lipophilic function might result in effective inhibitors, owing to the potential for nonpolar contacts in S4 and S5. Replacement of the guanidino group in **128** by a hydrophobic 3-pentylamino group gave **129**, a 100-fold more potent inhibitor of NA(A) ($IC_{50} = 0.048\,\mu M$), though an ineffective inhibitor of NA(B) ($IC_{50} = 104\,\mu M$). Analysis of the X-ray structure of the NA(A)–**129** complex showed conformational reorganization of Glu276 to form an intramolecular salt bridge with Arg224, as previously observed in the case of the C-6 carboxamide series of Neu5Ac2en analogues. This arrangement resulted in a small hydrophobic pocket allowing one branch of the pentyl chain to bind, while the other half interacts with the existing, extended lipophilic cleft of subsite S4 created by Ala246 and Ile222. It is believed that an entropy gain as a result of these hydrophobic interactions is mainly responsible for the enhanced NA(A) inhibitory activity.[116] For influenza B NA these rearrangements result in distortion of the protein backbone near Glu276, consequently lowering activity.[98]

Although a large number of candidates were synthesized, none of the resulting compounds demonstrated significantly increased activity against influenza virus sialidase. Moreover, the interactions of individual substituents on the benzene ring with the active site were not found to be additive. The overall interaction of the molecules with the active site of the enzyme was dependent upon the electronic and steric interaction of each unique substituent, which made the design of inhibitors difficult.[111] No compound of this family has proceeded to clinical trials.

5. Sialidase Inhibitors Based on a Cyclohexene Scaffold

a. Background.—The tremendous interest in synthesizing cyclohexene-based sialidase inhibitors was mainly ignited by the possibility to mimic Neu5Ac2en and its improved analogue Zanamivir, while at the same time overcoming the pharmacological shortcomings of the latter compounds regarding bioavailability. The similarity of the ring conformation in Neu5Ac2en (**5**) to the proposed ring-oxygen stabilized sialyl-cation transition-state (**4**) is thought to contribute heavily to the affinity of the enzyme for this molecule. However, a comparison of Neu5Ac2en and the putative intermediate **4** reveals that in fact these two structures are essentially positional isomers, which led to the evaluation of both isomers **130** and **131** in the context of a carbocyclic scaffold. While the Neu5Ac2en mimetic **130** exhibited a nearly 20-fold increase in IC_{50} (850 μM) relative to Neu5Ac2en itself (42 μM), the isomeric carbocycle **131** (20 μM) showed improved inhibition relative to Neu5Ac2en.[120]

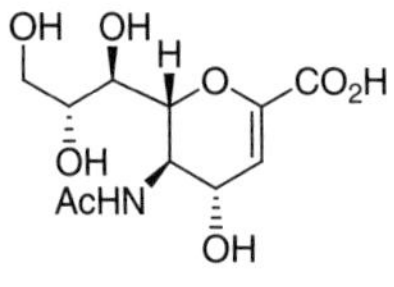

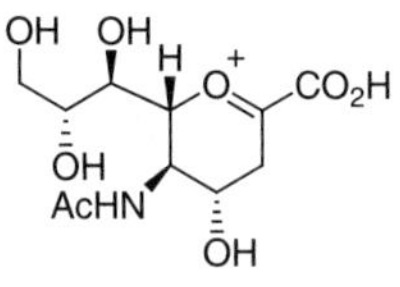

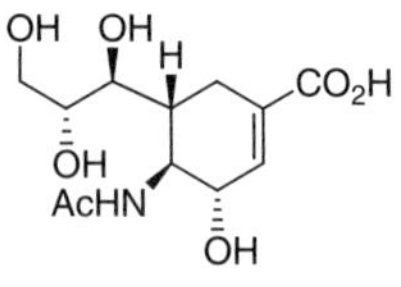

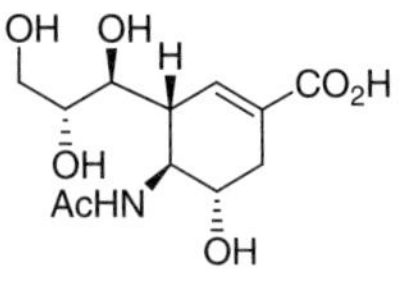

Comparative early work on the influenza virus sialidase inhibition of the cyclohexene isomers **132** and **133** revealed that **133** was a much more potent compound (IC_{50} = 6.3 μM) than its isomer **132**, which failed to demonstrate inhibition of the enzyme at concentrations up to 200 μM, supporting the significance of the position of the unsaturation.[121] Interestingly in seeming contrast to the results obtained for **132** and **133**, carbocyclic analogues of both 4-amino-4-deoxy-Neu5Ac2en and Zanamivir (**134** and **135**) bearing a double bond between C-2 and C-3 (Neu5Ac numbering) and a truncated glycerol side-chain demonstrated activity that approached or exceeded that of their truncated dihydropyran counterparts **87** and **90**.[68] Although the amine-substituted carbocycle **134** (K_i~900 μM) was an inferior inhibitor of sialidase activity relative to **135** (K_i~8 μM), it was the most potent of the four compounds **134**, **135**, **87**, and **90** in plaque-reduction assays.

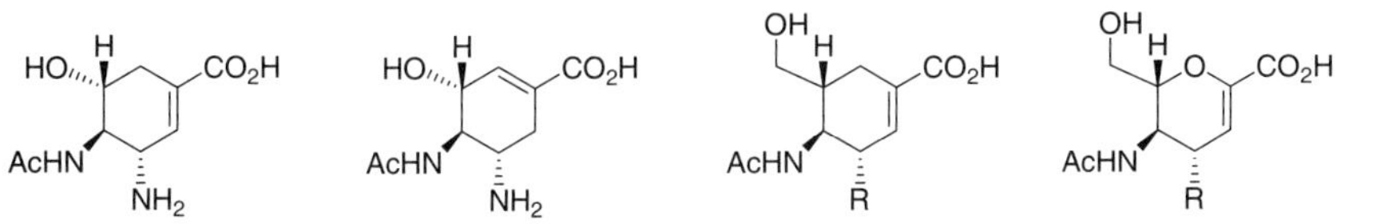

132 **133** **134** R = NH2 **135** R = NHC(NH)NH2 **87** R = NH2 **90** R = NHC(NH)NH2

b. Side-Chain Modification.—In an effort to introduce more hydrophobic character to the molecule with the hope of designing an inhibitor with oral bioavailability, the glycerol side-chain was replaced with an alkyl substituent, joined to the scaffold via an ether linkage for synthetic convenience. The strategy employed for synthesizing compounds such as **139** started with shikimic acid (**136**) and employed the aziridine derivative **137** as the key intermediate (Scheme 4). Exposure of the aziridine **137** to alcohols in the presence of $BF_3 \cdot Et_2O$ and subsequent amine acetylation gave azido derivative **138**. Reduction of the azide in **138** and ester saponification resulted in the required carbocylic derivatives **139**.

Systematic functionalization of the hydroxyl group of lead inhibitor **133** with various alkyl chains allowed for the introduction of hydrophobic groups designed to interact with nonpolar regions of subsites S4 and S5. In every reported case, alkyl ether derivatives were superior inhibitors to the alcohol.[121] Analysis of a range of inhibitors led to the conclusion that the length, branching, and

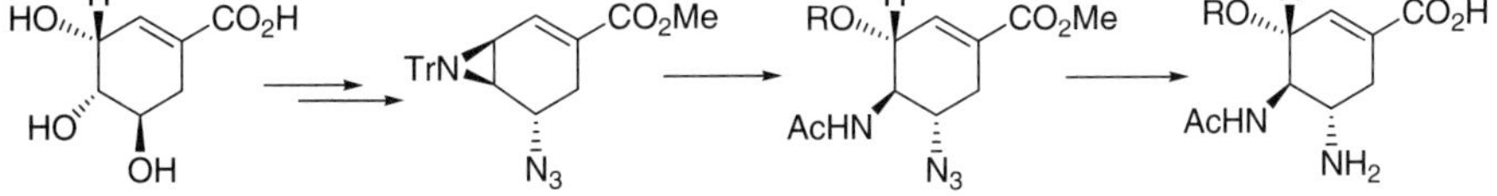

136 **137** **138** **139**

SCHEME 4. Synthesis of compounds such as **139** starting from shikimic acid (**136**) employing the aziridine derivative **137**.

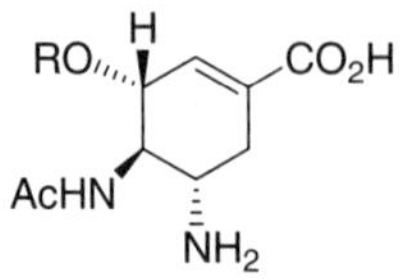

	R	
140	$-CH_3$	(3.7 μM)
141	$-CH_2CH_3$	(2.0 μM)
142	$-(CH_2)_2CH_3$	(0.2 μM)
143	$-(CH_2)_3CH_3$	(0.3 μM)
144	$-(CH_2)_5CH_3$	(0.2 μM)

Fig. 16. Influenza virus sialidase inhibitory activity of carbocyclic analogues with linear alkyl chains.

stereochemistry of the substituent defined the compound's activity. The activity of alkyl ethers **140–144** based on **133** was augmented with increasing length up to three carbon atoms (Fig. 16). Further extension of the alkyl chain beyond *n*-propyl length did not improve the influenza A activity and greatly decreased the NA(B) activity. The 20-fold increase in the NA activity of the *n*-propyl analogue **142** compared to that of the methyl derivative **140** implicated a significant hydrophobic interaction of the propyl group in the glycerol portion of the NA-binding pocket.

The impact of introduction of a branch in the side-chain was dependent on its position, as is evident from an analysis of four-carbon alkyl ethers **145–149**. Addition of a methyl substituent at the β-carbon of the *n*-propyl ether **142** yielded no appreciable change in inhibition (**145**). However, introduction of the same substituent at the α-carbon resulted in a 20-fold decrease in the IC_{50} (**146a** or **146b**). Addition of a single methyl group to **146** gave the isopentyl ether **147**, also known as GS4071, a remarkably potent inhibitor of NA from influenza A and B ($IC_{50} = 1$ nM against NA(A)). Further elongation of both alkyl chains (**148**) resulted in a decrease in potency ($IC_{50} = 16$ nM). However, the activity was restored when only one arm of the bifurcated side-chain was elongated (**149**, $IC_{50} = 1$ nM).

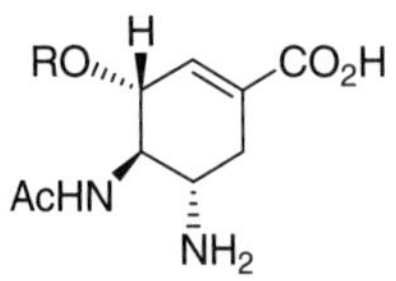

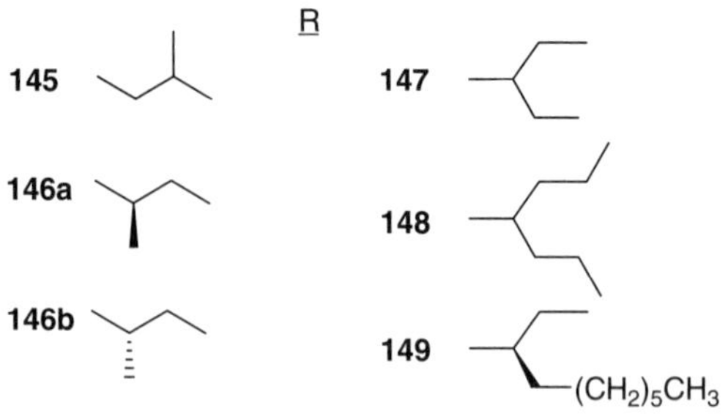

R

145

146a

146b

147

148

149 $-(CH_2)_5CH_3$

The crystal structures of GS4071 (**147**) and related molecules in complex with influenza virus sialidase have shed light on the origins of their observed high-affinity interactions and provided explanations for the variable success of the wide range of alkyl ethers examined.[121,122] The cyclohexene ring is oriented within the NA active site such that the carboxylate, acetamide, and C-4 substituent interact with the NA active site similar to those observed for Neu5Ac2en and derivatives, as expected. However, on binding of **147**, Glu276 adopts an alternative conformation that is stabilized by salt-bridge formation to Arg224, as reported for the carboxamide series of Neu5Ac2en inhibitors.[99] The two ethyl groups of **147** are accommodated by subsites S4 and S5, the latter of which is rendered considerably less polar by engagement of Glu276 with Arg224 (Fig. 17).

An examination of the inhibitory effects of numerous similar ethers containing larger alkyl groups and an asymmetric center has shed light on the active-site rearrangement and subsite geometry. Subsite S4 of influenza virus NA type A or B is able to accommodate smaller groups, such as an ethyl substituent. Larger substituents are bound in S5 by the enzyme from influenza A, but not influenza B. This finding is born out in the discrimination between stereoisomeric ethers by NA(A) and by the greatly diminished levels of inhibition of NA(B) observed when a large ether substituent is present. Both of these situations are illustrated by an examination of the inhibitory effects of diastereomeric inhibitors **150** and **151**. While these are both potent inhibitors of influenza A NA, the IC_{50} for the (*R*) isomer **151** is 40 times higher than for its (*S*) stereoisomer **150**. Crystallographic evidence indicates that the bulky phenethyl group is accommodated by S5 but not S4. Due to the side-chain asymmetry, **150** is better suited to this arrangement than **151**. Since Glu276 in NA(B) undergoes a much smaller conformational change upon binding of this series of inhibitors, S5 is not able to accommodate larger hydrophobic groups. As a consequence, substituents with one or more bulky group are selective inhibitors of influenza A virus NA. Even the less bulky 4-hex-1-ene ethers (**152**) display pronounced differences in their ability to inhibit the two enzyme subtypes.

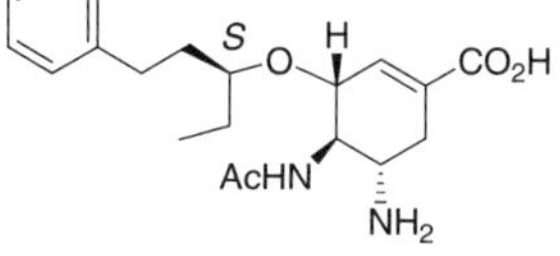

S O H CO_2H AcHN NH_2

150

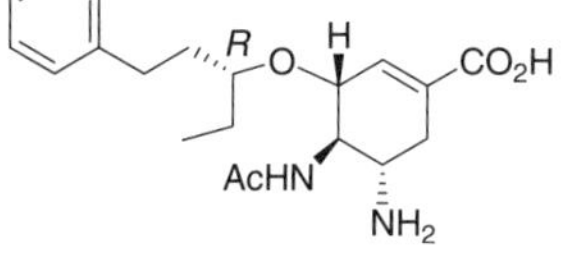

R O H CO_2H AcHN NH_2

151

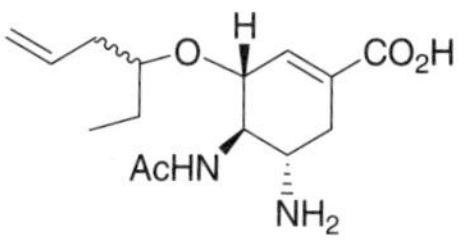

O H CO_2H AcHN NH_2

152

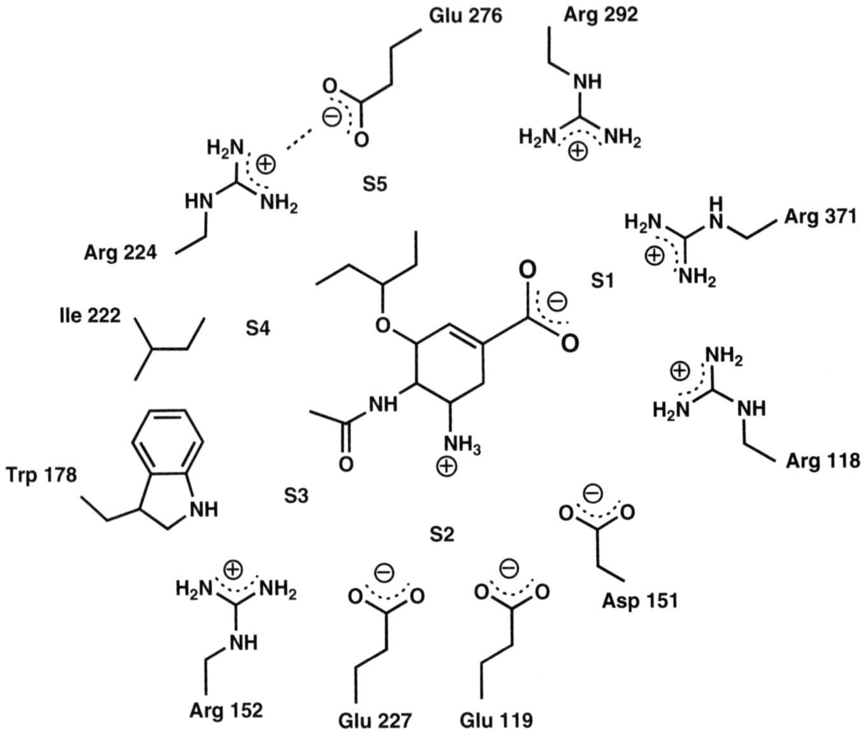

FIG. 17. A schematic representation of some key interactions of GS4071 **147** with conserved influenza virus A sialidase active-site residues.

The importance of the ether oxygen in **142** to its NA inhibitory properties was investigated via synthesis of the thia-, carba-, and aza-isosteres **153–157** (Fig. 18).[123,124] These new compounds produced similar levels of inhibition in enzyme assays against influenza A and B NA to the oxygen counterpart. This result confirms that the linker atom may not be part of the inhibitor interaction. The aza analogue (**156**) of GS4071 and a related tertiary amine **157** were also synthesized and evaluated as inhibitors of influenza virus NA. Although both compounds proved to be excellent inhibitors of enzyme activity, their potency was reduced by up to one order of magnitude relative to **147**.

The combined effects of a carbocyclic scaffold and replacement of the glycerol side-chain with an amide group were also investigated.[125] Neither GS4071-like

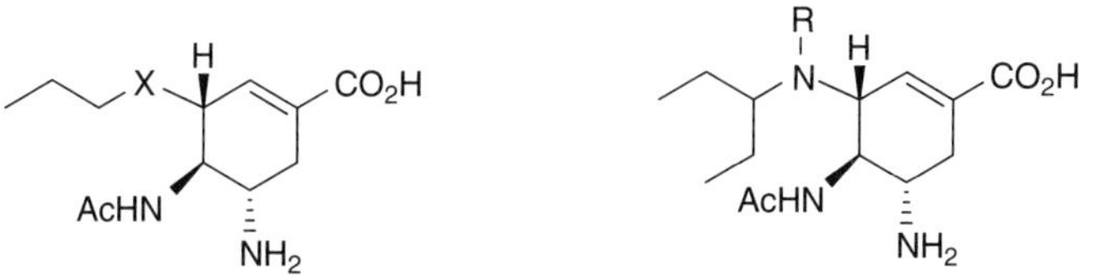

142 X = O (130 nM)
153 X = S (212 nM)
154 X = CH_2 (220 nM)
155 X = NH (200 nM)

156 R = H (11 nM)
157 R = Me (6 nM)

FIG. 18. Thia, carba, and aza analogues of cyclohexene ethers were assayed for inhibition of influenza A virus sialidase. IC_{50} values are reported in parentheses.

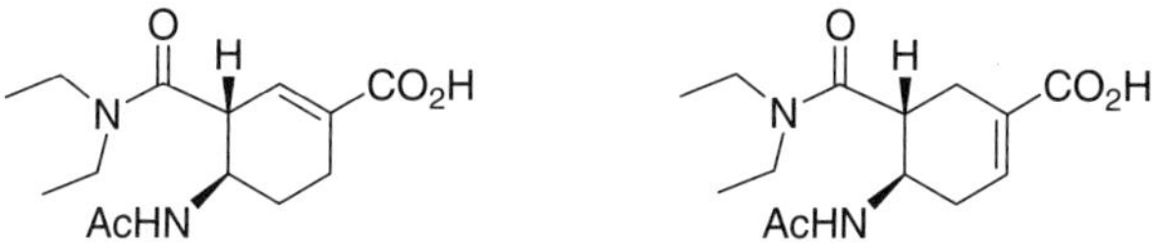

158
(0.017 μM, 23 μM)

159
(0.21 μM, 150 μM)

FIG. 19. Inhibition of influenza virus A and B sialidases by cyclohexene-based amides **158** and **159**. Numbers given in parentheses are IC_{50} values for NA(A) and NA(B), respectively. The IC_{50} values obtained for **1** under the same conditions were 0.005 μM for NA(A) and 0.004 μM for NA(B).

158 nor Zanamivir-like **159** were as effective against NA(A) as Zanamivir **1**, though these positional isomers lacked a substituent at C-4 for interaction with the S2-binding pocket (Fig. 19). Notably, the position of the double bond had a large impact on the effect of the inhibitor on NA activity, and both inhibitors exhibited selectivity for influenza A virus NA, as previously observed in the C-6 amide series.

c. C-5 Modification.—Functional-group modifications in the Neu5Ac2en series of NA inhibitors demonstrated the strict steric and electronic limitations imposed on the C-5 nitrogen substituent by the amino acid residues comprising subsite S3. Several acyl and sulfonamide derivatives of carbocycle **142** were synthesized and assayed for their ability to inhibit NA activity and viral replication *in vitro*.[126] The trifluoroacetamide derivative **160** showed moderate improvement over the acetamide in an enzymatic assay, but was a drastically less effective inhibitor of plaque formation. The increased steric demand of the

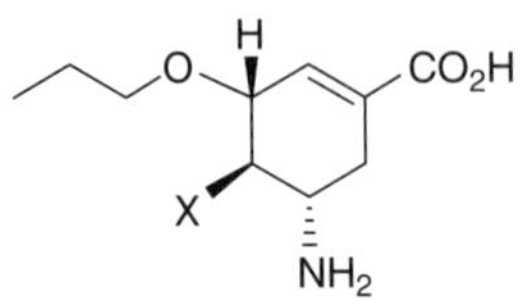

X
142 NHAc (130 nM)
160 $NHC(O)CF_3$ (100 nM)
161 $NHC(O)CH_2CH_3$ (1500 nM)
162 $NHSO_2CH_3$ (2500 nM)

FIG. 20. A comparison of influenza virus A sialidase inhibition by derivatives of **142** with various C-5 nitrogen substituents. IC_{50} values are given in parentheses.

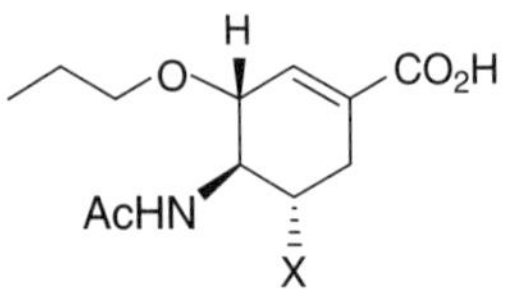

142 X = NH_2 (130 nM)
163 X = NHC(NH)H (140 nM)
164 X = $NHC(NH)NH_2$ (1.8 nM)
165 X = $NHC(NMe)NH_2$ (4.6 nM)

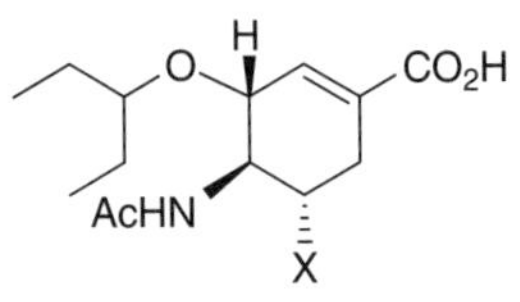

147 X = NH_2 (1.0 nM)
166 X = $NHC(NH)NH_2$ (0.5 nM)

FIG. 21. A comparison of influenza virus A sialidase inhibition by derivatives of **142** and **147** modified at C-4. IC_{50} values are given in parentheses.

propanamide **161** is a likely explanation for its reduction in activity by 10-fold relative to **142**. More surprising was the approximately 200-fold decrease in activity following replacement of the acetamide group with a methanesulfonamide (**162**), given the relatively small change in enzyme inhibition seen in the 4-amino-4-deoxy and 4-deoxy-4-guanidino-Neu5Ac2en series (**51**,**52**) (Fig. 20).

d. C-4 Modification.—In the Neu5Ac2en series, conversion of the 4-amino-4-deoxy function into a 4-deoxy-4-guanidino group imparted a 100-fold increased potency in NA inhibitory activity (**9** versus **1**).[62] Modification of the cyclohexene ether series by conversion of the amine to a guanidino group also produced affinity gains ranging from approximately 2-to 100-fold.[122,126] The effect of modifications at C-4 is illustrated by the propyl ether series in Fig. 21. Replacement of the amine with amidine (**163**), guanidino (**164**), or *N*-methylguanidino (**165**) groups resulted in a considerable enhancement in inhibition of

NA in the latter two cases. However, the guanidine analogue of GS4071 had only modestly improved inhibition (compare **166** with **147**). The origins of this differential improvement are unclear, but suggest that the nature of the alkyl ether component affects the hydrogen-bonding network in S2.

e. C-1 Modification.—Despite the intuitively apparent enhancement of lipophilicity accompanying replacement of the glycerol side-chain in the cyclohexene-ether series, both the guanidino derivative **166** and amine **147** displayed poor oral bioavailability, probably as a consequence of their zwitterionic character. However, conversion of **166** and **147** into the corresponding ethyl ester prodrug proved successful only in the latter case.[127] The ethyl ester of **147** was found to display considerably superior pharmacokinetic properties to the parent carboxylic acid. Oseltamivir **2**, the orally administered ethyl ester form of GS4071 (**147**), is now produced under the trade name Tamiflu™ by Roche.[128] The development of new synthetic routes for the production of Oseltamivir remains the subject of intensive research in academic and industrial settings, since demand currently outstrips production capacity.[129–132]

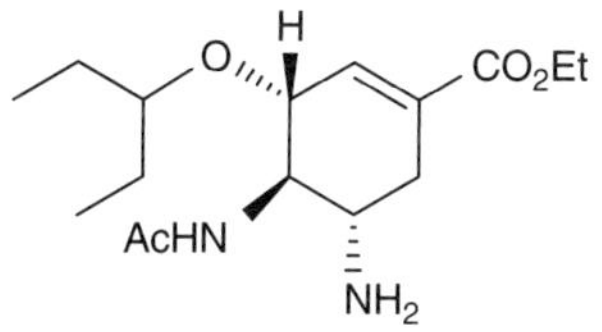

Oseltamivir, **2**

f. Miscellaneous Modifications to the Cyclohexene Scaffold.—Cyclohexene-based inhibitors, in contrast to the dihydropyran scaffolds, allow for addition of substituents at each position around the ring. Methyl and fluoro groups were appended to the carbon that replaces the oxygen ring of Neu5Ac derivatives, as shown in Fig. 22.[122] Methylation resulted in a dramatic decrease ($>$2000-fold) in the ability of this derivative (**167**) of **147** to interfere with catalysis. The less sterically demanding fluoro-substituted cyclohexene **168** retained its activity for the most part. A methyl group was appended to C-3 in an effort to capitalize potential hydrophobic contacts in this region.[122] Instead, inhibition by **169** was diminished by over three orders of magnitude relative to **147**. Cyclohexane-based substrate mimetics such as **170** have shown greatly reduced ability to inhibit NA relative to the cyclohexene transition-state analogues (Fig. 22).[133]

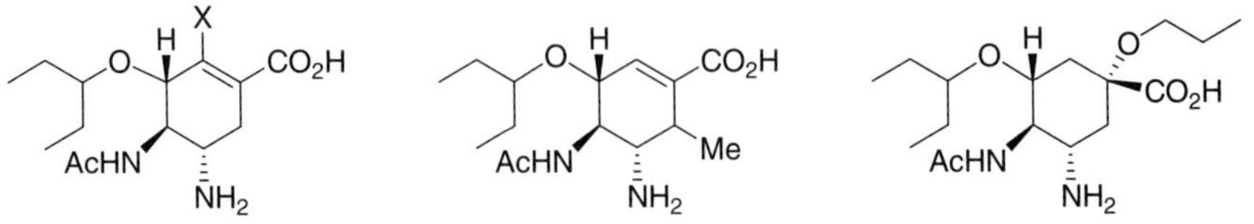

147 X = H (1 nM)
167 X = Me (2 μM)
168 X = F (3 nM)

169 (1.5 μM)

170 (75 μM)

FIG. 22. Inhibitory activity of GS4071 **147** and related compounds against influenza A virus sialidase. IC_{50} values are given in parentheses.

6. Sialidase Inhibitors Based on a Five-Membered Ring Scaffold

Whereas cyclohexene or benzene ring systems are intuitively apparent mimics of Neu5Ac2en and related inhibitors, it is less obvious that five-membered ring systems can play the same role. However, it has been demonstrated that compounds based on different five-membered cyclic scaffolds can engage each of the identified subsites (S1–S5) of influenza virus NA, affording potent inhibitory activity.

a. Cyclopentane Derivatives.—The nine-carbon furanose **171**, one of the first five-membered ring inhibitors of influenza virus NA described that demonstrated comparable activity to Neu5Ac2en, **5**.[134] Comparison of the crystal structures of **171** and Neu5Ac2en bound to N9 influenza virus sialidase revealed that the key functional groups (carboxylate, glycerol side-chain, *N*-acetyl group, and C-4 hydroxyl group) in both complexes shared similar orientations within the active site, despite the fact that the space occupied by the furanose ring differed from that filled by the dihydropyran backbone.[135] This observation reinforced the notion that the inhibition of influenza virus sialidases is highly dependent on the relative position of the pendant functional groups within the active site rather than the absolute position of the central ring.

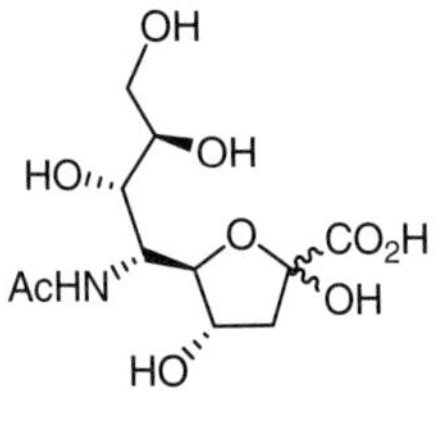

171 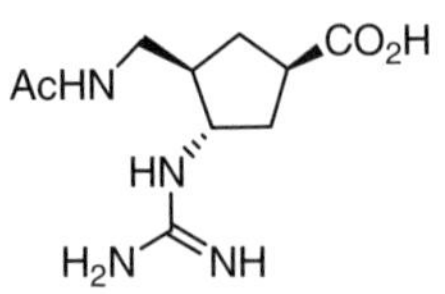 **172** 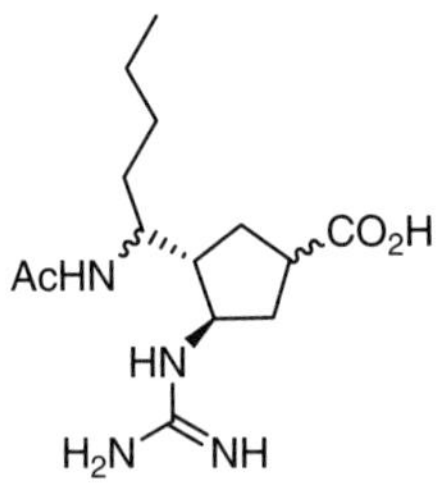 **173**

The encouraging NA inhibitory properties of nonulosonic acid **171** served as a starting point for the structure-based design of a new class of sialidase inhibitors at BioCryst Pharmaceuticals.[135] Evaluation of racemic **172** in complex with sialidase identified a single active isomer. As expected, the interactions of the guanidino group with the enzyme in S2 were found to parallel those observed for Zanamivir. This trisubstituted structure showed 10-fold greater inhibition of influenza A sialidase than did Neu5Ac2en. Further development efforts led to **173**, containing an *n*-butyl side-chain designed to generate additional contacts with the small hydrophobic surface in the active site formed by Ala246, Ile222, and the side-chain of Arg224 (S4). To determine the identity of the active form of **173**, a stereoisomeric mixture was soaked into crystals of N9 sialidase. Surprisingly, it was found that the absolute configuration at C-4 was *R* in compound **173** and *S* in compound **172**. As a consequence, the guanidino groups are arranged differently in the S2-binding subsite and the hydrophobic *n*-butyl side-chain of **173** adopts different positions in the active site of influenza virus sialidase A compared to B, as was reported for other Neu5Ac type inhibitors bearing hydrophobic groups in place of the glycerol side-chain.[99,121] Compound **173** showed IC_{50} values of 50 nM and 900 nM against NA(A) and NA(B), respectively.[135,136] The *n*-butyl side-chain of **173** was replaced by a 2′-ethylbutyl group, intended to take advantage of both the hydrophobic pocket created by the rearrangement of Glu276 and the hydrophobic surface of S4. Indeed the crystal structure of the enzyme bound to active isomer **174** revealed both of the desired interactions (Fig. 23). Once the isomer of highest binding affinity had been identified, the stereoselective synthesis of this compound was undertaken, as shown in Scheme 5.[136]

The final compound was synthesized starting from the commercially available lactam **175**, which was methanolyzed and N-protected followed by cycloaddition of 2-ethyl-butanonitrile oxide to give the isoxazoline **176**. After reduction and N-deprotection the amine was converted into the guanidine group and the ester was cleaved to give target compound **174**. This new cyclopentane-based compound BCX-1812 (**174**) was the first highly potent inhibitor of its class, inhibiting both influenza A and B virus sialidases at an nM level ($IC_{50} = 1.1$ nM and 0.2 nM against NA(A) and NA(B), respectively).

BCX-1812, also known as Peramivir and RWJ-270201, successfully completed animal studies and Phase I and Phase II clinical trials, in which the compound was administrated orally, showing neither major side effects nor toxicity.[137] However, in Phase III trial, Peramivir did not show statistical efficacy, presumably because of a lack of bioavailability. Recently the efficacy of

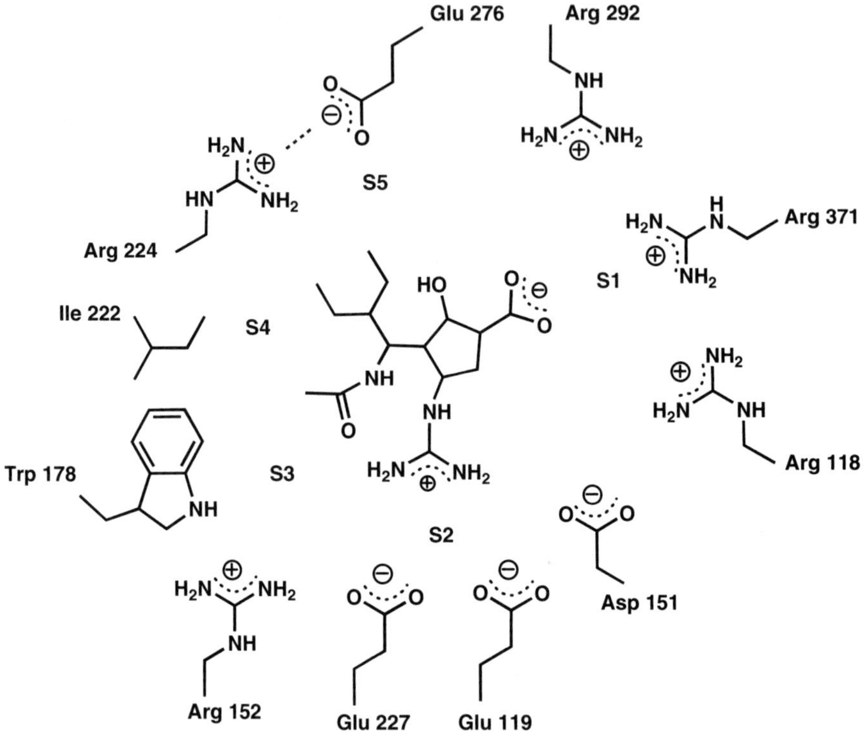

FIG. 23. A schematic representation of some key interactions of BCX-1812 **174** with conserved influenza virus A sialidase active-site residues.

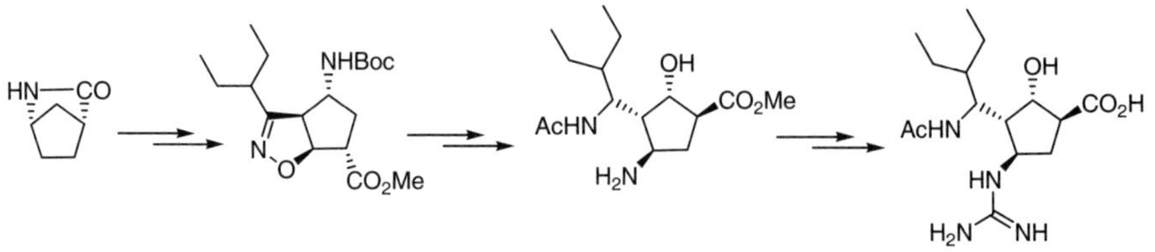

SCHEME 5. Synthesis of highly potent inhibitor BCX 1812 (**174**).

intramuscular injection of Peramivir in a mouse influenza model was demonstrated.[138]

The functionalized cyclopentane **178** (BCX-1827) displayed similar activity to **174**, exhibiting potent anti-influenza virus effects both *in vitro* in MDCK cells and *in vivo* in a mouse model.[139–141]

178

It is known that C-6-carboxamide derivatives of Zanamivir are very effective against NA(A) but only modestly active against NA(B).[96,97] An analogous modification was applied to cyclopentane-based inhibitors.[142] A series of multisubstituted cyclopentane amides such as **179–182** were prepared, including 14 alkyl or aralkyl monosubstituted amides, 13 alkyl or aralkyl disubstituted amides, and 12 examples of cycloaliphatic or substituted cycloaliphatic amides. The complexity of this library was increased by the initial synthesis and screening of stereoisomeric mixtures, arising from chiral centers at C-1 in the ring and C-1′ on the side-chain. Those diastereomeric mixtures that demonstrated significant inhibition of NA were synthesized separately in a stereoselective manner and reassayed in order to identify the most active component.

179 $R^1 = CH(CH_2CH_3)_2$ $R^2 = H$

180 $R^1 = CH_2CH_3$ $R^2 = CH_2CH_3$

181 $R^1 = (CH_2)_2CH_3$ $R^2 = (CH_2)_2CH_3$,

182 $R^1 = -(CH_2)_2O(CH_2)_2-$

In vitro test data for the best candidates **179–182** showed low μM activity for the inhibition of NA(A) but diminished activity towards NA(B), displayed IC_{50} values for **179** 0.08 μM and 3 μM, **180** 0.015 μM and 3 μM, **181** 0.06 μM and 9.2 μM, and **182** 0.04 μM and 7.8 μM, against NA(A) and NA(B), respectively.[142] These results were comparable to the results reported for the amides of Zanamivir type. The active stereoisomer of **180** was found to have the carboxylic acid and the guanidino group oriented *trans* to each other, while the carboxylic acid and the substituted amide group were *cis*. Thus, the absolute configuration at the carbon bearing the guanidino group (C-4) is *S* in **180**, whereas it is *R* in BCX-1827. Consequently, the guanidino group and the acetamido groups of **180** interact differently with the active-site residues compared to their counterparts in BCX-1827, however, similar to those observed in the dihydropyran series.[96,97]

b. Pyrrolidine Derivatives.—Researchers at Abbott Laboratories disclosed the use of pyrrolidine core to the design of influenza virus sialidase inhibitors.[49,143–145] Directed screening of selected chemical libraries using a sialidase biochemical assay, identified *cis*-*N*-*t*-butoxycarbonyl-3-aminopyrrolidine-4-carboxylic acid (**183**) as the lead compound (IC_{50} value of 58 μM against influenza virus A sialidase). The strategy for the optimization of this lead structure involved a combination of traditional medicinal chemistry including X-ray crystallographic analysis, computational modeling and chemical synthesis, and combinatorial chemistry (high-throughput parallel synthesis) allowing rapid and comprehensive exploration of a given scaffold. Docking studies of **183** with NA structures suggested addition of a substituent on the ring nitrogen to create additional contacts in both subsites S3 and S5. Thus, large numbers of trisubstituted pyrrolidine analogues such as amides (~250), carbamates (~25), sulfonamides (~25), and ureas (~250) were synthesized by automated solid-phase parallel synthesis. From within this pool of compounds, urea analogues with various alkyl groups demonstrated low-μM level inhibition. In particular, compound **184** was approximately 50-fold (1.1 μM against NA(A)) more active than the lead **183**. Analysis of the structures of **183** and **184** bound to influenza virus A sialidase showed that neither of the important subsites S3 (acetamide pocket of Zanamivir) and S2 (guanidino pocket of Zanamivir) were occupied properly and efforts to improve binding via variation of the exocyclic amine proved to be futile. As a consequence, a series of tetrasubstituted pyrrolidines was synthesized, and a method for high-throughput protein crystallographic X-ray analysis was established in order to verify design strategies. The rapid availability of structural information had a profound impact on this program, allowing for adjustment in response to unexpected binding modes of new inhibitors and for facile identification of active isomers from among diastereomeric mixtures. Of the tetrasubstituted pyrrolidine series of type **185–187**, only the acetamide **185** (7.5 μM against NA(A)) and the trifluoroacetamide **186** (0.28 μM against NA(A)) were active in the low-μM range, though selectivity for inhibition of NA(A) over NA(B) was pronounced. Evaluation of the NA–**185** complex confirmed the expected interaction of the carboxylate with the triarginyl cluster in S1, identified binding of the hydrophobic portion of the urea in S5 and the concomitant conformational reorganization of Glu276 observed previously,[99,127,135] and revealed that the exocyclic amine did not bind at S2 due to an unexpected rotation of the pyrrolidine ring. Carbocyclic analogue **187** exhibited similar binding but weaker activity (24 μM) against NA(A) and no activity against NA(B).

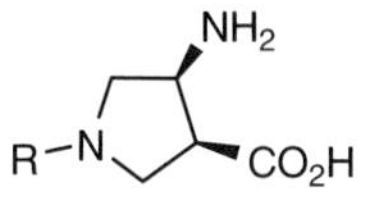

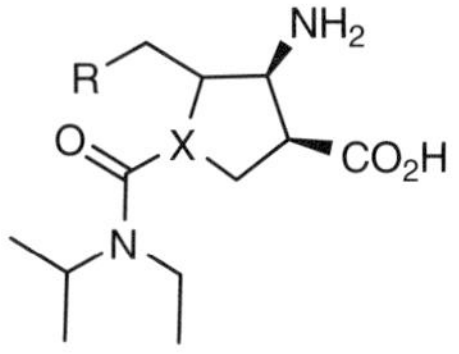

183 R =

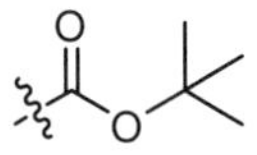

184 R =

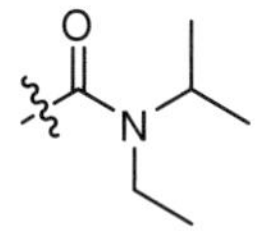

185 X = N, R = NHC(O)CH_3
186 X = N, R = NHC(O)CF_3
187 X = CH_2, R = NHC(O)CH_3

The breakthrough in the development of high-affinity inhibitors for influenza virus sialidase in this series was the surprising 180° reorientation (relative to **185–187**) of cyclopentane derivative **188** within the active site of the sialidase ($IC_{50} = 0.7$ μM against NA(A)).[146] Crystallographic analysis of **188** bound to NA showed an unpredicted hydrophobic interaction of the methyl ester with subsite S2, normally the site of charge–charge interactions. Compound **188** was considered as a three-site binder, since the exocyclic amine made no significant interaction with the protein. Four-site binders were obtained by introducing aliphatic alkyl groups on the side-chain methylene carbon of a pyrrolidine template, as illustrated by **189**, affording significant enhancement in potency against both NA(A) and NA(B) ($IC_{50} = 41$ nM against NA(A) and $IC_{50} = 56$ nM against NA(B)).[49,147] X-Ray crystallographic studies indicated that **189** retained the binding mode of **188**, with the methyl ester moiety interacting at S2[146] while the isobutyl group occupied S5. This new binding of the hydrophobic methyl ester arises mainly through strong van der Waals interaction with the side-chain methylene groups of Asp151 and Leu135 in subsite S2, as well as a potential π–π stacking with the C = O of Glu119. Additionally, the ester displaces two water molecules from S2, affording **189** an entropic advantage.[146]

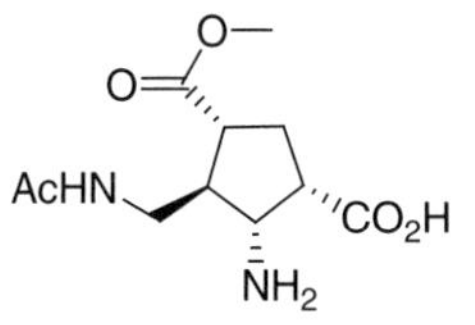

188

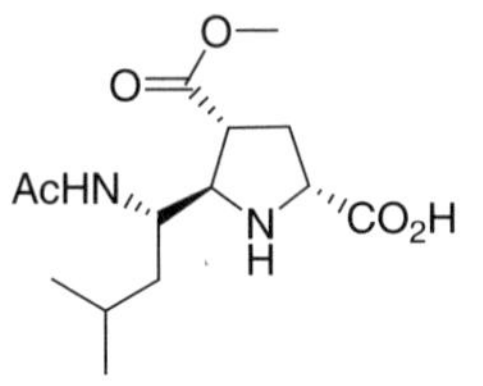

189

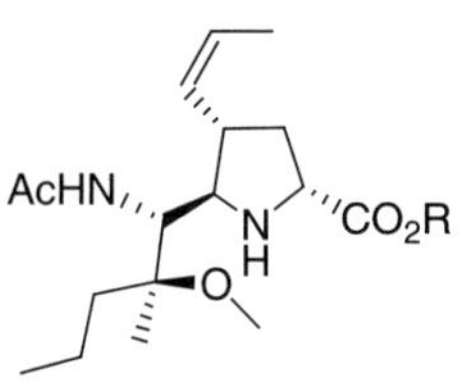

190 R = H
191 R = CH_2CH_3

Extensive efforts were devoted to replacement of the metabolically labile methyl ester in **189**. With the aid of molecular modeling and X-ray crystallography, a variety of suitable functionalities were identified including 2-pyrazolyl or short alkene residues such as *cis*-propenyl.[148] The latter alkene was incorporated into an inhibitor that ultimately led to development of the clinical candidate **190** (ABT-675, A-315675) after further optimization of the hydrophobic side-chain.[143,149,150] X-Ray crystallographic structures of ABT-675 bound to NA(A) and NA(B) confirmed the expected hydrophobic interactions of the *cis*-propenyl group with S2. The *n*-propyl group of the side-chain was observed to interact with S4, while the methyl and methoxy groups were associated with S5. In contrast to results obtained with other inhibitors that occupy both S4 and S5, Glu276 was observed to maintain its native conformation in the **190**–NA(B) complex and underwent conformational reorganization during binding by NA(A). This pyrrolidine-based inhibitor effectively inhibits both influenza virus A and B sialidase ($IC_{50} = 0.2\,nM$ against NA(A) and $IC_{50} = 0.1\,nM$ against NA(B)), and represents a 10^7-fold improvement in binding affinity relative to the original pyrrolidine screening hit **183**. Due to its zwitterionic character, ABT-675 has limited oral bioavailability.[143] Attempts to remedy this drawback via replacement of the pyrrolidine core with a furan or cyclopentane ring resulted in weaker inhibition of NA.[147]

A second approach involving use of the ethyl ester prodrug ABT-667 (**191**) has proven more successful, with this compound currently in clinical studies.

IV. Outlook

The worldwide spread of H5N1 avian influenza virus has raised the concern of its potential to emerge as a human-adapted virus. Three decades of intense research have yielded only two NA inhibitors, Relenza™ and Tamiflu™, that are currently approved for clinical use. Emergence of drug resistance further adds to the challenge and makes the development of anti-influenza drugs a priority. In early 2006, the threat of a pandemic, led the US FDA to grant a fast-track designation of the Biocryst injectable candidate Peramivir, and Biota and Sankyo have also announced their intention to market a new influenza virus sialidase inhibitor that is currently in clinical trials. Undoubtedly, the plethora of structural information and ongoing efforts will reveal new and more potent drugs. Recently, the crystal structure of N1 sialidase has been solved, which opens up a new avenue for structure-based drug design.[151]

From the sequence homology study, the nine sialidase subtypes can be subdivided into two distinct groups. Group-1 contains N1, N4, N5, and N8 subtypes whereas group-2 contains N2, N3, N6, N7, and N9.[151] Thus far, all X-ray crystallographic structural information that describes the sialidase active-site has been based on group-2 sialidases N2 and N9. Russell and coworkers have determined the crystal structures from three group-1 sialidases, N1, N4, and N8 subtypes, and of their complexes with the inhibitors Oseltamivir, Zanamivir, DANA, and Peramivir, to compare their active sites with those of the group-2 enzymes against which the current drugs were designed. These studies revealed

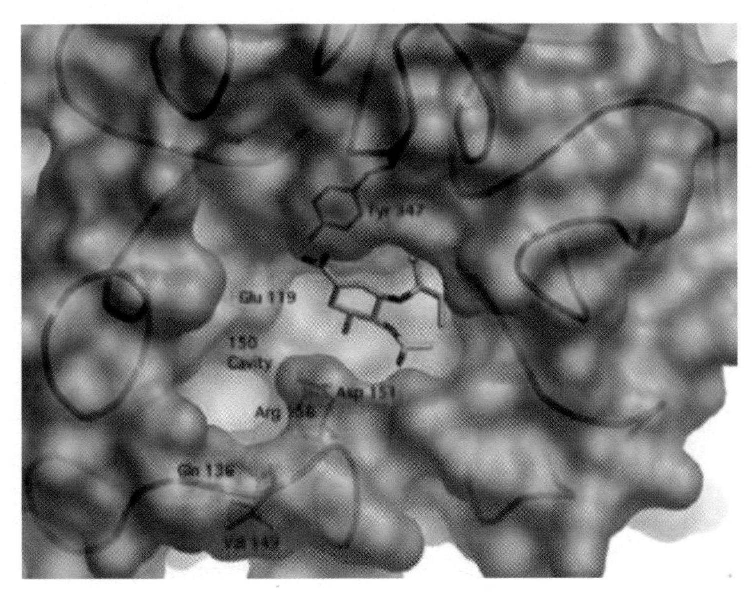

Fig. 24. Molecular surface of group-1 (N1) sialidase with bound Oseltamivir showing the 150-cavity.

that the active sites of both groups are virtually identical except for the presence of a large cavity, 150-loop (residues 147–152) in the group-1 enzymes adjacent to the subsite S2 of the active site (Fig. 24). This cavity is accessible from the active site due to a 1.5 Å difference in the side-chain position of the conserved Asp151 residue and the Glu119 residue which, in group-1 enzymes, adopts a conformation such that its carboxylate points in approximately the opposite direction from that seen in group-2 enzymes. The combined effect of the difference in position of these two acidic residues results in the 10 Å long and 5 Å wide and deep 150-cavity. This 150-loop appears to exist in this 'open' conformation, however, after a 30-minute soak of inhibitor into crystals of group-1 sialidase, a

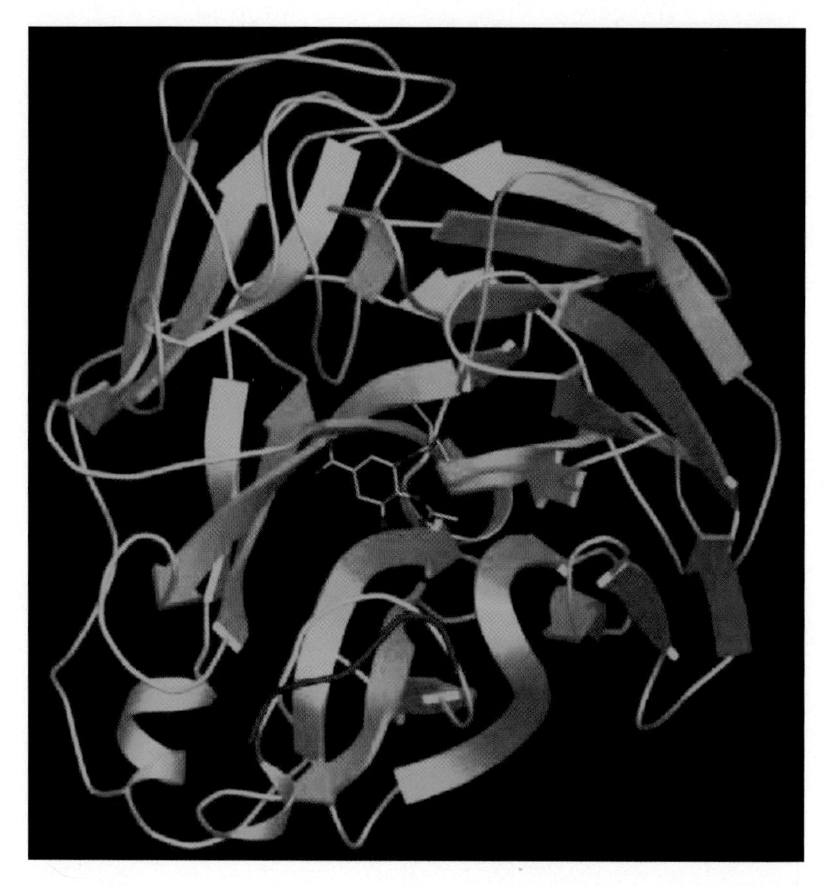

FIG. 25. Ribbon representation of the oseltamivir bound group-1 sialidase showing the 'open' (red) and 'close' (green) conformation of the 150-loop.

slow conformational change occurs, resulting in the 'closed' form, closely resembling the conformation observed in the group-2 sialidase structures (Fig. 25). On the basis of this discovery, new drugs against group-1 sialidases could be targeted by adding extra substituent moieties to the existing inhibitor skeleton. Although it would be ideal to prepare inhibitors that target against all virus subtypes, an effective group-specific inhibitor could be of considerable value against the avian H5N1 viruses.

Acknowledgments

Prof von Itzstein gratefully acknowledges the support of the Australian Research Council through the award of a Federation Fellowship and the National Health and Medical Research Council.

References

1. C. W. Potter, A history of influenza, *J. Appl. Microbiol.*, 91 (2001) 572–579.
2. A. W. Crosby, *America's Forgotten Pandemic*, Cambridge University Press, Cambridge, 1989.
3. R. G. J. Douglas, Drug therapy, prophylaxis and treament of influenza, *N. Engl. J. Med.*, 322 (1990) 443–450.
4. S. M. Wintermeyer and M. C. Nahata, Rimantidine: A clinical perspective, *Ann. Pharmacother.*, 29 (1995) 299–310.
5. L. H. Pinto, L. J. Holsinger, and R. A. Lamb, Influenza virus M2 protein has ion channel activity, *Cell*, 69 (1992) 517–528.
6. A. J. Hay, The action of adamantanamines against influenza A viruses: Inhibition of the M2 ion channel protein, *Semin. Virol.*, 3 (1992) 21–30.
7. A. J. Hay, C. A. Thompson, A. Geraghty, S. Hayhurst, S. Grambas, and M. S. Bennett, in C. Hannoun (Ed.), *Options for the Control of Influenza Virus*, Vol. II, Excerpta Medica, Amsterdam, 1993, pp. 281–288.
8. J. L. McKimm-Breschkin, Neuraminidase inhibitors for the treatment and prevention of influenza, *Expert Opin. Pharmacother.*, 3 (2002) 103–112.
9. L. V. Gubareva, L. Kaiser, and F. G. Hayden, Influenza virus neuraminidase inhibitors, *Lancet*, 355 (2000) 827–835.
10. J. Zhang and W. Xu, Recent advances in anti-influenza agents with neuraminidase as target, *Mini Rev. Med. Chem.*, 6 (2006) 429–448.
11. R. G. Webster, J. R. Schafer, J. Suss, W. J. Bean, and Y. Kawaoko, in Hannoun (Ed.), *Options for the control of influenza virus II*, Excerpta Medica, Amsterdam, 1993, pp. 177–185.
12. R. Wagner, M. Matrosovich, and H.-D. Klenk, Functional balance between haemagglutinin and neuraminidase in influenza virus infections, *Rev. Med. Virol.*, 12 (2002) 159–166.
13. J. S. Oxford, Influenza A pandemics of the 20th century with special reference to 1918: Virology, pathology and epidemiology, *Rev. Med. Virol.*, 10 (2000) 119–133.
14. G. Herrler, J. Hausmann, and H.-D. Klenk, in A. Rosenberg (Ed.), *Biology of the Sialic Acids*, Plenum Press, New York, 1995, pp. 315–336.

15. D. C. Wiley and J. J. Skehel, The structure and function of the hemagglutinin membrane glycoprotein of influenza virus, *Annu. Rev. Biochem.*, 56 (1987) 365–394.
16. J. N. Couceiro, J. C. Paulson, and L. G. Baum, Influenza virus strains selectively recognize sialyloligosaccharides on human respiratory epithelium; the role of the host cell in selection of hemagglutinin receptor specificity, *Virus Res.*, 29 (1993) 155–165.
17. Y. Suzuki, T. Ito, T. Suzuki, R. E. Holland Jr., T. M. Chambers, M. Kiso, H. Ishida, and Y. Kawaoka, Sialic acid species as a determinant of the host range of influenza A viruses, *J. Virol.*, 74 (2000) 11825–11831.
18. J. J. Skehel and D. C. Wiley, Receptor binding and membrane fusion in virus entry: The influenza hemagglutinin, *Annu. Rev. Biochem.*, 69 (2000) 531–569.
19. M. Matrosovich and H.-D. Klenk, Natural and synthetic sialic acid-containing inhibitors of influenza virus receptor binding, *Rev. Med. Virol.*, 13 (2003) 85–97.
20. E. Borrego-Diaz, M. E. Peeples, R. M. Markosyan, G. B. Melikyan, and F. S. Cohen, Completion of trimeric hairpin formation of influenza virus hemagglutinin promotes fusion pore opening and enlargement, *Virology*, 316 (2003) 234–244.
21. N. K. Sauter, M. D. Bednarski, B. A. Wurzburg, J. E. Hanson, G. M. Whitesides, J. J. Skehel, and D. C. Wiley, Hemagglutinins from two influenza virus variants bind to sialic acid derivatives with millimolar dissociation constants: A 500-MHz proton nuclear magnetic resonance study, *Biochemistry*, 28 (1989) 8388–8396.
22. P. A. Bullough, F. M. Hughson, A. C. Treharne, R. W. Ruigrok, J. J. Skehel, and D. C. Wiley, Crystals of a fragment of influenza haemagglutinin in the low pH induced conformation, *J. Mol. Biol.*, 236 (1994) 1262–1265.
23. P. M. Colman and C. W. Ward, Structure and diversity of influenza virus neuraminidase, *Curr. Top. Microbiol. Immunol.*, 114 (1985) 177–255.
24. J. N. Varghese, W. G. Laver, and P. M. Colman, Structure of the influenza virus glycoprotein antigen neuraminidase at 2.9 A resolution, *Nature*, 303 (1983) 35–40.
25. P. M. Colman, J. N. Varghese, and W. G. Laver, Structure of the catalytic and antigenic sites in influenza virus neuraminidase, *Nature*, 303 (1983) 41–44.
26. G. M. Air and W. G. Laver, The neuraminidase of influenza virus, *Proteins*, 6 (1989) 341–356.
27. P. M. Colman, Influenza virus neuraminidase: Structure, antibodies, and inhibitors, *Protein Sci.*, 3 (1994) 1687–1696.
28. A. Gottschalk, The structure of the prosthetic group of bovine submaxillary gland mucoprotein, *Biochim. Biophys. Acta*, 24 (1957) 649–650.
29. A. Gottschalk, Neuraminidase: its substrate and mode of action, *Adv. Enzymol. Relat. Subj. Biochem.*, 20 (1958) 135–146.
30. P. Palese, K. Tobita, M. Ueda, and R. W. Compans, Characterization of temperature sensitive influenza virus mutants defective in neuraminidase, *Virology*, 61 (1974) 397–410.
31. P. Palese and R. W. Compans, Inhibition of influenza virus replication in tissue culture by 2-deoxy-2,3-dehydro-*N*-trifluoroacetylneuraminic acid (FANA): Mechanism of action, *J. Gen. Virol.*, 33 (1976) 159–163.
32. A. P. Corfield and R. Schauer, in R. Schauer (Ed.), *Sialic Acids-Chemistry, Metabolism and Function*, Springer-Verlag, Wien, 1982, pp. 195–261.
33. A. Moscona, Neuraminidase inhibitors for influenza, *N. Engl. J. Med.*, 353 (2005) 1363–1373.
34. J. N. Watson, S. Newstead, A. A. Narine, G. Taylor, and A. J. Bennet, Two nucleophilic mutants of the *Micromonospora viridifaciens* sialidase operate with retention of configuration by two different mechanisms, *Chembiochem*, 6 (2005) 1999–2004.
35. J. N. Watson, S. Newstead, V. Dookhun, G. Taylor, and A. J. Bennet, Contribution of the active site aspartic acid to catalysis in the bacterial neuraminidase from *Micromonospora viridifaciens*, *FEBS Lett.*, 577 (2004) 265–269.

36. J. N. Watson, V. Dookhun, T. J. Borgford, and A. J. Bennet, Mutagenesis of the conserved active-site tyrosine changes a retaining sialidase into an inverting sialidase, *Biochemistry*, 42 (2003) 12682–12690.
37. C. A. Miller, P. Wang, and M. Flashner, Mechanism of Arthrobacter sialophilus neuraminidase: The binding of substrates and transition-state analogs, *Biochem. Biophys. Res. Commun.*, 83 (1978) 1479–1487.
38. J. N. Varghese, J. L. McKimm-Breschkin, J. B. Caldwell, A. A. Kortt, and P. M. Colman, The structure of the complex between influenza virus neuraminidase and sialic acid, the viral receptor, *Proteins*, 14 (1992) 327–332.
39. W. P. Burmeister, B. Henrissat, C. Bosso, S. Cusack, and R. W. Ruigrok, Influenza B virus neuraminidase can synthesize its own inhibitor, *Structure*, 1 (1993) 19–26.
40. A. K. Chong, M. S. Pegg, N. R. Taylor, and M. von Itzstein, Evidence for a sialosyl cation transition-state complex in the reaction of sialidase from influenza virus, *Eur. J. Biochem.*, 207 (1992) 335–343.
41. N. R. Taylor and M. von Itzstein, Molecular modeling studies on ligand binding to sialidase from influenza virus and the mechanism of catalysis, *J. Med. Chem.*, 37 (1994) 616–624.
42. C. S. Rye and S. G. Withers, Glycosidase mechanisms, *Curr. Opin. Chem. Biol.*, 4 (2000) 573–580.
43. H. D. Ly and S. G. Withers, Mutagenesis of glycosidases, *Annu. Rev. Biochem.*, 68 (1999) 487–522.
44. D. L. Zechel and S. G. Withers, Dissection of nucleophilic and acid-base catalysis in glycosidases, *Curr. Opin. Chem. Biol.*, 5 (2001) 643–649.
45. J. N. Varghese and P. M. Colman, Three-dimensional structure of the neuraminidase of influenza virus A/Tokyo/3/67 at 2.2 Å resolution, *J. Mol. Biol.*, 221 (1991) 473–486.
46. A. T. Baker, J. N. Varghese, W. G. Laver, G. M. Air, and P. M. Colman, Three-dimensional structure of neuraminidase of subtype N9 from an avian influenza virus, *Proteins*, 2 (1987) 111–117.
47. P. Bossart-Whitaker, M. Carson, Y. S. Babu, C. D. Smith, W. G. Laver, and G. M. Air, Three-dimensional structure of influenza A N9 neuraminidase and its complex with the inhibitor 2-deoxy 2,3-dehydro-*N*-acetyl neuraminic acid, *J. Mol. Biol.*, 232 (1993) 1069–1083.
48. W. P. Burmeister, R. W. Ruigrok, and S. Cusack, The 2.2 A resolution crystal structure of influenza B neuraminidase and its complex with sialic acid, *EMBO J.*, 11 (1992) 49–56.
49. V. Stoll, K. D. Stewart, C. J. Maring, S. Muchmore, V. Giranda, Y. G. Gu, G. Wang, Y. Chen, M. Sun, C. Zhao, A. L. Kennedy, D. L. Madigan, Y. Xu, A. Saldivar, W. Kati, G. Laver, T. Sowin, H. L. Sham, J. Greer, and D. Kempf, Influenza neuraminidase inhibitors: Structure-based design of a novel inhibitor series, *Biochemistry*, 42 (2003) 718–727.
50. M. J. Bamford, Neuraminidase inhibitors as potential anti-influenza drugs, *J. Enzyme Inhib.*, 10 (1995) 1–16.
51. T. H. Haskell, F. E. Peterson, D. Watson, N. R. Plessas, and T. Culbertson, Neuraminidase inhibition and viral chemotherapy, *J. Med. Chem.*, 13 (1970) 697–704.
52. J. D. Edmond, R. G. Johnston, D. Kidd, H. J. Rylance, and R. G. Sommerville, The inhibition of neuraminidase and antiviral action, *Br. J. Pharmacol. Chemother.*, 27 (1966) 415–426.
53. Y. Suzuki, K. Sato, M. Kiso, and A. Hasegawa, New ganglioside analogs that inhibit influenza virus sialidase, *Glycoconj. J.*, 7 (1990) 349–354.
54. T. Hagiwara, I. Kijima-Suda, T. Ido, H. Ohrui, and K. Tomita, Inhibition of bacterial and viral sialidases by 3-fluoro-*N*-acetylneuraminic acid, *Carbohydr. Res.*, 263 (1994) 167–172.
55. C. L. White, M. N. Janakiraman, W. G. Laver, C. Philippon, A. Vasella, G. M. Air, and M. Luo, A sialic acid-derived phosphonate analog inhibits different strains of influenza virus neuraminidase with different efficiencies, *J. Mol. Biol.*, 245 (1995) 623–634.

56. P. Meindl and H. Tuppy, 2-Deoxy-2,3-dehydrosialic acids. II. Competitive inhibition of Vibrio cholerae neuraminidase by 2-deoxy-2,3-dehydro-*N*-acylneuraminic acids, *Hoppe Seylers Z Physiol. Chem.*, 350 (1969) 1088–1092.
57. C. T. Holzer, M. von Itzstein, B. Jin, M. S. Pegg, W. P. Stewart, and W. Y. Wu, Inhibition of sialidases from viral, bacterial and mammalian sources by analogues of 2-deoxy-2,3-didehydro-*N*-acetylneuraminic acid modified at the C-4 position, *Glycoconj. J.*, 10 (1993) 40–44.
58. P. Meindl, G. Bodo, P. Palese, J. Schulman, and H. Tuppy, Inhibition of neuraminidase activity by derivatives of 2-deoxy-2,3-dehydro-*N*-acetylneuraminic acid, *Virology*, 58 (1974) 457–463.
59. U. Nohle, J. M. Beau, and R. Schauer, Uptake, metabolism and excretion of orally and intravenously administered, double-labeled *N*-glycoloylneuraminic acid and single-labeled 2-deoxy-2,3-dehydro-*N*-acetylneuraminic acid in mouse and rat, *Eur. J. Biochem.*, 126 (1982) 543–548.
60. P. Palese, J. L. Schulman, G. Bodo, and P. Meindl, Inhibition of influenza and parainfluenza virus replication in tissue culture by 2-deoxy-2,3-dehydro-*N*-trifluoroacetylneuraminic acid (FANA), *Virology*, 59 (1974) 490–498.
61. P. Palese and J. L. Schulman, Inhibitors of viral neuraminidase as potential antiviral drugs, *Chemoprophylaxis Virus Infect. Respir. Tract*, 1 (1977) 189–205.
62. M. von Itzstein, W. Y. Wu, G. B. Kok, M. S. Pegg, J. C. Dyason, B. Jin, T. Van Phan, M. L. Smythe, H. F. White, S. W. Oliver, P. M. Colman, J. N. Varghese, D. M. Ryan, J. M. Woods, R. C. Bethell, V. J. Hotham, J. M. Cameron, and C. R. Penn, Rational design of potent sialidase-based inhibitors of influenza virus replication, *Nature*, 363 (1993) 418–423.
63. P. J. Goodford, A computational procedure for determining energetically favorable binding sites on biologically important macromolecules, *J. Med. Chem.*, 28 (1985) 849–857.
64. M. von Itzstein, J. C. Dyason, S. W. Oliver, H. F. White, W. Y. Wu, G. B. Kok, and M. S. Pegg, A study of the active site of influenza virus sialidase: An approach to the rational design of novel anti-influenza drugs, *J. Med. Chem.*, 39 (1996) 388–391.
65. L.M. von Itzstein, W.Y. Wu, V. Phan Tho, B. Danylec, and B. Jin, *Preparation of Derivatives and Analogs of 2-Deoxy-2,3-Didehydro-N-Acetylneuraminic Acid as Antiviral Agents*, Biota Scientific Management Pty. Ltd., Australia, 1991, WO 9116320.
66. M. von Itzstein, W. Y. Wu, and B. Jin, The synthesis of 2,3-didehydro-2,4-dideoxy-4-guanidinyl-*N*-acetylneuraminic acid: A potent influenza virus sialidase inhibitor, *Carbohydr. Res.*, 259 (1994) 301–305.
67. M. von Itzstein, B. Jin, W. Y. Wu, and M. Chandler, A convenient method for the introduction of nitrogen and sulfur at C-4 on a sialic acid analog, *Carbohydr. Res.*, 244 (1993) 181–185.
68. M. Chandler, M. J. Bamford, R. Conroy, B. Lamont, B. Patel, V. K. Patel, I. P. Steeples, R. Storer, N. G. Weir, M. Wright, and C. Williamson, Synthesis of the potent influenza neuraminidase inhibitor 4-guanidino Neu5Ac2en. X-ray molecular structure of 5-acetamido-4-amino-2,6-anhydro-3,4,5-trideoxy-D-*erythro*-L-*gluco*-nonionic acid, *J. Chem. Soc. Perkin Trans.*, 1 (1995) 1173–1180.
69. J. M. Woods, R. C. Bethell, J. A. Coates, N. Healy, S. A. Hiscox, B. A. Pearson, D. M. Ryan, J. Ticehurst, J. Tilling, S. M. Walcott, et al., 4-Guanidino-2,4-dideoxy-2,3-dehydro-*N*-acetylneuraminic acid is a highly effective inhibitor both of the sialidase (neuraminidase) and of growth of a wide range of influenza A and B viruses *in vitro*, *Antimicrob. Agents Chemother.*, 37 (1993) 1473–1479.
70. A. C. Wallace, R. A. Laskowski, and J. M. Thornton, LIGPLOT: A program to generate schematic diagrams of protein-ligand interactions, *Protein Eng.*, 8 (1995) 127–134.
71. R. W. Sidwell and D. F. Smee, *In vitro* and *in vivo* assay systems for study of influenza virus inhibitors, *Antiviral. Res.*, 48 (2000) 1–16.

72. M. Flashner, J. Kessler, and S. W. Tanenbaum, The interaction of substrate-related ketals with bacterial and viral neuraminidases, *Arch. Biochem. Biophys.*, 221 (1983) 188–196.
73. P.-A. Driguez, B. Barrere, G. Quash, and A. Doutheau, Synthesis of transition-state analogues as potential inhibitors of sialidase from influenza virus, *Carbohydr. Res.*, 262 (1994) 297–310.
74. K. Ikeda, K. Sano, M. Ito, M. Saito, K. Hidari, T. Suzuki, Y. Suzuki, and K. Tanaka, Synthesis of 2-deoxy-2,3-didehydro-*N*-acetylneuraminic acid analogues modified at the C-4 and C-9 positions and their behaviour towards sialidase from influenza virus and pig liver membrane, *Carbohydr. Res.*, 330 (2001) 31–41.
75. K. Ikeda, F. Kimura, K. Sano, Y. Suzuki, and K. Achiwa, Chemoenzymatic synthesis of an *N*-acetylneuraminic acid analogue having a carbamoylmethyl group at C-4 as an inhibitor of sialidase from influenza virus, *Carbohydr. Res.*, 312 (1998) 183–189.
76. J. Li, M. Zheng, W. Tang, P. L. He, W. Zhu, T. Li, J. P. Zuo, H. Liu, and H. Jiang, Syntheses of triazole-modified zanamivir analogues via click chemistry and anti-AIV activities, *Bioorg. Med. Chem. Lett.*, 16 (2006) 5009–5013.
77. J. C. Wilson, R. J. Thomson, J. C. Dyason, P. Florio, K. J. Quelch, S. Abo, and M. von Itzstein, The design, synthesis and biological evaluation of neuraminic acid-based probes of Vibrio cholerae sialidase, *Tetrahedron Asymmetry*, 11 (2000) 53–73.
78. I. D. Starkey, M. Mahmoudian, D. Noble, P. W. Smith, P. Cherry, P. D. Howes, and S. L. Sollis, Synthesis and influenza virus sialidase inhibitory activity of the 5-desacetamido analog of 2,3-didehydro-2,4-dideoxy-4-guanidinyl-*N*-acetylneuraminic acid (GG167), *Tetrahedron Lett.*, 36 (1995) 299–302.
79. E. Schreiner, E. Zbiral, R. G. Kleineidam, and R. Schauer, 2,3-Didehydro-2-deoxysialic acids structurally varied at C-5 and their behaviour towards the sialidase from *Vibrio cholerae*, *Carbohydr. Res.*, 216 (1991) 61–66.
80. P. W. Smith, I. D. Starkey, P. D. Howes, S. L. Sollis, S. P. Keeling, P. C. Cherry, M. von Itzstein, W. Y. Wu, and B. Jin, Synthesis and influenza virus sialidase inhibitory activity of analogs of 4-guanidino-Neu5Ac2en (GG167) with modified 5-substituents, *Eur. J. Med. Chem.*, 31 (1996) 143–150.
81. T. Honda, T. Masuda, S. Yoshida, M. Arai, Y. Kobayashi, and M. Yamashita, Synthesis and anti-influenza virus activity of 4-guanidino-7-substituted Neu5Ac2en derivatives, *Bioorg. Med. Chem. Lett.*, 12 (2002) 1921–1924.
82. T. Honda, T. Masuda, S. Yoshida, M. Arai, S. Kaneko, and M. Yamashita, Synthesis and anti-influenza virus activity of 7-*O*-alkylated derivatives related to zanamivir, *Bioorg. Med. Chem. Lett.*, 12 (2002) 1925–1928.
83. D. M. Andrews, P. C. Cherry, D. C. Humber, P. S. Jones, S. P. Keeling, P. F. Martin, C. D. Shaw, and S. Swanson, Synthesis and influenza virus sialidase inhibitory activity of analogues of 4-guanidino-Neu5Ac2en (Zanamivir) modified in the glycerol side-chain, *Eur. J. Med. Chem.*, 34 (1999) 563–574.
84. J. L. McKimm-Breschkin, P. M. Colman, B. Jin, G. Y. Krippner, M. McDonald, P. A. Reece, S. P. Tucker, L. Waddington, K. G. Watson, and W. Y. Wu, Tethered neuraminidase inhibitors that bind an influenza virus: A first step towards a diagnostic method for influenza, *Angew. Chem. Int. Ed. Engl.*, 42 (2003) 3118–3121.
85. T. Honda, S. Yoshida, M. Arai, T. Masuda, and M. Yamashita, Synthesis and anti-influenza evaluation of polyvalent sialidase inhibitors bearing 4-guanidino-Neu5Ac2en derivatives, *Bioorg. Med. Chem. Lett.*, 12 (2002) 1929–1932.
86. T. Masuda, S. Yoshida, M. Arai, S. Kaneko, M. Yamashita, and T. Honda, Synthesis and anti-influenza evaluation of polyvalent sialidase inhibitors bearing 4-guanidino-Neu5Ac2en derivatives, *Chem. Pharm. Bull. (Tokyo)*, 51 (2003) 1386–1398.

87. K. G. Watson, R. Cameron, R. J. Fenton, D. Gower, S. Hamilton, B. Jin, G. Y. Krippner, A. Luttick, D. McConnell, S. J. MacDonald, A. M. Mason, V. Nguyen, S. P. Tucker, and W. Y. Wu, Highly potent and long-acting trimeric and tetrameric inhibitors of influenza virus neuraminidase, *Bioorg. Med. Chem. Lett.*, 14 (2004) 1589–1592.
88. S. J. Macdonald, R. Cameron, D. A. Demaine, R. J. Fenton, G. Foster, D. Gower, J. N. Hamblin, S. Hamilton, G. J. Hart, A. P. Hill, G. G. Inglis, B. Jin, H. T. Jones, D. B. McConnell, J. McKimm-Breschkin, G. Mills, V. Nguyen, I. J. Owens, N. Parry, S. E. Shanahan, D. Smith, K. G. Watson, W. Y. Wu, and S. P. Tucker, Dimeric zanamivir conjugates with various linking groups are potent, long-lasting inhibitors of influenza neuraminidase including H5N1 avian influenza, *J. Med. Chem.*, 48 (2005) 2964–2971.
89. S. J. Macdonald, K. G. Watson, R. Cameron, D. K. Chalmers, D. A. Demaine, R. J. Fenton, D. Gower, J. N. Hamblin, S. Hamilton, G. J. Hart, G. G. Inglis, B. Jin, H. T. Jones, D. B. McConnell, A. M. Mason, V. Nguyen, I. J. Owens, N. Parry, P. A. Reece, S. E. Shanahan, D. Smith, W. Y. Wu, and S. P. Tucker, Potent and long-acting dimeric inhibitors of influenza virus neuraminidase are effective at a once-weekly dosing regimen, *Antimicrob. Agents Chemother.*, 48 (2004) 4542–4549.
90. G. B. Kok, M. Campbell, B. Mackey, and M. von Itzstein, Synthesis and biological evaluation of sulfur isosteres of the potent influenza virus sialidase inhibitors 4-amino-4-deoxy- and 4-deoxy-4-guanidino-Neu5Ac2en, *J. Chem. Soc. Perkin Trans.*, 1 (1996) 2811–2815.
91. B. J. Smith, P. M. Colman, M. Von Itzstein, B. Danylec, and J. N. Varghese, Analysis of inhibitor binding in influenza virus neuraminidase, *Protein Sci.*, 10 (2001) 689–696.
92. M. Murakami, K. Ikeda, and K. Achiwa, Chemoenzymatic synthesis of neuraminic acid analogs structurally varied at C-5 and C-9 as potential inhibitors of the sialidase from influenza virus, *Carbohydr. Res.*, 280 (1996) 101–110.
93. M. R. Lentz, R. G. Webster, and G. M. Air, Site-directed mutation of the active site of influenza neuraminidase and implications for the catalytic mechanism, *Biochemistry*, 26 (1987) 5351–5358.
94. M. J. Bamford, J. C. Pichel, W. Husman, B. Patel, R. Storer, and N. G. Wier, Synthesis of 6-, 7- and 8-carbon sugar analogs of potent anti-influenza 2,3-didehydro-2,3-dideoxy-*N*-acetylneuraminic acid derivatives, *J. Chem. Soc. Perkin Trans.*, 1 (1995) 1181–1187.
95. S. L. Sollis, P. W. Smith, P. D. Howes, P. C. Cherry, and R. C. Bethell, Novel inhibitors of influenza sialidase related to GG167. Synthesis of 4-amino and guanidino-4H-pyran-2-carboxylic acid-6-propylamides; selective inhibitors of influenza A virus sialidase, *Bioorg. Med. Chem. Lett.*, 6 (1996) 1805–1808.
96. P. W. Smith, S. L. Sollis, P. D. Howes, P. C. Cherry, I. D. Starkey, K. N. Cobley, H. Weston, J. Scicinski, A. Merritt, A. Whittington, P. Wyatt, N. Taylor, D. Green, R. Bethell, S. Madar, R. J. Fenton, P. J. Morley, T. Pateman, and A. Beresford, Dihydropyrancarboxamides related to zanamivir: A new series of inhibitors of influenza virus sialidases. 1. Discovery, synthesis, biological activity, and structure-activity relationships of 4-guanidino- and 4-amino-4*H*-pyran-6-carboxamides, *J. Med. Chem.*, 41 (1998) 787–797.
97. P. G. Wyatt, B. A. Coomber, D. N. Evans, T. I. Jack, H. E. Fulton, A. J. Wonacott, P. Colman, and J. Varghese, Sialidase inhibitors related to zanamivir. Further SAR studies of 4-amino-4*H*-pyran-2-carboxylic acid-6-propylamides, *Bioorg. Med. Chem. Lett.*, 11 (2001) 669–673.
98. P. W. Smith, S. L. Sollis, P. D. Howes, P. C. Cherry, K. N. Cobley, H. Taylor, H. R. Whittington, R. C. Bethell, N. Taylor, J. N. Varghese, P. M. Colman, O. Singh, T. Slkarzynski, A. Cleasby, and A. J. Wonacott, Novel inhibitors of influenza sialidases related to GG167. Structure-activity, crystallographic and molecular dynamics studies with 4*H*-pyran-2-carboxylic acid 6-carboxamides, *Bioorg. Med. Chem. Lett.*, 6 (1996) 2931–2936.

99. N. R. Taylor, A. Cleasby, O. Singh, T. Skarzynski, A. J. Wonacott, P. W. Smith, S. L. Sollis, P. D. Howes, P. C. Cherry, R. Bethell, P. Colman, and J. Varghese, Dihydropyrancarboxamides related to zanamivir: A new series of inhibitors of influenza virus sialidases. 2. Crystallographic and molecular modeling study of complexes of 4-amino-4*H*-pyran-6-carboxamides and sialidase from influenza virus types A and B, *J. Med. Chem.*, 41 (1998) 798–807.
100. P. W. Smith, J. E. Robinson, D. N. Evans, S. L. Sollis, P. D. Howes, N. Trivedi, and R. C. Bethell, Sialidase inhibitors related to zanamivir: Synthesis and biological evaluation of 4*H*-pyran 6-ether and ketone, *Bioorg. Med. Chem. Lett.*, 9 (1999) 601–604.
101. P. W. Smith and A. R. Whittington, Novel inhibitors of influenza sialidases related to zanamivir. Heterocyclic replacements of the glycerol sidechain, *Bioorg. Med. Chem. Lett.*, 7 (1997) 2239–2242.
102. M. C. Mann, R. J. Thomson, J. C. Dyason, S. McAtamney, and M. von Itzstein, Modelling, synthesis and biological evaluation of novel glucuronide-based probes of *Vibrio cholerae* sialidase, *Bioorg. Med. Chem.*, 14 (2006) 1518–1537.
103. M. C. Mann, T. Islam, J. C. Dyason, P. Florio, C. J. Trower, R. J. Thomson, and M. von Itzstein, Unsaturated *N*-acetyl-D-glucosaminuronic acid glycosides as inhibitors of influenza virus sialidase, *Glycoconj. J.*, 23 (2006) 127–133.
104. M. C. Mann, R. J. Thomson, and M. von Itzstein, An efficient approach to *N*-acetyl-D-glucosaminuronic acid-based sialylmimetics as potential sialidase inhibitors, *Bioorg. Med. Chem. Lett.*, 14 (2004) 5555–5558.
105. P. Florio, R. J. Thomson, A. Alafaci, S. Abo, and M. von Itzstein, Synthesis of δ 4-β-D-glucopyranosiduronic acids as mimetics of 2,3-unsaturated sialic acids for sialidase inhibition, *Bioorg. Med. Chem. Lett.*, 9 (1999) 2065–2068.
106. P. Florio, R.J. Thomson, B. Smith, P.M. Colman, and M. von Itzstein, unpublished data.
107. K. Sasaki, Y. Nishida, M. Kambara, H. Uzawa, T. Takahashi, T. Suzuki, Y. Suzuki, and K. Kobayashi, Design of *N*-acetyl-6-sulfo-beta-d-glucosaminide-based inhibitors of influenza virus sialidase, *Bioorg. Med. Chem.*, 12 (2004) 1367–1375.
108. K. Sasaki, Y. Nishida, H. Uzawa, and K. Kobayashi, *N*-Acetyl-6-sulfo-D-glucosamine as a promising mimic of *N*-acetyl neuraminic acid, *Bioorg. Med. Chem. Lett.*, 13 (2003) 2821–2823.
109. M. J. Jedrzejas, S. Singh, W. J. Brouillette, W. G. Laver, G. M. Air, and M. Luo, Structures of aromatic inhibitors of influenza virus neuraminidase, *Biochemistry*, 34 (1995) 3144–3151.
110. S. Singh, M. J. Jedrzejas, G. M. Air, M. Luo, W. G. Laver, and W. J. Brouillette, Structure-based inhibitors of influenza virus sialidase: A benzoic acid lead with novel interaction, *J. Med. Chem.*, 38 (1995) 3217–3225.
111. P. Chand, Y. S. Babu, S. Bantia, N. Chu, L. B. Cole, P. L. Kotian, W. G. Laver, J. A. Montgomery, V. P. Pathak, S. L. Petty, D. P. Shrout, D. A. Walsh, and G. M. Walsh, Design and synthesis of benzoic acid derivatives as influenza neuraminidase inhibitors using structure-based drug design, *J. Med. Chem.*, 40 (1997) 4030–4052.
112. E. A. Sudbeck, M. J. Jedrzejas, S. Singh, W. J. Brouillette, G. M. Air, W. G. Laver, Y. S. Babu, S. Bantia, P. Chand, N. Chu, J. A. Montgomery, D. A. Walsh, and M. Luo, Guanidinobenzoic acid inhibitors of influenza virus neuraminidase, *J. Mol. Biol.*, 267 (1997) 584–594.
113. P. D. Howes, A. Cleasby, D. N. Evans, H. Fielden, P. W. Smith, S. L. Sollis, N. Taylor, and A. J. Wonacott, 4-Acetylamino-3-(imidazol-1-yl)-benzoic acids as novel inhibitors of influenza sialidase, *Eur. J. Med. Chem.*, 34 (1999) 225–234.
114. W. J. Brouillette, V. R. Atigadda, F. Duarte, M. Luo, G. M. Air, Y. S. Babu, and S. Bantia, Design of benzoic acid inhibitors of influenza neuraminidase containing a cyclic substitution for the *N*-acetyl grouping. (Erratum to document cited in CA131:266560), *Bioorg. Med. Chem. Lett.*, 9 (1999) 3259.

115. J. B. Finley, V. R. Atigadda, F. Duarte, J. J. Zhao, W. J. Brouillette, G. M. Air, and M. Luo, Novel aromatic inhibitors of influenza virus neuraminidase make selective interactions with conserved residues and water molecules in the active site, *J. Mol. Biol.*, 293 (1999) 1107–1119.
116. V. R. Atigadda, W. J. Brouillette, F. Duarte, S. M. Ali, Y. S. Babu, S. Bantia, P. Chand, N. Chu, J. A. Montgomery, D. A. Walsh, E. A. Sudbeck, J. Finley, M. Luo, G. M. Air, and G. W. Laver, Potent inhibition of influenza sialidase by a benzoic acid containing a 2-pyrrolidinone substituent, *J. Med. Chem.*, 42 (1999) 2332–2343.
117. M. Williams, N. Bischofberger, S. Swaminathan, and C. U. Kim, Synthesis and influenza neuraminidase inhibitory activity of aromatic analogs of sialic acid, *Bioorg. Med. Chem. Lett.*, 5 (1995) 2251–2254.
118. P. Chand, P. L. Kotian, P. E. Morris, S. Bantia, D. A. Walsh, and Y. S. Babu, Synthesis and inhibitory activity of benzoic acid and pyridine derivatives on influenza neuraminidase, *Bioorg. Med. Chem.*, 13 (2005) 2665–2678.
119. W. J. Brouillette, V. R. Atigadda, M. Luo, G. M. Air, Y. S. Babu, and S. Bantia, Design of benzoic acid inhibitors of influenza neuraminidase containing a cyclic substitution for the N-acetyl grouping, *Bioorg. Med. Chem. Lett.*, 9 (1999) 1901–1906.
120. S. Vorwerk and A. Vasella, Carbocyclic analogs of *N*-acetyl-2,3-didehydro-2-deoxy-D-neuraminic acid (Neu5Ac2en, DANA): Synthesis and inhibition of viral and bacterial neuraminidases, *Angew. Chem. Int. Ed.*, 37 (1998) 1732–1734.
121. C. U. Kim, W. Lew, M. A. Williams, H. Liu, L. Zhang, S. Swaminathan, N. Bischofberger, M. S. Chen, D. B. Mendel, C. Y. Tai, W. G. Laver, and R. C. Stevens, Influenza neuraminidase inhibitors possessing a novel hydrophobic interaction in the enzyme active site: Design, synthesis, and structural analysis of carbocyclic sialic acid analogues with potent anti-influenza activity, *J. Am. Chem. Soc.*, 119 (1997) 681–690.
122. C. U. Kim, W. Lew, M. A. Williams, H. Wu, L. Zhang, X. Chen, P. A. Escarpe, D. B. Mendel, W. G. Laver, and R. C. Stevens, Structure–activity relationship studies of novel carbocyclic influenza neuraminidase inhibitors, *J. Med. Chem.*, 41 (1998) 2451–2460.
123. W. Lew, M. A. Williams, D. B. Mendel, P. A. Escarpe, and C. U. Kim, C3-Thia and C3-carba isosteres of a carbocyclic influenza neuraminidase inhibitor, (3R,4R,5R)-4-acetamido-5-amino-3-propoxy-1-cyclohexene-1-carboxylic acid, *Bioorg. Med. Chem. Lett.*, 7 (1997) 1843–1846.
124. W. Lew, X. Chen, and C. U. Kim, Discovery and development of GS 4104 (oseltamivir): An orally active influenza neuraminidase inhibitor, *Curr. Med. Chem.*, 7 (2000) 663–672.
125. S. A. Kerrigan, P. W. Smith, and R. J. Stoodley, Synthesis of (4R*,5R*)-4-acetylamino-5-diethylcarbamoylcyclohex-1-ene-1-carboxylic acid and (3R*,4R*)-4-acetylamino-3-diethylcarbamoylcyclohex-1-ene-1-carboxylic acid: New inhibitors of influenza virus sialidases, *Tetrahedron Lett.*, 42 (2001) 4709–4712.
126. M. A. Williams, W. Lew, D. B. Mendel, C. Y. Tai, P. A. Escarpe, W. G. Laver, R. C. Stevens, and C. U. Kim, Structure–activity relationships of carbocyclic influenza neuraminidase inhibitors, *Bioorg. Med. Chem. Lett.*, 7 (1997) 1837–1842.
127. W. Li, P. A. Escarpe, E. J. Eisenberg, K. C. Cundy, C. Sweet, K. J. Jakeman, J. Merson, W. Lew, M. Williams, L. Zhang, C. U. Kim, N. Bischofberger, M. S. Chen, and D. B. Mendel, Identification of GS 4104 as an orally bioavailable prodrug of the influenza virus neuraminidase inhibitor GS 4071, *Antimicrob. Agents Chemother.*, 42 (1998) 647–653.
128. K. E. Doucette and F. Y. Aoki, Oseltamivir: A clinical and pharmacological perspective, *Expert Opin. Pharmacother.*, 2 (2001) 1671–1683.
129. J. A. McCullers, Antiviral therapy of influenza, *Expert Opin. Investig. Drugs*, 14 (2005) 305–312.

130. Y. Y. Yeung, S. Hong, and E. J. Corey, A short enantioselective pathway for the synthesis of the anti-influenza neuramidase inhibitor oseltamivir from 1,3-butadiene and acrylic acid, *J. Am. Chem. Soc.*, 128 (2006) 6310–6311.
131. P. J. Harrington, J. D. Brown, T. Foderaro, and R. C. Hughes, Research and development of a second-generation process for oseltamivir phosphate, prodrug for a neuraminidase inhibitor, *Org. Process Res. Dev.*, 8 (2004) 86–91.
132. S. Abrecht, P. Harrington, H. Iding, M. Karpf, R. Trussardi, B. Wirz, and U. Zutter, The synthetic development of the anti-influenza neuraminidase inhibitor oseltamivir phosphate (Tamiflu): A challenge for synthesis & process research, *Chimia*, 58 (2004) 621–929.
133. J. W. Jeong, J. K. Kim, B. W. Woo, B. J. Song, D. C. Ha, and N. Z, A substrate mimetic approach for influenza neuraminidase inhibitors, *Bull. Korean Chem. Soc.*, 25 (2004) 1575–1577.
134. T. Yamamoto, H. Kumazawa, K. Inami, T. Teshima, and T. Shiba, Syntheses of sialic acid isomers with inhibitory activity against neuraminidase, *Tetrahedron Lett.*, 33 (1992) 5791–5794.
135. Y. S. Babu, P. Chand, S. Bantia, P. Kotian, A. Dehghani, Y. El-Kattan, T. H. Lin, T. L. Hutchison, A. J. Elliott, C. D. Parker, S. L. Ananth, L. L. Horn, G. W. Laver, and J. A. Montgomery, BCX-1812 (RWJ-270201): Discovery of a novel, highly potent, orally active, and selective influenza neuraminidase inhibitor through structure-based drug design, *J. Med. Chem.*, 43 (2000) 3482–3486.
136. P. Chand, P. L. Kotian, A. Dehghani, Y. El-Kattan, T. H. Lin, T. L. Hutchison, Y. S. Babu, S. Bantia, A. J. Elliott, and J. A. Montgomery, Systematic structure-based design and stereoselective synthesis of novel multisubstituted cyclopentane derivatives with potent antiinfluenza activity, *J. Med. Chem.*, 44 (2001) 4379–4392.
137. R. W. Sidwell and D. F. Smee, Peramivir (BCX-1812, RWJ-270201): Potential new therapy for influenza, *Expert Opin. Investig. Drugs*, 11 (2002) 859–869.
138. S. Bantia, C. S. Arnold, C. D. Parker, R. Upshaw, and P. Chand, Anti-influenza virus activity of peramivir in mice with single intramuscular injection, *Antiviral Res.*, 69 (2006) 39–45.
139. S. Bantia, C. D. Parker, S. L. Ananth, L. L. Horn, K. Andries, P. Chand, P. L. Kotian, A. Dehghani, Y. El-Kattan, T. Lin, T. L. Hutchison, J. A. Montgomery, D. L. Kellog, and Y. S. Babu, Comparison of the anti-influenza virus activity of RWJ-270201 with those of oseltamivir and zanamivir, *Antimicrob. Agents Chemother.*, 45 (2001) 1162–1167.
140. R. W. Sidwell, D. F. Smee, J. H. Huffman, D. L. Barnard, K. W. Bailey, J. D. Morrey, and Y. S. Babu, *In vivo* influenza virus-inhibitory effects of the cyclopentane neuraminidase inhibitor RJW-270201, *Antimicrob. Agents Chemother.*, 45 (2001) 749–757.
141. D. F. Smee, J. H. Huffman, A. C. Morrison, D. L. Barnard, and R. W. Sidwell, Cyclopentane neuraminidase inhibitors with potent *in vitro* anti-influenza virus activities, *Antimicrob. Agents Chemother.*, 45 (2001) 743–748.
142. P. Chand, Y. S. Babu, S. Bantia, S. Rowland, A. Dehghani, P. L. Kotian, T. L. Hutchison, S. Ali, W. Brouillette, Y. El-Kattan, and T. H. Lin, Syntheses and neuraminidase inhibitory activity of multisubstituted cyclopentane amide derivatives, *J. Med. Chem.*, 47 (2004) 1919–1929.
143. W. M. Kati, D. Montgomery, R. Carrick, L. Gubareva, C. Maring, K. McDaniel, K. Steffy, A. Molla, F. Hayden, D. Kempf, and W. Kohlbrenner, *In vitro* characterization of A-315675, a highly potent inhibitor of A and B strain influenza virus neuraminidases and influenza virus replication, *Antimicrob. Agents Chemother.*, 46 (2002) 1014–1021.
144. G. T. Wang, Y. Chen, S. Wang, R. Gentles, T. Sowin, W. Kati, S. Muchmore, V. Giranda, K. Stewart, H. Sham, D. Kempf, and W. G. Laver, Design, synthesis, and structural analysis of influenza neuraminidase inhibitors containing pyrrolidine cores, *J. Med. Chem.*, 44 (2001) 1192–1201.

145. W. M. Kati, D. Montgomery, C. Maring, V. S. Stoll, V. Giranda, X. Chen, W. G. Laver, W. Kohlbrenner, and D. W. Norbeck, Novel α- and β-amino acid inhibitors of influenza virus neuraminidase, *Antimicrob. Agents Chemother.*, 45 (2001) 2563–2570.
146. G. T. Wang, Recent advances in the discovery and development of anti-influenza drugs, *Expert Opin. Ther. Patents*, 12 (2002) 845–861.
147. G. T. Wang, S. Wang, R. Gentles, T. Sowin, C. J. Maring, D. J. Kempf, W. M. Kati, V. Stoll, K. D. Stewart, and G. Laver, Design, synthesis, and structural analysis of inhibitors of influenza neuraminidase containing a 2,3-disubstituted tetrahydrofuran-5-carboxylic acid core, *Bioorg. Med. Chem. Lett.*, 15 (2005) 125–128.
148. C. J. Maring, V. S. Stoll, C. Zhao, M. Sun, A. C. Krueger, K. D. Stewart, D. L. Madigan, W. M. Kati, Y. Xu, R. J. Carrick, D. A. Montgomery, A. Kempf-Grote, K. C. Marsh, A. Molla, K. R. Steffy, H. L. Sham, W. G. Laver, Y. G. Gu, D. J. Kempf, and W. E. Kohlbrenner, Structure-based characterization and optimization of novel hydrophobic binding interactions in a series of pyrrolidine influenza neuraminidase inhibitors, *J. Med. Chem.*, 48 (2005) 3980–3990.
149. D. A. DeGoey, H. J. Chen, W. J. Flosi, D. J. Grampovnik, C. M. Yeung, L. L. Klein, and D. J. Kempf, Enantioselective synthesis of antiinfluenza compound A-315675, *J. Org. Chem.*, 67 (2002) 5445–5453.
150. S. Hanessian, M. Bayrakdarian, and X. Luo, Total synthesis of A-315675: A potent inhibitor of influenza neuraminidase, *J. Am. Chem. Soc.*, 124 (2002) 4716–4721.
151. R. J. Russell, L. F. Haire, D. J. Stevens, P. J. Collins, Y. P. Lin, G. M. Blackburn, A. J. Hay, S. J. Gamblin, and J. J. Skehel, The structure of H5N1 avian influenza neuraminidase suggests new opportunities for drug design, *Nature*, 443 (2006) 45–49.

ADVANCES IN CARBOHYDRATE CHEMISTRY AND BIOCHEMISTRY, VOL. 61

CHEMOSELECTIVE NEOGLYCOSYLATION

BY FRANCESCO NICOTRA, LAURA CIPOLLA, FRANCESCO PERI, BARBARA LA FERLA and CRISTINA REDAELLI

Department of Biotechnology and Bioscience, University of Milano Bicocca, Piazza della Scienza 2, I-20126 Milano, Italy

I. INTRODUCTION

Carbohydrates are key components of glycoproteins, glycolipids, and proteoglycans. They are involved in various recognition processes, such as cell adhesion, migration and signaling, and in folding of proteins.[1] Many recognition processes of biological relevance are mediated by carbohydrate–protein interactions that are essential for normal tissue growth and repair, and are an

ISBN: 978-0-12-373920-9
DOI: 10.1016/S0065-2318(07)61007-5

integral part of many pathological processes, such as bacterial and viral invasion of the host,[2] as well as tumor-cell motility and progression.[3] Understanding how and why these highly specific interactions come about is a hard task because of the difficulties associated with the "microheterogenity" of glycoconjugates, in particular glycoproteins, considering that glycosylation is a post-transcriptional process.

The possible variations in anomeric configuration, regiochemistry of bonding, states of oxidation (or reduction), and substitution of the monosaccharide composing oligosaccharides generate a wide range of topographies, which provide an extremely large and sophisticated vocabulary to enable the development of a large number of molecular-recognition events. A precise understanding of the relationship between sugars and biological activity remains limited by the availability of convenient and effective glycosylation tools.

A great demand for synthetic glycoconjugates (glycoproteins, glycolipids) has been created by the needs of both basic research and pharmacological interest. Recent chemical advances, such as improved synthetic methods, including solid-phase synthesis,[4] and methods for enzymatic synthesis,[5] have opened new and exciting possibilities for obtaining homogeneous natural glycoconjugates or glycoconjugate mimetics, which are termed "neoglycoconjugates".[6]

The development of neoglycoconjugates in the place of natural glycoconjugates, in particular in the field of glycopeptides and glycoproteins, lies in the fact that selective glycosylation of specific units of serine or threonine in O-linked glycopropeins, and of asparagine in N-linked glycoproteins, is a very difficult task. This is due to the multifunctionality of these compounds, which requires extensive and time-consuming protection/deprotection steps, and due to the complexity in manipulation of some aglycons, such as proteins, in the experimental condition for glycosylation reactions. To overcome these difficulties, a variety of unnatural linkages between the reducing end of the sugar and the aglycon have been evaluated, including the use of spacers with appropriate functional groups, to perform selective ligations and to render this process more efficient and selective.

To better understand carbohydrate–protein interactions, and to define at the molecular level normal and pathological processes, particularly promising are such carbohydrate-based diagnostic and therapeutic tools as arrays, gold nanoparticles, and microspheres, as well as naturally occurring glycoconjugates and mimetics. These glyco-tools not only permit qualitative and quantitative *in vitro* studies on the interaction phenomena under investigation, but also offer promise for therapeutic purposes and for *in vivo* diagnosis.[7]

Finally, the cell-adhesion properties of carbohydrates can be exploited in tissue engineering for the generation of polymeric biocompatible materials allowing the regeneration of the tissue on the polymeric scaffold. Such diverse biomedical applications, such as drug delivery, cell targeting, and adhesion, have been suggested for sugar-containing polymeric materials and nanoparticles,[8] thanks to the characteristics conferred by the carbohydrate residue.

The interaction between carbohydrates and proteins is generally weak. Three models have been suggested for glycoconjugate interactions: (i) the Carbohydrate Recognition Model—only single oligosaccharide motifs are ligands, (ii) the Cluster Model—clusters of many carbohydrate motifs are ligands, and (iii) the Carbohydrate–Protein Recognition Model—the ligand consists in the carbohydrate motif *and* a given region of protein that supports it. To take advantage of the cluster model, a great variety of dendrimeric ligands, which allow multiple saccharidic structures to cluster, have been developed and tested for adhesion properties.

This chapter describes the interest in generating neoglycoconjugates of different types: glycopeptides, glycoproteins, glycolipids, glycochips, glyconanoparticles, and glyco-biomaterials; and reports on the different methods that permit conjugation of the sugar, with or without spacers, to different aglycons. Particular attention is devoted to those procedures that perform the process chemoselectively, avoiding protection and deprotection steps as much as possible, and allowing the process to be performed in aqueous solution.

II. Neoglycopeptides and Neoglycoproteins

It is well recognized that glycoconjugates, and in particular glycopeptides and glycoproteins, act as mediators of an enormous variety of cellular events. With respect to structural diversity, they have the capacity to far exceed proteins and nucleic acids as agents encoding information for specific molecular recognition, and they act as determinants of protein folding, stability, and pharmacokinetics. The structural diversity of oligosaccharide expression patterns on proteins and cells makes it extremely difficult to correlate structure with function. Some technical challenges are analytical in nature: determination of the oligosaccharide sequence on a specific glycoconjugate is still far from routine. Others originate from glycoconjugate biosynthesis, which is neither template-driven nor under direct transcriptional control. Oligosaccharides are assembled in stepwise fashion, primarily in the endoplasmic reticulum and Golgi apparatus, a process that affords significant product microheterogeneity.[9] Unlike the biosynthesis of proteins and nucleic acids, there appears to be no associated mechanism for proofreading and correcting differently glycosylated biomolecules. Glycoproteins therefore occur naturally in a number of forms (glycoforms) that possess the same peptide backbone, but differ in both the nature and site of glycosylation. The different properties exhibited[10] by each component within these microheterogeneous mixtures present regulatory difficulties and problems in determining exact function through structure–activity relationships. It has even been suggested that these naturally occurring mixtures of glycoforms provide a spectrum of activities that can be biased in one direction or another as a mean of fine-tuning.

As a result, it is difficult to obtain homogeneous and chemically defined glycoconjugates from biological sources. Access to structurally defined oligosaccharides and glycoconjugates is a prerequisite for discovering their function. Chemical routes for the production of oligosaccharides are, therefore, essential. Advances on this front are now providing materials for the assessment of glycan function, the establishment of those structural features important for function, the elucidation of biosynthetic pathways, the creation of carbohydrate-based vaccines, the production of non-natural glycosylated antibiotics, and the generation of inhibitors of glycoconjugate function.

In general, glycoconjugate assembly, that is the synthesis of glycosylated peptide fragments and full-length proteins, suffers from serious limitations.[11] The synthesis of merely the core fragments, the regions in closest proximity to the polypeptide backbone, can require complex orthogonal protection schemes to form branched oligosaccharide structures. The methods for such transformations that are currently at our disposal are not compatible with the complex functionality of the polypeptide backbone, precluding the chemical elaboration of the oligosaccharide from the polypeptide scaffold. Despite these challenges, in recent years several groups have conquered the assembly of glycopeptide fragments and some impressive structures have been reported.[12] In addition, advances in enzymatic glycosylation methods[13] and in recombinant DNA technologies[14] promise to relieve the strain of protecting-group manipulations in glycoprotein syntheses.

Typically, syntheses of glycoconjugates adopt one of two strategies. The first is that of linear assembly, using protected glycosylated amino acids as building blocks for automated solid-phase peptide synthesis (SPPS) to generate glycosylated peptide fragments.[15] Fmoc-based peptide synthetic methods are sufficiently mild for the oligosaccharide to remain intact throughout the synthesis.

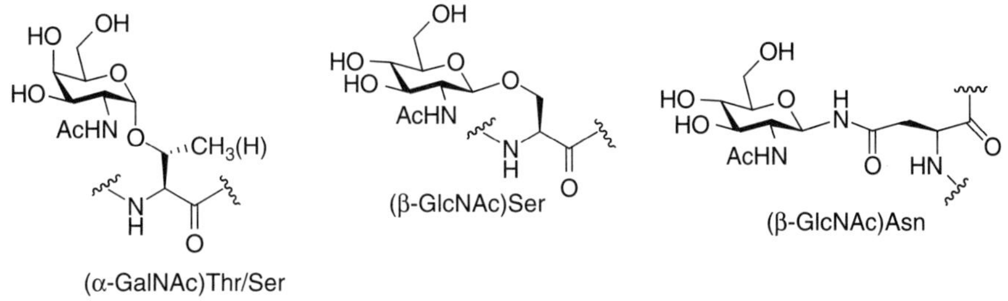

(α-GalNAc)Thr/Ser (β-GlcNAc)Ser (β-GlcNAc)Asn

FIG. 1. Common O- and N-linked glycosides in protein glycosylation.

The linear assembly of glycosylated amino acids in particular usually provides well-defined products; following this approach, the required carbohydrate structure is attached to an amino acid residue, typically serine and threonine for O-linked glycopeptides and asparagine for N-linked glycopeptides (Fig. 1).[16]

The second is a convergent approach that relies on the formation of the link late in the synthesis, once the protein scaffold for its presentation is in place. In this case, site-selective glycosylation must be in order to have chemically defined structures.[17]

Given the instability associated with the glycosidic bond,[18] and the requirements for protection that need to be considered in the use of glycosylated building blocks, the latter approach has often seemed the most attractive option. However, the convergent assembly of separately prepared oligosaccharides and proteins is also marked by several obstacles. In addition, the necessary syntheses of oligosaccharides have to continue hand-in-hand with methods for their conjugation to proteins and peptides. Two factors limit the work: first, the need not only for extensive carbohydrate protection but also for amino acid protection and, second, the acid- and base-lability of glycosylated amino acid residues. In addition, the extension of these methods to full-length glycoproteins has proved more troublesome, largely because of limitations inherent in linear, stepwise SPPS. Peptides larger than 50–60 residues are difficult to obtain using conventional methods as a result of poor yields and accumulation of by-products.

To surpass the size limits inherent in linear SPPS, the convergent condensation of unprotected or partially protected peptide/glycopeptide segments has emerged as an alternative approach for the total synthesis of glycoproteins. As already mentioned, it is too difficult to construct the glycosidic linkage between an oligosaccharide and a large protein. In response, chemists have begun to explore various approaches for the construction of glycopeptide/protein

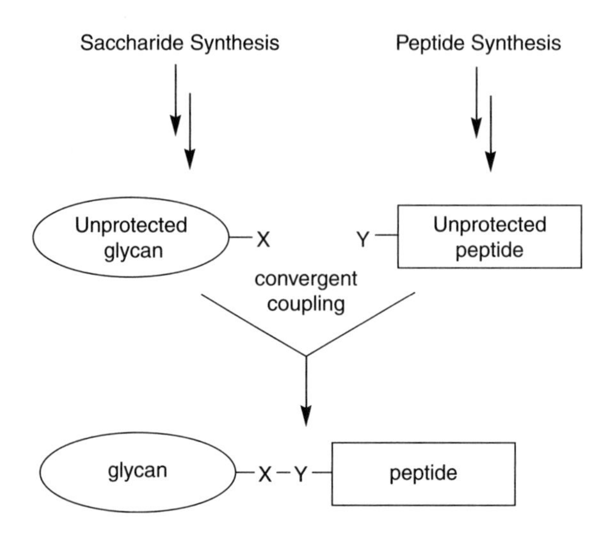

FIG. 2. Schematic representation of the chemoselective method for glycopeptide synthesis.

mimetics that have superior properties for therapeutic applications, or lend themselves to a more-facile convergent assembly from proteins and synthetic oligosaccharides. Hence, in an attempt to increase the selectivity and predictability of protein glycosylation, various novel approaches have been described, all of which exploit the chemoselectivities of different enzymatic and traditional methods (Fig. 2).[19]

Chemoselective ligation, first described by protein chemists as the coupling of two mutually and uniquely reactive groups in an aqueous environment, provides access to complex glycoconjugates in an elegant and convergent manner. In this technique, uniquely reactive pairs of functional groups (generally an electrophile and a nucleophile) are introduced into the peptide and sugar fragments, giving rise to selective covalent-bond formation, even in the presence of an array of other unprotected functional groups (see Section VII).

Although the resulting linkages do not precisely replicate natural connections, the simplified procedures allow rapid, convenient access to a wide array of neoglycopeptides that have been shown by some researchers to maintain their native biological activity.[20]

Changes in glycosylation are often a hallmark of disease states.[21] The structures of glycans, which decorate all eukaryotic-cell surfaces, change, for example, with the onset of cancer[22] and inflammation.[23] Synthetic mimics of the complex glycocopeptides and glycoproteins found on cell surfaces can modulate cellular interactions and are under development as therapeutic agents. The next section outlines different applications of neoglycosylation of peptides.

1. Neoglycopeptides in the Diagnosis and Treatment of Cancer

Malignant cells are commonly characterized by incomplete glycosylation and a large number of tumor-associated carbohydrate antigens (TACA) expressed on glycolipids and glycoproteins have been identified.[24] Tumor immunotherapy is based on the theory that tumor-associated antigens (TAA) become immunogenic if presented to a properly trained immunosystem.[25] Many of the tumor antigens are of carbohydrate constitution. In particular, the monosaccharide α-GalNAc, also known as the Tn antigen, has been extensively studied. It is expressed on mucin-type glycoproteins by the majority of human adenocarcinomas as a consequence of aberrant glycosylation, whereas it is hidden in normal cells.[26] An immune response directed against carbohydrate antigens results in the induction of antibodies that could eradicate the micrometastases and the circulating tumor cells in the blood stream, thus providing protection against the tumor. However, carbohydrate-based vaccines have so far been unsuccessful in inducing detectable T-cell immunity.[27] To overcome this limitation, and in order to induce an immunological memory toward a carbohydrate epitope, a suitable vaccine has to provide a helper T-cell response for B-cell induction and IgG production. According to this concept, semisynthetic vaccines have been prepared by conjugation of a carbohydrate B-epitope with a peptide by chemoselective methods.[28] Neoglycopeptides **1–3** (Fig. 3) have been synthesized as potential small-molecular weight antitumor vaccines, formed by covalently linking a short peptide, as the T-epitope, to the B-epitope, the Tn antigen. To achieve increased stability toward *in vivo* degradation, the two moieties are connected by a *C*-glycosylic bond in place of the natural *O*-glycosidic linkage. In addition, the concept of chemoselective ligation has been applied to the site-specific attachment of the sugar moiety to peptides. Chemoselectivity was achieved by coupling an aminoxy group suitably introduced in the peptide backbone with a keto function present on the *C*-glycosyl analogue of the Tn antigen, affording the corresponding oxime.

A current chemoselective approach to neoglycopeptides involves the assembly of multiepitopic glycoconjugates bearing clustered Tn antigen as synthetic anticancer vaccines.[29] In this study, the use of regioselectively addressable functionalized templates (RAFTs),[30] as new scaffolds for the design of anticancer vaccines was investigated. RAFTs are topological templates, which

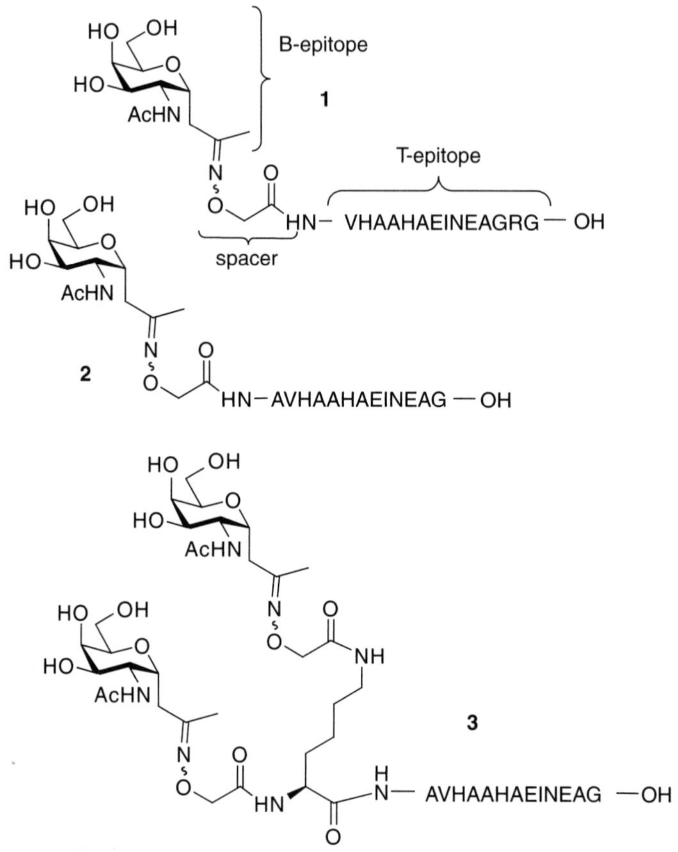

Fig. 3. Structure of neoglycopeptides as potential cancer vaccines.

are commonly composed of a backbone-cyclized decapeptide containing two proline–glycine units as β-turns inducers that stabilize their conformation in solution. These cyclic templates provide a suitable tool for the multimeric presentation of B- and T-epitopes. They display via lysinyl side chains two independent functional faces, one dedicated to the attachment of the TACA moieties in a multivalent manner and the other to the T-cell peptide epitopes. These conjugates exhibit the clustered Tn analogue as TACA (B-cell epitope) and the CD4+ helper T-cell peptide from the type 1 poliovirus. The saccharidic and peptidic epitopes were both synthesized separately and combined regioselectively to the RAFT core, using a sequential oxime-bond formation strategy (Fig. 4).

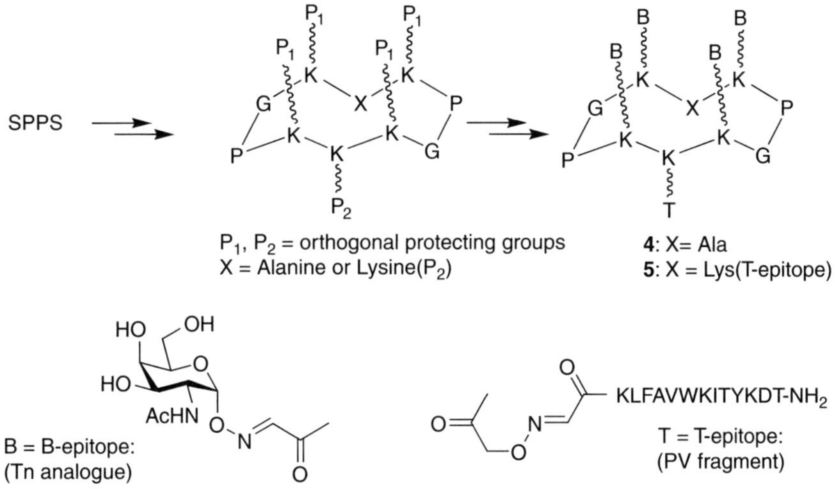

FIG. 4. Strategy for the synthesis of RAFT-derived vaccines by chemoselective ligation.

An efficient method for the convergent assembly of O-linked mucin-type glycopeptides bearing sulfated and sialylated oligosaccharides in a chemoselective fashion has also been proposed.[31] Interest in the synthesis of sulfated and sialylated oligosaccharides is motivated in part by their unique expression in tissues of certain disease states, including chronic inflammation and many types of cancer. Glycosyl amino acid **6** (Fig. 5), in which a protected thiol group substitutes for the hydroxyl group, is used for the synthesis of glycopeptide analogues by thiol alkylation.[32]

Building block **6** contains an unnatural 3-thio-GalNAc residue that can undergo chemoselective condensation with *N*-bromoacetamido sugars, following its incorporation into a peptide by Fmoc-based SPPS. The thiol group can be alkylated selectively in preference to all natural amino acids except cysteine. The resulting product contains an unnatural thioether linkage at the C-3 position, a common branch-point of mucin-type oligosaccharides, but retains the native sugar–peptide linkage. In this manner, carbohydrate ligands have been appended chemoselectively to native mucin-like scaffolds. This methodology was applied to the synthesis of thioether-linked analogues of the mucin-related 2,3-sialyl-TF (STF) and MECA-79 antigens (Fig. 6). The STF antigen is present on breast tumors, and thus glycopeptide fragments containing this structure are of interest in the preparation of anticancer vaccines.[25] The

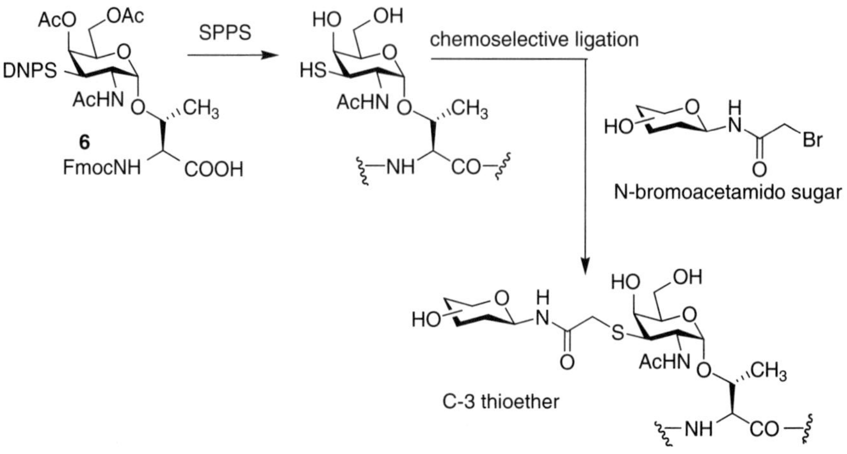

FIG. 5. Synthesis of glycopeptide analogues by thioalkylation (DNP = dinitrophenyl).

MECA-79 antigen is of interest because of its unique expression at sites of chronic inflammation.[33]

An interesting evolution of chemical ligation is a process termed "metabolic oligosaccharide engineering";[19b,c,34] wherein unnatural sugars bearing bioorthogonal functional groups can be metabolically introduced into cellular glycans. Bioorthogonal chemistry offers a means to visualize such metabolites as sugars, lipids, and cofactors non-invasively *in vivo*. To achieve this in practice requires a pair of functional groups that are mutually reactive in a physiological environment and, at the same time, inert to the biological milieu. One of these functional groups is appended to the biomolecule of interest (for example, a monosaccharide) and the other is attached to a reporter (for example, a magnetic resonance imaging (MRI) contrast-reagent or an affinity purification tag). A reaction that fulfils these criteria is the Staudinger ligation between azides and a specific class of phosphines.[35] The azide is essentially unreactive with biological nucleophiles, yet readily condenses with exogenously delivered phosphine reagents. The ability of the Staudinger ligation to probe glycosylation *in vivo* has significant implications for imaging disease-associated glycans. This methodology can be applied for cancer diagnostics based on cancer glycans; in particular (Fig. 7), azide-containing sialic acid residues (SiaNAz) which have been metabolically incorporated into membrane glycoconjugates in mice are treated with the azido sugar precursor *N*-α-(2-azidoacetyl)mannosamine (ManNAz). Because cancer cells overexpress sialic acids and are more metabolically active than

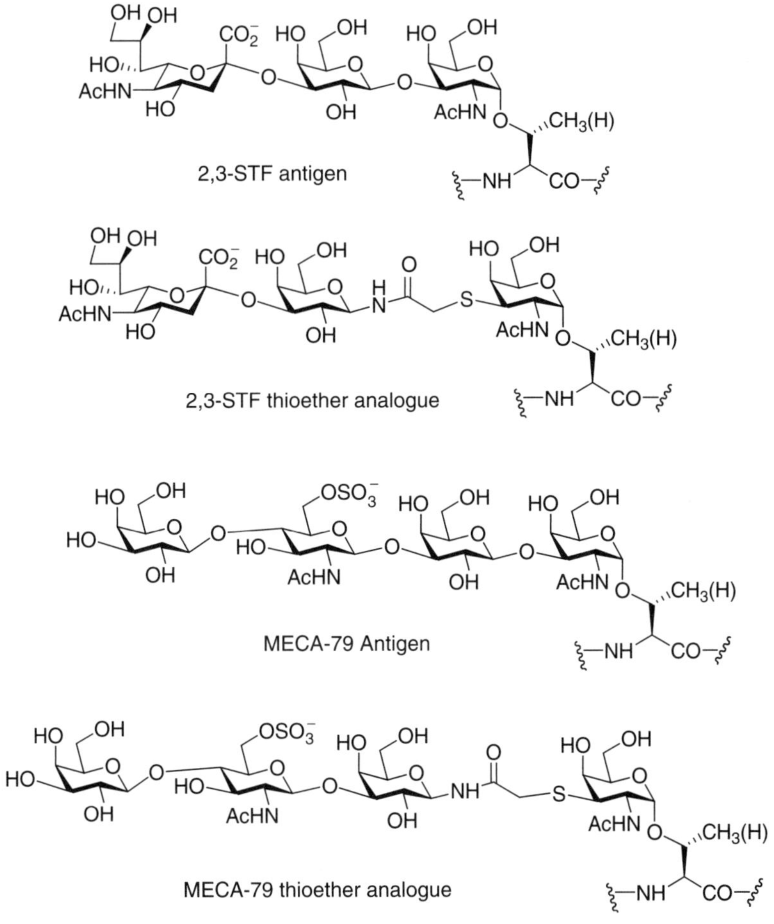

2,3-STF antigen

2,3-STF thioether analogue

MECA-79 Antigen

MECA-79 thioether analogue

FIG. 6. Structure of 2,3-STF and MECA-79 antigens and their thioether analogues.

normal cells, it is possible that these cells would incorporate disproportionately high amounts of SiaNAz into their cell-surface glycans. Subsequent reaction between azide-modified cancer cells and phosphine-conjugated imaging probes would then allow discrimination of cancerous tissue from the surrounding healthy tissue, using non-invasive diagnostic techniques such as MRI.[36]

More broadly, the azide might serve as an *in vivo* reporter of glycan expression. In principle, any sugar could metabolically be labeled with an azide if the biosynthetic enzymes are tolerant of azido substrates.

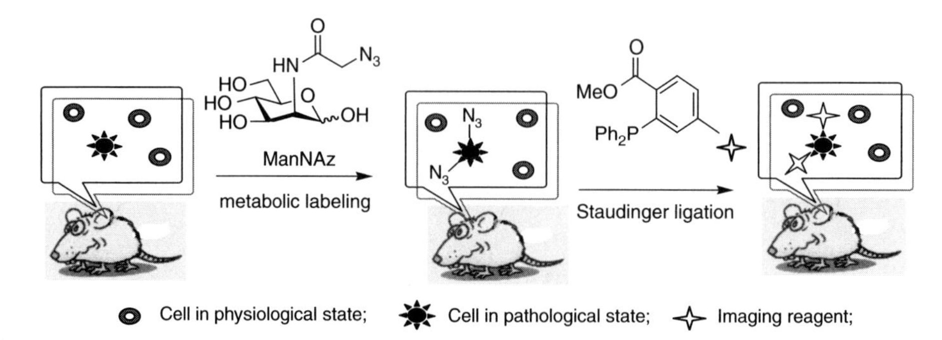

FIG. 7. Metabolic oligosaccharide engineering via bioorthogonal Staudinger ligation.

FIG. 8. Synthetic vaccines against *Haemophilus influenzae*; protein = HSA or TT, $n \sim 7$.

2. Neoglycoproteins as Vaccines Against *Haemophilus influenzae* Type b

A sulfur-based chemoselective method was applied to the synthesis of a conjugate polysaccharide vaccine against *Haemophilus influenzae* type b.[37] The polysaccharide **7** was conjugated to human serum albumin (HSA) and to the tetanus toxoid (TT) through the thiolation of lysine ε-amino groups (Fig. 8).

3. Neoglycopeptides as Inhibitors of Oligosaccharyl Transferase

Linear hexapeptides featuring the asparagine mimetics alanine-β-hydrazide **9**, alanine-β-hydroxylamine **10**, and 1,3-diaminobutanoic acid **11** have been synthesized as oligosaccharyl transferase (OT)-substrate mimetics and chemoselectively N-glycosylated to obtain the corresponding neoglycopeptides as OT-product mimetics as enzymatic inhibitors (Fig. 9).[38] These residues have been chosen as asparagine mimics since the side chain functionality reacts chemoselectively with reducing sugars, without the need for protecting groups or activating agents, to afford N-linked glycoconjugates in highly convergent synthetic routes.

An *N′*-methyl-amino-oxy amino acid has also been incorporated into peptides using standard SPPS procedures. Reaction of these peptides with native reducing sugars yields neoglycopeptides via a chemoselective reaction with the aminooxy side chains, maintaining the linked sugars in their cyclic conformations and close to the peptide backbone.[39]

Neoglycosylation of proteins was also applied to the modification of the psychrophilic Atlantic cod trypsin; the applicability of psychrophilic enzymes is limited because of their lower thermodynamic stability, despite their higher catalytic rate. It has been shown that the thermodynamic stability could be enhanced appreciably by covalent chemical modification with an oxidized sucrose polymer without affecting the enzymatic activity. The acquired stability of cod trypsin was found to be on par with the mesophilic porcine trypsin.[40]

III. Neoglycolipids

Glycosylation of lipids is known to influence many biological processes and the pathogenesis of infectious agents including prions, viruses, and bacteria, giving rise to a wide range of diseases, ranging from rare congenital disorders to diabetes and cancer.[1e,14b,41] Glycolipids are, indeed, natural receptors for hormones, matrix proteins, viruses, and bacteria or toxins, such as cholera toxin, shiga toxin, and verotoxin of *Escherichia coli*.[42] Furthermore, as a component of important antigen systems and membrane receptors, glycolipids play a central role in many important intracellular signaling mechanisms controlling cell growth, survival, and development,[43] as has been well catalogued since the first

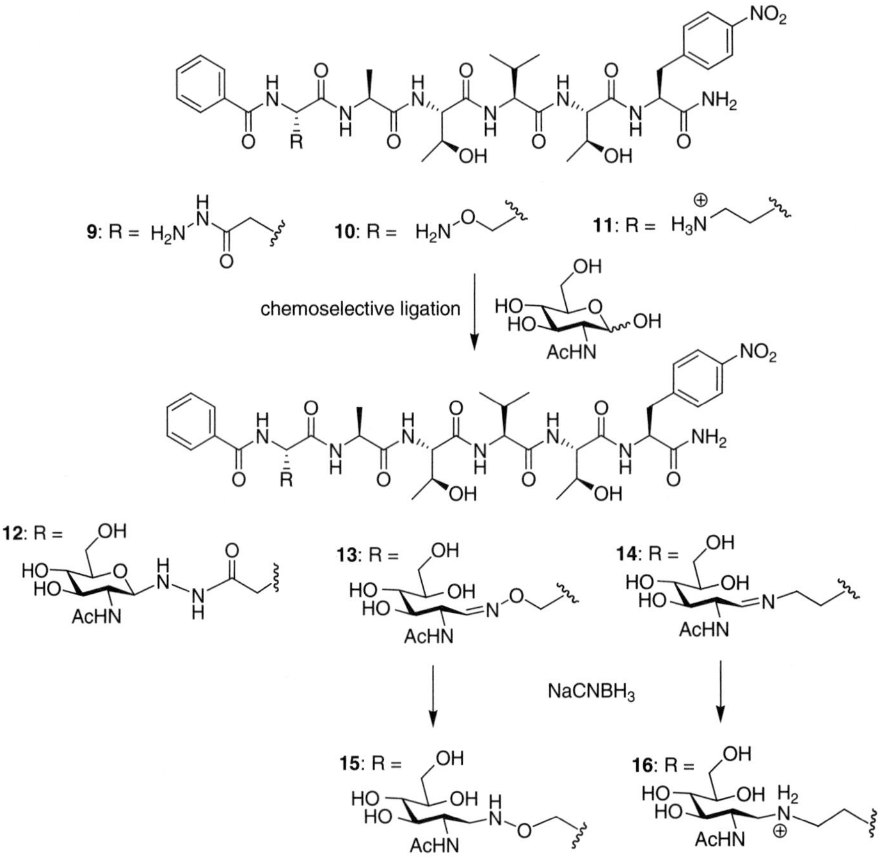

FIG. 9. Substrate mimetics and chemoselectively N-glycosylation to the corresponding neoglycopeptides as OT inhibitors.

review on glycolipids in 1937.[44] In addition, such glycolipids as GM2, GD2, GD3, 9-*O*-acetyl GD3, fucosyl GM1 (gangliosides), Lewis^y (Le^y), and globo-H, attached to the lipid bilayer at the cell surface by hydrophobic forces through the ceramide moiety, are found upregulated on the cell surface of tumor cells.[45] Common types of cancers such as melanomas, sarcomas, and neuroblastomas express a broad range of carbohydrate antigens as glycolipids.

Despite the small size of glycolipids, their complexity has still driven efforts for synthesis of simpler mimetics in an attempt to identify the minimal structural features required for the biological activity. To this end, many synthetic methods have been adopted in glycolipids synthesis, based on traditional glycosylation procedures for the formation of the glycan–lipid bond.[46] However, in order to satisfy the rapidly increasing demand for these glycolipid mimetics, an alternative, speedy, and more efficient methodology is desirable. Chemoselective coupling reactions have become very popular and useful in organic chemistry, and therefore can also constitute a good approach for neoglycolipid synthesis. The notable advantage of such an approach is the use of unprotected sugars and mild reaction conditions.

However, despite the evident advantages of chemoselective techniques, only a limited number of such reactions have, to the best of our knowledge, found application in the formation of the sugar–lipid bond.

1. Neoglycolipids for the Functionalization of Liposomes for Drug or Gene Delivery

Current pharmaceutical interest in glycolipid research is more and more focused on the possibility of targeting active drugs specifically to their site of action (drug targeting). While many therapeutic agents are discovered every year, their clinical application is often limited by their toxicity at non-target sites and by their failure to reach the site of action. To this aim, drug targeting would not only reduce systemic toxicity but would also potentiate the pharmacological activity by concentrating the drug in the target cells or tissues.

Liposomes constitute a particularly promising vector system.[47] In particular, glycoliposomes comprise a special type of carrier for drug delivery; the presence of the glycan moiety allows improved targeting of tissues or cells,[48] because of the crucial roles played by glycans in recognition events.[1] Neoglycolipids have been proposed for preparation of glycoliposomes and present many advantages, such as ease of handling, quantitative controls, and definitive identification of the saccharide determinant. Many examples have been reported, for instance the syntheses of five biantennary oligomannose neoglycolipids that could be incorporated into cationic liposome systems for the active targeting of liposomes to macrophages in gene-therapy strategies.[49] A recent report describes the syntheses of *N*-acetyl-α-D-galactosamine-containing neoglycolipids which incorporated into liposomes can be recognized by the ST6GalNAc tranferase involved in the cancer process.[50]

Gene therapy constitutes a novel form of drug delivery.[51] To employ their full potential as therapeutic agents, genes have to be delivered in a very specific and efficient way to their cellular targets. In the field of gene therapy, the problems and high risks associated with the use of virus particles for efficient delivery of transgenes to their cellular targets is often overcome by the use of synthetic non-viral alternatives, such as histone proteins,[52] dendrimers,[53] polymers[54] or lipids,[55] and liposomes.[47e,56] In this context, Miller and co-workers developed a simple methodology to generate valuable neoglycolipids for drug or gene delivery, which enhanced transfection efficacies in comparison to the first-generation ternary liposome:mu:DNA (LMD) vector system.[57] Chemoselective neoglycosylation was applied in the syntheses of novel carbohydrate-based agents for the stabilization of LMD non-viral vector systems.[58] LMD is a ternary nucleic acid delivery system, built around the condensed core complex of the cationic adenoviral peptide μ (mu, M) associated with the plasmid DNA (pDNA, D), and complexed with extruded cationic liposomes (L). The general structure of these neoglycolipids consists of a completely unprotected sugar, such as D-mannose, D-glucose, D-galactose, D-glucuronic acid, maltose, lactose, maltotriose, maltotetraose, or maltoheptaose, coupled with a cholesterol-based aminoxy lipid (Fig. 10). The desired neoglycolipids **17** were constructed, as shown in Fig. 10, by chemoselective coupling of the aminoxy functional group of the lipid moiety with the aldehyde group present in the open-chain form of

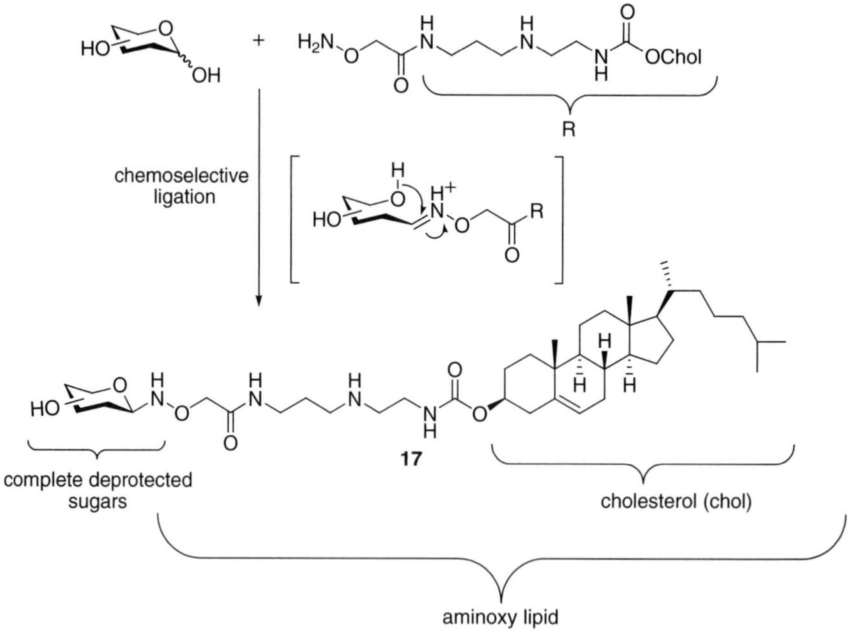

FIG. 10. Synthesis of cholesterol-based neoglycolipids.

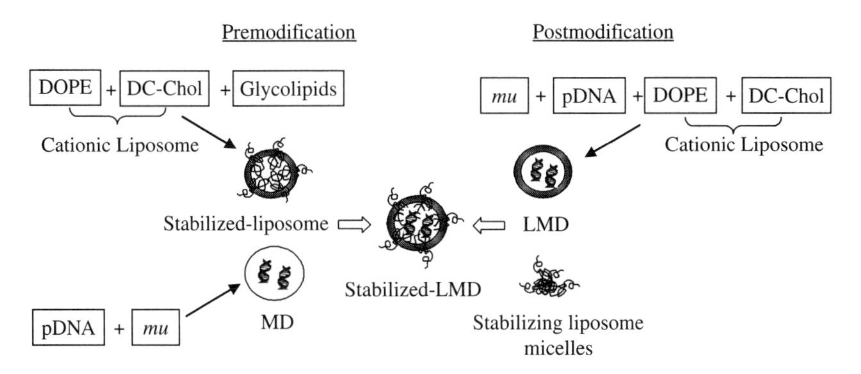

FIG. 11. Stabilized LMD preparations.

glycan residues (reducing termini). Coupling reactions were performed under mild conditions, making use of a solvent–buffer system comprising *N*,*N*-dimethylformamide (DMF) and aqueous acetic acid (1:1, v/v), pH 4, selected to optimize simultaneously both the combined solubilities of sugars and aminoxy lipid, together with aminoxy functional-group reactivity. Both α and β anomers of the aminoxy glycopyranose residues were formed, although ^{1}H NMR revealed a major configurational preference for the β anomer. Stabilization of the LMD systems was achieved both with post- and pre-modification principles (Fig. 11). In post-modification preparations, neoglycolipids micelles were incubated with unmodified LMD. In pre-modification preparations, the first incubations of defined molar percents of the neoglycolipid with DC-Chol (3β-[*N*-(*N*′,*N*′-dimethylaminoethane)carbamoyl]cholesterol) and the neutral lipid dioleoyl DOPE (L-α-phosphatidylethanolamine) were followed by the standard preparation of liposome and LMD.

2. Neoglycolipids as Probes for Investigating Biological Recognition Phenomena

Inclusion of oligo- or poly-saccharides in cell membrane models, such as lipid monolayers or liposomes, provides a basis for understanding the mechanisms that govern specific recognition phenomena.[59] For example, glycosphingolipids mimics, such as non-labile L-glucocerebrosides, were synthesized for the study of lipid accumulation found in Gaucher disease,[60] while a large number of derivatives of the globo series of glycolipids have been proposed to map the interactions with lectins in the pathogens *Escherichia coli* and *Streptococcus suis*.[61] Fluorescent neoglycolipids, consisting of aminolipids linked to several neutral or acidic oligosaccharides, have been designed as probes for biological and structural studies.[62] These lipid-linked oligosaccharide probes, obtained by reductive amination between the carbohydrate moiety and the lipid chain *N*-aminoacetyl-*N*-(9-anthracenylmethyl)-1,2-dihexadecyl-*sn-glycero*-3-phospho-ethanolamine (ADHP), should be powerful tools for studying the roles of specific oligosaccharides in cell signaling, as well as in the elucidation of oligosaccharide-binding specificities of antibodies and endogenous carbohydrate-binding proteins, and in microbial adhesion studies.[63]

Lactosylceramide (LacCer) is a pivotal intermediate in the degradation and syntheses of many complex glycosphingolipids, such as ganglioside GM3,[64] globotriosylceramide, and sulfatides, involved in many biological events, such as cell–cell and cell–matrix interactions.[65] Moreover its important role as a lipid second messenger is well known in signaling events linked to cell differentiation, development, apoptosis, and oncogenesis,[3,66] and as receptors for viruses and bacteria.[67] The difficulties encountered in large-scale synthesis of LacCer in purified form led to the development of neoglycolipids that can also be used as molecular carriers to the target site, and as inhibitors of cell adhesion. Furthermore, LacCer mimics should be useful tools for the identification of bioactive oligosaccharides.

To improve the rational design of glycoliposomes as probes for investigating biological recognition phenomena, Harada and co-workers provided efficient enzymatic and chemoenzymatic procedures for preparing unnatural "double-tailed" lactosylated neoglycolipids (**18**–**20**) as LacCer-like structures (Fig. 12).[68] The mimics are formed by a disaccharide glycoside and phospholipid moieties, connected by a suitable linker. The hexadecanoylphosphatidyl moiety (DPPC) was selected as the lipid part, as it is one of the most widely distributed lipids in biological systems. The resulting neoglycolipids were readily transformed into glycoliposomes, wherein the glycan is exposed on the liposomal surface, as showed by NMR experiments and lectin-binding assays. Syntheses of the neoglycolipids 1,2-dihexadecanoylphosphatidylethanolaminyl β-lactoside (Lac-DPPE, **18**), 1,2-hexadecanoylphosphatidylhexyl β-lactoside (Lac-DPPA, **19**), and *N*-acetyl-β-lactosaminide (LacNAc-DPPA, **20**) were achieved by an endo-β-glycosidase that is a cellulase from *Trichoderma reesi* that catalyzes transglycosylation and condensation reactions (Fig. 12).[69] The cellulase-mediated condensation is chemoselective for the hemiacetal hydroxyl group and proceeds in satisfactorily high yields. In Lac-DPPA (**19**) and LacNAc-DPPA (**20**), the phospholipid moieties were regioselectively linked to the primary hydroxyl group of the hydroxyhexyl aglycon moiety via enzymatic transphosphatidylation mediated by phospholipase

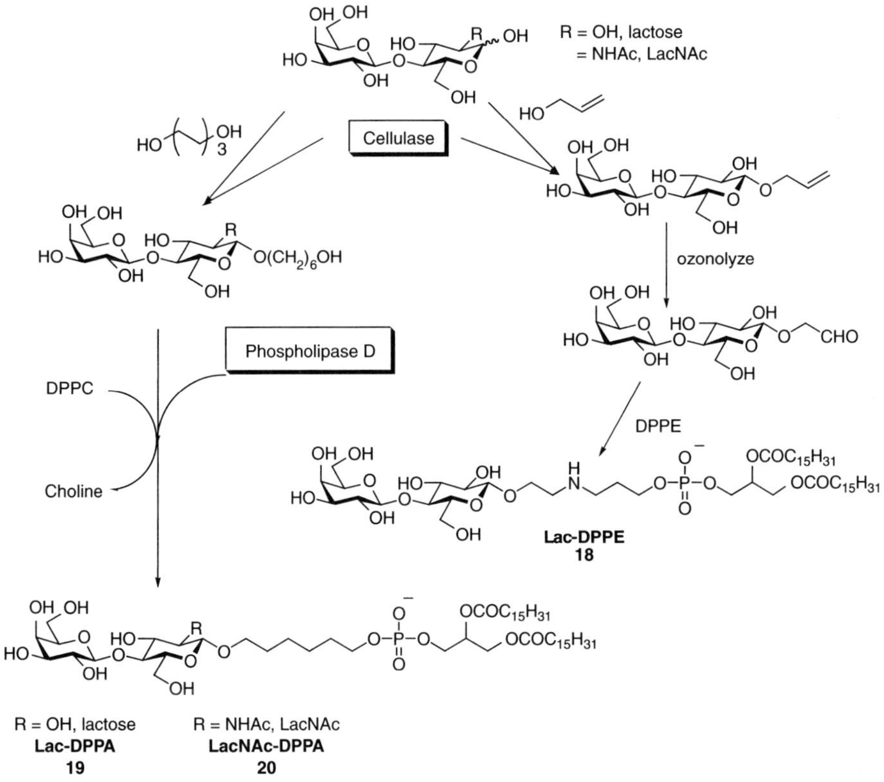

FIG. 12. Synthesis of lactosylated neoglycolipids.

D from *Streptomyces* sp. (PLD, Fig. 12). This enzyme catalyzes transfer of the phosphatidyl group from phospholipids to primary alcohols.[70] An alternative route was used for synthesis of Lac-DPPE (**18**): reductive amination of a suitable sugar-aldehyde with 1,2-dihexadecanolyphosphatidylethanolamine (DPPE) allowed conjugation of the lipid moiety to the disaccharide glycoside (Fig. 12). The ability of these neoglycolipids to detect lectins at very low concentrations and to trap calcein, a chelating derivative of fluorescein, in multilamellar vesicles makes them particularly interesting as probes for elucidating the mechanisms that govern specific cellular recognition.

Finally, an interesting example has been reported in the syntheses of new photoreactive neoglycolipid probes carrying carbohydrate determinants specific for some murine mammary adenocarcinoma cells for the study of lectin receptors on tumor cell surfaces.[71]

3. Neoglycolipids in Anticancer Treatments

The ability of liposomes to carry drugs to active sites makes them also interesting vectors in vaccine design.[72] In the field of antitumor treatments, Boons and co-workers reported the synthesis of a fully synthetic lipidated glycopeptide (**21**) as an anticancer vaccine, where the lipid moiety serves as a "built-in" adjuvant.[73] The title compound is composed of the tumor-associated Tn antigen, as the B-epitope, the peptide YAFKYARHANVGRNAFELFL (YAF) as the T-epitope, and the lipopolypeptide *S*-[(*R*)-2,3-dihexadecanoyloxypropyl]-*N*-palmitoyl-(*R*)-cysteine (Pam_3Cys) (Fig. 13). The Tn antigen, which serves as the B-epitope, is overexpressed on the surface of human epithelial tumor cells of the breast, colon, and prostate. To overcome the T-cell independent properties of the carbohydrate antigen, the YAF peptide was incorporated. This 20-amino acid peptide sequence is derived from an outer-membrane protein of *Neisseria meningitides*, and has been identified as an MHC class II-restricted site for human T-cells. However, the combined B-cell and helper T-cell epitope lacks the ability to provide appropriate "danger signals" for dendritic cell (DC) maturation. Therefore, the lipopeptide Pam_3Cys, which is derived from the immunologically active N-terminal sequence of the principal lipoprotein of *E. coli*, was incorporated.

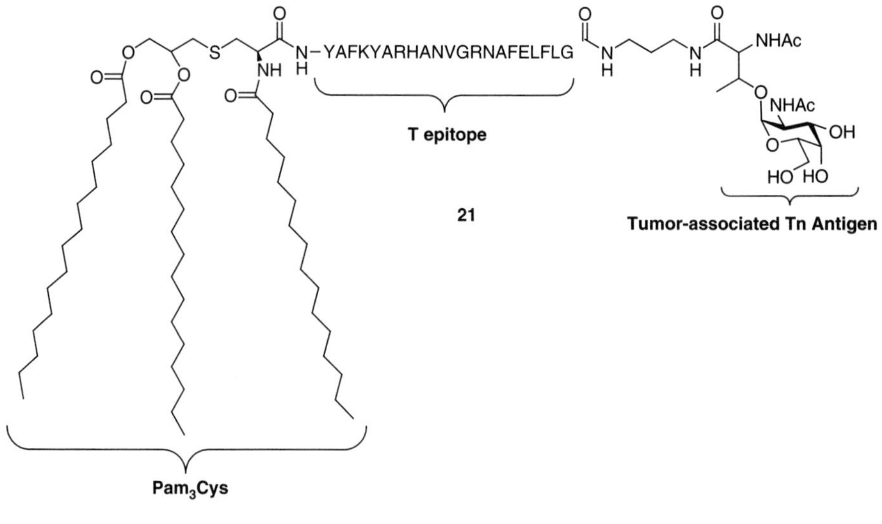

FIG. 13. Structure of the fully synthetic lipidated glycopeptide.

The vaccine was incorporated into phospholipid-based liposomes prior to administration; the mice that were immunized with the liposome preparations elicited IgM and IgG antibodies against the Tn antigen.

Synthesis of the vaccine candidate **21** was achieved in high yield by condensation of the C-terminal carboxylic acid of lipopolypeptide (**24**) and the amine of spacer-containing Tn antigen (**25**), followed by deprotection of the amino acid side chain. The fully protected lipopolypeptide **24** was produced by coupling the lipid *S*-[2,3-bis(hexadecanoyloxy)-propyl]-*N*-Fmoc-Cys ($Pam_2FmocCys$ (**3**)) to the N-terminal amine of the resin-bound peptide (**22**), and by subsequent linking of the hexadecanoic acid to the free amine of the resin-bound lipopolypeptide (Fig. 14).

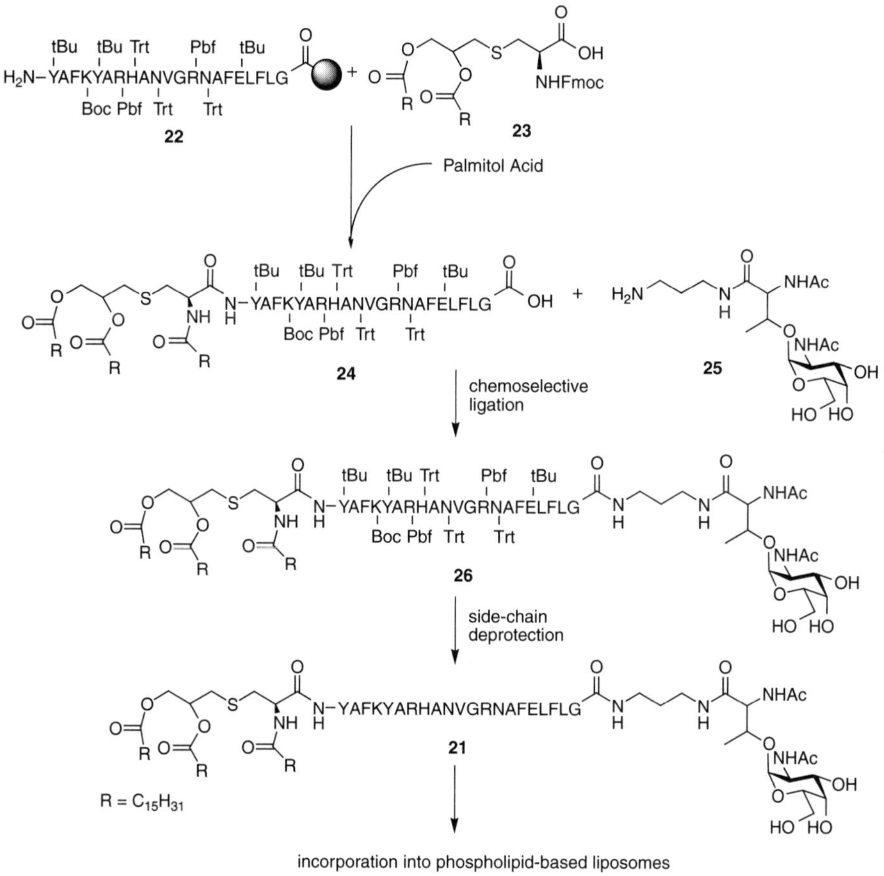

tBu tBu Trt Pbf tBu
H_2N–YAFKYARHANVGRNAFELFLG
Boc Pbf Trt Trt
22
NHFmoc
OH
23
Palmitol Acid
24
25
chemoselective ligation
26
side-chain deprotection
N–YAFKYARHANVGRNAFELFLG
21
$R = C_{15}H_{31}$
NHAc
HO HO
incorporation into phospholipid-based liposomes

FIG. 14. Synthesis of the lipidated glycopeptide.

IV. Glycoarrays

The role that carbohydrates exert in recognition phenomena of biological and pharmaceutical relevance has stimulated the quest for tools that can monitor and even quantitatively evaluate these interactions. The potential diagnostic applications of such tools is evident, and the possible development of commercial carbohydrate microarrays, such as those established for oligonucleotides, and very recently peptides, is of high interest.

The microarray approach is ideal for screening the interaction with proteins of a large number of glycoforms that are available in only small quantities. Few approaches have yet been developed for the generation of carbohydrate microarrays,[74] in contrast to gene-chip technology, a procedure now well established. Once carbohydrate microarrays have been perfected for selected applications, as for example, for immunodiagnostic purposes, they will be amenable to large-scale production and commercialization.

The generation of such tools, based on the conjugation of saccharide entities to different and heterogeneous materials, requires not only the availability of a wide variety of glycans of biological relevance (obtained by isolation from natural sources or by synthesis), but more importantly and specifically, the availability of efficient methods for linking them to supports of different types and size.

In addition to the chips technologies, other interesting materials, such as gold microparticles grafted with carbohydrates[75] or glycocluster nanoparticles,[76] and paramagnetic microspheres (Fig. 15), are under investigation and their applications in biomedicine is promising.

In general, thin layers, beads, micro-, and nanoparticles made of ceramics, glasses, nitrocellulose, polystirene, and gold, have been exploited, and the concept of multivalency has been widely applied.[77]

V. Glycodendrimers

Owing to the weakness of carbohydrate receptor–protein interactions, in order to attain biological meaningful affinities for the receptor, carbohydrates very often need to be clustered and expressed in multiple copies. For this purpose glycodendrimers, which are multivalent glycoconjugates with well-defined chemical structures, have received recent attention for their considerable potential as tools for studying cell-surface protein–carbohydrate interactions, because of the affinity enhancement obtained by multivalency.

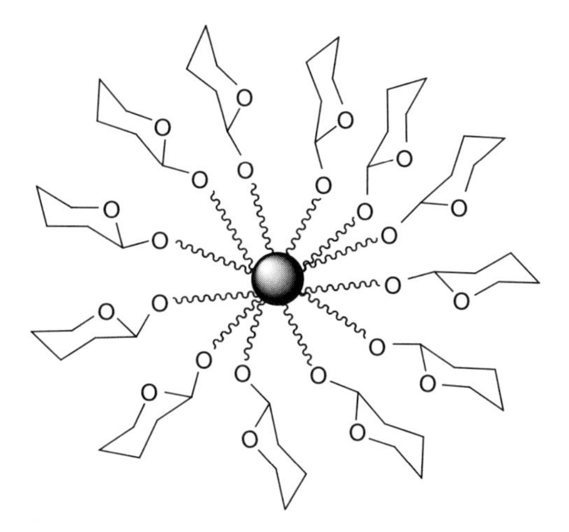

FIG. 15. Representation of a glycosylated nanoparticle.

The importance of these structures is clear by observing the numerous biomedical applications[77b] that have been suggested for such compounds: including inhibition of viral and bacterial infections, tumor eradication by stimulating the immune response, preventing toxin binding, and many others. Selected examples have been chosen to illustrate these potentialities.

Glyco-coated dendrimers possess a core structure of varied chemical nature, with a number of peripheral groups to which carbohydrate moieties are attached. The most common core structures are based on polyaminoamides (PAMAM) or poly(propyleneimine) (Astramol^{TM}) with a primary amine at the periphery, and these are commercially available. Polyamides based on *N*,*N*-bis(3-aminopropyl)glycine and *N*,*N*-bis(3-aminopropyl)succinic acid have also been prepared.[78]

Examples of glyco-coated dendrimers are the sialylated dendrimers[79] that have been developed to prevent infection by the influenza virus. This infection is mediated by the interaction of viral glycoprotein hemagglutinin and the sialic acid residues present on cell-surface glycoproteins and glycolipids.

A G4 PAMAM dendrimer conjugate[80] (Fig. 16) has been synthesized, and *in vivo* studies have demonstrated its ability to protect against experimental infection by influenza A X-31 H3N2 virus in mice. In this case the glycodendrimer not only showed increased binding affinity due to the multivalency, but also facilitated the delivery of sialic acid, preventing enzymatic breakdown *in vivo*, and diminishing the cytotoxicity.

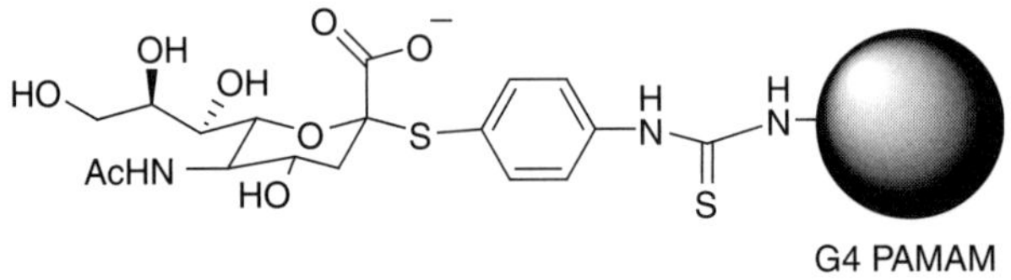

FIG. 16. A tetravalent sialylated PAMAM dendrimer.

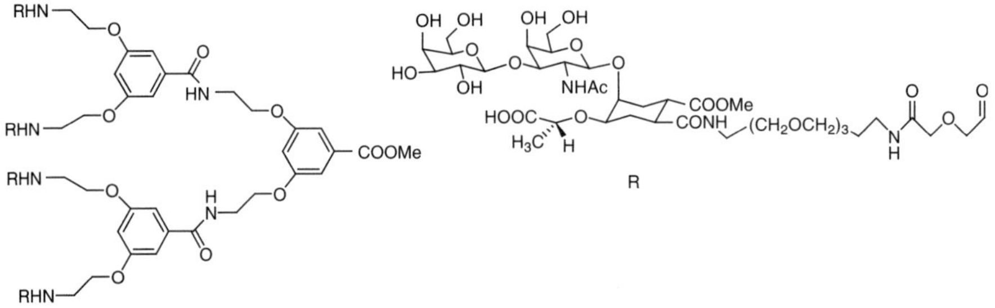

FIG. 17. A tetravalent dendrimer of a GM1 mimic.

The GM1 or GM1 mimics glycodendrimers comprise another interesting example of oligosaccharide-derivatized glycodendrimers. The oligosaccharidic part of GM1 ganglioside (βGal1→3βGalNAc1→4[αSia2→3]-βGal1→4Glc) is responsible for the adhesion of cholera toxin on the cell surface of the intestinal tract. This interaction is crucial for toxin uptake and disease development, and is multivalent since five subunits are present on the binding site. Dendrimers based on a poly(propylene)imine or polyamidoamine core decorated with GM1 oligosaccharides showed up to 1000-fold higher inhibition properties than the free oligosaccharide.[81] More-accessible mimics (Fig. 17) of these complex oligosaccharides have also been prepared and attached to a dendrimer core based on the 3,5-bis-(2-aminoethoxy)benzoic acid branching unit.[82]

The binding affinities of the tetramer with respect to the monomer showed a 442-fold enhancement.

A series of mannosylated dendrimers have been prepared and evaluated for their biological properties, in particular as inhibitors of plant lectins.[83] Subsequently, mannosylated PAMAM, and a dendrimer based on a 3,5-bis-(2-aminoethoxy)benzoic acid branching unit, have been tested for their ability to inhibit bacterial infection by interfering with the binding of uropathogenic

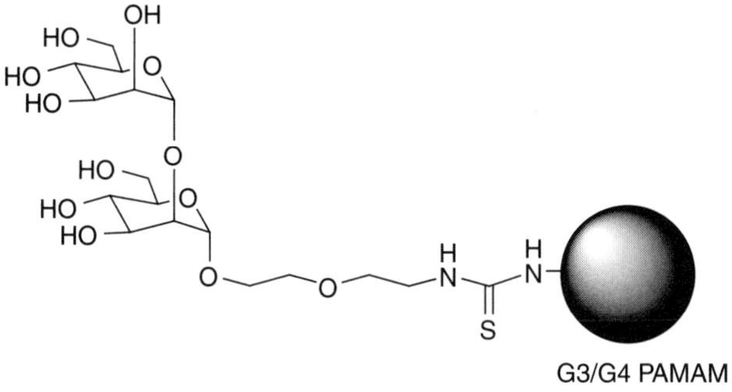

FIG. 18. Dimannosylated PAMAM dendrimer.

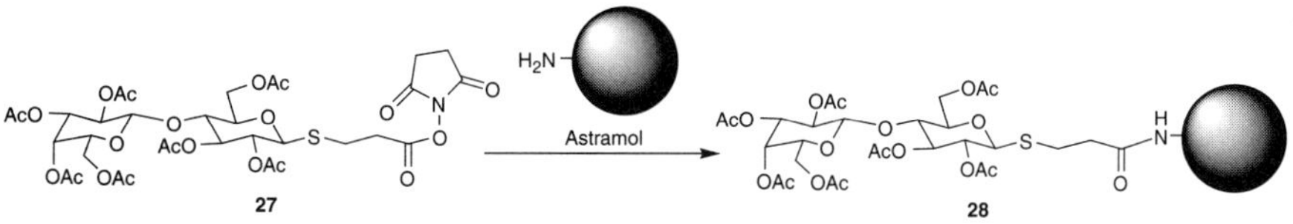

FIG. 19. Coupling of *N*-hydroxysuccinimide-activated thiolactoside **27** to Astramol™.

E. coli. All multivalent compounds showed enhanced affinity as compared to mannose, with IC_{50} values in the micromolar range.[84]

Mannosylated PAMAM dendrimers have also been used in inhibition studies on HIV infection.[85] In this case, dimannosylated generation three and four of the dendrimer (Fig. 18) were prepared and tested for their binding to cianovirin-N (CV-N), an HIV-inactivating protein that blocks virus-to-cell fusion through a high mannose-mediated interaction. These dendrimers were found to be effective for the recruitment of CV-N.

As shown in the previous examples, the sugar moieties can be attached to the core structures by various approaches. More often the carbohydrate moieties are attached to the core through a linker introduced at the anomeric position of the sugar,[86] as for example in compound **27** (Fig. 19). The linker is bound to the anomeric center of the sugar through an *O*-glycosidic or *S*-glycosidic linkage, performed through classical glycosylation methods that require laborious protection/deprotection steps.[83a,87]

Dendrimers bearing primary amino groups at the periphery have also been directly glycosylated. The linkage is effected by reacting the amino group

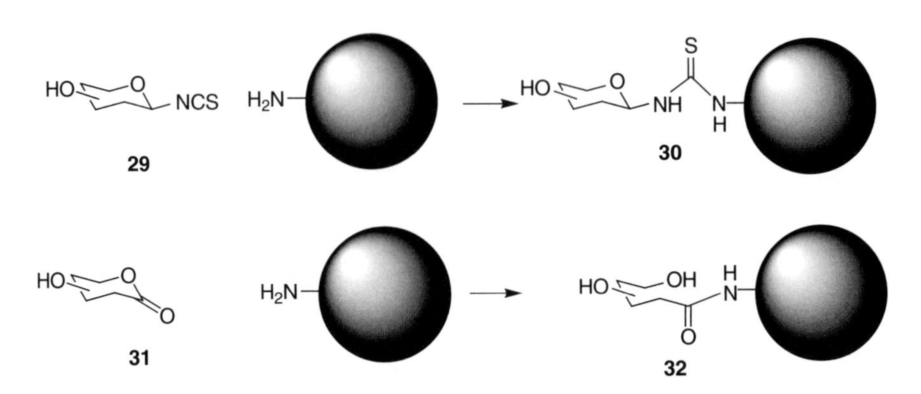

FIG. 20. Direct coupling of glyconolactones (**31**) and sugars activated as isothiocyanates (**29**) to dendrimers bearing primary amino groups.

with glyconolactones, affording an amide **31**[88] (Fig. 20), while reaction with an isothiocyanate derivative **29** (Fig. 20) forms a thiourea bond.[83,89]

These approaches have the advantage of utlilizing unprotected carbohydrates, but the preparation of the saccharide component remains in general a laborious task.

An alternative approach consists in functionalizing the dendrimer for a chemoselective linkage employing glycosyl thiols.[90] Synthesis of the α-thiosialo dendrimer **33** (Fig. 21) is an interesting example.[91] Here the 2-thioglycosyl component is linked to a 3,3′-iminobis(propylamine)-based core structure, where the terminal amino groups are functionalized by chloroacetyl groups.

The chemoselective character of this reaction permits the use of unprotected glycosyl thiols.[92] With a similar strategy, an unprotected sugar derivative has been linked to 2-mercaptoacetyl-functionalized dendrimer forming a disulfide bond.[93]

Instead of a single carbohydrate moiety, a small cluster can be attached to each terminal functional group of the dendrimer **36** (Fig. 22),[86a] or used in a convergent approach wherein the cluster is attached to branching units, which in turn are finally linked to a suitable core.[83a,87a,b]

Completely unprotected carbohydrate clusters have been successfully synthesized following this approach,[94] overcoming the problems of steric hindrance inside the glycocluster and during the coupling reactions.

Glycodendrimers are useful for generating the multivalency important for protein–carbohydrate interactions and for many biomedical applications. Nevertheless, the formation of these structures is most often a laborious task. The

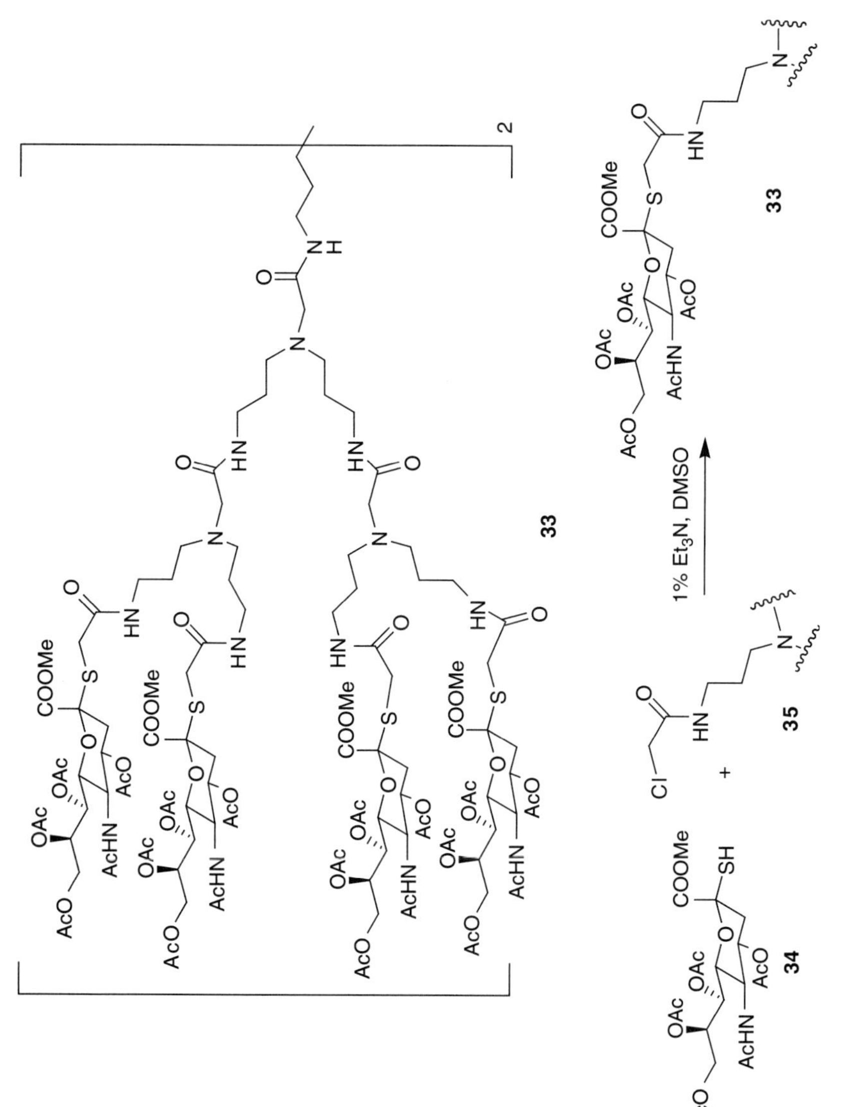

FIG. 21. α-Thiosialo dendrimer obtained by reaction of the 2-thioglycoside with a chloroacetyl-functionalized dendrimer.

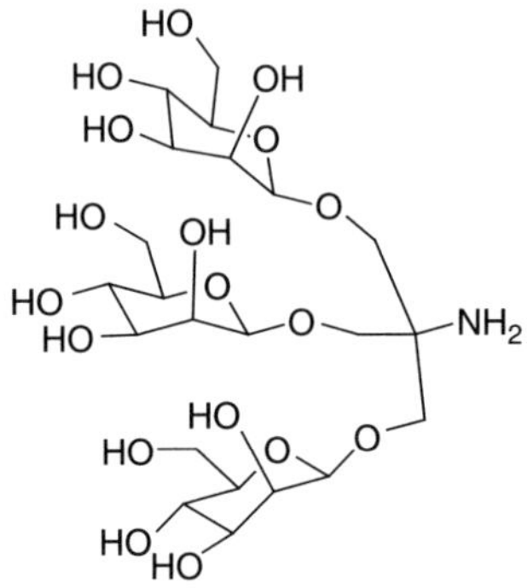

Fig. 22. Example of a clustered carbohydrate moiety.

potential for chemoselective procedures to generate glycodendrimer structures with less chemical manipulation of the carbohydrate moieties is therefore welcome.

VI. Glycosylated Polymers

Glycopolymers, consisting of sugar residues attached to a polymer backbone, have emerged as important tools for the investigation of sugar–protein interactions.[8a,77j,95]

Various biomedical applications (such as drug delivery, cell targeting, adhesion, tumor diagnosis and detection, trapping of viruses and bacterial toxins, and the like) have been suggested for these sugar polymers[8a–c,95b,c,96] based on the properties conferred by the carbohydrate residue. Their utility is mainly ascribed to their strong interaction with receptor proteins through the multivalency effect of the binding.

Particular interest has been directed to glycopolymers carrying such cell-surface oligosaccharides as sialyl Lewis^{x} (sLe^{x})[97] and globosyl oligosaccharides[98] and their mimics. An example of the former is the poly(acrylamide-co-3′-sulfo-Lewis^{x}-Glc) obtained by copolymerization of the acryloylated monomer **37** (Fig. 23) with acrylamide. This polymer is a promising antagonist of cell-adhesion molecules, since it is a good inhibitor of L- and E-selectins, displaying IC_{50} values in the micromolar range.

Other glycopolymers carrying a sialyl Lewis^{x} mimic were obtained by copolymerization of functionalized carbohydrate monomers with acrylamide; for

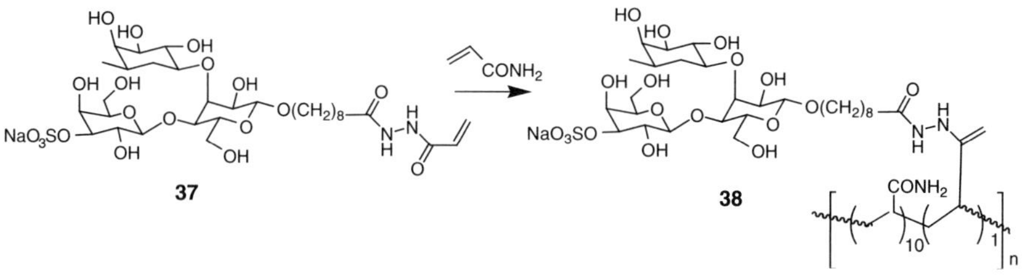

FIG. 23. Glycopolymer containing a sialyl Lewis^x mimic.

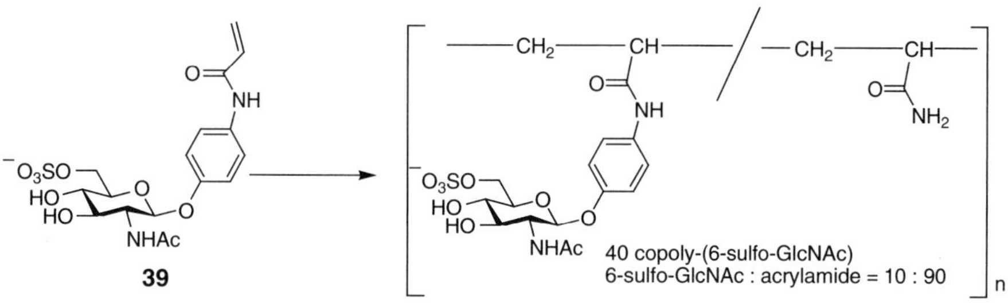

FIG. 24. Polymer containing a 6-sulfo-GlcNAc component.

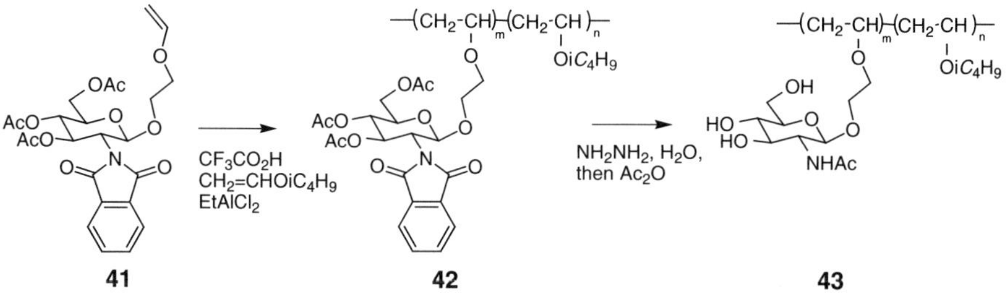

FIG. 25. Preparation of the GlcNAc-carrying polymer by living-cationic polymerization.

example the glycopolymer **40**[99] (Fig. 24) carrying a 6-sulfo-GlcNAc cluster was obtained from monomer **39**.

An array of synthetic methods for the polymerization of sugar-based monomers has been developed, including ionic polymerization, controlled radical polymerization, and atom-transfer radical polymerization (ATRP). The GlcNAc-carrying polymer **43**[100] (Fig. 25) is an illustrative example of controlled-structure sugar polymers. It was prepared by living-cationic polymerization, starting from the vinyl ether building block **41** (Fig. 25). The resultant copolymer showed a significant increase in binding affinity toward WGA lectin as compared to GlcNAc and its oligomers.

Functionalized biopolymers are also of great interest in the field of tissue engineering, where they are used to repair and regenerate tissue. They serve to support, reinforce, and organize regenerating tissue and in some cases they also release bioactive substances. Many polymers presently used in biomedicine are such natural polysaccharides as chitosan, hyaluronate, alginate, and agarose, which show adequate biocompatibility, biodegradability, and the possibility of being derivatized.[101]

Specific carbohydrate ligands may be linked to these polymers in order to enhance the cell-adhesion properties of the biomaterial or in order to modify their physicochemical properties. An example is a biomimetic surfactant polymer of a poly(vinylamine) backbone bearing pendant maltoheptose units to increase local hydration and to inhibit non-specific adsorption.[102]

The ability to prepare such biopolymers directly from available carbohydrates and matrices, adopting chemoselective-neoglycosylation methods offers promise for valuable improvements in this field.

VII. Neoglycosylation Procedures

This section summarizes the different methods of neoglycosylation, most of which have been applied to the synthesis of the various molecular targets described in the previous sections. Depending on the nature of the aglycon, different ligation technologies can be used. The covalent ligation methodologies can be divided roughly into three categories: (a) methods that effect direct linkage of the sugar chain to the polymer, exploiting the anomeric center of the reducing sugar (the "true" neoglycosylation methods); (b) methods whereby appropriately functionalized sugars are linked to a spacer that increases the distance of the carbohydrate entity from the solid support (different spacers with a variety of functional groups for the ligation have been exploited); and (c) methods in which multiple carbohydrate entities are clustered on dendrimeric linkers in order to effect stronger interactions with the sugar receptors, a phenomenon termed "multivalency".

1. Direct Linkage

The direct linkage of saccharides to different aglycons has been widely applied in glycosylation of proteins, peptides, and other materials. The most widely used

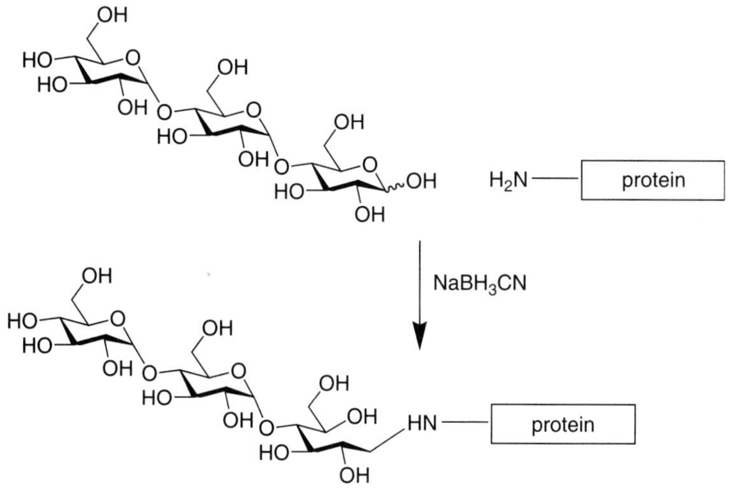

FIG. 26. Direct linkage of saccharides to different aglycons by reductive amination.

procedure is that of reductive amination, in which the reducing end of an oligosaccharide reacts with a primary amine (in the protein, a lysine residue), thereby losing its cyclic structure and generating a stable linkage R-NH-R′ with the aglycon.[103] The reaction has the advantage of being chemoselective, no protection/deprotection steps are required, and it can be performed in aqueous (physiological) media. On the other hand, the structure of the oligosaccharide is partially modified, the sugar involved in the linkage being no longer cyclic. Nevertheless, if the sugar chain is long enough, this structural modification should not interfere in the recognition phenomenon (Fig. 26).

An interesting alternative has been proposed that allows the chemoselective ligation of a "reducing sugar" without affecting its cyclic structure.

The chemoselective synthesis of neoglycoconjugates has benefitted from the techniques previously developed for the assembly of protein mimics, starting from unprotected fragment peptides.[4,30] These techniques were based on the formation of oxime, hydrazone, hydrazide, thioether, thioester, and thiazolidine bonds. The formation of these linkages has been exploited for chemoselective glycosylation as depicted in Fig. 27. Well-known organic reactions, such as nucleophilic addition to carbonyls, nucleophile displacement by sulfur nucleophiles, and Diels-Alder (4 + 2) or azide–alkyne (3 + 2) cycloadditions, have been revisited in the light of chemoselective ligation chemistry.

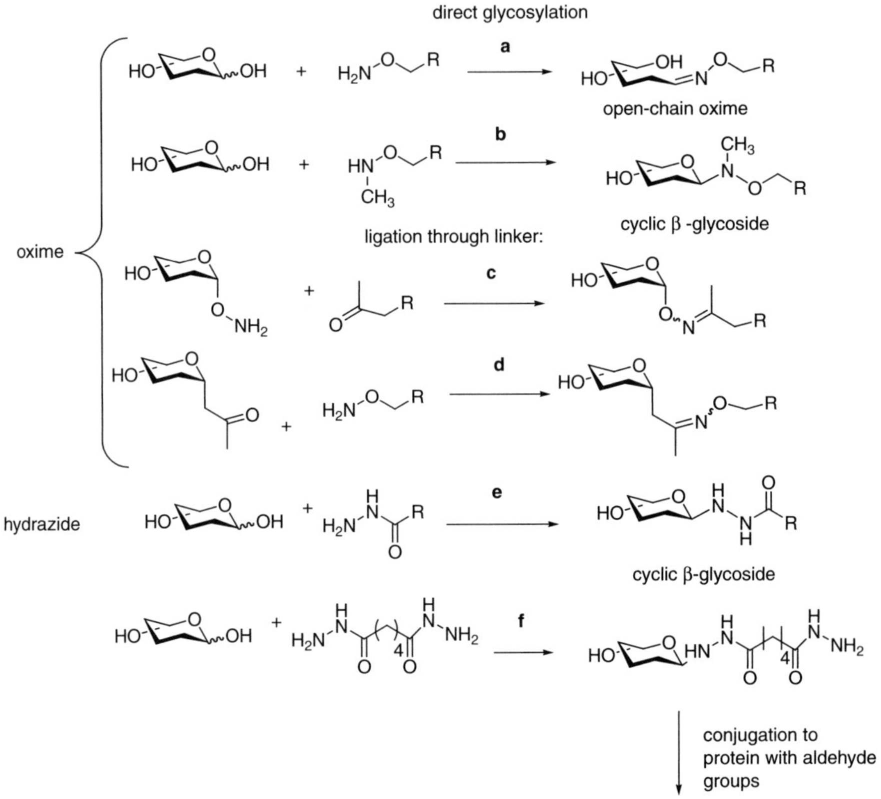

FIG. 27. Chemoselective glycosylation methods based on carbonyl chemistry.

The absence of aldehydes and ketones on the side chains of the naturally occurring amino acids inspired the development of a set of chemoselective reactions for the synthesis of neoglycopeptides and proteins. These are based on the condensation of the anomeric carbon of a reducing sugar (aldehyde group masked as a cyclic hemiacetal) with a variety of non-natural nucleophile groups introduced into the peptide chain (Fig. 27a–e). An amino-oxyacetylated unprotected peptide was glycosylated in a chemoselective way after SPPS with a variety of unprotected mono- and oligosaccharides, affording oxime-linked neoglycopeptides having increased bioavailability (Fig. 27a).[104] The same method, leading to open-chain oxime glycoconjugates, was subsequently used for the functionalization of surfaces with sugars for SPR studies.[105] The main drawback of this method is the formation of the open-chain oxime isomers of the first attached sugar: this form of the sugar is quite different from the natural cyclic (pyranose or furanose) form. This problem has been addressed in our laboratories by using an *N*-methylated amino-oxy group on the peptide moiety, which permitted formation of the closed-ring neoglycosides (Fig. 27b).[39,106] The glycosidic linkage to the aglycon was obtained exclusively in the β-configuration, making these mimetics structurally quite close to the natural N-linked glycoproteins. The methyl amino-oxy group can be attached to the peptide through a linker, or be connected directly on the side chain of asparagine mimetics. This method was extended by Thorson and co-workers to the synthesis of a library of neoglycosteroids (cardiac glycosides) obtained through the reaction of reducing sugars with secondary hydroxylamine-containing steroids,[107] and by ourselves in the chemoselective synthesis of $N(OCH_3)$ disaccharide mimetics.[108]

In addition to these appealing direct glycosylation methods, other methods have been explored based on the formation of neoglycoconjugates in which the oxime bond is part of a spacer. The first involves the ligation of a sugar bearing an amino-oxy group in the anomeric position to a peptide having a ketone or an aldehyde group (Fig. 27c).[109,30] The second is based on the reaction of a *C*-glycosyl derivative with a keto group and an aminoxy-peptide (Fig. 27d). Linear and branched neoglycopeptides, active as synthetic vaccines targeted to DC, have been chemoselectively assembled by this method.[23]

A direct chemoselective glycosylation method was developed by Imperiali *et al.* using an asparagine surrogate bearing a hydrazide functionality on the side chain (Fig. 27e).[38] In this instance the native cyclic form of the sugar is likewise maintained in the neoglycopeptide. A method for the one-step generation of glycosylhydrazides by reaction of the (bifunctional) adipic dihydrazide with unprotected mono- and oligosaccharides has been described (Fig. 27f).[110] The resultant sugar-hydrazides were used for chemoselective conjugation to bovine serum albumin (BSA) decorated with aldehyde functionalities.

The superior nucleophilic character of thiols inspired the use of these functional groups in combination with a series of electrophiles for chemoselective formation of neoglycoconjugates. The natural amino acid cysteine is a chemoselective attachment-point for site-selective reactions. Davis and co-workers showed that site-directed mutagenesis combined with chemical modification provides, in principle, a general method that allows both regio- and sugar-specific glycosylation of proteins. The serine protease subtilisin, from *Bacillus lentus* (SBL), which is not naturally glycosylated, was mutated in order to introduce cysteine residues in precise sites. Mutants were then treated with protected and unprotected monosaccharides bearing a methanethiosulfonate or an ethyl-tethered methanethiosulfonate as

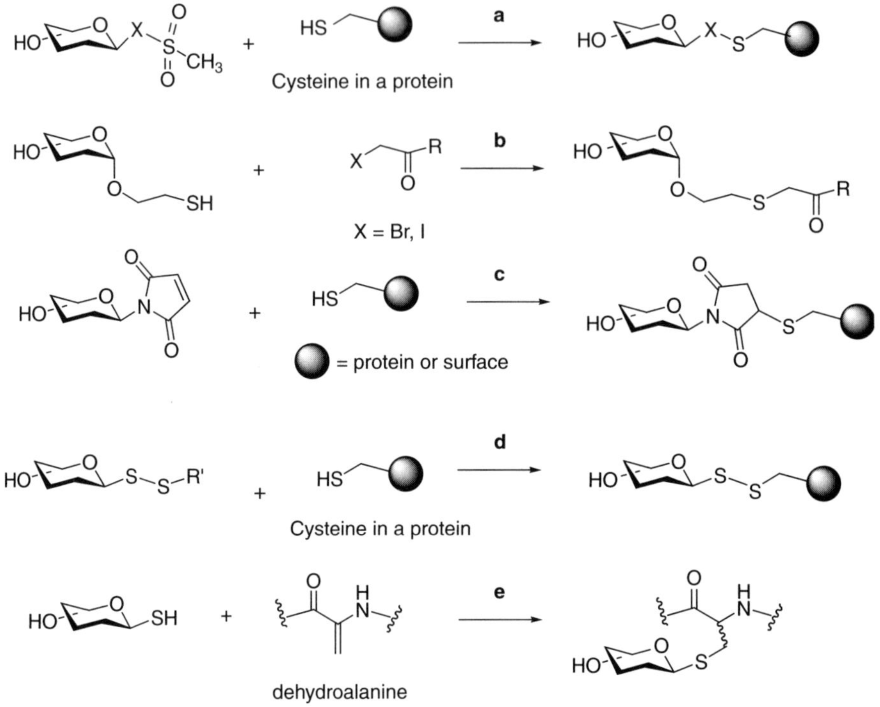

FIG. 28. Sulfur-based chemoselective glycosylation methods.

selective thiol electrophiles at the anomeric position, thus generating disulfide linkages (Fig. 28a).[111] The glycoprotein hormone erythropoietin (EPO) is naturally glycosylated and N-linked glycans at asparagine residues 24, 38, and 83 are essential for *in vitro* activity. Flitsch and co-workers undertook a systematic and elegant study on this protein through Asn→Cys mutation at the natural N-linked glycosylation sites.[112] The three mutants N24C, N38C, and N83C were refolded in the native form, and then allowed to react with an excess of glycosyl iodoacetamide, affording neoglycoproteins selectively glycosylated in the desired regions. It was also demonstrated that, despite the non-natural linker between sugar and protein, the stability of neoglycoproteins was similar to that of wild-type EPO. Bertozzi and co-workers employed sugars functionalized with bromoacetamido groups, which reacted chemoselectively with thioglycosides (Fig. 28b).[32] In model studies with short peptides, cysteines incorporated during SPPS underwent chemoselective reaction with sugars bearing maleimide (Fig. 28c)[113] and disulfide (Fig. 28d)[114] electrophiles.

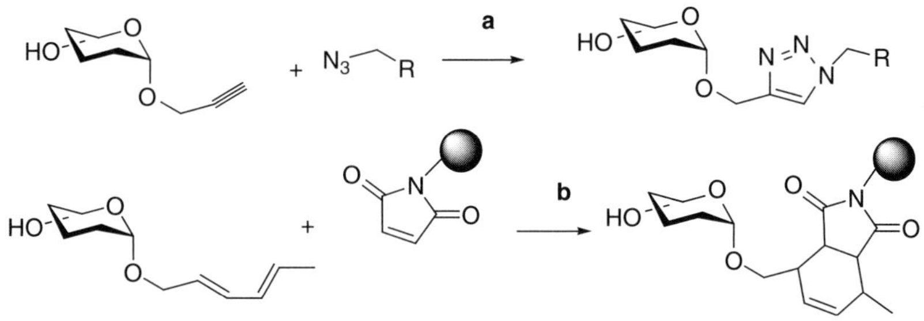

FIG. 29. Cycloaddition-based chemoselective glycosylation methods.

Sugars having maleimide groups linked to the anomeric position underwent selective reaction with maleimides tethered on surfaces for the preparation of sugar microarrays.[115]

Anomeric thioglycoslydes reacted with tripeptides containing dehydroalanine, derived from selenocysteine by oxidative elimination, thus affording S-linked neoglycopeptides through chemoselective Michael addition (Fig. 28e).[116]

Azide–alkyne (3+2) cycloadditions (Fig. 29a)[117] and Diels–Alder (4+2) cycloadditions (Fig. 29b)[118] were used for the generation of unprotected glycodendrimers and for conjugating sugars to proteins, respectively.

Melnyk and co-workers used thiol-based ligation in combination with aldehyde/hydrazide chemistry to build up complex dendrimeric synthetic vaccines targeted to DCs. An array of synthetic vaccines was generated, presenting different epitopes linked to the poly-mannose cluster.[119] This synthetic approach, based on the combination of orthogonal hydrazone and thioether methods, is an elegant and rare example of double ligation strategy, one reaction being used for sugar–peptide, the other for peptide–peptide conjugation.

An exciting extension of the chemoselective ligation chemistry to macromolecular and cellular systems was demonstrated by Bertozzi and co-workers. Glycoproteins anchored to the cell membrane were decorated by chemoselective covalent reactions in the cellular environment. Through the addition of chemically modified metabolic precursors of sialic acid, namely *N*-levulinoylmannosamine, and *N*-(2-azidoacetyl)mannosamine, cell-surface glycoproteins can be made to display ketones and azides which may be extended by selective covalent reactions. In the case of ketones, such complementary nucleophiles as hydrazides and amino-oxy groups react, forming stable adducts. This technology was applied to the remodeling of cells with alternative glycoforms,[120] to the selective targeting of toxins and diagnostic agents to tumor cells, and to modulate the cell-surface immunoreactivity.[36] The azide group on the other hand was shown to undergo a Staudinger reaction with a triphenylphosphine, and a prototypical decoration of Jurkat cells with water-soluble biotinylated phosphine has been demonstrated.[35]

References

1. (a) R. A. Dwek, Glycobiology: Toward understanding the function of sugars, *Chem. Rev.*, 96 (1996) 683–720; (b) M. Fukuda, Carbohydrate-dependent cell adhesion, *Bioorg. Med. Chem.*, 3 (1995) 205–215; (c) A. Varki, Biological roles of oligosaccharides: All of the theories are correct, *Glycobiology*, 3 (1993) 97–130; (d) T. Feizi, Progress in deciphering the information content of the 'glycome'—a crescendo in the closing years of the millennium, *Glycoconjugate J.*, 17 (2000) 553–565; (e) C. Bertozzi and L. L. Kiessling, Chemical glycobiology, *Science*, 291 (2001) 2357–2364; (f) A. Varki, R. Cummings, J. Esko, H. Freeze, G. Hart, and J. Marth, *Essentials of Glycobiology*, Cold Spring Harbor Laboratory Press, Cold Spring Harbor, NY, 1999.
2. K. A. Karlsson, Microbial recognition of target-cell glycoconjugates, *Curr. Opin. Struct. Biol.*, 5 (1995) 622–635.
3. S. Hakomori and Y. Zhang, Glycosphingolipid antigens and cancer therapy, *Chem. Biol.*, 4 (1997) 97–104.
4. O. J. Plante, E. R. Palmacci, and P. H. Seeberger, Automated solid-phase synthesis of oligosaccharides, *Science*, 291 (2001) 1523–1527.
5. X. Chen, Z.-Y. Liu, J.-B. Zhang, W. Zheng, P. Kowal, and P. G. Wang, Reassembled biosynthetic pathway for large-scale carbohydrate synthesis: α-Gal epitope producing superbug, *ChemBioChem*, 3 (2002) 47–53.
6. M. Meldal, Glycopeptide synthesis, in Y. C. Lee and R. T. Lee (Eds.), *Neoglycoconjugates: Preparation and Application*, Academic Press, San Diego, CA, 1994, pp. 145–198.
7. D. M. Ratner, E. W. Adams, M. D. Disney, and P. H. Seeberger, Tools for glycomics: Mapping interactions of carbohydrates in biological systems, *ChemBioChem*, 5 (2004) 1375–1383.
8. (a) Y. C. Lee and R. T. Lee, Neoglycoconjugates, Part A: Synthesis, in J. N. Abelson, M. I. Simon, Y. C. Lee, and R. T. Lee (Eds.), *Methods Enzymology*, Vol. 242, Academic Press, San Diego, CA, 1994; (b) K. Kobayashi and A. Tshuchida, A new type of artificial glycoconjugate polymer: A convenient synthesis and its interaction with lectins, *Macromolecules*, 30 (1997) 2016–2020; (c) A. Kobayashi, M. Goto, K. Kobayashi, and T. Akaike, Receptor-mediated regulation of differentiation and proliferation of hepatocytes by a synthetic polymer model of asialoglycoprotein, *J. Biomater. Sci. Polymer Edn.*, 6 (1994) 325–342.
9. N. Jenkins, R. B. Parekh, and D. C. James, Getting the glycosylation right: Implications for the biotechnology industry, *Nat. Biotechnol.*, 14 (1996) 975–981.
10. (a) P. M. Rudd, H. C. Joao, E. Coghill, P. Fiten, M. R. Saunders, G. Opdenakker, and R. A. Dwek, Glycoforms modify the dynamic stability and functional activity of an enzyme, *Biochemistry*, 33 (1994) 17–22; (b) R. B. Parekh, R. A. Dwek, P. M. Rudd, J. R. Thomas, T. W. Rademacher, T. Warren, T. C. Wun, B. Hebert, B. Reitz, M. Palmier, T. Ramabhadran, and D. C. Tiemeier, N-Glycosylation and in vitro enzymic activity of human recombinant tissue plasminogen activator expressed in Chinese hamster ovary cells and a murine cell line, *Biochemistry*, 28 (1989) 7670–7679.
11. M. R. Pratt and C. R. Bertozzi, Synthetic glycopeptides and glycoproteins as tools for biology, *Chem. Soc. Rev.*, 34 (2005) 58–68.

12. (a) N. Mathieux, H. Paulsen, M. Meldal, and K. Bock, Synthesis of glycopeptide sequences of repeating units of the mucins MUC 2 and MUC 3 containing oligosaccharide side-chains with core 1, core 2, core 3, core 4 and core 6 structure, *J. Chem. Soc. Perkin Trans. 1*, 6 (1997) 2359–2368; (b) D. Sames, X.-T. Chen, and S. Danishefsky, Convergent total synthesis of a tumour-associated mucin motif, *Nature*, 389 (1997) 587–591; (c) S. J. Danishefsky, S. Hu, P. F. Cirillo, M. Eckhardt, and P. Seeberger, A highly convergent total synthetic route to glycopeptides carrying a high-mannose core pentasaccharide domain N-linked to a natural peptide motif, *Chem. Eur. J.*, 3 (1997) 1617–1628; (d) Z.-G. Wang, X.-F. Zhang, Y. Ito, Y. Nakahara, and T. Ogawa, Stereocontrolled syntheses of O-glycans of core class 2 with a linear tetrameric lactosamine chain and with three lactosamine branches, *Carbohydr. Res.*, 295 (1996) 25–39; (e) J. Seifert and C. Unverzagt, Synthesis of a core-fucosylated, biantennary octasaccharide as a precursor for glycopeptides of complex N-glycans, *Tetrahedron Lett.*, 37 (1996) 6527–6530; (f) E. Meinjohanns, M. Meldal, H. Paulsen, R. A. Dwek, and K. Bock, Novel sequential solid-phase synthesis of N-linked glycopeptides from natural sources, *J. Chem. Soc. Perkin Trans.*, 1 (1998) 549–560; (g) V. Y. Dudkin, J. S. Miller, and S. J. Danishefsky, Chemical synthesis of normal and transformed PSA glycopeptides, *J. Am. Chem. Soc.*, 126 (2004) 736–738; (h) M. Mandal, V. Y. Dudkin, X. Geng, and S. J. Danishefsky, In pursuit of carbohydrate-based HIV vaccines, Part 1: The total synthesis of hybrid-type gp120 fragments, *Angew. Chem. Int. Ed.*, 43 (2004) 2557–2561; (i) X. Geng, V. Y. Dudkin, M. Mandal, and S. J. Danishefsky, In pursuit of carbohydrate-based HIV vaccines, Part 2: The total synthesis of high-mannose-type gp120 fragments-evaluation of strategies directed to maximal convergence, *Angew. Chem. Int. Ed.*, 43 (2004) 2562–2575; (j) M. R. Pratt and C. R. Bertozzi, Syntheses of 6-sulfo sialyl Lewis X glycans corresponding to the L-selectin ligand sulfoadhesin, *Organic Lett.*, 6 (2004) 2345–2348.
13. (a) C. Unverzagt, Chemoenzymatic synthesis of a sialylated diantennary N-glycan linked to asparagines, *Carbohydr. Res.*, 305 (1998) 423–431; (b) O. Seitz and C.-H. Wong, Chemoenzymatic solution- and solid-phase synthesis of *O*-glycopeptides of the mucin domain of MAdCAM-1. A general route to *O*-LacNAc, *O*-sialyl-LacNAc, and *O*-sialyl-Lewis-X peptides, *J. Am. Chem. Soc.*, 119 (1997) 8766–8776.
14. (a) B. G. Davis, Mimicking posttranslational modifications of proteins, *Science*, 303 (2004) 480–482; (b) Z. Zhang, J. Gildersleeve, Y.-Y. Yang, R. Xu, J. A. Loo, S. Uryu, C.-H. Wong, and P. G. Schultz, A new strategy for the synthesis of glycoproteins, *Science*, 303 (2004) 371–373.
15. H. Herzner, T. Reipen, M. Schultz, and H. Kunz, Synthesis of glycopeptides containing carbohydrate and peptide recognition motifs, *Chem. Rev.*, 100 (2000) 4495–4538.
16. (a) C. M. Taylor, Glycopeptides and glycoproteins: Focus on the glycosidic linkage, *Tetrahedron*, 54 (1998) 11317–11362; (b) G. Arsequell and G. Valencia, O-Glycosyl α-amino acids as building blocks for glycopeptide synthesis, *Tetrahedron: Asymmetry*, 8 (1997) 2839–2876.
17. For recent reviews on glycopeptide/glycoprotein synthesis see for example; (a) B. G. Davis, Synthesis of glycoproteins, *Chem. Rev.*, 102 (2002) 579–601; (b) A. Hölemann and P. H. Seeberger, Carbohydrate diversity: Synthesis of glycoconjugates and complex carbohydrates, *Curr. Opin. Biotechnol.*, 15 (2004) 615–622; (c) L. Liu, C. S. Bennett, and C.- H. Wong, Advances in glycoprotein synthesis, *Chem. Commun.* (2006) 21–33.
18. H. Kunz, Synthesis of glycopeptides, partial structures of biological recognition components [new synthetic methods (67)], *Angew. Chem. Int. Ed. Engl.*, 26 (1987) 294–308.
19. (a) L. A. Marcaurelle and C. R. Bertozzi, New directions in the synthesis of glycopeptide mimetics, *Chem. Eur. J.*, 5 (1999) 1384–1390; (b) G. A. Lemieux and C. R. Bertozzi, Chemoselective ligation reactions with proteins, oligosaccharides and cells, *Trends Biotechnol.*, 16 (1998) 506–513; (c) H. C. Hang and C. R. Bertozzi, Chemoselective approaches to glycoprotein assembly, *Acc. Chem. Res.*, 34 (2001) 727–736; (d) F. Peri, Extending chemoselective

ligation to sugar chemistry: Convergent assembly of bioactive neoglycoconjugates, *Mini Rev. Med. Chem.*, 3 (2003) 651–658; (e) F. F. Peri, Nicotra chemoselective ligation in glycochemistry, *Chem. Commun.*, 6 (2004) 623–627.
20. L. A. Marcaurelle, E. C. Rodriguez, and C. R. Bertozzi, Synthesis of an oxime-linked neoglycopeptide with glycosylation-dependent activity similar to its native counterpart, *Tetrahedron Lett.*, 39 (1998) 8417–8420.
21. H. J. Gabius, Biological information transfer beyond the genetic code: The sugar code, *Natur Wissenschaften*, 87 (2000) 108–121.
22. E. Meezan, H. C. Wu, P. H. Black, and P. W. Robbins, Comparative studies on carbohydrate-containing membrane components of normal and virus-transformed mouse fibroblasts. Separation of glycoproteins and glycopeptides by sephadex chromatography, *Biochemistry*, 8 (1969) 2518–2524.
23. (a) G. A. Turner, *N*-Glycosylation of serum-proteins in disease and its investigation using lectins, *Clin. Chim. Acta*, 208 (1992) 149–171; (b) J. S. Axford, Glycosylation and rheumatic disease, *Biochim. Biophys. Acta*, 1455 (1999) 219–229; (c) A. Mackiewicz and K. Mackiewicz, Glycoforms of serum α1-acid glycoprotein as markers of inflammation and cancer, *Glycoconjugate J.*, 12 (1995) 241–247.
24. (a) S. Hakomori, Aberrant glycosylation in tumors and tumor-associated carbohydrate antigens, *Adv. Cancer Res.*, 52 (1989) 257–331; (b) T. Toyokuni and A. K. Singhal, Synthetic carbohydrate vaccines based on tumour-associated antigens, *Chem. Soc. Rev.*, 25 (1995) 231–242.
25. S. J. Danishefsky and J. R. Allen, From the laboratory to the clinic: A retrospective on fully synthetic carbohydrate-based anticancer vaccines, *Angew. Chem. Int. Ed.*, 39 (2000) 836–863.
26. G. F. Springer, T and Tn, general carcinoma autoantigens, *Science*, 224 (1984) 1198–1206.
27. M. J. Francis, in G. Gregoriadis, A. C. Allison, and G. Poste (Eds.), *Vaccines: Recent Trends and Progress*, Plenum Press, New York, 1991.
28. (a) F. Peri, L. Cipolla, M. Rescigno, B. La Ferla, and F. Nicotra, Synthesis and biological evaluation of an anticancer vaccine containing the C-glycoside analogue of the Tn epitope, *Bioconj. Chem.*, 12 (2001) 325–328; (b) L. Cipolla, M. Rescigno, A. Leone, F. Peri, B. La Ferla, and F. Nicotra, Novel Tn antigen containing neoglycopeptides: Synthesis and evaluation as antitumour vaccines, *Bioorg. Med. Chem.*, 10 (2002) 1639–1646.
29. S. Grigalevicius, S. Chierici, O. Renaudet, R. Lo-Man, E. Dériaud, C. Leclerc, and P. Dumy, Chemoselective assembly and immunological evaluation of multiepitopic glycoconjugates bearing clustered Tn antigen as synthetic anticancer vaccines, *Bioconjugate Chem.*, 16 (2005) 1149–1159.
30. P. Dumy, I. M. Eggleston, S. Cervigni, U. Sila, X. Sun, and M. Mutter, A convenient synthesis of cyclic peptides as regioselectively addressable functionalized templates (RAFT), *Tetrahedron Lett.*, 36 (1995) 1255–1258.
31. L. A. Marcaurelle, M. R. Pratt, and C. R. Bertozzi, Synthesis of thioether-linked analogues of the 2,3-sialyl-TF and MECA-79 antigens: Mucin-type glycopeptides associated with cancer and inflammation, *ChemBioChem*, 2–3 (2003) 224–228.
32. L. A. Marcaurelle and C. R. Bertozzi, Chemoselective elaboration of O-linked glycopeptide mimetics by alkylation of 3-thioGalNAc, *J. Am. Chem. Soc.*, 123 (2001) 1587–1595.
33. J. Yeh, N. Hiraoka, B. Petryniak, J. Nakayama, L. G. Ellies, D. Rabuka, O. Hindsgaul, J. D. Marth, J. B. Lowe, and M. Fukuda, Novel sulfated lymphocyte homing receptors and their control by a core1 extension β-1,3-*N*-acetylglucosaminyltransferase, *Cell*, 105 (2001) 957–969.
34. D. H. Dube and C. R. Bertozzi, Glycans in cancer and Inflammation. Potential for therapeutics and diagnostics, *Nature Rev. Drug Discov.*, 4 (2005) 477–488.

35. E. Saxon and C. R. Bertozzi, Cell surface engineering by a modified Staudinger reaction, *Science*, 287 (2000) 2007–2010.
36. (a) G. A. Lemieux, K. J. Yarema, C. L. Jacobs, and C. R. Bertozzi, Exploiting differences in sialoside expression for selective targeting of MRI contrast reagents, *J. Am. Chem. Soc.*, 121 (1999) 4278–4279; (b) G. A. Lemieux and C. R. Bertozzi, Modulating cell surface immunoreactivity by metabolic induction of unnatural carbohydrate antigens, *Chem. Biol.*, 8 (2001) 265–275.
37. V. Verez-Bencomo, V. Fernandez-Santana, E. Hardy, M. E. Toledo, M. C. Rodrı guez, L. Heynngnezz, A. Rodriguez, A. Baly, L. Herrera, M. Izquierdo, A. Villar, Y. Valdes, K. Cosme, M. L. Deler, M. Montane, E. Garcia, A. Ramos, A. Aguilar, E. Medina, G. Torano, I. Sosa, I. Hernandez, R. Martinez, A. Muzachio, A. Carmenates, L. Costa, F. Cardoso, C. Campa, M. Diaz, and R. Roy, A synthetic coniugate polysaccharide vaccine against *Haemophilus influenzae* type b, *Science*, 305 (2004) 522–525.
38. (a) S. Peluso and B. Imperiali, Asparagine surrogates for the assembly of *N*-linked glycopeptide mimetics by chemoselective ligation, *Tetrahedron Lett.*, 42 (2001) 2085–2087; (b) S. Peluso, M. de L. Ufret, M. K. O'Reilly, and B. Imperiali, Neoglycopeptides as inhibitors of oligosaccharyl transferase: Insight into negotiating product inhibition, *Chem. Biol.*, 9 (2002) 1323–1328.
39. (a) M. R. Carrasco, M. J. Nguyen, D. R. Burnell, M. D. MacLaren, and S. M. Hengel, Synthesis of neoglycopeptides by chemoselective reaction of carbohydrates with peptides containing a novel N′-methyl-aminooxy amino acid, *Tetrahedron Lett.*, 43 (2002) 5727–5729; (b) M. R. Carrasco, R. T. Brown, I. M. Serafimova, and O. Silva, Synthesis of *N*-Fmoc-*O*-(*N*′-Boc-*N*′-methyl)-aminohomoserine, an amino acid for the facile preparation of neoglycopeptides, *J. Org. Chem.*, 68 (2003) 195–197.
40. R. Venkatesh, S. Srimathi, A. Yamuna, and G. Jayaraman, Enhanced catalytic and conformational stability of Atlantic cod trypsin upon neoglycosylation, *Biochim. Biophys. Acta*, 1722 (2005) 113–115.
41. (a) P. Sears and C.-H. Wong, Toward automated synthesis of oligosaccharides and glycoproteins, *Science*, 291 (2001) 2344–2350; (b) J. A. Prescher, D. H. Dube, and C. R. Bertozzi, Chemical remodelling of cell surfaces in living animals, *Nature*, 430 (2004) 873–877.
42. (a) W. I. Weis and K. Drickamer, Structural basis of lectin-carbohydrate recognition, *Annu. Rev. Biochem.*, 65 (1996) 441–473; (b) S. Olofsson and T. Bergstroem, Glycoconjugate glycans as viral receptors, *Ann. Med.*, 37 (2005) 154–172; V. C. Chu and R. Gary, Influenza virus entry and infection require host cell N-linked glycoprotein, *Proc. Natl. Acad. Sci. USA*, 101 (2004) 18153–18158; (c) T. G. Boyce, D. L. Swerdlow, and P. M. Griffin, *Escherichia coli* O157:H7 and the hemolytic-uremic syndrome, *New Engl. J. Med.*, 333 (1995) 364–368; (d) J. E. Brown, P. Echeverria, and A. A. Lindberg, Digalactosyl-containing glycolipids as cell surface receptors for shiga toxin of *Shigella dysenteriae* 1 and related cytotoxins of *Escherichia coli*, *Rev. Infect. Dis.*, 13 (Suppl. 4) (1991) 298–303.
43. (a) R. Harel and A. Futerman, Inhibition of sphingolipid synthesis affects axonal outgrowth in cultured hippocampal neurons, *J. Biol. Chem.*, 268 (1993) 14476–14481; (b) S. Furuya, J. Mitoma, A. Makino, and Y. Hirabayashi, Ceramide and its interconvertible metabolite sphingosine function as indispensable lipid factors involved in survival and dendritic differentiation of cerebellar Purkinje cells, *J. Neurochem.*, 71 (1998) 366–377.
44. E. Klenk and K. Schuwirth, The chemistry of the lipins, *Annu. Rev. Biochem.*, 6 (1937) 115–138.
45. S. Hakomori and Y. Zhang, Glycosphingolipid antigens and cancer therapy, *Chem. Biol.*, 4 (1997) 97–104.
46. B. G. Davis, Recent developments in glycoconjugates, *J. Chem. Soc., Perkin Trans.*, 1 (1999) 3215–3237.

47. (a) C. Le Bec and A. M. Douar, Gene therapy progress and prospects-vectorology: Design and production of expression cassettes in AAV vectors, *Gene Ther.*, 13 (2006) 805–813; (b) K. Ewert, N. L. Slack, A. Ahmad, H. M. Evans, A. J. Lin, C. E. Samuel, and C. R. Safinya, Cationic lipid-DNA complexes for gene therapy: Understanding the relationship between complex structure and gene delivery pathways at the molecular level, *Curr. Med. Chem.*, 11 (2004) 133–149; (c) K. Okamoto, M. Mizuno, and J. Yoshida, Application of liposomes for human gene therapy, *Drug Del. Syst.*, 15 (2000) 443–446; (d) X. Gao and L. Huang, Cationic liposome-mediated gene transfer, *Gene Ther.*, 2 (1995) 710–722; (e) A. D. Miller, Cationic liposomes for gene therapy, *Angew. Chem. Int. Ed.*, 37 (1998) 1768–1785.
48. (a) F. Yan, J. Xue, J. Zhu, R. E. Marchant, and Z. Guo, Synthesis of a lipid conjugate of SO3Lea and its enhancement on liposomal binding to activated platelets, *Bioconjugate Chem.*, 16 (2005) 90–96; (b) R. Zeisig, R. Stahn, K. Wenzel, D. Behrens, and I. Fichtner, Effect of sialyl Lewis X-glycoliposomes on the inhibition of E-selectin-mediated tumour cell adhesion in vitro, *Biochim. Biophys. Acta, Biomembranes*, 1660 (2004) 31–40.
49. A. Düffels, L. G. Green, S. V. Ley, and A. D. Miller, Synthesis of high-mannose type neoglycolipids: Active targeting of liposomes to macrophages in gene therapy, *Chem. Eur. J.*, 6 (2000) 1416–1430.
50. N. Laurent, D. Lafont, and P. Boullanger, Syntheses of α-D-galactosamine neoglycolipids, *Carbohydr. Res.*, 341 (2006) 823–835.
51. (a) I. M. Verma and M. D. Weitzman, Gene therapy: Twenty-first century medicine, *Annu. Rev. Biochem.*, 74 (2005) 711–738; (b) A. D. Miller, Human gene therapy comes of age, *Nature*, 357 (1992) 455–460; (c) W. F. Anderson, Human gene therapy, *Science*, 256 (1992) 808–813.
52. (a) D. K. Hoganson, L. A. Chandler, G. A. Fleurbaaij, W. Ying, M. E. Black, J. Doukas, G. F. Pierce, A. Baird, and B. A. Sosnowski, Targeted delivery of DNA encoding cytotoxic proteins through high-affinity fibroblast growth factor receptors, *Hum. Gene Therapy*, 9 (1998) 2565–2575; (b) T. Boulikas and F. Martin, Histones, protamine, and polylysine but not poly(E:K) enhance transfection efficiency, *Int. J. Oncol.*, 10 (1997) 317–322.
53. (a) A. A. Gurtovenko, S. V. Lyulin, M. Karttunen, and I. Vattulainen, Molecular dynamics study of charged dendrimers in salt-free solution: Effect of counterions, *J. Chem. Phys*, 124 (2006) 94904; (b) H. R. Ihre, O. L. P. De Jesus, F. C. Szoka, and J. M. J. Frechet, Polyester dendritic systems for drug delivery applications: Design, synthesis, and characterization, *Bioconjugate Chem.*, 13 (2002) 443–452; (c) A. U. Bielinska, C. L. Chen, J. Johnson, and J. R. Baker, DNA complexing with polyamidoamine dendrimers: Implications for transfection, *Bioconjugate Chem.*, 10 (1999) 843–850.
54. (a) K. Regnstroem, E. G. E. Ragnarsson, M. Fryknaes, M. Koeping-Hoeggard, and P. Artursson, Gene expression profiles in mouse lung tissue after administration of two cationic polymers used for nonviral gene delivery, *Pharma. Res.*, 23 (2006) 475–482; (b) S. Dincer, M. Tuerk, and E. Piskin, Intelligent polymers as nonviral vectors, *Gene Therapy*, 12 (Suppl. 1) (2005) 139–145; (c) L. Pitkaenen, M. Ruponen, J. Nieminen, and A. Urtti, Vitreous is a barrier in nonviral gene transfer by cationic lipids and polymers, *Pharm. Res.*, 20 (2003) 576–583; (d) A. Bragonzi, A. Boletta, A. Biffi, A. Muggia, G. Sersale, S. H. Cheng, C. Bordignon, B. M. Assael, and M. Conese, Comparison between cationic polymers and lipids in mediating systemic gene delivery to the lungs, *Gene Ther.*, 6 (1999) 1995–2004.
55. (a) K. Ewert, A. Ahmad, H. M. Evans, H.-W. Schmidt, and C. R. Safinya, Efficient synthesis and cell-transfection properties of a new multivalent cationic lipid for nonviral gene delivery, *J. Med. Chem.*, 45 (2002) 5023–5029; (b) C.-Y. Huang, T. Uno, J. E. Murphy, S. Lee, J. D. Hamer, J. A. Escobedo, F. E. Cohen, R. Radhakrishnan, V. Dwarki, and R. N. Zuckermann, Lipitoids-novel cationic lipids for cellular delivery of plasmid DNA in

vitro, *Chem. Biol.*, 5 (1998) 345–354; (c) C. J. Wheeler, P. L. Felgner, Y. J. Tsai, J. Marshall, L. Sukhu, S. G. Doh, J. Hartikka, J. Nietupski, M. Manthorpe, M. Nichols, M. Plewe, X. W. Liang, J. Norman, A. Smith, and S. H. Cheng, A novel cationic lipid greatly enhances plasmid DNA delivery and expression in mouse lung, *Proc. Natl. Acad. Sci. USA*, 93 (1996) 11454–11459.

56. (a) J. Van Zanten, B. Doornbos-van der Meer, S. Audouy, R. J. Kok, and L. de Leij, A nonviral carrier for targeted gene delivery to tumor cells, *Cancer Gene Ther.*, 11 (2004) 156–164; (b) I. Moret-Tatay, J. Diaz, F. M. Marco, A. Crespo, and S. F. Alino, Complete tumor prevention by engineered tumor cell vaccines employing nonviral vectors, *Cancer Gene Ther.*, 10 (2003) 887–897; (c) M. Cemazar, G. Sersa, J. Wilson, G. M. Tozer, S. L. Hart, A. Grosel, and G. U. Dachs, Effective gene transfer to solid tumors using different nonviral gene delivery techniques: Electroporation, liposomes, and integrin-targeted vector, *Cancer Gene Ther.*, 9 (2002) 399–406; (d) Y. Yamazaki, M. Nango, M. Matsuura, Y. Hasegawa, M. Hasegawa, and N. Oku, Polycation liposomes, a novel nonviral gene transfer system, constructed from cetylated polyethylenimine, *Gene Ther.*, 7 (2000) 1148–1155; (e) R. G. Cooper, C. J. Etheridge, L. Stewart, J. Marshall, S. Rudginsky, S. H. Cheng, and A. D. Miller, Cationic lipids for gene therapy. 1. Polyamine analogs of 3β-[N-(N′,N′-dimethylaminoethane)carbamoyl]cholesterol (DC-chol) as agents for gene delivery, *Chem. Eur. J.*, 4 (1998) 137–151.
57. T. Tagawa, M. Manvell, N. Brown, M. Keller, E. Perouzel, K. D. Murray, R. P. Harbottle, M. Tecle, F. Booy, M. C. Brahimi-Horn, C. Coutelle, N. R. Lemoine, E. W. Alton, and A. D. Miller, Characterisation of LMD virus-like nanoparticles self-assembled from cationic liposomes, adenovirus core peptide mu and plasmid DNA, *Gene Ther.*, 9 (2002) 564–576.
58. E. Perouzel, M. R. Jorgensen, M. Keller, and A. D. Miller, Synthesis and formulation of neoglycolipids for the functionalization of liposomes and lipoplexes, *Bioconjugate Chem.*, 14 (2003) 884–898.
59. L. Berthelot, V. Rosilio, M. L. Costa, S. Chierici, G. Albrecht, P. Boullanger, and A. Baszkin, Behavior of amphiphilic neoglycolipids at the air:solution interface. Interaction with a specific lectin, *Colloids Surf., B Biointerfaces*, 11 (1998) 239–248.
60. A. E. Gal, P. G. Pentchev, J. M. Massey, and R. O. Brady, L-Glucosylceramide: Synthesis, properties, and resistance to catabolism by glucocerebrosidase in vitro, *Proc. Natl. Acad. Sci. USA*, 76 (1979) 3083–3086.
61. J. Kihlberg and G. Magnusson, Use of carbohydrates and peptides in studies of adhesion of pathogenic bacteria and in efforts to generate carbohydrate-specific T cells, *Pure Appl. Chem.*, 68 (1996) 2119–2128.
62. M. S. Stoll, T. Feizi, W. Loveless, W. Chai, A. M. Lawson, and C.-T. Yuen, Fluorescent neoglycolipids Improved probes for oligosaccharide ligand discovery, *Eur. J. Biochem.*, 267 (2000) 1795–1804.
63. T. Feizi and R. A. Childs, Neoglycolipids: Probes in structure/function assignments to oligosaccharides, *Meth. Enzymol.*, 242 (1994) 205–217.
64. J.-I. Inokuchi, K. Kabayama, S. Uemura, and Y. Igarashi, Glycosphingolipids govern gene expression, *Glycoconjugate J.*, 20 (2004) 169–178.
65. G. Van Mer and J. C. M. Holthuis, Sphingolipid transport in eukaryotic cells, *Biochim. Biophys. Acta*, 1486 (2000) 145–170.
66. (a) S. Uemura, K. Kabayama, M. Noguchi, Y. Igarashi, and J.-I. Inokuchi, Sialylation and sulfation of lactosylceramide distinctly regulate anchorage-independent growth, apoptosis, and gene expression in 3LL Lewis lung carcinoma cells, *Glycobiology*, 13 (2003) 207–216; (b) E. M. Schmelz, P. C. Roberts, E. M. Kustin, L. A. Lemonnier, M. C. Sullards, D. L. Dillehay, and A. H. Merrill Jr., Modulation of intracellular β-catenin localization and intestinal tumorigenesis in vivo and in vitro by sphingolipids, *Cancer Res.*, 61 (2001) 6723–6729;

(c) T. Iwamoto, S. Fukumoto, K. Kanaoka, E. Sakai, M. Shibata, E. Fukumoto, J. J. Inokuchi, K. Takamiya, K. Furukawa, K. Furukawa, Y. Kato, and A. Mizuno, Lactosylceramide is essential for the osteoclastogenesis mediated by macrophage-colony-stimulating factor and receptor activator of nuclear factor-kB ligand, *J. Boil. Chem.*, 276 (2001) 46031–46038.

67. (a) R. Komagome, H. Sawa, T. Suzuki, Y. Suzuki, S. Tanaka, W. J. Atwood, and K. Nagashima, Oligosaccharides as receptors for JC virus, *J. Virol.*, 76 (2002) 12992–13000; (b) V. I. Razinkov and F. S. Cohen, Sterols and sphingolipids strongly affect the growth of fusion pores induced by the hemagglutinin of influenza virus, *Biochemistry*, 39 (2000) 13462–13468; (c) S. Teneberg, I. Leonardsson, H. Karlsson, P.-Å. Jovall, J. Ångström, D. Danielsson, I. Näslund, Å. Ljungh, T. Wadström, and K.-A. Karlsson, Lactotetraosylceramide, a novel glycosphingolipid receptor for Helicobacter pylori, present in human gastric epithelium, *J. Biol. Chem.*, 277 (2002) 19709–19719; (d) S. Teneberg, J. Ångström, and Å. Ljungh, Carbohydrate recognition by enterohemorrhagic *Escherichia coli*: Characterization of a novel glycosphingolipid from cat small intestine, *Glycobiology*, 14 (2004) 187–196.
68. Y. Harada, T. Murata, K. Totani, T. Kajimoto, S. Md. Masum, Y. Tamba, M. Yamazaki, and T. Usui, Design and facile synthesis of neoglycolipids as lactosylceramide mimetics and their transformation into glycoliposomes, *Biosci. Biotachnol. Biochem.*, 69 (2005) 166–178.
69. (a) N. Yasutake, K. Totani, Y. Harada, S. Haraguchi, T. Murata, and T. Usui, Efficient synthesis of glyceroyl beta-lactoside and its derivatives through a condensation reaction by cellulose, *Biochim. Biophys. Acta*, 1620 (2003) 252–258; (b) N. Yasutake, K. Totani, H. S. Ohi, T. Murata, and T. Usui, Enzymatic synthesis of aliphatic β-lactosides as mimic units of glycosphingolipids by use of trichoderma reesei cellulase, *Arch. Biochem. Biophys*, 385 (2001) 70–77.
70. P. Wang, M. Schuster, Y.-F. Wang, and C.-H. Wong, Synthesis of phospholipid-inhibitor conjugates by enzymatic transphosphatidylation with phospholipase D, *J. Am. Chem. Soc.*, 115 (1993) 10487–10491.
71. E. L. Vodovozova, G. V. Pazynina, A. B. Tuzikov, I. V. Grechishnikova, and J. G. Molotkovsky, The synthesis of photoaffinity neoglycolipid probes as tools for studying membrane lectins, *Russian J. Bioorg. Chem.*, 30 (2004) 154–160.
72. (a) A. Roth, S. Espuelas, C. Thumann, B. Frisch, and F. Schuber, Synthesis of thiol-reactive lipopeptide adjuvants. Incorporation into liposomes and study of their mitogenic effect on mouse splenocytes, *Bioconjugate Chem.*, 15 (2004) 541–553; (b) G. F. A. Kersten and D. J. A. Crommelin, Liposomes and ISCOMS as vaccine formulations, *Biochim. Biophys. Acta Rev. Biomembranes*, 1241 (1995) 117–138.
73. T. Buskas, S. Ingale, and G.-J. Boons, Towards a fully synthetic carbohydrate-based anticancer vaccine: Synthesis and immunological evaluation of a lipidated glycopeptide containing the tumor-associated Tn antigen, *Angew. Chem. Int. Ed.*, 44 (2005) 5985–5988.
74. (a) D. Wang, S. Liu, B. J. Trummer, C. Deng, and A. Wang, Carbohydrate microarrays for the recognition of cross-reactive molecular markers of microbes and host cells, *Nature Biotechnol.*, 20 (2002) 275–281; (b) B. T. Houseman and M. Mrksich, Carbohydrate arrays for the evaluation of protein binding and enzymatic modifications, *Chem. Biol.*, 9 (2002) 443–454; (c) S. Fukui, T. Feizi, C. Galustian, A. M. Lawson, and W. Chai, Oligosaccharide microarrays for the high-throughput detection and specificity assignments of carbohydrate-protein interactions, *Nature Biotechnol.*, 20 (2002) 1011–1017; (d) F. Fazio, M. C. Bryan, O. Blixt, J. C. Paulson, and C-H. Wong, Synthesis of sugar arrays in microtiter plates, *J. Am. Chem. Soc.*, 124 (2002) 14397–14402; (e) K. R. Love and P. Seeberger, Carbohydrate arrays as tools for glycomics, *Angew. Chem. Int. Ed.*, 41 (2002) 3583–3586; (f) T. Feizi, F. Fazio, W. Chai, and C-H. Wong, Carbohydrate microarray—a new set of technologies at frontiers of glycomics, *Curr. Opin. Struct. Biol.*, 13 (2003) 637–645; (g) T. Feizi and W. Chai, Oligosaccharide microarrays to decipher the glyco code, *Nat. Rev. Mol. Cell Biol.*, 5 (2004) 582–588; (h) O. Blixt, S. Head,

T. Mondala, C. Scanlan, M. E. Huflejt, R. Alvarez, M. C. Bryan, F. Fazio, D. Calarese, J. Stevens, N. Razi, D. J. Stevens, J. J. Skehel, D. Van, I. D. R. Burton, I. A. Wilson, R. Cummings, N. Bovin, C. H. Wong, and J. C. Paulson, Printed covalent glycan array for ligand profiling of diverse glycan binding proteins, *Proc. Natl. Acad. Sci. U.S.A.*, 101 (2004) 17033–17038.

75. (a) M. J. Hernaiz, J. M. de la Fuente, A. G. Barrientos, and S. Penades, A model system mimicking glycosphingolipid clusters to quantify carbohydrate self-interactions by surface plasmon resonance, *Angew. Chem. Int. Ed.*, 41 (2002) 1554–1557; (b) A. G. Barrientos, J. M. de la Fuente, T. C. Rojas, A. Fernandez, and S. Penades, Gold glyconanoparticles: Synthetic polyvalent ligands mimicking glycocalyx-like surfaces as tools for glycobiological studies, *Chem. Eur. J.*, 9 (2003) 1909–1921; (c) A. Carvalho de Souza, K. M. Halkes, J. D. Meeldijk, A. J. Verkleij, J. F. G. Vliegenthart, and J. P. Kamerling, Synthesis of gold glyconanoparticles: Possible probes for the exploration of carbohydrate-mediated self-recognition of marine sponge cells, *Eur. J. Org. Chem.*, 21 (2004) 4323–4339.

76. Y. Aoyama, T. Kanamori, T. Nakai, T. Sasaki, and S. Horiuchi, Artificial viruses and their application to gene delivery. Size-controlled gene coating with glycocluster nanoparticles, *J. Am. Chem. Soc.*, 125 (2003) 3345–3455.

77. (a) W. B. Turnbull and J. F. Stoddart, Design and synthesis of glycodendrimers, *Rev. Mol. Biotechnol.*, 90 (2002) 231–255; (b) K. Bezousoeka, Design, functional evaluation and biomedical applications of carbohydrate dendrimers (glycodendrimers), *Rev. Mol. Biotechnol.*, 90 (2002) 269–290; (c) T. K. Lindhorst, Artificial multivalent sugar ligands to understand and manipulate carbohydrate-protein interactions, *Top. Curr. Chem.*, 218 (2002) 201–235; (d) L. L. Kiessling, J. E. Gestwicki, and L. E. Strong, Synthetic multivalent ligands in the exploration of cell-surface interactions, *Curr. Opin. Chem. Biol.*, 4 (2000) 696–703; (e) N. N. Jayaraman, S. A. Nepogodiev, and J. F. Stoddart, Synthetic carbohydrate dendrimers. Synthetic carbohydrate-containing dendrimers, *Chem. Eur. J.*, 3 (1997) 1193–1199; (f) R. Roy, Recent developments in the rational design of multivalent glycoconjugates, *Top. Curr. Chem.*, 187 (1997) 241–274; (g) J. J. Lundquist and E. J. Toone, The cluster glycoside effect, *Chem. Rev.*, 102 (2002) 555–578; (h) R. T. Lee and Y. C. Lee, Affinity enhancement by multivalent lectin-carbohydrate interaction, *Glycoconjugate J.*, 17 (2000) 543–551; (i) M. Mammen, S.-K. Choi, and G. M. Whitesides, Polyvalent interactions in biological systems: Implications for design and use of multivalent ligands and inhibitors, *Angew. Chem. Int. Ed.*, 37 (1998) 2754–2794; (j) Y. C. Lee and R. T. Lee, Carbohydrate–protein interactions: Basis of glycobiology, *Acc. Chem. Res.*, 28 (1995) 321–327.

78. D. Zanini and R. Roy, Novel dendritic α-sialosides: Synthesis of glycodendrimers based on a 3,3′-iminobis(propylamine) core, *J. Org. Chem.*, 61 (1996) 7348–7354.

79. (a) R. Roy, R. A. Pon, F. D. Tropper, and F. O. Andersson, Michael addition of poly-L-lysine to N-acryloylated sialosides syntheses of influenza A virus haemagglutinin inhibitor and Group B meningococcal polysaccharide vaccines, *J. Chem. Soc. Chem. Commun.* (1993) 264–265; (b) J. D. Reuter, A. Myc, M. M. Hayes, Z. Gan, R. Roy, D. Qin, R. Yin, L. T. Piehler, R. Esfand, D. A. Tomalia, and J. R. Baker Jr., Inhibition of viral adhesion and infection by sialic acid conjugated dendritic polymers, *Bioconj. Chem.*, 10 (1999) 271–278.

80. J. J. Landers, Z. Cao, I. Lee, L. T. Piehler, P. P. Myc, A. Myc, T. Hamouda, A. T. Galecki, and J. R. Baker Jr., Prevention of influenza pneumonitis by sialic acid-conjugated dendritic polymers, *J. Infect. Dis.*, 186 (2002) 1222–1230.

81. J. P. Thompson and C. L. Schengrund, Oligosaccharide derivatized dendrimers: Defined multivalent inhibitors of the adherence of the cholera toxin B subunit and the heat labile enterotoxin of *Escherichia coli* to GM1, *Glycoconjugate J.*, 14 (1997) 837–845.

82. D. Arosio, I. Vrasidas, P. Valentini, R. M. Liskamp, R. J. Pieters, and A. Bernardi, Synthesis and cholera toxin binding properties of multivalent GM1 mimics, *Org. Biomol. Chem.*, 2 (2004) 2113–2124.

83. (a) P. R. Ashton, E. F. Hounsell, N. Jayaraman, T. M. Nielsen, N. Spencer, J. F. Stoddart, and M. Young, Synthesis and biological studies of α-D-mannopyranoside-containing dendrimers, *J. Org. Chem.*, 63 (1998) 3429–3437; (b) D. Pagé and R. Roy, Synthesis and biological properties of mannosylated starburst poly(amidoamine) dendrimers, *Bioconj. Chem.*, 8 (1997) 714–723; (c) S. L. Mangold and M. J. Cloninger, Binding of monomeric and dimeric Concavalin A to mannose-functionalized dendrimers, *Org. Biomol. Chem.*, 4 (2006) 2458–2465.
84. C. C. M. Appeldoorn, J. A. Joosten, F. Ait el Maate, U. Dobrindt, J. Hacker, R. M. J. Liskamp, A. S. Khan, and R. J. Pieters, Novel multivalent mannose compounds and their inhibition of the adhesion of type 1 fimbriated uropathogenic *E. coli*, *Tetrahedron: Asymmetry*, 16 (2005) 361–372.
85. S. L. Mangold, J. R. Morgan, G. C. Strohmeyer, A. M. Gronenborn, and M. J. Cloninger, Cianovirin-N binding Manα1-2Man functionalized dendrimers, *Org. Biomol. Chem.*, 3 (2005) 2354–2358.
86. (a) P. R. Ashton, S. E. Boyd, C. L. Brown, S. A. Nepogodiev, E. W. Meijer, H. W. I. Peerlings, and J. F. Stoddart, Synthesis of glycodendrimers by modification of poly(propylene imine) dendrimers, *Chem. Eur. J.*, 3 (1997) 974–983; (b) H. W. I. Peerlings, S. A. Nepogodiev, J. F. Stoddart, and E. W. Meijer, Synthesis of spacer-armed glycodendrimers based on modification of poly(propylene imine) dendrimers, *Eur. J. Org. Chem.*, 9 (1998) 1879–1886.
87. (a) P. R. Ashton, S. E. Boyd, C. L. Brown, N. Jayaraman, S. A. Nepogodiev, and J. F. Stoddart, A convergent synthesis of carbohydrate containing dendrimers, *Chem. Eur. J.*, 2 (1996) 1115–1128; (b) P. R. Ashton, S. E. Boyd, C. L. Brown, N. Jayaraman, and J. F. Stoddart, A convergent synthesis of carbohydrate containing dendrimers, *Angew. Chem. Int. Ed. Engl.*, 36 (1997) 732–735; (c) W. B. Turnbull, A. R. Pease, and J. F. Stoddart, Towards the synthesis of large oligosaccharide-based dendrimers, *ChemBiochem*, 1 (2000) 70–74.
88. K. Aoi, K. Itoh, and M. Okada, Globular carbohydrate macromolecule sugar balls. 1. Syntheis of novel sugar-persubstituted poly(amido amine) dendrimers, *Macromolecules*, 28 (1995) 5391–5393.
89. (a) T. K. Lindhorst and C. Kieburg, Glycocoating of oligovalent amines: Synthesis of thiourea-bridged cluster glycosides from glycosyl isothiocyanates, *Angew. Chem. Int. Ed. Engl.*, 35 (1996) 1953–1956; (b) D. Pagè, S. Aravind, and R. Roy, Synthesis and lectin binding properties of dendritic mannopyranoside, *Chem. Commun.* (1996) 1913–1914; (c) C. Kieburg and T. K. Lindhorst, Glycodendrimer synthesis without using protecting groups, *Tetrahedron Lett.*, 38 (1997) 3885–3888.
90. D. Zanini and R. Roy, Chemoenzymatic synthesis and lectin binding properties of dendritic N-acetyllactosamines, *Bioconjugate Chem.*, 8 (1997) 187–192.
91. D. Zanini and R. Roy, Synthesis of new α-thiosialodendrimers and their specific binding properties to the sialic acid specific lectin from *Limax flavus*, *J. Am. Chem. Soc.*, 119 (1997) 2088–2095.
92. C. Grandjean, R. Rommens, H. Gras-Masse, and O. Melnyk, One-pot synthesis of antigen-bearing, lysine-based cluster mannosides using two orthogonal chemoselective ligation reactions, *Angew. Chem. Int. Ed. Engl.*, 39 (2000) 1068–1072.
93. B. G. Davis, The controlled glycosylation of a protein with a bivalen glycan: Towards a new class of glycoconjugates, glycodendriproteins, *Chem. Commun.* (2001) 351–352.
94. N. Jayaraman and J. F. Stoddart, Synthesis of carbohydrate containing dendrimers. 5. Preparation of dendrimers using unprotected carbohydrates, *Tetrahedron Lett.*, 38 (1997) 6767–6770.
95. (a) L. L. Kiessling and N. L. Pohl, Strength in numbers: Non-natural polyvalent carbohydrate derivatives, *Chem. Biol.*, 3 (1996) 71–77; (b) R. Roy, Syntheses and some applications of chemically defined multivalent glycoconjugates, *Curr. Opin. Struct. Biol.*, 6 (1996) 692–702;

(c) Y. C. Lee and R. T. Lee, Neoglycoconjugates, Part B: Synthesis, in Y. C. Lee and R. T. Lee (Eds.), *Methods Enzymology*, Vol. 247, Academic Press, San Diego, CA, 1994.

96. N. V. Bovin and H. J. Galbius, Polymer-immobilized carbohydrate ligands: Versatile chemical tools for biochemistry and medical sciences, *Chem. Soc. Rev.*, 24 (1995) 413–422.
97. (a) R. Roy, W. K. C. Park, O. P. Srivastava, and C. Foxall, Combined glycomimetic and multivalent strategies for the design of potent selectin antagonists, *Bioorg. Med. Chem. Lett.*, 6 (1996) 1399–1402; (b) H. Miyauchi, M. Tanaka, H. Koike, N. Kawamura, and M. Hayashi, Synthesis and inhibitory effect of a sialyl Lewis x-acrylamide homopolymer at preventing cell adhesion, *Bioorg. Med. Chem. Lett.*, 7 (1997) 985–988; (c) E. J. Gordon, W. J. Sanders, and L. L. Kiessling, Synthetic ligands point to cell surface strategies, *Nature*, 392 (1998) 30–31; (d) G. Thoma, J. T. Patton, J. L. Magnani, B. Ernst, R. Öhrlein, and R. O. Duthaler, Versatile functionalization of polylysine: Synthesis, characterization, and use of neoglycoconjugates, *J. Am. Chem. Soc.*, 121 (1999) 5919–5929.
98. (a) H. Dohi, Y. Nishida, M. Mizuno, M. Shinkai, T. Kobayashi, T. Takeda, H. Uzawa, and K. Kobayashi, Synthesis of an artificial glycoconjugate polymer carrying P^k-antigenic trisaccharide and its potent neutralization activity against Shiga-like toxin, *Bioorg. Med. Chem.*, 7 (1999) 2053–2062; (b) P. I. Kitov, J. M. Sadowska, G. Mulvey, G. D. Armstrong, H. Ling, N. S. Pannu, R. J. Read, and D. R. Bundle, Shiga-like toxins are neutralized by tailored multivalent carbohydrate ligands, *Nature*, 403 (2000) 669–672.
99. K. Sasaki, Y. Nishida, T. Tsurumi, H. Uzawa, H. Kondo, and K. Kobayashi, Facile assembly of cell surface oligosaccharide mimics by copolymerization of carbohydrate modules, *Angew. Chem. Int. Ed.*, 41 (2002) 4463–4467.
100. K. Yamada, M. Minoda, and T. Miyamoto, Controlled synthesis of amphiphilic block copolymers with pendant N-acetyl-D-glucosamine residues by living cationic polymerization and their interaction with WGA lectin, *Macromolecules*, 32 (1999) 3553–3558.
101. I. Velter, B. La Ferla, and F. Nicotra, Carbohydrate-based molecular scaffolding, *J. Carbohydr. Chem.*, 25 (2006) 97–138.
102. S. M. Sagnella, F. Kligman, E. H. Anderson, J. E. King, G. Murugesan, R. E. Marchant, and K. Kottke-Marchant, Human microvascular endothelial cell growth and migration on biomimetic surfactant polymers, *Biomaterials*, 25 (2004) 1249–1259.
103. (a) P. Wang, T. G. Hill, C. A. Wartchow, M. E. Huston, L. M. Oehler, M. Bradley Smith, M. D. Bednarski, and M. R. Callstrom, New carbohydrate-based materials for the stabilisation of proteins, *J. Am. Chem. Soc.*, 60 (1992) 2216–2226; (b) C. A. Wartchow, P. Wang, M. D. Bednarski, and M. R. Callstrom, Carbohydrate protease conjugates: Stabilized proteases for peptide synthesis, *J. Org. Chem.*, 60 (1995) 2216–2226.
104. S. E. Cervigni, P. Dumy, and M. Mutter, Synthesis of glycopeptides and lipopeptides by chemoselective ligation, *Angew. Chem. Int. Ed.*, 35 (1996) 1230–1232.
105. M. Vila-Perello, R. Gutierréz-Gallego, and D. Andrei, A simple approach to well-defined sugar-coated surfaces for intercation studies, *ChemBioChem*, 6 (2005) 1831–1838.
106. F. Peri, P. Dumy, and M. Mutter, Chemo- and stereoselective glycosilation of hydroxylamino derivatives: A versatile approach to glycoconjugates, *Tetrahedron*, 54 (1998) 12269–12278.
107. J. M. Langheran, N. R. Peters, I. A. Guzei, F. M. Hoffmann, and J. S. Thorson, Enhancing the anticancer properties of cardiac glycosides by neoglycorandomization, *Proc. Natl. Acad. Sci. USA*, 102 (2005) 12305–12310.
108. (a) F. Peri, A. Deutman, B. La Ferla, and F. Nicotra, Solution and solid-phase chemoselective synthesis of amino(methoxy) di- and trisaccharide analogues, *Chem. Commun.* (2002) 1504–1505; (b) F. Peri, J. Jiménez-Barbero, V. García-Aparicio, I. Tvaroska, and F. Nicotra, Synthesis and conformational analysis of novel $N(OCH_3)$ disaccharide analogues, *Chem. Eur. J.*, 10 (2004) 1433–1444.

109. (a) E. C. Rodrigues, K. A. Winans, D. S. King, and C. R. Bertozzi, A strategy for the chemoselective synthesis of O-linked glycopeptides with the native sugar-peptide linkages, *J. Am. Chem. Soc.*, 119 (1997) 9905–9906; (b) E. C. Rodriguez, L. A. Marcaurelle, and C. R. Bertozzi, Aminooxy-, hydrazide-, and thiosemicarbazide-functionalized saccharides: Versatile reagents for glycoconjugate synthesis, *J. Org. Chem.*, 63 (1998) 7134–7135; (c) O. Renaudet and P. Dumy, Chemoselectively template-assembled glcoconjugates as mimics for multivalent presentation of carbohydrates, *Org. Lett.*, 5 (2003) 243–246.
110. N. S. Flinn, M. Quibell, T. P. Monk, K. Ramjee, and C. J. Urch, A single-step method for the production of sugar hydrazides: Intermediates for the chemoselective preparation of glycoconjugates, *Bioconjugate Chem.*, 16 (2005) 722–728.
111. B. G. Davis, R. C. Lloyd, and J. Bryan Jones, Controlled site-selective glycosylation of proteins by a combined site-directed mutagenesis and chemical modification approach, *J. Org. Chem.*, 63 (1998) 9614–9615.
112. D. Macmillan, R. M. Bill, K. A. Sage, D. Fern, and S. Flitsch, Selective in vitro glycosylation of recombinant proteins: Semi-synthesis of novel homogeneous glycoforms of human erythropoietin, *Chem. Biol.*, 8 (2001) 133–145.
113. I. Shin, H.-J. Jung, and M.-R. Lee, Chemoselective ligation of maleimidosugars to peptides/protein for the preparation of neoglycopeptides/neoglycoprotein, *Tetrahedron Lett.*, 42 (2001) 1325–1328.
114. W. M. Macindoe, A. H. van Oijen, and G. -J. Boons, A unique and highly facile method for synthesizing disulfide linked neoglycoconjugates: a new approach for remodelling of peptides and proteins, *Chem. Commun.* (1998) 847–848.
115. S. Park, M.-R. Lee, S.-J. Pyo, and I. Shin, Carbohydrate chips for studying high-throughput carbohydrate-protein interactions, *J. Am. Chem. Soc.*, 126 (2004) 4812–4819.
116. M. D. Gieselman, Y. Zhu, H. Zhou, D. Galonic, and W. A. van der Donk, Selenocysteine derivatives for chemoselective ligations, *ChemBioChem*, 3 (2002) 709–716.
117. E. Fernandez-Megia, J. Correa, I. Rodriguez-Meizoso, and R. Rigueira, A click approach to unprotected glycodendrimers, *Macromolecules*, 39 (2006) 2113–2120.
118. V. Pozsgay, N. E. Vieira, and A. Yergey, A method for bioconjugation of carbohydrates using Diels–Alder cycloaddition, *Org. Lett.*, 19 (2002) 3191–3194.
119. (a) G. Angyalosi, C. Grandjean, M. Lamirand, C. Auriault, H. Gras-Masse, and O. Melnyk, Synthesis and mannose receptor-mediated uptake of clustered glycomimetics by human dendritic cells: Effect of charge, *Bioorg. Med. Chem.*, 12 (2002) 2723–2727; (b) C. Grandjean, V. Santraine, J.-S. Fruchart, O. Melnyk, and H. Gras-Masse, Combined thioether/hydrazone chemoselective ligation reactions for the synthesis of glycocluster-antigen peptide conjugates, *Bioconjugate Chem.*, 13 (2002) 887–892.
120. J. Kevin, L. K. Yarema, R. E. Mahal, E. C. Bruehl, and C. R. Bertozzi, Metabolic delivery of ketone groups to sialic acid residues. Application to cell surface glycoform engineering, *J. Biol. Chem.*, 273 (1998) 31168–31179.

AUTHOR INDEX

Page numbers in roman type indicate that the listed author is cited on that page; page numbers in italic denote the page where the literature citation is given.

A

E

G

H

I

J

K

M

P

X

SUBJECT INDEX

T